Pattern Recognition
AND SIGNAL ANALYSIS IN MEDICAL IMAGING

SECOND EDITION

Pattern Recognition and Signal Analysis in Medical Imaging

SECOND EDITION

ANKE MEYER-BÄESE
Department of Scientific Computing
Florida State University
Tallahassee, USA

VOLKER SCHMID
Department of Statistics
Ludwig-Maximilians-University Munich
Munich, Germany

AMSTERDAM • BOSTON • HEIDELBERG • LONDON
NEW YORK • OXFORD • PARIS • SAN DIEGO
SAN FRANCISCO • SINGAPORE • SYDNEY • TOKYO
Academic Press is an imprint of Elsevier

Academic Press is an imprint of Elsevier
The Boulevard, Langford Lane, Kidlington, Oxford OX5 1GB, UK
225 Wymann Street, Waltham, MA 02451, USA

First edition 2004
Second edition 2014

British Library Cataloguing in Publication Data
A catalogue record for this book is available from the British Library

Library of Congress Cataloging-in-Publication Data
A catalog record for this book is available from the Library of Congress

ISBN–: 978-0-12-810116-2

For information on all Academic Press publications
visit our website at store.elsevier.com

Printed and bound in the US
14 15 16 17 18 10 9 8 7 6 5 4 3 2 1

DEDICATION

To my daughter Lisa
To my son Konrad and my daughter Rosa

CONTENTS

FOREWORD

Recent advances in medical imaging have transformed it from a primarily morphological assessment tool, to a versatile modality, offering detailed quantitative information for predictive modeling, disease stratification, and personalised therapy. Throughout the evolution of medical imaging, pattern recognition and signal analysis have played a key role in clinical decision support, whether for low-level processing such as filtering and segmentation, or high-level analysis including classification and outcome prediction.

In this book, Meyer-Bäese and Schmid have provided a comprehensive reference of both traditional and modern approaches to medical image computing. Differing from many other books in medical image analysis, it focuses on the underlying information content of data and explains how feature extraction, transformation, and spatio-temporal models can be used for computer-aided diagnosis.

From machine learning and signal processing perspectives, topics related to feature selection and extraction, wavelet transform, and neural networks are familiar topics to the medical imaging community. The authors, however, have managed to provide a seamless link to this diverse range of topics, supplemented with practical examples and exercises. These can help to create a solid foundation for those entering the field or serve as a valuable reference for those who have already embarked on a research career in medical image computing.

The topics on spatio-temporal models in functional and perfusion imaging and time-series analysis address some of the new approaches to deriving functional data from image sequences. Different types of temporal models including linear, nonlinear, and nonparametric models are described and issues related to local versus global smoothing are addressed. Example approaches to cerebral time-series analysis for contrast-enhanced perfusion MRI are provided, allowing a detailed insight into the underlying processing steps involved and stimulating further considerations of improved approaches for addressing this clinically challenging problem. The authors also attempt to provide an integrated approach for computer-aided diagnosis for breast lesions in dynamic contrast-enhanced MRI by linking some of the key concepts that are described in the early part of the book.

It is difficult to unite the wide range of topics that are covered in this book in a seamless fashion and the effort by the authors is laudable. Medical imaging research is intrinsically multidisciplinary and direct application of existing methods in machine learning and signal processing is unlikely to yield clinically significant results. From their extensive experience in clinical collaboration, the authors have demonstrated the importance of focusing on theoretical innovation with a strong emphasis on clinical translation and direct patient benefit. I hope the readers of this book will find these views resonating and applaud the great effort by the authors.

Guang-Zhong Yang
The Hamlyn Center
Imperial College London
February, 2014

PREFACE

Medical imaging is today becoming one of the most important visualization and interpretation methods in biology and medicine. The past decade has witnessed a tremendous development of new, powerful instruments for detecting, storing, transmitting, analyzing, and displaying images. These instruments are greatly amplifying the ability of biochemists, biologists, medical scientists, and physicians to see their objects of study and to obtain quantitative measurements to support scientific hypotheses and medical diagnoses. An awareness of the power of computer-aided analytical techniques, coupled with a continuing need to derive more information from medical images, has led to a growing application of digital processing techniques for the problems of medicine. The most challenging aspect in medical imaging lies in the development of integrated systems for the use in the clinical sector. Design, implementation, and validation of complex medical systems require not solely medical expertise but also a tight collaboration between physicians and biologists, on the one hand, and engineers and physicists, on the other. It is well known that it was the interdisciplinary collaboration between a physicist, G. N. Hounsfield, and a neuroradiologist, J. Ambrose, that led in the late 1960s to the development of the first computer tomographic scanner. Noise, artifacts, and weak contrast are the cause of a decrease in image quality and make the interpretation of medical images very difficult. These sources of interference, which are of a different nature for mammograms than for ultrasound images, are responsible for the fact that conventional or traditional analysis and detection algorithms are not always successful. The biomedical scene is one of the most difficult to cope with since we have to deal with non-Gaussian, nonstationary, and nonlinear processes (transients, bursts, ruptures) but also with mixtures of components interacting in a quite complicated form. Therefore much of the research done today is geared toward improvement of the reduced quality of the available biosignal material. The very recent years have proclaimed spatio-temporal approaches as the future of image analysis in MRI since they combine temporal aspects with local spatial information and thus retain sharp features and borders of lesions or of myocardial tissue areas. The standard assumption of global spatial smoothness proved to be unsuitable for medical imaging and novel motion compensation and segmentation approaches as well as feature extraction techniques have been developed to overcome these new challenges. All these methods emphasize local image information and local adaptive smoothing. The goal of this new edition is to respond to the new demands in medical imaging and to present a complete range of proven and new methods, which play a leading role in the improvement of the biomedical signal analysis and interpretation as well as presentation of intelligent and automated CAD systems with application to spatio-temporal medical images.

The goal of the present book is to present a complete range of proven and new methods, which play a leading role in the improvement of image quality, as well as analysis and interpretation, in the modern medical imaging of this decade. These methods offer solutions to a vast number of problems, for which the classical methods provide only insufficient solutions. Chapter I provides an overview of the foundations of medical imaging. Imaging with ionization radiation, magnetic resonance imaging, ultrasound and ultrasonic imaging, and biomagnetic imaging play a central role in the present book and are described in detail. Chapter II contains a description of methods for feature selection and extraction. Feature selection methods presented are nontransforming and transforming signal characteristics, graphical and structural descriptors, and texture. Methods for feature extraction are exhaustive search, branch and bound algorithm, max-min feature selection, and Fisher's linear discriminant function. Wavelets, for example, are leaders in edge extraction, compression, noise cancellation, feature extraction, image enhancement, and image fusion and occupy, therefore, a central role in this book. Novel feature extraction techniques are added such as local and velocity moments to describe spatio-temporal phenomena in medical image sequences, as well as Minkowski functionals and Writhe number as descriptors for tumor morphology. In addition, Gaussian Markov Random Field and Markov Chain Monte Carlo are defined and applied to medical imaging. Two chapters are dedicated for discussion of wavelets: A mathematical basic part, Chapter III, and an application part, Chapter IV, regarding the application of the wavelets to the above-mentioned problems. Another basic feature extraction method having its roots in evolution theory is genetic algorithms, discussed in Chapter V. Both genetic algorithms and neural networks are among the few approaches for large-scale feature extraction providing an optimal solution for extraction of relevant parameters. Chapters VI–X describe cognitive and noncognitive classification methods relevant for medical imaging. Chapter VI develops the traditional statistical classification methods, presenting both parametric and nonparametric estimation methods, and the less known syntactic or structural approach. Novel statistical pattern recognition techniques such as Bayesian networks and the Bayesian Information Criterion are added. The potential of the methods presented in Chapters II–VI is illustrated by means of relevant applications in radiology, digital mammography, and fMRI. Neural networks have been an emerging technique since the early 1980s and have established themselves as an effective parallel processing technique in pattern recognition. The foundations of these networks are described in Chapter VII. Chapter VIII reviews neural implementations of principal and independent component analysis and presents their application in medical image coding and exploratory data analysis in functional MRI. Besides neural networks, fuzzy logic methods represent one of the most recent techniques applied to data analysis in medical imaging. They are always of interest when we have to deal with imperfect knowledge, when precise modeling of a system is difficult, and when we have to cope with both uncertain and imprecise knowledge. Chapter IX develops the foundations of fuzzy logic and that of several fuzzy clustering algorithms and their application in radiology, fMRI, and MRI. Chapter X details the emerging complex neural architectures for medical imaging. Specialized architectures such as invariant neural networks, context-based neural networks, optimization networks, and elastic contour models are very detailed. The chapter also includes the application of convolutional neural networks, hierarchical pyramidal neural networks, neural networks with receptive fields, and modified Hopfield networks to almost all types of medical images. Principal component analysis and independent component analysis for fMRI data analysis based on self-organizing neural networks are also shown as a comparative procedure. Compression of radiological images based on neural networks is compared to JPEG and SPHIT wavelet compression. Chapter XI describes spatio-temporal models in functional and perfusion imaging and covers spatial approaches for three different types of temporal models: linear, nonlinear, and nonparametric models. Assuming a global spatial smoothness is typically not appropriate for medical images and locally adaptive smoothing allows to retain sharp features and borders in the images. Chapter XII addresses the cerebral time series analysis in contrast-enhanced perfusion MRI time series. Chapter XIII describes integrated complex computer-aided diagnosis systems for medical imaging and shows the application of modern spatio-temporal and local feature selection and classification methods from the previous chapters.

The emphasis of the book lies in the compilation and organization of a breadth of new approaches, modeling, and applications from pattern recognition relevant to medical imaging and aims to respond to novel challenges in spatio-temporal medical image processing. Many references are included and are the basis of an in-depth study. Only basic knowledge of digital signal processing, linear algebra, and probability is necessary to fully appreciate the topics considered in this book. Therefore, we hope that the book will receive widespread attention in an interdisciplinary scientific community.

ACKNOWLEDGMENTS

A book does not just "happen," but requires a significant commitment from its author as well as a stimulating and supporting environment. The author has been very fortunate in this respect. The environment in the Department of Scientific Computing was also conducive to this task. My thanks to the Chair, Max Gunzburger. I would like to thank my graduate students, who used earlier versions of the notes and provided both valuable feedback and continuous motivation.

I am deeply indebted to Prof. Heinrich Werner, Thomas Martinetz, Tim Nattkemper, Fabian Theis, Axel Wismüller, Joachim Weickert, Bernhard Burgeth, Uwe Meyer-Bäse, Andrew Laine, Marek Ogiela, Carla Boetes, Marc Lobbes, Thomas Schlossbauer, and Joachim Wildberger.

The efforts of the professional staff at Elsevier Science, especially Jonathan Simpson and Cari Owen, deserve special thanks. Finally, watching my daughter Lisa-Marie laugh and play rewarded me for the many hours spent with the manuscript.

My thanks to all, many unmentioned, for their help.

Funding for this "ScholarlyWorks" project was made possible by grant G13LM009832 from the National Library of Medicine, NIH, DHHS. The views expressed in any written publication, or other media, do not necessarily reflect the official policies of the Department of Health and Human Services; nor does mention by trade names, commercial practices, or organizations imply endorsement by the U.S. Government.

Anke Meyer-Bäese

I am indebted to my co-author for the opportunity to take part in this book project. I would also like to thank the staff at Elsevier Science for their support.

Thanks go to the Department of Statistics and the graduate students in my group for their support and input while writing this manuscript. I am also grateful for the support of Leonhard Held and Brandon Whitcher, who helped me to develop the necessary skills for my research.

My special thanks go to my wife Stefanie Volz, who provided me with time and moral support for finishing the manuscript.

Volker J. Schmid

LIST OF SYMBOLS

a	Scale parameter
$a_t(i)$	Activity of neuron i at time t
a_{ij}	Connection between the ith and the jth neuron of the same layer
$\mathbf{a}_i$	Vector of lateral connections of the ith neuron
$\mathbf{A}$	Mixing matrix
b	Shift parameter
$c_{m,n}$	Scaling coefficient at scale m and translation n
C	Fuzzy set
$d_{m,n}$	Wavelet coefficient at scale m and translation n
$d(\mathbf{x}, \mathbf{y})$	Distance between data vectors $\mathbf{x}$ and $\mathbf{y}$
$d(\mathbf{x}, \mathbf{m^i}, \mathbf{K^i})$	Mahalanobis distance of the ith receptive field
$D(\mathbf{x_j}, \mathbf{L_i})$	Dissimilarity between a data point $\mathbf{x_j}$ and a prototype $\mathbf{L_i}$
E_p	Error function due to the pth feature vector
f_i	Evaluation associated with the ith string
f_v	Approximation signal
f_w	Detail signal
$\bar{f}$	Average evaluation of all strings in the population
$f_i(x_i)$	Output function of the ith neuron for the current activity level x_i
$\mathbf{f}(\mathbf{x}(t))$	Vector of output functions
$\mathbf{F_X}$	Input field of a neural network
$\mathbf{F_Y}$	Output field of a neural network
$g_i(n)$	Filter coefficients in the synthesis part
$\mathbf{G}$	Grammar
$\mathbf{G}(\|\|\|\|)$	Radial basis function
$h_i(n)$	Filter coefficients in the analysis part
$H(\mathbf{y})$	(Differential) entropy of stochastic signal $\mathbf{y}$
$H(y_s, \mathbf{d})$	Probability that a given difference y_s occurs at distance $\mathbf{d}$
$I(y_1, \ldots, y_m)$	mutual information between m random variables y_i
$I(A_i, \mathbf{L_i})$	Inadequacy between a fuzzy class A_i and its prototype $\mathbf{L}_i$
J	Criterion function
$J(\mathbf{P}, \mathbf{L})$	Criterion function (inadequacy) for partition $\mathbf{P}$ and representation $\mathbf{L}$ for GFNM algorithm
$J(\mathbf{P}, \mathbf{L}, \mathbf{M})$	Criterion function for partition $\mathbf{P}$ and representation $\mathbf{L}$ and shape matrix $\mathbf{M}$ for GAFNM algorithm
$J(\mathbf{P}, \mathbf{V}, R)$	Criterion function for partition $\mathbf{P}$ and hypersphere prototypes $\mathbf{L}$ described by set of radii R and centers $\mathbf{V}$ for GFNS algorithm

$J(\mathbf{P}, \mathbf{V}, R, \mathbf{M})$	Criterion function for partition **P** and hyperellipsoidal prototypes **L** described by set of radii R and centers **V** and shape matrix **M** for GAFNS algorithm
$J(\mathbf{y})$	Negentropy of stochastic signal **y**
$\mathbf{K^i}$	Inverse covariance matrix of the ith receptive field
$K(f\|g)$	Kullback divergence between the pdfs f and g
L	Language or log–likelihood function in ICA
$L(X)$	Family of all fuzzy sets on universe X
$\mathbf{L}_i$	Prototype of fuzzy class A_i
$\mathbf{L}_i(\mathbf{v}_i, r_i)$	Hyperspherical shell prototype of the fuzzy class A_i
$\mathbf{L}_i(\mathbf{v}_i, r_i, \mathbf{M}_i)$	Hyperellipsoidal shell prototype of the fuzzy class A_i
$\mathbf{m}_i$	Centroid of the ith receptive field
M	Factor of the sampling rate change (either decimation or interpolation)
M	Feedforward weight matrix between field $\mathbf{F_X}$ and field $\mathbf{F_Y}$ or shape matrix in fuzzy logic
N	Feedback weight matrix between field $\mathbf{F_Y}$ and field $\mathbf{F_X}$
$p(\mathbf{x})$	Probability density function of **x**
$p(\mathbf{x}\|\omega_i)$	Likelihood function of ω_i with respect to **x**
p	Feature vector
P	Set of production rules or fuzzy partition in fuzzy logic
P	Intraconnection matrix in field $\mathbf{F_X}$
$P(\omega_i)$	A-priori probability of feature class ω_i
$P(\omega_i \mathbf{x})$	A-posteriori probability for a pattern **x** to belong to class ω_i
$P_m f$	Lowpass content of f in the subspace V_m
$\mathbf{q}_i$	ith eigenvector
$Q_m f$	Highpass content of f in the subspace W_m
Q	Eigenvector matrix
r_i	Radius of hypersphere of prototype $\mathbf{L}_i$
R	Set of radii for prototypes **L**
R	Correlation matrix
s	Source signal
$\mathbf{S}_i$	Within-class scatter matrix of fuzzy class A_i
S	Start variable
T_1	Time constant for the molecules to recover their orientation
T_2	Decay time constant
u_{ij}	Membership value of jth input vector to ith fuzzy class
U	Membership matrix
v	Chromosome (binary vector)
$\mathbf{v}_i$	Center of the hypersphere of the prototype $\mathbf{L}_i$

V	Auxiliary alphabet or set of centers in fuzzy logic
V_m	Subspace of the approximation signals
w_{ij}	Feedforward connection between the *i*th and the *j*th neuron
$\mathbf{w}_i$	Feedforward weight vector for the *i*th neuron
$\mathbf{W}$	Demixing matrix
W_m	Orthogonal complement of V_m
$x(n)$	Input signal
$x'(n)$	Intermediate signal
$\mathbf{x}$	Input vector (either feature vector or neural activity vector) or sensor signal
X	Remaining feature set or universe of discourse in fuzzy logic
$y(n)$	Output signal
$\mathbf{y}$	Output vector
Y	Initial available feature set
Z	Alphabet
η	Learning rate
$\phi[\mathbf{y}]$	Contrast function of output $\mathbf{y}$
σ^i_j	*j*th standard deviation of the *i*th receptive field
$\psi(t)$	Wavelet function
$\phi(t)$	Scaling function
$\gamma(\mathbf{x}, \mathbf{m})$	Potential function
λ_i	*i*th eigenvalue
ω_i	*i*th feature class
ϱ	Proton density
$\oplus$	Orthogonal sum
$\perp$	Orthogonal complement

CHAPTER ONE

Introduction*

Contents

Medical imaging deals with the interaction of all forms of radiation with tissue and the design of technical systems to extract clinically relevant information, which is then represented in image format. Medical images range from the simplest such as a chest X-ray to sophisticated images displaying temporal phenomena such as the functional magnetic resonance imaging (fMRI).

The past decades have witnessed a tremendous development of a new, powerful technology for detecting, storing, transmitting, analyzing, and displaying digital medical images. This technology is helping biochemists, biologists, medical scientists, and physicians to obtain quantitative measurements, which facilitate the validation of scientific hypothesis and accurate medical diagnosis.

This chapter gives an overview of image analysis and describes the basic model for computer-aided systems as a common basis enabling the study of several problems of medical-imaging-based diagnostics.

1.1. MODEL FOR MEDICAL IMAGE PROCESSING

The analysis and interpretation of medical images represent two of the most responsible and complex tasks and usually consist of multiple processing steps. However, it is not difficult to generalize this procedure for all medical imaging modalities, and the resulting three-level processing model is shown in Fig. 1.1. Image formation represents the bottom

* This chapter contains material reprinted from chapter 1 of Biomedical Signal Analysis: Contemporary Methods and Applications, by Fabian Theis and Anke Meyer-Base, published by The MIT Press. Reprinted with permission from MIT Press.

Pattern Recognition and Signal Analysis in Medical Imaging
http://dx.doi.org/10.1016/B978-0-12-409545-8.00001-7

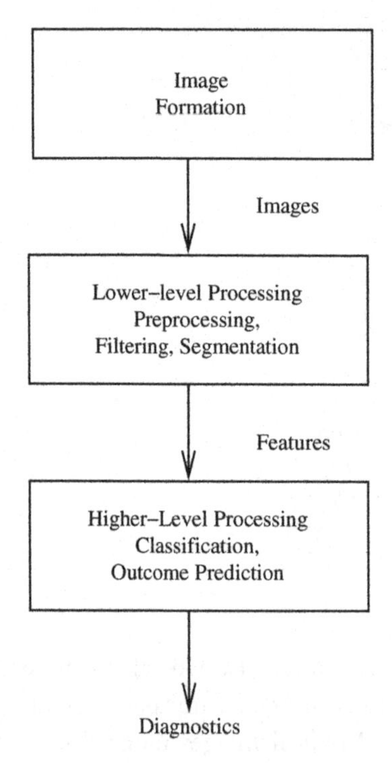

Figure 1.1 Model for diagnostic system using medical images [213].

level of a diagnostic system. Some imaging modalities, such as conventional X-ray, do not rely on any computation, while others such as single-photon emission computed tomography employ image reconstruction as an image processing technique. Image processing is performed in two steps: a lower-level and a higher-level step. The former performs filtering, image enhancement and segmentation, feature extraction, and selection, directly on the raw pixels, while the latter uses the preprocessed pixel data and provides a medical diagnosis based on it. The most important tasks associated with this processing level are feature classification, tumor detection, and, in general, diagnosis for several diseases.

The basic image processing operations can be classified into five categories:

- *Preprocessing:* Preprocessing serves to better visualize object contours exhibiting a low resolution. The most common techniques include motion image registration, histogram transformation, filters, or Laplace-operators.
- *Filtering:* Filtering includes enhancement, deblurring, and edge detection. Enhancement techniques consist of linear or nonlinear, local or global filters, or are wavelet-based. Deblurring techniques may consist of inverse or Wiener filters. Edge-detection techniques include the Haar transform, local operators, prediction, and/or classification methods.
- *Segmentation:* Segmentation can be both region-based and curve-based. There are several different kinds of segmentation algorithms including the classical region growers,

clustering algorithms, and line and circular arc detectors. A critical issue in medical imaging is whether or not segmentation can be performed for many different domains using general bottom-up methods that do not use any special domain knowledge.

- *Shape modeling:* Shape modeling is performed based on features that can be used independently of, or in combination with, size measurements. For medical images, it is sometimes useful to describe the shape of an object in more detail than that offered by a single feature but more compactly than is reflected in the object image itself. A shape descriptor represents in such cases a more compact representation of an object's shape.
- *Classification:* Classification is based on feature selection, texture characterization, and a decision regarding the feature class. Each abnormality or disease is recognized as belonging to a particular class, and the recognition is implemented as a classification process.

1.2. MEDICAL IMAGE ANALYSIS

Medical imaging techniques, mostly noninvasive, play an important role in several disciplines such as medicine, psychology, and linguistics. The four main medical imaging signals are: (1) X-ray transmission, (2) γ-ray transmission, (3) ultrasound echoes, and (4) nuclear magnetic resonance induction. This is illustrated in Table 1.1 where US means ultrasound and MR means magnetic resonance.

The most frequently used medical imaging modalities are illustrated in Fig. 1.2.

Figure 1.2a and b illustrate the concept of ionizing radiation. Projection radiography and computed tomography are based on X-ray transmission through the body and the selective attenuation of these rays by the body's tissue to produce an image. Since they transmit energy through the body they belong to transmission imaging modalities contrary to emission imaging modalities found in nuclear medicine where the radioactive sources are localized within the body. They are based on injecting radioactive compounds into the body which finally move to certain regions or body parts which then emit gamma rays of intensity proportional to the local concentration of the compounds.

Magnetic resonance imaging is visualized in Fig. 1.2c and is based on the property of nuclear magnetic resonance. This means that protons tend to align themselves with this field. Regions within the body can be selectively excited such that these protons tip away from the magnetic field direction. The returning of the protons back to alignment with

Table 1.1 Range of applications of the most important radiologic imaging modalities [248].

X-rays	Breast, lung, bone
γ-rays	Brain, organ parenchyma, heart function
MR	Soft tissue, disks, brain
US	Fetus, pathological changes, internal organs

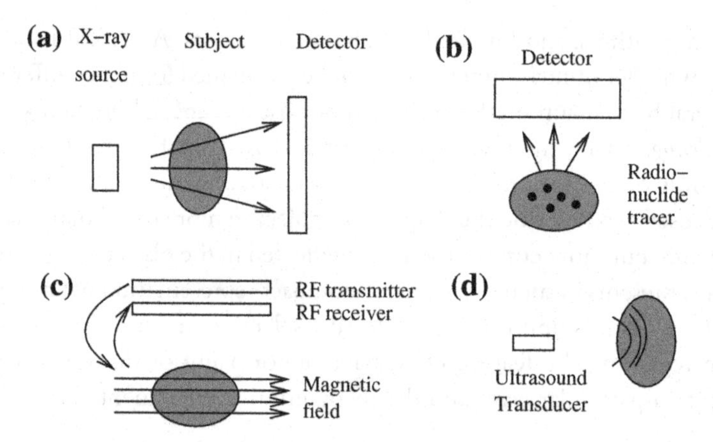

Figure 1.2 Schematic schemes of the most frequently used medical imaging modalities [213] (a) X-ray imaging, (b) radionuclide imaging, (c) MRI, and (d) ultrasound.

the field causes a precession. This produces a radio-frequency electromagnetic signature which can be detected by an antenna.

Figure 1.2d represents the concept of ultrasound imaging: high-frequency acoustic waves are sent into the body and the received echoes are used to create an image.

In this chapter, we discuss the four main medical imaging signals introduced in Fig. 1.2. Both the medical physics behind these imaging modalities will be presented as well as the image analysis challenges. Since the goal of medical imaging is to be automated as much as possible, we will give an overview about computer-aided diagnosis systems in Section 1.3. Their main component, the workstation, is very detailedly described.

For further details on medical imaging, readers are referred to [59,228,384].

1.2.1 Imaging with Ionizing Radiation

X-ray is the widest-spread medical imaging modality, discovered by W.C. Röntgen in 1895. X-rays represent a form of ionizing radiation with a typical energy range between 25 keV and 500 keV for medical imaging. A conventional radiographic system contains an X-ray tube that generates a short pulse of X-rays that travels through the human body. Those X-ray photons that are not absorbed or scattered reach the large area detector creating an image on a film. The attenuation has a spatial pattern in function of the linear attenuation coefficient distribution in the body. This energy and material-dependent effect is captured by the basic imaging equation

$$I_d = \int_0^{E_{max}} S_0(E) E \exp\left[-\int_0^d \mu(s; E) ds\right] dE \tag{1.1}$$

where $S_0(E)$ is the X-ray spectrum and $\mu(s; E)$ represents the linear attenuation coefficient along the line between the source and detector. s is the distance from the origin and d is the source-to-detector distance.

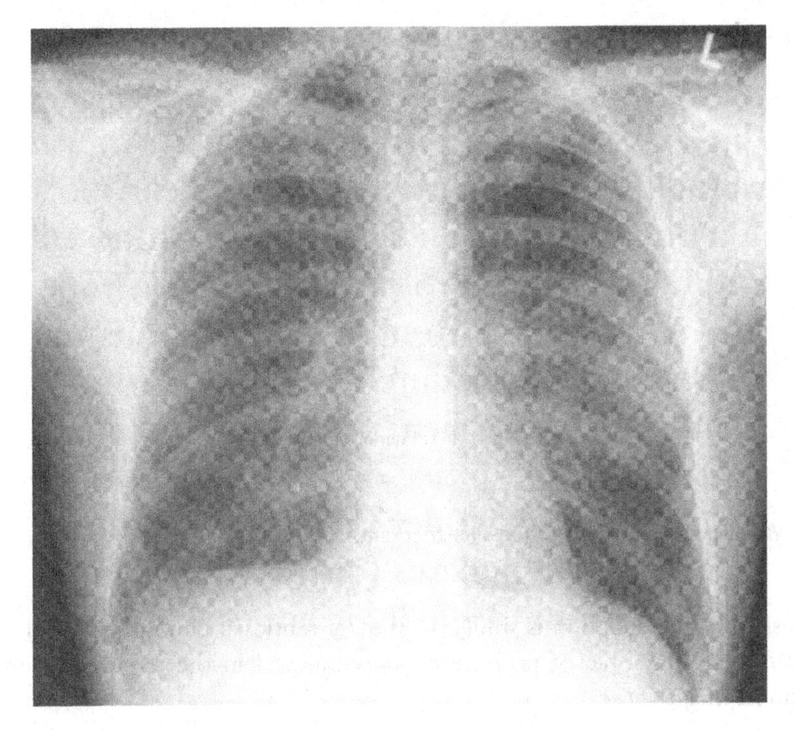

Figure 1.3 Thorax X-ray. *(Courtesy of Publicis-MCD-Verlag.)*

The image quality is influenced by both the noise stemming from the random nature of the X-rays or their transmission. Figure 1.3 displays a thorax X-ray.

A popular imaging modality is computed tomography (CT), introduced by Hounsfield in 1972, eliminates the artifacts stemming from overlaying tissues and thus hampering a correct diagnosis. In CT, X-ray projections are collected around the patient. It can be visualized as a series of conventional X-rays taken as the patient is rotated slightly around an axis.

The films show a 2-D projection at different angles of a 3-D body. A horizontal line in a film visualizes a 1-D projection of a 2-D axial cross-section of the body. The collection of horizontal lines stemming from films at the same height represents an one axial cross-section. The two-dimensional cross-sectional slices of the subject are reconstructed from the projection data based on the Radon transform [59], an integral transform introduced by J. Radon in 1917. This transformation collects 1-D projections of a 2-D object over many angles and the reconstruction is based on a filtered backpropagation, which is the most employed reconstruction algorithm. The projection-slice theorem forms the basis of the reconstruction: it states that a 1-D Fourier transform of a projection is a slice of the 2-D Fourier transform of the object. Figure 1.4 visualizes this aspect.

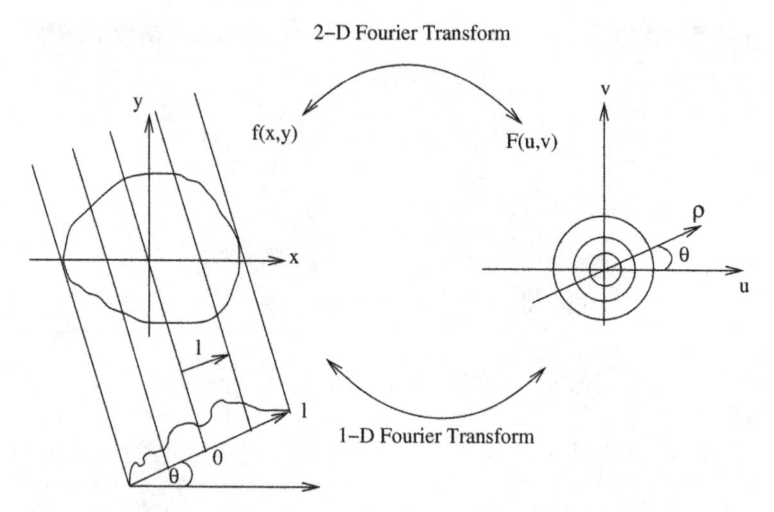

Figure 1.4 Visualization of the projection-slice theorem.

The basic imaging equation is similar to the conventional radiography with the sole difference that an ensemble of projections are employed in the reconstruction of the cross-sectional images

$$I_d = I_0 \exp\left[-\int_0^d \mu(s;\bar{E})ds\right]dE \tag{1.2}$$

where I_0 is the reference intensity and $\bar{E}$ the effective energy.

The major advantages of CT over projection radiography are: (1) eliminating the superposition of images of structures outside the region of interest, (2) providing a high-contrast resolution such that differences between tissues of physical density of less than 1% become visible, and (3) as a tomographic and potentially three-dimensional method allowing the analysis of isolated cross-sectional visual slices of the body. The most common artifacts in CT images are aliasing and beam hardening. CT represents an important tool in medical imaging used to provide additional information than X-rays or ultrasound. It is mostly employed in the diagnosis of cerebrovascular diseases, acute and chronic changes of the lung parenchyma, supporting ECG, and for a detailed diagnosis of abdominal and pelvic organs. An example of a CT image is shown in Fig. 1.5.

Nuclear medicine began in the late 1930s and many of its procedures use radiopharmaceuticals. Its beginning marked the use of radioactive iodine to treat thyroid disease. Like X-ray imaging, nuclear medicine imaging developed from projection imaging to tomographic imaging. Nuclear medicine is based on ionizing radiation, and image generation is similar to an X-ray's but with an emphasis on the physiological function rather than anatomy. However, in nuclear medicine radiotracers and thus the source of emission are introduced into the body. This technique is a functional imaging modality: the physiology

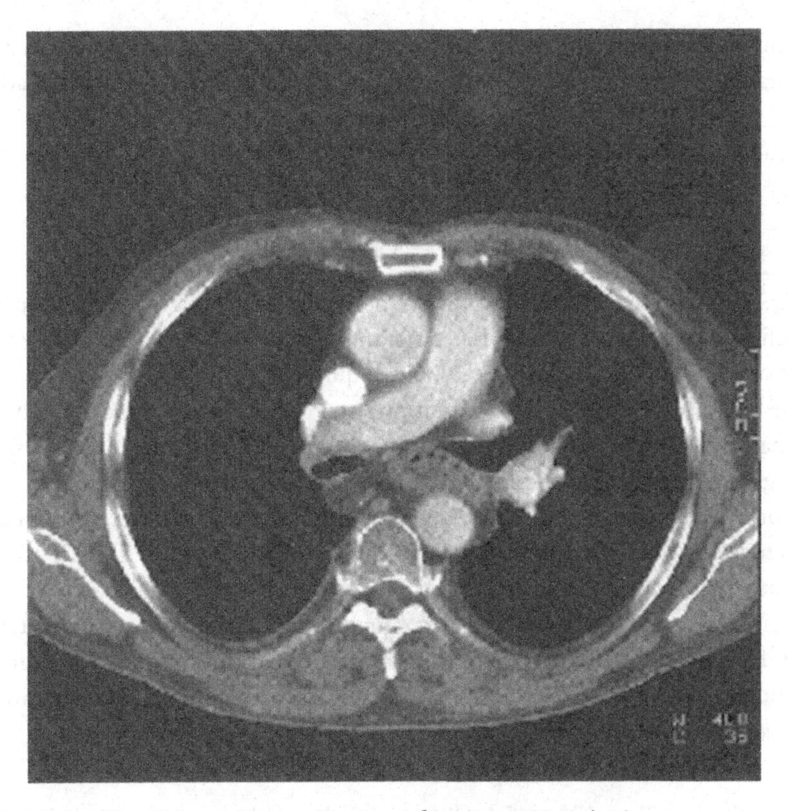

Figure 1.5 CT of mediastinum and lungs. *(Courtesy of Publicis-MCD-Verlag.)*

and biochemistry of the body determine the spatial distribution of measurable radiation of the radiotracer. In nuclear medicine, different radiotracers visualize different functions and thus provide different information. In other words, a variety of physiological and biochemical functions can be visualized by different radiotracers. The emissions stemming from a patient are recorded by scintillation cameras (external imaging devices) and converted either into a planar, 2-D image, or cross-sectional images.

Nuclear medicine is relevant for clinical diagnosis and treatment covering a broad range of applications: tumor diagnosis and therapy, acute care, cardiology, neurology, and renal and gastrointestinal disorders.

Based on the method of radiopharmaceutical disintegration, the three basic imaging modalities in nuclear medicine are usually divided into two main areas: (1) planar imaging and single-photon emission computed tomography (SPECT) using gamma emitters as radiotracers and (2) positron emission tomography (PET) using positrons as radiotracers. Projection imaging, also called planar scintigraphy, uses the Anger scintillation camera, an electronic detection instrumentation. This imaging modality is based on the detection and estimation of the position of individual scintillation events on the face of an Anger camera.

The fundamental imaging equation contains two important components: activity as the desired parameter, and attenuation as an undesired but extremely important additional part.

The fundamental imaging equation is given below:

$$\phi(x, y) = \int_{\infty}^{0} \frac{A(x, y, z)}{4\pi z^2} \exp\left(-\int_{z}^{0} \mu(x, y, z'; E)dz'\right) dz \tag{1.3}$$

where $A(x, y, z)$ represents the activity in the body and E the energy of the photon. The image quality is determined mainly by camera resolution and noise stemming from the sensitivity of the system, injected activity and acquisition time.

On the other hand, SPECT uses a rotating Anger scintillation camera to obtain projection data from multiple angles. Single-photon emission uses nuclei that disintegrate by emitting a single γ photon, which is measured with a gamma camera system. SPECT is a slice-oriented technique, in the sense that the obtained data are tomographically reconstructed to produce a 3-D data set or thin (two-dimensional) slices. This imaging modality can be viewed as a collection of projection images where each is a conventional planar scintigram. The basic imaging equation contains two inseparable terms, the activity and attenuation. Before giving the imaging equation, we need some geometric considerations: if x and y are rectlinear coordinates in the plane, the line equation in the plane is given as

$$L(l, \theta) = \{(x, y)|x \cos\theta + y \sin\theta = l\} \tag{1.4}$$

with l being the lateral position of the line and θ the angle of a unit normal to the line. Figure 1.6 visualizes this aspect.

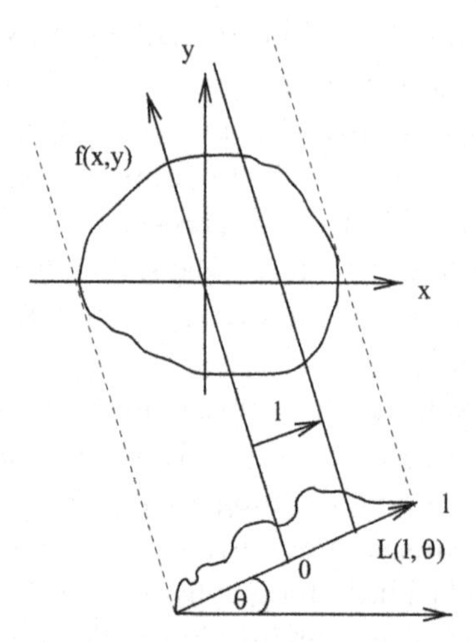

Figure 1.6 Geometric representations of lines and projections.

This yields the following parametrization for the coordinates $x(s)$ and $y(s)$

$$x(s) = l\cos\theta - s\sin\theta \tag{1.5}$$

$$y(s) = l\sin\theta + s\cos\theta \tag{1.6}$$

Thus, the line integral of a function $f(x, y)$ is given as

$$g(l, \theta) = \int_{-\infty}^{\infty} f(x(s), y(s))ds \tag{1.7}$$

For a fixed angle $\theta, g(l, \theta)$ represents a projection while for all l and θ it is called the 2-D Radon Transformation of $f(x, y)$.

The imaging equation for SPECT ignoring the effect of the attenuation term is given below:

$$\phi(l, \theta) = \int_{-\infty}^{\infty} A(x(s), y(s))ds \tag{1.8}$$

where $A(x(s), y(s))$ describes the radioactivity within the 3-D body and represents the inverse 2-D Radon transform of $\phi(l, \theta)$. Therefore, there is no closed-form solution for attenuation correction in SPECT. SPECT represents an important imaging technique by providing an accurate localization in the 3-D space and is used to provide functional images of organs. Its main applications are in functional cardiac and brain imaging. Figure 1.7 shows an image of a brain SPECT study.

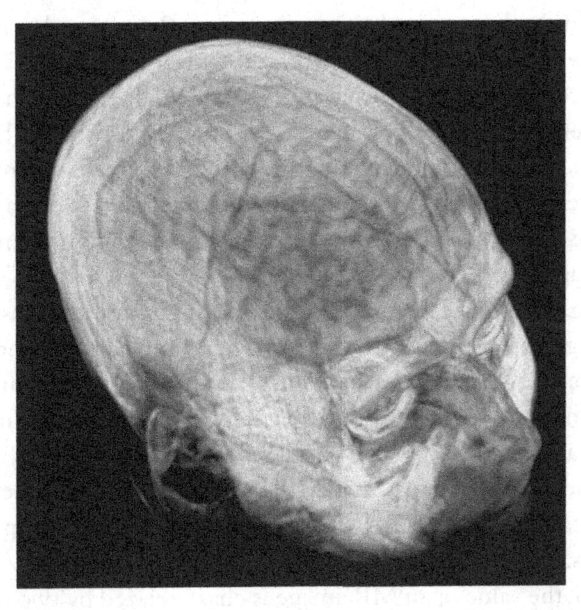

Figure 1.7 SPECT brain study. *(Image courtesy Dr. A. Wismüller, Dept. of Radiology, University of Munich.)*

PET is a particular technique having no analogy with other imaging modality. The radionuclides employed for PET emit positrons instead of γ-rays. These positrons, antiparticles of electrons, are measured and their positions are computed. The reconstruction is produced by using algorithms of filtered backprojection. The imaging equation in PET is similar to that in SPECT with one single difference: the limits of integration for the attenuation term span the entire body because of the coincidence detection of paired γ-rays, the so-called annihilation photons. The imaging equation is given as

$$\phi(l,\theta) = K\int_{-R}^{R} A(x(s), y(s))ds \tag{1.9}$$

where K represents a constant that includes the constant factors, such as detector area and efficiency, that influence ϕ. The image quality in both SPECT and PET is limited by resolution, scatter, and noise. PET has its main clinical application in oncology, neurology, and psychiatry. An important area represents neurological disorders such as early detection of Alzheimer disease, dementia, and epilepsy.

1.2.2 Magnetic Resonance Imaging

Magnetic resonance imaging (MRI) represents noninvasive imaging methods used to render images of the inside of the body. During the past 30 years, it became one of the key bioimaging modalities in medicine. It provides pathological and physiological changes of body's tissues like nuclear medicine, in addition to structural details of organs like CT.

The MRI signal stems from the nuclear magnetism of hydrogen atoms located in the fat and water of the human body and is based on the physical principle of nuclear magnetic resonance (NMR). NMR is concerned with the charge and angular momentum possessed by certain nuclei. Nuclei are of positive charge and have in case of an odd atomic number or mass number an angular momentum Φ. By having spin, these nuclei are NMR-active. Each nucleus that has a spin also has a microscopic magnetic field. When an external electric field is applied, the spins tend to align with the applied magnetic field. This property is called nuclear magnetism. Thus, the spin systems become macroscopically magnetized.

In MR imaging, we look at the macroscopic magnetization by considering a specific spin system (hydrogen atoms) within a sample. The "sample" represents a small volume of tissue—i.e., a voxel. By applying a static magnetic field $\mathbf{B}_0$, the spin system becomes magnetized and can be modeled by a bulk magnetization vector $\mathbf{M}$. In the undisturbed state, $\mathbf{M}$ will reach an equilibrium value $\mathbf{M}_0$ parallel to the direction of $\mathbf{B}_0$, see Fig. 1.9a.

It is very important to mention that $\mathbf{M}(\mathbf{r}, t)$ is a function of time and of the 3-D coordinate $\mathbf{r}$ that can be manipulated spatially by external radio-frequency excitations and magnetic fields.

At a given voxel, the value of an MR image is characterized by two important factors: the tissue properties and the scanner imaging protocol. The most relevant tissue properties are the relaxation parameters T_1 and T_2 as well as the proton density. The proton density is defined as the number of targeted nuclei per unit volume. The scanner software and hardware are manipulating the magnetization vector **M** over time and space based on the so-called pulse-sequence.

In the following, we will focus on a particular voxel and give the equations of motion for $\mathbf{M}(t)$ as a function of time t. These equations are based on the Bloch equations and describe a precession of the magnetization vector around the external applied magnetic field with a frequency ω_0 which is known as the resonance or Larmor frequency.

The magnetization vector $\mathbf{M}(t)$ has two components:

1. The longitudinal magnetization given by $M_z(t)$, the z-component of $\mathbf{M}(t)$.
2. The transverse magnetization vector $M_{xy}(t)$, a complex quantity, which combines two orthogonal components

$$M_{xy}(t) = M_x(t) + jM_y(t) \tag{1.10}$$

where ϕ is the angle of the complex number M_{xy}, known as phase angle, given as

$$\phi = \tan^{-1} \frac{M_x}{M_y} \tag{1.11}$$

Since $\mathbf{M}(t)$ is a magnetic moment, it will have a torque if an external time-varying magnetic field $\mathbf{B}(t)$ is applied. If this field is static and oriented parallel to z-direction, then $\mathbf{B}(t) = \mathbf{B}_0$.

The magnetization vector **M** precesses if it is initially oriented away from the $\mathbf{B}_0$. The spin system can also be excited by using RF signals such that RF signals are produced as output by the stimulated system. This RF excitation is achieved by applying $\mathbf{B}_1$ at the Larmor frequency rather than keeping it constant and allows a tracking of the position of $\mathbf{M}(t)$. However, the precession is not perpetual and we will show that there are two independent mechanisms to dampen the motion and cause the received signal to vanish: the longitudinal and transversal relaxation.

By the RF excitation, $\mathbf{M}(t)$ is pushed down at an angle α toward the xy-plane if $\mathbf{B}_1$ is along the direction of the y-axis. At $\alpha = 0$, we have $M_z = 0$ and the magnetization vector rotates in the xy-plane with a frequency equal to the Larmor frequency. The $\mathbf{B}_1$ pulse needed for an angle $\alpha = \pi/2$ is called the 90 pulse. The magnetization vector returns to its equilibrium state and the relaxation process is described by

$$M_z(t) = M_0 \left[1 - \exp\left(-\frac{t}{T_1} \right) \right] \tag{1.12}$$

and depends on the longitudinal or spin-lattice relaxation time (T_1), see Fig. 1.8.

Transverse or spin-spin relaxation is the effect of perturbations caused by other neighboring spins as they change their phase relative to others. This dephasing leads to a loss of the signal in the receiver antenna. The resulting signal is named free induction decay

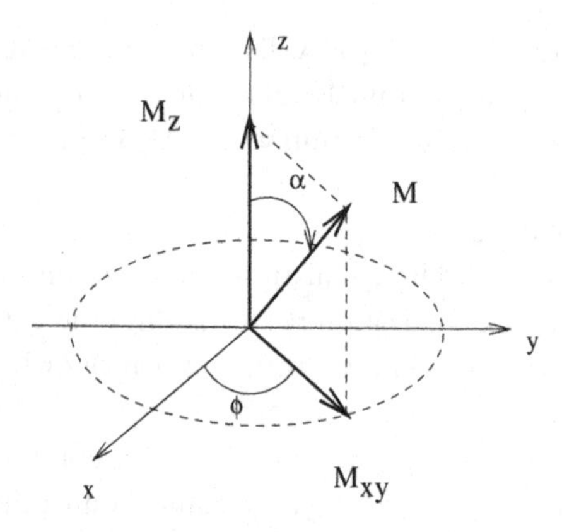

Figure 1.8 The magnetization vector **M** precesses about the *z*-axis.

(FID). The return of the transverse magnetization $\mathbf{M}_{xy}$ to the equilibrium is described by

$$M_{xy}(t) = M_{x_0 y_0} \exp\left(-\frac{t}{T_2}\right) \tag{1.13}$$

where T_2 is the spin-spin relaxation time. T_2 is tissue-dependent and produces the contrast in MR images. However, the received signal decays faster than T_2. Local perturbations in the static field $\mathbf{B}_0$ give rise to a faster time constant T_2^* where $T_2^* < T_2$. Figure 1.9b visualizes this situation. The decay associated with the external fields effects is modeled by the time constant T_2'. The relationship between the three transverse relaxation constants is modeled by

$$\frac{1}{T_2^*} = \frac{1}{T_2} + \frac{1}{T_2'} \tag{1.14}$$

It is important to mention that both T_1 and T_2 are tissue-dependent and for all material $T_2 \leq T_1$.

Valuable information is obtained from measuring the temporal course of the T1/T2 relaxation process after applying an RF-pulse sequence. This measured time course is converted from the time- to the frequency-domain based on the Fourier transform. The amplitude in the spectrum appears at the resonance frequency of hydrogen nucleons in water, see Fig. 1.10.

A contrast between tissues can be seen if the measured signal is different in those tissues. In order to achieve this, two possibilities are available: the intrinsic NMR properties such as P_D, T_1, and T_2 and the characteristics of the externally applied excitations. It is possible to control the tip angle α and to use sophisticated pulse-sequences such as the spin-echo-sequence. A 90°-pulse has a period of TR seconds (repetition time) and is followed by an 180°-pulse after TE seconds (echo time). This second pulse rephases partially the spins and produces an echo signal.

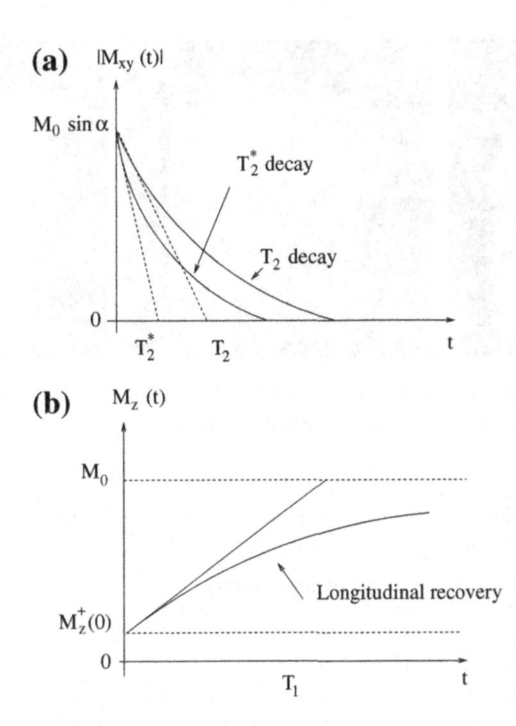

Figure 1.9 (a) Transverse and (b) longitudinal relaxation.

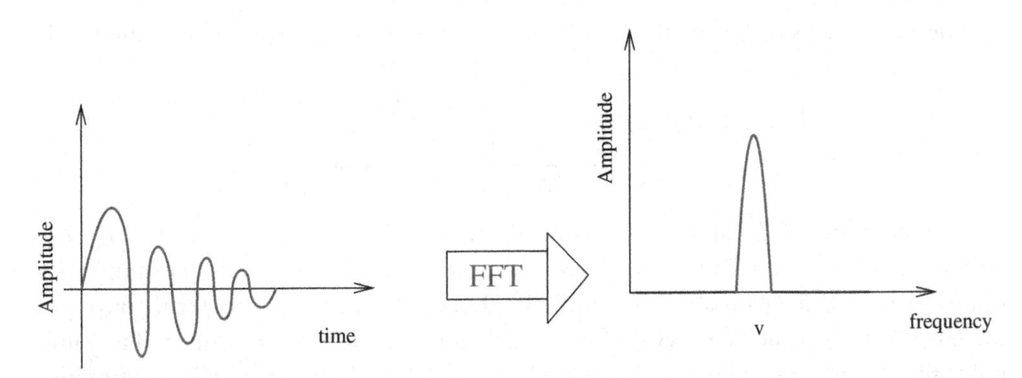

Figure 1.10 Frequency-domain transformation of the measured temporal course. The amplitude in the spectrum is exhibited at the Larmor frequency.

Figure 1.11 shows a brain scan as T_1-weighted, T_2-weighted, and hydrogen density-weighted images.

By weighted is meant that the differences in intensity observed between different tissues are mainly caused by the differences in T_1, T_2, and P_D, respectively, of the tissues. The basic way to create contrast based on the above parameters is visualized in Table 1.2.

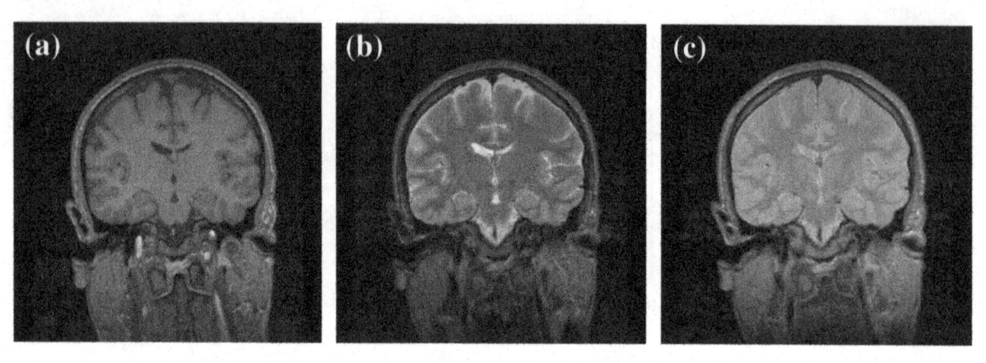

Figure 1.11 Brain MRI showing (a) T_1, (b) T_2, and (c) hydrogen density-weighted images. *(Image courtesy Dr. A. Wismüller, Dept. of Radiology, University of Munich.)*

Table 1.2 Basic way to create contrast depending on P_D, T_1, and T_2.

Contrast	Scanner parameters
P_D	Long T_R, read FID or use short T_E
T_2	Long T_R, $T_E \approx T_2$
T_1	Read FID or use short T_E, $T_R \approx T_1$

The pixel intensity $I(x, y)$ of an MR image obtained using a spin-echo sequence is given by

$$I(x,y) \propto P_D(x,y) \underbrace{\left(1 - \exp\left[-\frac{T_R}{T_1}\right]\right)}_{\text{T1-weighting}} \underbrace{\exp\left[-\frac{T_E}{T_2}\right]}_{\text{T2-weighting}} \tag{1.15}$$

Varying the values of T_R and T_E will control the sensitivity of the signal to the T_1/T_2-relaxation process and will produce different weighted contrast images. If for example, T_R is much larger than T_1 for all tissues in the ROI, then the T_1-weighting term converges to zero and there is no sensitivity of the signal to the T_1-relaxation process. The same holds for T_E much smaller than T_2 for all tissues. When both T_1 and T_2 sensitivities decrease, then the pixel density depends only on the proton density $P_D(x, y)$.

The MR image quality depends not only on contrast but also on sampling and noise. Summarizing, the advantages of MRI as an imaging tool are: (1) excellent contrasts between the various organs and tumors essential for image quality, (2) three-dimensional nature of the image, and (3) the contrast provided by the T_1 and T_2 relaxation mechanism as one of the most important imaging modalities.

An important technique in MRI represents multispectral magnetic resonance imaging. A sequence of 3-D MRI images of the same ROI is recorded assuming that the images are correctly registered. This imaging type enables the discrimination of different tissue types.

To further enhance the contrast between tissue types, contrast agents (CA) are used to manipulate the relaxation times. CAs are intravenously administrated, and during that time a signal enhancement is achieved for tissue with increased vascularity.

Functional magnetic resonance imaging (fMRI) represents a novel noninvasive technique for the study of cognitive functions of the brain [267]. The basis of this technique represents the fact that the MRI signal is susceptible to the changes of hemodynamic parameters, such as blood flow, blood volume, and oxygenation, that arise during neural activity. The most commonly used fMRI signal is the blood oxygenation level-dependent (BOLD) contrast. The BOLD temporal response changes when the local deoxyhemoglobin concentration decreases in an area of neuronal activity. This fact is reflected in T_2^*- and T_2-weighted MR images.

The two underlying characteristics of hemodynamic effects are spatial and temporal. While vasculature is mainly responsible for spatial effects, the temporal effects are responsible for the caused delay of the detected MR signal changes in response to neural activity and a longer duration of the dispersion of the hemodynamic changes. The temporal aspects impose two different types of fMRI experiments: "block" designs and "event-related" designs. The block designs are characterized by an experimental task performed in an alternating sequence of 20–60 s blocks. In event-related designs multiple stimuli are presented randomly and the corresponding hemodynamic response to each is measured. The main concept behind this type of experiment represents the almost linear response to multiple stimulus presentations. fMRI with high temporal and spatial resolution represents a powerful technique for visualizing rapid and fine activation patterns of the human brain. The functional localization is based on the evident correlation between neuronal activities and MR signal changes. As is known from both theoretical estimations and experimental results [266], an activated signal variation appears very low on a clinical scanner. This motivates the application of analysis methods to determine the response waveforms and associated activated regions.

The main advantages of this technique are: (1) noninvasively recording of brain signals without any risk of radiation like CT, (2) excellent spatial and temporal resolution, and (3) integration of fMRI with other techniques such as MEG and EEG to study the human brain.

fMRI's main feature is to image brain activity in vivo. Therefore its applications lie in the diagnosis, interpretation, and treatment evaluation of clinical disorders of cognitive brain functions. The most important clinical application lies in preoperative planning and risk-assessment in intractable focal epilepsy. In pharmacology, fMRI is a valuable tool in determining how the brain is responding to a certain drug. Besides in clinical applications, the importance of fMRI in understanding neurological and psychiatric disorders and refining the diagnosis is continuously growing.

1.2.3 Ultrasound and Acoustic Imaging

Ultrasound is a leading imaging modality and has been extensively studied since the early 1950. It represents a noninvasive imaging modality which produces oscillations of 1–10 MHz when passing soft tissues and fluid.

The cost effectiveness and the portability of the modality made this technique extremely popular. Its importance in diagnostic radiology is unquestionable enabling imaging of pathological changes of inner organs, blood vessels, and supporting breast cancer detection.

The principle of the ultrasonic imaging is very simple: the acoustic wave launched by a transducer into the body interacts with tissue and blood and some of the energy, that is not absorbed, returns to the transducer and is detected by the transducer. As a result, so-called "ultrasonic signatures" emerge from the interaction of ultrasound energy with different tissue types that are subsequently used for diagnosis.

The speed of sound in tissue is a function of tissue type, temperature, and pressure. Table 1.3 gives some examples of acoustic properties of some materials and biological tissues. Because of scattering, absorption, or reflection, an attenuation of the acoustic wave is observed. The attenuation is described by an exponential function of the distance, described by $A(x) = A_0 \exp(-\alpha x)$, where A is the amplitude, A_0 a constant, α the attenuation factor, and x the distance. The important characteristics of the returning signal such as amplitude and phase provide pertinent information about the interaction and the type of the crossed medium. The basic imaging equation is the pulse-echo equation which gives a relation between the excitation pulse, the transducer face, and object reflectivity, and received signal.

In ultrasound, we have the following imaging modes:

- A-mode or amplitude mode: most simple method that displays the envelope of pulse-echoes versus time. Mostly used in ophthalmology to determine the relative distances between different regions of the eye, and in localization of the brain midline, or myocardium infarction. Figure 1.12 visualizes this aspect.

Table 1.3 Acoustical properties of some materials and biological tissues.

Medium	Speed of sound (m/s)	Impedance (10^6 kg/m^{2}s)	Attenuation (dB/cm at 1 MHz)
Air	344	0.0004	12
Water	1480	1.48	0.0025
Fat	1410	1.38	0.63
Muscle	1566	1.70	1.2–3.3
Liver	1540	1.65	0.94
Bone	4080	7.80	20.0

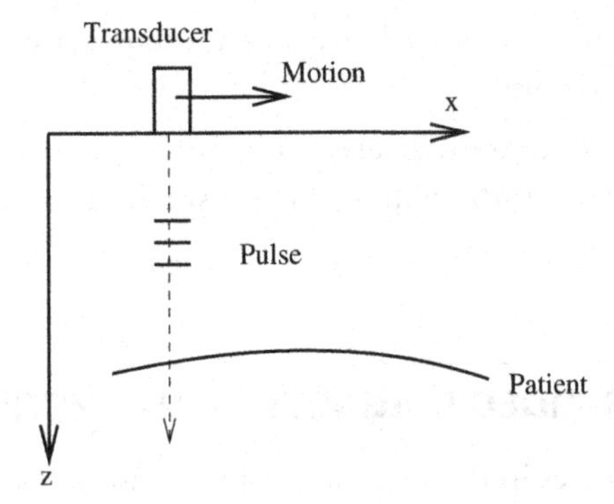

Figure 1.12 A-mode display.

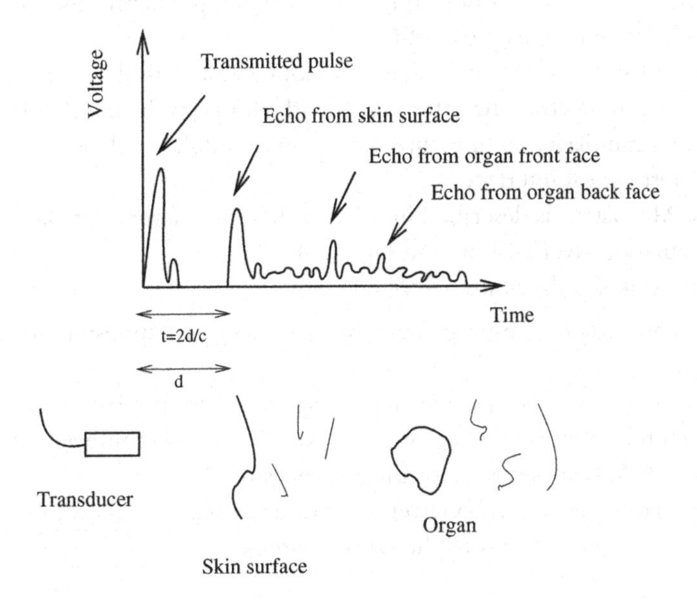

Figure 1.13 B-mode scanner.

- B-mode or brightness mode: produced by scanning the transducer beam in a plane as shown in Fig. 1.13. It can be used for both stationary and moving structures such as cardiac valve motion.
- M-mode or motion mode: it displays the A-mode signal corresponding to repeated pulses in a separate column of a 2-D image. It is mostly employed in conjunction with ECG for motion of the heart valves.

The two basic techniques used to achieve a better sensitivity of the echoes along the dominant (steered) direction are:

- Beamforming: increases the transducer's sensitivity to a particular direction.
- Dynamic focusing: increases the transducer's sensitivity to a particular point in space at a particular time.

1.3. COMPUTER-AIDED DIAGNOSIS (CAD) SYSTEMS

The important advances in the field of computer vision paired with artificial intelligence techniques and data mining have been facilitating the development of automatic medical image analysis and interpretation. Computer-aided diagnosis (CAD) systems are the result of these research endeavors and provide a parallel second opinion in order to assist clinicians in the process of detecting abnormalities, predicting the disease progress and obtaining a differential diagnosis of lesions.

Modern CAD systems are becoming very sophisticated tools with a user-friendly graphical interface supporting the interactions with clinicians during the diagnostic process. They have a multilayer architecture with many modules such as image processing, databases, and a graphical interface.

A typical CAD system is described in [294]. It has three layers: data layer, application layer, and presentation layer as shown in Fig. 1.14.

The functions of each layer are described below:

- Data layer: has a database management system which is responsible for archival and distribution of data.
- Application layer: has a management application server for database access and presentation to graphical user interface, a www server to ensure remote access to the CAD system, and a CAD workstation for image processing.
- Presentation layer: has the web viewer to allow a fast remote access to the system and at the user site it grants access to the whole system.

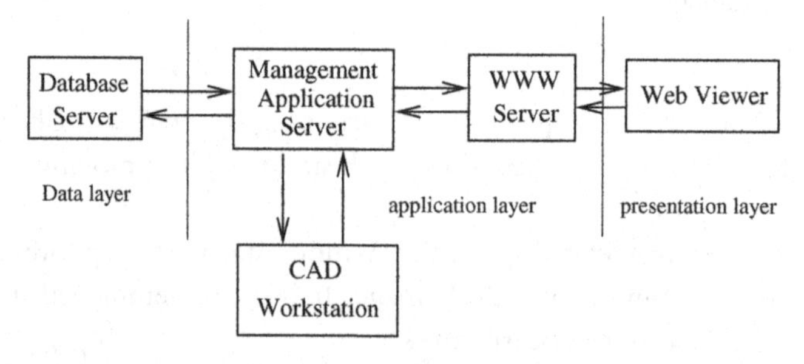

Figure 1.14 Multilayer structure of a CAD system [294].

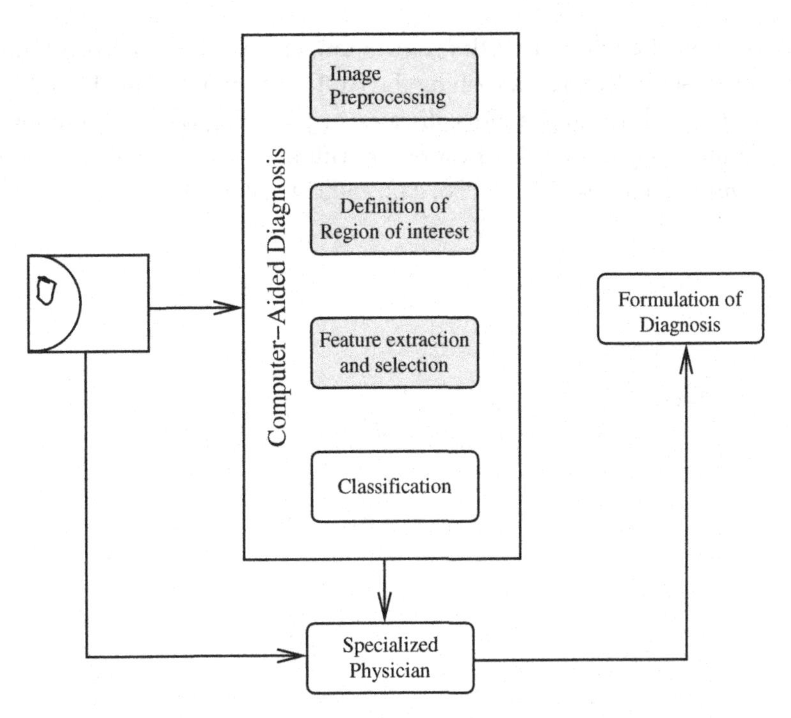

Figure 1.15 Typical architecture of a CAD workstation.

1.3.1 CAD Workstation

A typical CAD system's architecture is shown in Fig. 1.15. It has four important components: (i) image preprocessing, (ii) definition of a region of interest (ROI), (iii) extraction and selection of features, and (iv) classification of the selected ROI.

These basic components are described in the following:

- Image preprocessing: the goal of it is to improve the quality of the image based on image registration, denoising, and enhancing the edges of the image or its contrast. This task is crucial for subsequent tasks.
- Definition of an ROI: the ROIs are mostly determined by seeded region growing and by active contour models that approximate correctly the shapes of organ boundaries.
- Extraction and selection of features: are crucial for the subsequent classification and are based on finding mathematical methods for reducing the size of measurements of medical images. Feature extraction is typically carried out in the spectral or spatial domain and considers the whole image content and maps it onto a lower dimensional feature space. On the other hand, feature selection considers only the necessary information in order to achieve a robust and accurate classification. The employed methods for removing redundant information are either exhaustive, heuristic, or nondeterministic.

- Classification of the selected ROI: classification, either supervised or unsupervised, assigns a given set of features describing the ROI to its proper class. These classes can be used in medical imaging tumors, diseases, or physiological signal groups. Several supervised and unsupervised classification algorithms have been applied in the context of breast tumor diagnosis [58, 87, 226, 242, 250, 292, 360, 403].

CHAPTER TWO

Feature Selection and Extraction

Contents

Pattern Recognition and Signal Analysis in Medical Imaging
http://dx.doi.org/10.1016/B978-0-12-409545-8.00002-9

2.1. INTRODUCTION

Pattern recognition tasks require the conversion of pattern in features describing the collected sensor data in a compact form. Ideally, this should pertain only to relevant information. Feature selection methods can be either classical methods (statistic or of syntactic nature) or biologically oriented (neural or genetic algorithm based) methods.

Feature extraction and selection in pattern recognition are based on finding mathematical methods for reducing dimensionality of pattern representation. A lower-dimensional representation based on pattern descriptors is a so-called feature. It plays a crucial role in determining the separating properties of pattern classes. The choice of features, attributes, or measurements has an important influence on: (1) accuracy of classification, (2) time needed for classification, (3) number of examples needed for learning, and (4) cost of performing classification.

The cost of performing classification is very important in medical diagnosis where patterns are described by using both observable symptoms and results of diagnostic tests. Different diagnostic tests have different costs as well as risks associated with them. For example, needle biopsy is more expensive than X-ray mammography.

Generally speaking, for each pattern recognition problem there are many solution strategies. Table 2.1 illustrates this aspect. Deciding on one specific solution is not always simple.

This chapter gives an overview of the most relevant feature selection and extraction methods for biomedical image processing. Feature extraction methods encompass, besides the traditional transformed and nontransformed signal characteristics and texture, structural and graph descriptors. The feature selection methods described in this chapter are the exhaustive search, branch and bound algorithm, max–min feature selection, sequential forward and backward selection, and Fisher's linear discriminant. Advanced feature representation methods are becoming necessary when it comes to dealing with the local image content or with spatio-temporal characteristics or with the statistical image content.

2.2. ROLE OF FEATURE SELECTION AND EXTRACTION

Dimensionality reduction plays an important role in classification performance. A recognition system is designed using a finite set of inputs. While the performance of this system increases if we add additional features, at some point a further inclusion leads to

Table 2.1 Standard approaches in pattern recognition.

Measured features	**Transformation of features**	**Structural features**
Amplitude	Polynomials	Peaks
Bias	Harmonic analysis	Derivatives
Duration	Fourier transform	Lines
Phase	Wavelet transform	Edges
Energy	Haar transform	LPC coefficients
Moments	Karhunen-Loeve transform	Parametric models
Singular values		
Karhunen-Loeve eigenvalues		
Feature selection	**Classifiers**	**Clustering methods**
Discriminant analysis	Euclidian distance	Isodata algorithm
Chernoff bound	Mahalanobis distance	Fisher's linear discriminant
Bhattacharya	Linear discriminant functions	Parsing
Divergence	Bayesian linear classifier	
Exhaustive search	Maximum likelihood	
Dynamic programming	Production rules	
	Density functions	
	Parzen estimator	
	k-NN algorithm	
	Histogram	

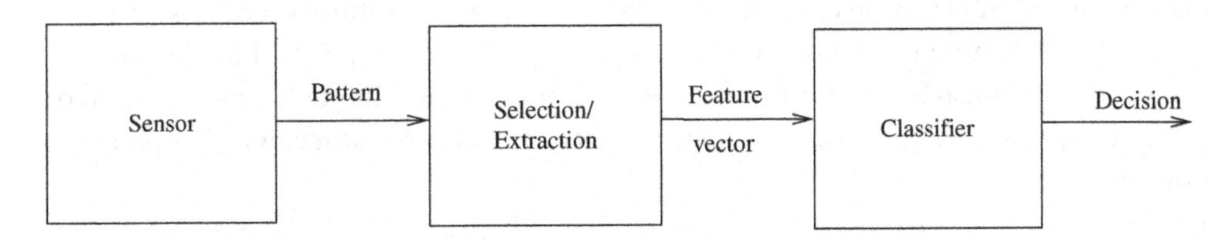

Figure 2.1 Pattern recognition system including feature selection and extraction.

a performance degradation. Thus a dimensionality reduction may not always improve a classification system.

A model of the pattern recognition system including the feature selection and extraction stages is shown in Fig. 2.1.

The sensor data are subject to a feature extraction and selection process for determining the input vector for the subsequent classifier. This makes a decision regarding the class associated with this pattern vector.

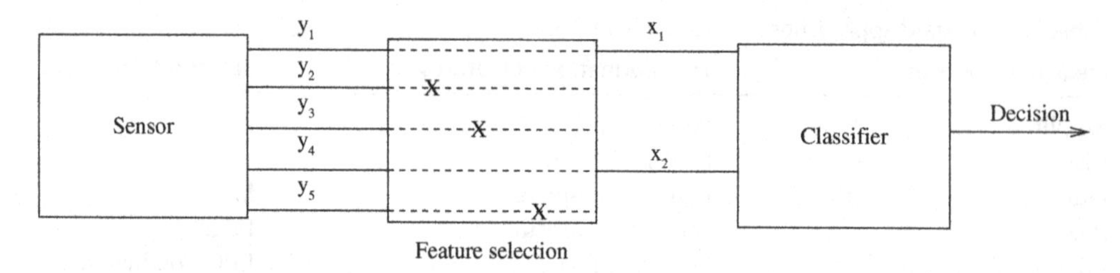

Figure 2.2 Dimensionality reduction based on feature selection.

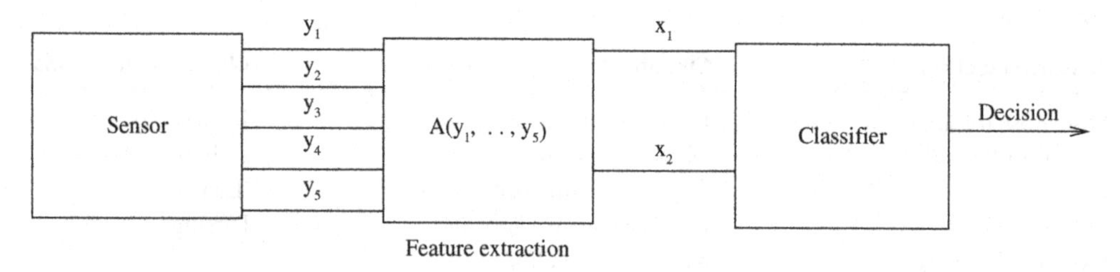

Figure 2.3 Dimensionality reduction based on feature extraction.

Dimensionality reduction is accomplished based on either feature selection or feature extraction. Feature selection is based on omitting those features from the available measurements which do not contribute to class separability. In other words, redundant and irrelevant features are ignored. This is illustrated in Fig. 2.2.

Feature extraction, on the other hand, considers the whole information content and maps the useful information content into a lower dimensional feature space. This is shown in Fig. 2.3. In feature extraction, the mapping type A has to be specified beforehand.

We see immediately that for feature selection or extraction the following is required: (1) feature evaluation criterion, (2) dimensionality of the feature space, and (3) optimization procedure.

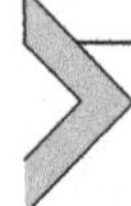

2.3. PRELIMINARY NOTATIONS FOR FEATURE SELECTION AND EXTRACTION

We will introduce the necessary notations to explain feature selection and extraction methods.

The available sensor outputs are given either in the form of a vector component $y_k, k = 1, 2, \ldots, D$ or in the form of a vector $\mathbf{y}$ with $\mathbf{y} = [y_1, \ldots, y_D]^T$. Each pattern vector $\mathbf{y}$ is associated with one of the possible M classes $\omega_i, i = 1, \ldots, M$. A set of candidate features $\zeta_j, j = 1, \ldots, d$, is described by χ while $X = \{\mathbf{x}_j | j = 1, \ldots, d\}$ describes the optimized set based on some criterion function J.

For feature selection optimization of all possible candidate feature sets have to be considered:

$$J(X) = \overset{max}{\chi} J(\chi) \tag{2.1}$$

whereas for feature extraction the optimization concerns only all admissible mappings

$$J(A) = \overset{max}{A} J(A(\mathbf{y})) \tag{2.2}$$

where A represents an optimal feature extractor. Based on this specification, we can determine the feature vector $\mathbf{x} = [x_1, \ldots, x_d]^T$ based on the transformation $\mathbf{x} = A(\mathbf{y})$.

2.4. FEATURE EXTRACTION METHODS

A good feature should remain unchanged if variations take place within a class, and it should reveal important differences when discriminating between patterns of different classes. In other words, patterns are described with as little loss as possible of pertinent information.

There are four known categories in the literature for extracting features [63]:

1. Nontransformed structural characteristics: moments, power, amplitude information, energy, etc.
2. Transformed structural characteristics: frequency and amplitude spectra, subspace transformation methods, etc.
3. Structural descriptions: formal languages and their grammars, parsing techniques, and string matching techniques.
4. Graph descriptors: attributed graphs, relational graphs, and semantic networks.

One of the most frequent problems when applying any classification technique to a pattern recognition problem is the so-called curse of dimensionality [84]. Methods that are adequate for a low-dimensional feature space might be completely impracticable for a high-dimensional space (number of features > 50). This shows that techniques for dimensionality reduction in the feature space had to be developed in order to obtain a more manageable problem.

We will tackle the problem of feature extraction applied to medical image analysis. Our goal will be for a given image, or a region within an image, to generate the features that will be the input to a classifier, which has to assign this image or region of an image to one of the possible classes.

A digital monochrome image is obtained by sampling the corresponding continuous image function $I(x, y)$ and storing the discretized values for x and y in form of a two-dimensional $N_x \times N_y$ image matrix $I(m, n)$ with $m = 0, 1, \ldots, N_x - 1$ and $n = 0, 1, \ldots, N_y - 1$. Every (m, n) element is called a pixel whose gray value is determined by $I(m, n)$. There are N_g quantization levels leading to 2^{N_g} distinct gray levels. To achieve a smooth perception of the image, we need a resolution of $N_g = 8$ bits per pixel.

A small 128 × 128 image has 16,384 pixels, and it is evident that we cannot use this raw information for classification purposes. Therefore, we have to generate new features from the available image matrix $I(m, n)$, and those features should extract precisely the relevant information contained in the original image.

This section will review the most important feature extraction methods in biomedical image analysis.

2.4.1 Nontransformed Signal Characteristics

To describe the properties of random variables and processes, we need statistical parameters that are obtained from these random variables and processes. Although to describe a Gaussian process second-order statistics (mean value and variance) is perfectly sufficient, signal separation problems usually need higher-order statistics [285].

2.4.1.1 Moments

Moments represent extracted features that are derived from raw measurements. For two-dimensional signals, such as images, they can be used to achieve rotation, scale, and translation invariant.

By $I(x, y)$ we define a continuous image function. The geometric moment of order $p + q$ is given by

$$m_{pq} = \int_{-\infty}^{\infty} \int_{-\infty}^{\infty} x^p y^q I(x, y) dx\, dy \tag{2.3}$$

The geometric moment provides important information for the purpose of image reconstruction [285]. In other words, every single coefficient contributes to providing image information.

Invariance in geometric transformation is required for many classification tasks [364]. Moments defined by eq. (2.3) do not provide the desired feature invariance. However, a remedy can be found by defining moments that are invariant to

- *Translation:*

$$x' = x + a, \quad y' = y + b \tag{2.4}$$

- *Scaling:*

$$x' = \alpha x, \quad y' = \alpha y \tag{2.5}$$

- *Rotations:*

$$\begin{bmatrix} x' \\ y' \end{bmatrix} = \begin{bmatrix} \cos\theta & \sin\theta \\ -\sin\theta & \cos\theta \end{bmatrix} \begin{bmatrix} x \\ y \end{bmatrix} \tag{2.6}$$

We also can show that central moments are invariant to translations

$$\mu_{pq} = \int \int I(x, y)(x - \bar{x})^p (y - \bar{y})^q dx\, dy \tag{2.7}$$

where

$$\bar{x} = \frac{m_{10}}{m_{00}}, \quad \bar{y} = \frac{m_{01}}{m_{00}} \tag{2.8}$$

On the other hand, normalized central moments η_{pq} are both scaling and translation invariant [364].

$$\eta_{pq} = \frac{\mu_{pq}}{\mu_{00}^{\gamma}}, \quad \gamma = \frac{p+q+2}{2} \tag{2.9}$$

The preceding equations represent the definitions regarding continuous image functions. Similarly, we can define for a digital image $I(x, y)$ with $i = 0, 1, \ldots, N_x - 1, j = 0, 1, \ldots, N_y - 1$ the corresponding moments by just replacing integrals by summations

$$m_{pq} = \sum_i \sum_j I(x, y) i^p j^q \tag{2.10}$$

2.4.1.2 Parametric Modeling

Parametric modeling is based on representing a signal as a weighted combination of previous samples. For a detailed introduction, see [347]. By applying this method to digital images, we can obtain a useful feature set to be used in conjunction with a subsequent classifier. For this, we will assume that $I(m, n)$ is a real nondiscrete random variable.

There are two ways to proceed. One thinks of an image as a successive sequence of rows or columns and assumes it is a one-dimensional random process $I(n)$, while the other considers a two-dimensional random process $I(m, n)$ or a so-called random field.

The basic idea is that a random sequence can be generated at the output of a linear, causal, stable, time-invariant system with impulse response $h(n)$, whose input is a white noise sequence [285]. Let $I(n)$ be a random stationary sequence with the autocorrelation $R(k)$:

$$R(k) = E[I(n)I(n-k)] \tag{2.11}$$

With $\mu(n)$ representing a white noise sequence, we obtain equivalently for $I(n)$

$$I(n) = \sum_{k=0}^{\infty} h(k)\mu(n-k) \tag{2.12}$$

Such a process is referred to as an autoregressive (AR) process, and is generated recursively by

$$I(n) = \sum_{k=1}^{p} a(k)I(n-k) + \mu(n) \tag{2.13}$$

We immediately see that the random sequence $I(n)$ represents a linear combination of past values of $I(n-k)$ and an additive constant $\mu(n)$. p is the order of the AR model. The coefficients $a(k)$ with $k = 1, 2, \ldots, p$ are the parameters of the AR model and at the same time the predictor parameters of the sequence $I(n)$. In other words, they represent the weighting terms of the past sampled values $I(n-1), \ldots, I(n-p)$ and serve as a prediction of the actual value $I(n)$

$$\hat{I}(n) = \sum_{k=1}^{p} a(k)I(n-k) \equiv \mathbf{a}^T \mathbf{I}(n-1) \tag{2.14}$$

with $\mathbf{I}^T(n-1) \equiv [I(n-1), \ldots, I(n-p)]$ and $\mu(n)$ the prediction error. The unknown parameter vector $\mathbf{a}^T = [a(1), a(2), \ldots, a(p)]$ can optimally be estimated by minimizing the mean-square prediction error:

$$E[\mu^2(n)] = E[(I(n) - \hat{I}(n))^2] = E[(I(n) - \mathbf{a}^T\mathbf{I}(n-1))^2] \tag{2.15}$$

The unknown parameters can be determined from

$$E[\mathbf{I}(n-1)\mathbf{I}^T(n-1)]\mathbf{a} = E[I(n)\mathbf{I}(n-1)] \tag{2.16}$$

This leads to the following equation system

$$\begin{bmatrix} R(0) & R(-1) & \cdots & R(-p+1) \\ R(1) & R(0) & \cdots & R(-p+2) \\ \vdots & \vdots & \vdots & \vdots \\ R(p-2) & R(p-3) & \cdots & R(-1) \\ R(p-1) & R(p-2) & \cdots & R(0) \end{bmatrix} \begin{bmatrix} a(1) \\ a(2) \\ \vdots \\ a(p-1) \\ a(p) \end{bmatrix} = \begin{bmatrix} r(1) \\ r(2) \\ \vdots \\ r(p-1) \\ r(p) \end{bmatrix} \tag{2.17}$$

or equivalently in matrix notation to

$$\tilde{\mathbf{R}}\mathbf{a} = \mathbf{R} \tag{2.18}$$

with $\mathbf{r} \equiv [r(1), \ldots, r(p)]^T$. The variance of the prediction error σ_μ^2 is determined based on eqs. (2.15) and (2.18):

$$\sigma_\mu^2 = E[\mu^2(n)] = R(0) - \sum_{k=1}^{p} a(k)R(k) \tag{2.19}$$

The resulting autocorrelation matrix is Töplitz. This property is very desirable because it provides, based on the Levinson-Durbin-algorithm [285], a solution of the linear system described in eq. (2.18).

The AR-parameters can also serve as features to distinguish between different classes.

Example 2.4.1 Let the AR random sequence of order $p = 2$ be

$$I(n) = \sum_{k=1}^{2} a(k)I(n-k) + \mu(k) \tag{2.20}$$

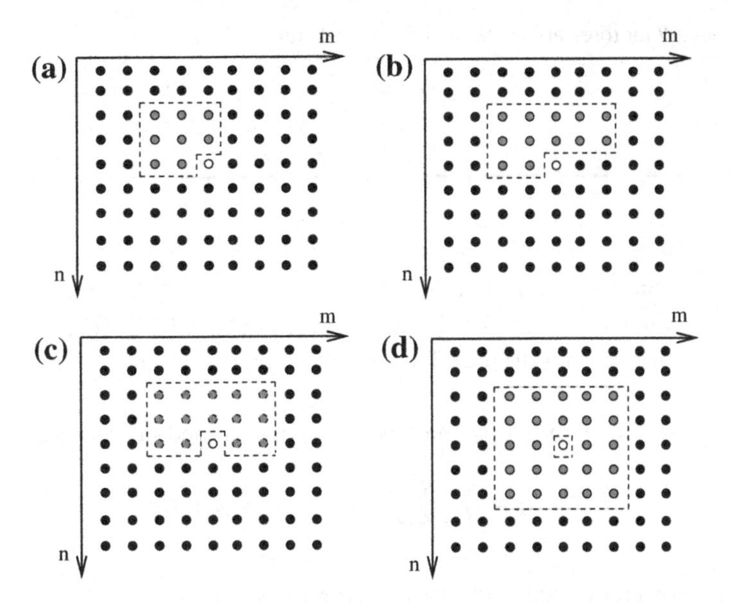

Figure 2.4 Different types of 2-D predictor models. The dotted polygon corresponds to the region W. (a) Strict causal, (b) causal, (c) semicausal, and (d) noncausal.

with $R(0) = 1, R(1) = 0.3,$ *and* $R(2) = 0.55$. The AR parameters can be computed from eq. (2.18):

$$\begin{bmatrix} 1 & 0.3 \\ 0.3 & 1 \end{bmatrix} \begin{bmatrix} a(1) \\ a(2) \end{bmatrix} = \begin{bmatrix} 0.3 \\ 0.55 \end{bmatrix} \tag{2.21}$$

We obtain $a(1) = 0.15$ and $a(2) = 0.5$.

As a generalization we also give the two-dimensional AR random sequence $I(m, n)$

$$\hat{I}(m, n) = \sum_k \sum_l a(k, l) I(m - k, n - l), \quad (k, l) \in W \tag{2.22}$$

and

$$I(m, n) = \hat{I}(m, n) + \mu(m, n) \tag{2.23}$$

Figure 2.4 illustrates the pixel domain W, which leads to the prediction of $I(m, n)$. The figure illustrates four different two-dimensional predictor models [364]. Figure 2.4a illustrates a strict causal predictor model. The coordinates of every single pixel in the contributing frame are smaller than the coordinates m, n of the predicted pixel represented by an unshaded circle. The corresponding frame is given by $W_1 = \{0 \leq k \leq p, 0 \leq l \leq q, (k, l) \neq (0, 0)\}$. Besides causal predictors, there are also noncausal predictors defined by

$$I(m, n) = \sum_{k=-p}^{p} \sum_{l=-q}^{q} a(k, l) I(m - k, n - l) + \mu(m, n) \tag{2.24}$$

Table 2.2 Domain W for three application relevant predictors.

Causal:	$W_2 = \{(-p \leq k \leq p, 1 \leq l \leq q) \cup (1 \leq k \leq p, l = 0)\}$
Semicausal:	$W_3 = \{-p \leq k \leq p, 0 \leq l \leq q, (k,l) \neq (0,0)\}$
Noncausal:	$W_4 = \{-p \leq k \leq p, -q \leq l \leq q, (k,l) \neq (0,0)\}$

Figure 2.4d illustrates the corresponding frame for $p = q = 2$. Figure 2.4c shows a semicausal predictor, while Fig. 2.4b shows a causal predictor. In most applications, strict causality is not of primary importance. Thus, the other three relevant cases are summarized in Table 2.2.

In the 2-D case, the predictor parameters can be determined similarly to the 1-D case:

$$\hat{I}(m,n) = \sum_k \sum_l a(k,l) I(m-k, n-l) \tag{2.25}$$

From the orthogonality condition between error and signal, we obtain

$$R(i,j) = \sum_k \sum_l a(k,l) R(i-k, j-l), \quad (i,j) \in W \tag{2.26}$$

$R(i,j)$ is a 2-D autocorrelation of the image array $I(m,n)$. The system of equations shown in (2.26) is linear, and its solution provides an estimation of $a(k,l)$. By choosing a causal window with $p = q$, we obtain from the symmetry property of the window that for every single index pair (i,j) there also exists an associated index pair $(-i,-j)$. The variance of the prediction error is determined by

$$\sigma_\mu^2 = R(0,0) - \sum_k \sum_l a(k,l) R(k,l) \tag{2.27}$$

It is important to note that the correlation $R(k,l)$ of a homogeneous and direction invariant (isotropic) image depends only on the relative distance between pixels:

$$R(k,l) = R\left(\sqrt{k^2 + l^2}\right) \tag{2.28}$$

This means that the resulting autocorrelation matrix is symmetric and Töplitz,

$$\tilde{\mathbf{R}} = \begin{bmatrix} \tilde{\mathbf{R}}_0 & \tilde{\mathbf{R}}_1 & \cdots & \tilde{\mathbf{R}}_{2p} \\ \tilde{\mathbf{R}}_1 & \tilde{\mathbf{R}}_0 & \cdots & \tilde{\mathbf{R}}_{2p-1} \\ \vdots & \vdots & \vdots & \vdots \\ \tilde{\mathbf{R}}_{2p} & \tilde{\mathbf{R}}_{2p-1} & \cdots & \tilde{\mathbf{R}}_0 \end{bmatrix} \tag{2.29}$$

while each matrix element again represents a matrix which is also Töplitz:

$$\tilde{\mathbf{R}}_{\mathbf{i}} = \begin{bmatrix} R(i,0) & \cdots & R(i,2p) \\ \vdots & \vdots & \vdots \\ R(i,2p) & \cdots & R(i,0) \end{bmatrix} \tag{2.30}$$

By choosing a symmetric frame W, we get symmetric AR parameters, $a(k,l) = a(-k,-l)$, and the system can be solved based on the Levinson-Durbin algorithm.

The AR parameters represent useful features used for many image processing classification tasks [168,234].

Example 2.4.2 Let $R(k,l)$ be the autocorrelation sequence for a medical image

$$R(k,l) = 0.4^{\sqrt{k^2+l^2}}$$

Estimate the AR parameters for a noncausal $p = q = 1$ window.

From eq. (2.22) we obtain

$$\begin{aligned} \hat{I}(m,n) = & \, a(1,1)I(m-1,n-1) + a(1,0)I(m-1,n) \\ & + a(1,-1)I(m-1,n+1) + a(0,1)I(m,n-1) \\ & + a(0,-1)I(m,n+1) + a(-1,1)I(m+1,n-1) \\ & + a(-1,0)I(m+1,n) + a(-1,1)I(m+1,n+1) \end{aligned}$$

The resulting matrix $\tilde{\mathbf{R}}$ is a $(2p+1) \times (2p+1)$ matrix, whose elements are 3×3 matrices

$$\tilde{\mathbf{R}} = \begin{bmatrix} \tilde{\mathbf{R}}_0 & \tilde{\mathbf{R}}_1 & \tilde{\mathbf{R}}_2 \\ \tilde{\mathbf{R}}_1 & \tilde{\mathbf{R}}_0 & \tilde{\mathbf{R}}_1 \\ \tilde{\mathbf{R}}_2 & \tilde{\mathbf{R}}_1 & \tilde{\mathbf{R}}_0 \end{bmatrix}$$

with

$$\tilde{\mathbf{R}}_0 = \begin{bmatrix} R(0,0) & R(0,1) & R(0,2) \\ R(0,1) & R(0,0) & R(0,1) \\ R(0,2) & R(0,1) & R(0,0) \end{bmatrix}$$

$$\tilde{\mathbf{R}}_1 = \begin{bmatrix} R(1,0) & R(1,1) & R(1,2) \\ R(1,1) & R(1,0) & R(1,1) \\ R(1,2) & R(1,1) & R(1,0) \end{bmatrix}$$

and

$$\tilde{\mathbf{R}}_2 = \begin{bmatrix} R(2,0) & R(2,1) & R(2,2) \\ R(2,1) & R(2,0) & R(2,1) \\ R(2,2) & R(2,1) & R(2,0) \end{bmatrix}$$

We obtain the following AR parameters: $a(1,1) = a(-1,-1) = -0.0055$, $a(1,0) = a(-1,0) = -0.125$, $a(1,-1) = a(-1,1) = -0.0055$, $a(0,1) = a(0,-1) = -0,125$, and $\sigma_\mu^2 = 0.585$.

2.4.2 Transformed Signal Characteristics

The basic idea employed in transformed signal characteristics is to find such transform-based features with a high information density of the original input and a low redundancy. To understand this aspect better, let us consider a mammographic image. The pixels (input samples) at the various positions have a large degree of correlation. By using only the pixels, irrelevant information for the subsequent classification is also considered. By using, for example, the Fourier transform we obtain a feature set based on the Fourier coefficients which retains only the important image information residing in low-frequency coefficients. These coefficients preserve the high correlation between the pixels.

There are several methods for obtaining transformed signal characteristics. For example, Karhunen-Loeve transform and singular value decomposition are problem-dependent and the result of an optimization process [84,364]. They are optimal in terms of decorrelation and information concentration properties, but at the same time too computationally expensive. On the other hand, transforms which use fixed basis vectors (images) such as the Fourier and wavelet transform, exhibit low computational complexity while being suboptimal in terms of decorrelation and redundancy.

In this section, we will review the most important methods for obtaining transformed signal characteristics, such as principal component analysis, the discrete Fourier transform, and the discrete cosine and sine transform.

2.4.2.1 Principal Component Analysis (PCA)

Principal component analysis is a basic technique used for data reduction in pattern recognition, signal processing, and bioengineering. It has been widely used in data analysis [79,165,238,271,272,356] and compression [55,160,375]. PCA is also referred to as Karhunen-Loeve transformation or the Hotelling transform.

The idea is that similar input patterns belong to the same class. Thus, the input data can be normalized within the unit interval and then chosen based on their variances. In this sense, the larger the variances, the better discriminatory properties the input features have.

Sometimes, combining two features provides a better recognition result than either one alone.

PCA involves a mathematical procedure that transforms a number of (possibly) correlated variables into a (smaller) number of uncorrelated variables called principal components. This is a highly desirable property, since besides being optimally uncorrelated, the redundancy in data information is removed.

If the data are concentrated in a linear subspace, this provides a way to compress them without losing much information and simplifying the representation. By selecting the eigenvectors having the largest eigenvalues we lose as little information as possible in the mean-square sense. A fixed number of eigenvectors and their respective eigenvalues can be chosen to obtain a consistent representation of the data.

Let $\mathbf{x} = [x_1, \ldots, x_m]^T$ be a random vector generated by a stationary stochastic process. The correlation matrix $\mathbf{R}$ of this vector is defined as

$$\mathbf{R} = E[\mathbf{x}\mathbf{x}^T] \tag{2.31}$$

Also, let $\mathbf{q}_i$ be the ith eigenvector and λ_i the corresponding ith eigenvalue of the matrix $\mathbf{R}$. The eigenvalues λ_i are also known as singular values. Let $\mathbf{Q}$ be an $m \times m$ eigenvector matrix

$$\mathbf{Q} = [\mathbf{q}_1, \cdots, \mathbf{q}_m] \tag{2.32}$$

and $\mathbf{\Lambda}$ be the $m \times m$ diagonal matrix given by the eigenvalues of matrix $\mathbf{R}$

$$\mathbf{\Lambda} = \text{diag}[\lambda_1, \ldots, \lambda_m] \tag{2.33}$$

Without loss of generality, we may assume that the eigenvalues are arranged in decreasing order:

$$\lambda_1 > \lambda_2 \cdots \lambda_m \tag{2.34}$$

such that the first eigenvalue represents the maximal eigenvalue, $\lambda_1 = \lambda_{max}$.

We immediately see that matrix $\mathbf{Q}$ is an orthogonal matrix because its column vectors satisfy the conditions of orthonormality:

$$\mathbf{q}_i^T \mathbf{q}_i = \begin{cases} 1, & j = i \\ 0, & \text{otherwise} \end{cases} \tag{2.35}$$

In matrix representation, this becomes

$$\mathbf{Q}^T\mathbf{Q} = \mathbf{I} \tag{2.36}$$

Thus, the correlation matrix $\mathbf{R}$ can be rewritten in terms of the eigenvalues as

$$\mathbf{R} = \sum_{i=1}^{m} \lambda_i \mathbf{q}_i \mathbf{q}_i^T \tag{2.37}$$

This is known as the spectral theorem. By selecting the eigenvectors $\mathbf{q}_i$ as new basis vectors of the data space, we can rewrite the original data vector $\mathbf{x}$ in the following form:

$$\mathbf{x} = \sum_{i=1}^{m} \mathbf{q}_i a_i = \mathbf{Q}\mathbf{a} \tag{2.38}$$

where $\mathbf{a} = [a_1, \ldots, a_m]^T$ is a coefficient vector. The coefficients a_i represent the projections of $\mathbf{x}$ onto the principal directions represented by the vectors $\mathbf{q}_i$. In other words, they are produced by the linear transformation from the vector space of the original data vector $\mathbf{x}$. They are determined based on the inner product of the transposed eigenvector matrix $\mathbf{Q}$ with the data vector $\mathbf{x}$

$$\mathbf{a} = \mathbf{Q}^T\mathbf{x} \tag{2.39}$$

or

$$a_j = \mathbf{q}_j^T\mathbf{x} = \mathbf{x}^T\mathbf{q}_j, \quad j = 1, 2, \ldots, m \tag{2.40}$$

The a_i represent the principal components and have the same physical dimensions as the original data vector $\mathbf{x}$.

The importance of PCA lies in the dimensionality reduction. The number of features of the input vector can be reduced by eliminating those linear combinations in eq. (2.38) that have small variances and keeping those with large variances.

Let us assume that the eigenvalues $\lambda_1, \ldots, \lambda_l$ have larger variances than the others $\lambda_{l+1}, \ldots, \lambda_m$. By omitting the latter we obtain an approximation $\widehat{\mathbf{x}}$ of vector $\mathbf{x}$ as

$$\widehat{\mathbf{x}} = \sum_{i=1}^{l} \mathbf{q}_i a_i = \sum_{i=1}^{l} \mathbf{q}_i \mathbf{q}_i^T \mathbf{x} = \tilde{\mathbf{Q}}\tilde{\mathbf{Q}}^T\mathbf{x} = [\mathbf{q}_1, \mathbf{q}_2, \ldots, \mathbf{q}_l] \begin{bmatrix} a_1 \\ a_2 \\ \vdots \\ a_l \end{bmatrix}, \quad l \neq m \tag{2.41}$$

where $\tilde{\mathbf{Q}} = [\mathbf{q}_1, \ldots, \mathbf{q}_l]$. Equation (2.41) is the orthogonal projection of vector $\mathbf{x}$ onto the subspace spanned by the principal components with the l largest eigenvalues. Based on eq. (2.40), we are able to determine the set of the principal components with the l largest eigenvalues:

$$\begin{bmatrix} a_1 \\ a_2 \\ \vdots \\ a_l \end{bmatrix} = \begin{bmatrix} \mathbf{q}_1^T \\ \mathbf{q}_2^T \\ \vdots \\ \mathbf{q}_l^T \end{bmatrix} \mathbf{x}, \quad l \leq m \tag{2.42}$$

These l largest eigenvalues correspond to the most relevant features of the input vector $\mathbf{x}$. It is evident that the elimination of features yields an approximation error $\mathbf{e}$ that is given by the difference between the original data vector $\mathbf{x}$ and the new reduced-order data vector $\tilde{\mathbf{x}}$. The approximation error is given by

$$\mathbf{e} = \mathbf{x} - \widehat{\mathbf{x}} = \sum_{i=l+1}^{m} a_i \mathbf{q}_i \tag{2.43}$$

The total variance of the considered l components of the vector $\mathbf{x}$ is

$$\sum_{i=1}^{l} \sigma_i^2 = \sum_{i=1}^{l} \lambda_i \tag{2.44}$$

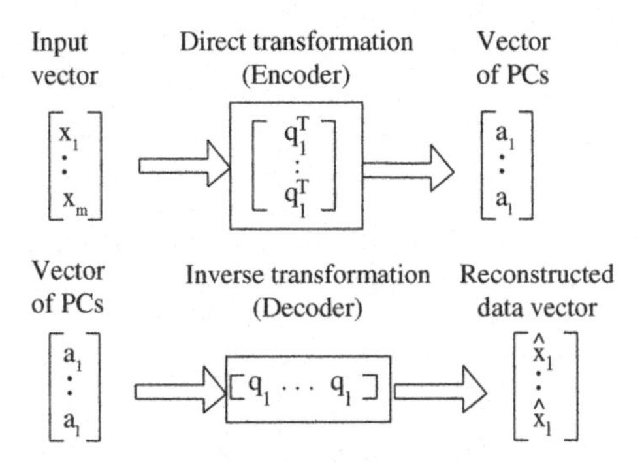

Figure 2.5 Coding and decoding mechanism of PCA.

while the total variance of the remaining $(l - m)$ components is

$$\sum_{i=l+1}^{m} \sigma_i^2 = \sum_{i=l+1}^{m} \lambda_i \tag{2.45}$$

The smallest eigenvalues $\lambda_{l+1}, \ldots, \lambda_m$ of the correlation matrix $\mathbf{R}$ are thus discarded from eq. (2.38). The most effective dimensionality reduction is achieved if the discarded eigenvalues are close to zero, and represent only redundant information. Thus, the strategy for dimensionality reduction based on PCA is to determine the eigenvalues and eigenvectors of the correlation matrix of the input data vector, and then project the data orthogonally onto the subspace that belongs to the largest eigenvalues. This technique is known as subspace decomposition [272].

By analyzing eqs. (2.42) and (2.41), we observe an interesting aspect: eq. (2.42) describes a mapping from the data space $\mathcal{R}^m$ to a reduced-order feature space $\mathcal{R}^l$ and represents an encoder for the approximate representation of the data vector $\mathbf{x}$. Equation (2.41) describes the inverse transformation as a projection from the reduced-order feature space $\mathcal{R}^l$ to the original data space $\mathcal{R}^m$ and represents a decoder for the reconstruction of the data vector $\mathbf{x}$. Figure 2.5 illustrates this.

It is important to point out the major difference between PCA and a related transformation, singular value decomposition (SVD). SVD is related only to a single set of samples, whereas PCA is related to an ensemble of them. PCA is playing a major role as a preprocessing technique in pattern recognition and in context with other transformations such as independent component analysis.

There are also some failures of PCA when used for feature extraction. Figure 2.6 illustrates the case when the first principal component does not contribute toward class separability but only the second principal component does.

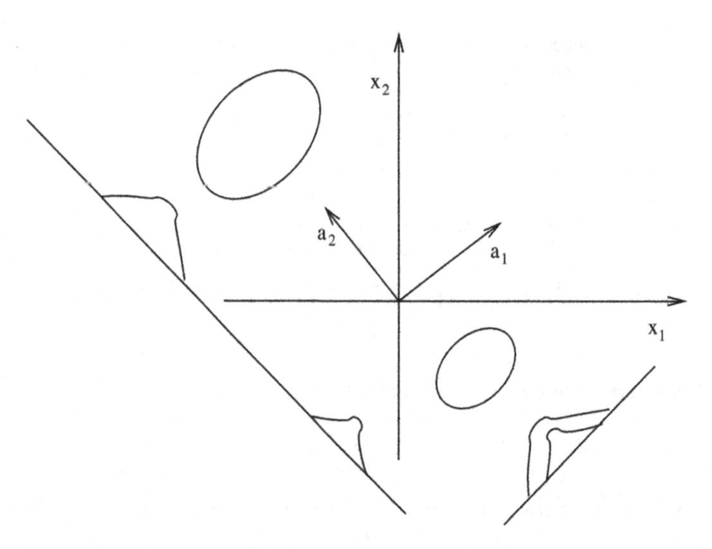

Figure 2.6 PCA is not always associated with class separability. Different contributions of the first two principal components to class separability are shown.

2.4.2.2 Discrete Fourier Transform

In the following we will review the discrete Fourier transform (DFT) for both one- and two-dimensional cases.

Let us consider N sampled values $x(0), \ldots, x(N-1)$. Their DFT is given by

$$y(k) = \frac{1}{\sqrt{N}} \sum_{n=0}^{N-1} x(n) \exp\left(-j\frac{2\pi}{N}kn\right), \quad k = 0, 1, \ldots, N-1 \tag{2.46}$$

while the corresponding inverse transform is

$$x(n) = \frac{1}{\sqrt{N}} \sum_{k=0}^{N-1} y(k) \exp\left(j\frac{2\pi}{N}kn\right), \quad n = 0, 1, \ldots, N-1 \tag{2.47}$$

with $j \equiv \sqrt{-1}$. All $x(n)$ and $y(k)$ can be concatenated in the form of two $N \times 1$ vectors. Let us also define

$$W_N \equiv \exp\left(-j\frac{2\pi}{N}\right) \tag{2.48}$$

such that eqs. (2.46) and (2.47) can be written in matrix form

$$\mathbf{y} = \tilde{\mathbf{W}}\mathbf{x}, \quad \mathbf{x} = \mathbf{W}\mathbf{y} \tag{2.49}$$

with

$$\tilde{\mathbf{W}} = \frac{1}{\sqrt{N}} \begin{bmatrix} 1 & 1 & 1 & \cdots & 1 \\ 1 & W_N & W_N^2 & \cdots & W_N^{N-1} \\ \vdots & \vdots & \vdots & \vdots & \vdots \\ 1 & W_N^{N-1} & W_N^{2(N-1)} & \cdots & W_N^{(N-1)(N-1)} \end{bmatrix} \tag{2.50}$$

where $\mathbf{W}$ is an unitary and symmetric matrix

$$\mathbf{W}^{-1} = \tilde{\mathbf{W}} = \mathbf{W}^*$$

Let us choose as an example the case $N = 2$.

Example 2.4.3 We then obtain for $N = 2$

$$\mathbf{W} = \frac{\sqrt{2}}{2} \begin{bmatrix} 1 & 1 \\ 1 & -1 \end{bmatrix}$$

We see that the columns of $\mathbf{W}$ correspond to the basis vectors

$$\mathbf{w_0} = \frac{1}{\sqrt{2}}[1, 1]^T$$
$$\mathbf{w_1} = \frac{1}{\sqrt{2}}[1, -1]^T$$

and based on them we can reconstruct the original signal:

$$\mathbf{x} = \sum_{i=0}^{1} y(i)\mathbf{w_i}$$

The two-dimensional DFT for an $N \times N$-image is defined as

$$Y(k, l) = \frac{1}{N} \sum_{m=0}^{N-1} \sum_{n=0}^{N-1} X(m, n) W_N^{km} W_N^{ln} \tag{2.51}$$

while its inverse DFT is given by

$$X(m, n) = \frac{1}{N} \sum_{k=0}^{N-1} \sum_{l=0}^{N-1} Y(k, l) W_N^{-km} W_N^{-ln} \tag{2.52}$$

The corresponding matrix representation yields

$$\mathbf{Y} = \tilde{\mathbf{W}}\mathbf{X}\tilde{\mathbf{W}}, \quad \mathbf{X} = \mathbf{W}\mathbf{Y}\mathbf{W} \tag{2.53}$$

We immediately see that the two-dimensional DFT represents a separable transformation with the basis images $\mathbf{w_i}\mathbf{w_j}^{\mathbf{T}}, i, j = 0, 1, \ldots, N-1$.

2.4.2.3 Discrete Cosine and Sine Transform

Another very useful transformation is the discrete cosine transform (DCT) which plays an important role in image compression. In the following, we will review the DCT for both the one- and two-dimensional cases.

For N given input samples the DCT is defined as

$$y(k) = \alpha(k) \sum_{n=0}^{N-1} x(n) \cos\left(\frac{\pi(2n+1)k}{2N}\right), \quad k = 0, 1, \ldots, N-1 \tag{2.54}$$

Its inverse transform is given by

$$x(n) = \sum_{k=0}^{N-1} \alpha(k) y(n) \cos\left(\frac{\pi(2n+1)k}{2N}\right), \quad n = 0, 1, \ldots, N-1 \tag{2.55}$$

with

$$\alpha(0) = \sqrt{\frac{1}{N}}, \quad k = 0 \quad \text{and} \quad \alpha(k) = \sqrt{\frac{2}{N}}, \quad 1 \le k \le N-1 \tag{2.56}$$

The vector form of the DCT is given by

$$\mathbf{y} = \mathbf{C}^{\mathrm{T}} \mathbf{x} \tag{2.57}$$

while we have for the elements of the matrix **C**

$$C(n,k) = \sqrt{\frac{1}{N}}, \quad k = 0, \quad 0 \le n \le N-1$$

and

$$C(n,k) = \sqrt{\frac{2}{N}} \cos\left(\frac{\pi(2n+1)k}{2N}\right)$$
$$1 \le k \le N-1, \quad 0 \le n \le N-1$$

C represents an orthogonal matrix with real numbers as elements:

$$\mathbf{C}^{-1} = \mathbf{C}^{\mathrm{T}}$$

In the two-dimensional case the DCT becomes

$$\mathbf{Y} = \mathbf{C}^{\mathrm{T}} \mathbf{X} \mathbf{C}, \quad \mathbf{X} = \mathbf{C} \mathbf{Y} \mathbf{C}^{\mathrm{T}} \tag{2.58}$$

Unlike the DFT, the DCT is real valued. It has found application in image compression.

Another orthogonal transform is the discrete sine transform (DST) defined as

$$S(k,n) = \sqrt{\frac{2}{N+1}} \sin\left(\frac{\pi(n+1)(k+1)}{N+1}\right), \quad k, n = 0, 1, \ldots, N-1 \tag{2.59}$$

Both DCT and DST have excellent information concentration properties since they concentrate most of the energy in a few coefficients.

Other important transforms are the Haar, wavelet, Hadamard, and Walsh transforms [52,364]. Because of the powerful properties of the wavelet transform and its extensive application opportunities in biomedical imaging, Chapters III and IV are dedicated to the wavelet transform and its applications.

2.4.3 Advanced Techniques for Nontransformed Signal Characteristics and Transformed Signal Characteristics

Many bioimaging applications require a detailed yet compact description of the morphological characteristics of a lesion. Most standard transformed and nontransformed techniques cannot accurately capture blurred and missing information or quantize nonsmooth surfaces.

In this section, we will present advanced methods to describe the local moment characteristics such as Krawtchouk and Zernike moments and the spatio-temporal moments such as Zernike velocity moments that are able to track the temporal changes of the shape parameter. In addition, we will introduce the Minkowski functionals and the Writhe number to describe the geometric structure without considering the gray values.

2.4.3.1 Krawtchouk Moments

Global and local shape description represents an important field in 3D medical image analysis. For lesion classification, there is a stringent need to describe properly the huge data volumes stemming from 3D images by a small set of parameters which captures the morphology (shape) well. Krawtchouk moments represent powerful global and local shape descriptors. Krawtchouk moments represent a set of orthonormal polynomials associated with the binomial distribution [400]. The nth order Krawtchouk classical polynomials can be expressed as a hyper-geometric function:

$$K_n(x; p, N) = \sum_{k=0}^{N} a_{k,n,p} x^k = {}_2F_1\left(-n, -x; -N; \frac{1}{p}\right) \tag{2.60}$$

with $x, n = 0, 1 \cdots N$; $N > 0$; $p \in (0, 1)$ and the hypergeometric function ${}_2F_1$ is defined as:

$${}_2F_1(a, b; c; z) = \sum_{k=0}^{\infty} \frac{(a)_k (b)_k}{(c)_k} \frac{z^k}{k!} \tag{2.61}$$

and with $(a)_k$ being the Pochhammer symbol

$$(a)_k = a(a+1) \cdots (a+k-1) = \frac{\Gamma(a+k)}{\Gamma(a)} \tag{2.62}$$

The set of the Krawtchouk polynomials $S = \{K_n(x; p, N), n = 0 \dots N\}$ has $N + 1$ elements. This corresponds to a set of discrete basis functions with the weight function

$$w(x; p, N) = \binom{N}{x} p^x (1-p)^{N-x} \tag{2.63}$$

and

$$\rho(n; p, N) = (-1)^n \left(\frac{1-p}{p}\right)^n \frac{n!}{(-N)_n} \tag{2.64}$$

We assume that $f(x, y, z)$ is a three-dimensional function defined in a discrete field $A = \{(x, y, z) : x, y, z \in N, x = [0 \cdots N-1], y = [0 \cdots M-1], z = [0 \cdots L-1]\}$. The weighted three-dimensional moments of order $(n + m + l)$ of f are given as:

$$\begin{aligned} \tilde{Q}_{mnl} = & \sum_{x=0}^{N-1} \sum_{y=0}^{M-1} \sum_{z=0}^{L-1} \bar{K}_n(x; p_x, N-1) \\ & \cdot \bar{K}_m(y; p_y, M-1) \bar{K}_l(z; p_z, L-1) \\ & \cdot f(x, y, z) \end{aligned} \tag{2.65}$$

where $p_x, p_y, p_z \in (0, 1)$ enable based on appropriate selection local features to be extracted by the low-order Krawtchouk moments. $\bar{K}_n(x; p, N)$ is given as

$$\bar{K}_n(x; p, N) = K_n(x; p, N) \sqrt{\frac{w(x; p, N)}{\rho(n; p, N)}} \tag{2.66}$$

$\bar{K}_m(y; p_y, M-1)$ and $\bar{K}_l(z; p_z, L-1)$ are defined correspondingly. Thus, every three-dimensional function $f(x, y, z)$ in a three-dimensional field can be decomposed into weighted three-dimensional Krawtchouk moments $\tilde{Q}_{nml}$.

The tumor can be represented by Krawtchouk moments since it is expressed as a function $f(x, y, z)$ in a discrete space $[0 \dots N-1] \times [0 \dots M-1] \times [0 \dots L-1]$ [309].

Weighted 3-D Krawtchouk moments have several advantages compared to other known methods: (1) they are defined in the discrete field and thus do not introduce any discretization error like Spherical Harmonics defined in a continuous field and (2) low-order moments can capture abrupt changes in the shape of an object. The weighted 3-D Krawtchouk moments [229] form a very compact descriptor of a tumor, achieved in a very short computational time. Every tumor can be represented by Krawtchouk moments since it is expressed as a function $f(x, y, z)$ in a discrete grid $[0 \dots N-1] \times [0 \dots M-1] \times [0 \dots L-1]$.

2.4.3.2 Zernike Moments

Zernike moments are compared to geometric moments less susceptible to noise and superior in terms of information redundancy and reconstruction capability. They are constructed based on an orthogonal basis set and this introduces less correlation between the determined moments and eventually leads to better classification results. However, a drawback is the computational effort.

Their computation is based on a transformation of the coordinate system to a polar one and re-scale the object such that it fits entirely into the unit disk. This mapping onto the unit circle ensures the translation and scaling invariance of these moments. A two-dimensional image is consequently described by the radius r and an angle φ. The discrete Zernike moments are then defined as follows:

$$\Omega_{mn} = \sum_{r=0}^{1} \sum_{\varphi=0}^{2\pi} f(r,\varphi)[V_{mn}(r,\varphi)]^* \tag{2.67}$$

The value $m = 0, 1, \ldots, \infty$ specifies the order of the moments, $*$ denotes the complex conjugate, and the angular dependence is steered by $n \in N$ having the constraints $|n| \leq m$ and $m - |n|$ being an even number. The Zernike polynomials V_{mn} thereby serve as basis functions with

$$V_{mn}(r,\varphi) = \frac{m+1}{\pi} R_{mn}(r) \exp(jn\varphi) \tag{2.68}$$

where $j = \sqrt{-1}$ is the imaginary unit and the radial polynomial R is defined as:

$$R_{mn}(r) = \sum_{s=0}^{\frac{m-|n|}{2}} (-1)^s \frac{(m-s)!}{s!\left(\frac{m+|n|}{2} - s\right)!\left(\frac{m-|n|}{2} - s\right)!} r^{m-2s} \tag{2.69}$$

Rotation invariance can be achieved by taking the norm $|A_{mn}|$ as proposed in [340].

In order to apply the moments to MRI images, a 3-D representation is needed. The image is again expressed in polar coordinates and is shifted and scaled to fit into the unit ball. In [48] the 3-D Zernike moments have been computed as:

$$\Omega_{mn}^{l} = \sum_{r=0}^{1} \sum_{\varphi=0}^{\pi} \sum_{\vartheta=0}^{2\pi} f(r,\varphi,\vartheta) \left[R_{mn}(r) S_n^l(\varphi,\vartheta)\right]^* \tag{2.70}$$

The radial polynomial R_{mn} as well as the parameters m, n are the same as before. The additional parameter l takes values in the set $\{-n, \ldots, n\}$. S_n^l depicts the spherical harmonics with

$$S_n^l(\varphi,\vartheta) = N_n^l P_n^l(\cos\varphi) \exp(jl\vartheta) \tag{2.71}$$

with P_n^l being the Legendre functions and N_n^l a normalization factor defined as:

$$N_n^l = \sqrt{\frac{2n+1}{4\pi} \frac{(n-l)!}{(n+l)!}} \tag{2.72}$$

This sophisticated formulation has been simplified in [264] into a linear combination of geometric moments:

$$\Omega_{mn}^{l} = \frac{3}{4\pi} \sum_{r+s+t \leq m} [\chi_{mnl}^{rst}]^{*} \, m_{rst} \tag{2.73}$$

The definition of the factors χ_{mnl}^{rst} is described in [264]. Differently from the 2D Zernike moments, the three-dimensional ones are not as easy to make rotationally invariant. One possibility described in [264] is to consider the vector $\Omega_{mn} = (\Omega_{mn}^{-n}, \Omega_{mn}^{-n+1}, \ldots, \Omega_{mn}^{n})^{\top}$. The norm of this vector then represents the rotationally invariant 3-D Zernike descriptors F_{mn}:

$$F_{mn} := \|\Omega_{mn}\| \tag{2.74}$$

The advantage is that the number of Zernike descriptors is the same as the number of the 2-D Zernike moments for the same order.

2.4.3.3 Zernike Velocity Moments

Velocity moments (Zernike or Cartesian) are based around the statistical center of mass (COM) of an image and are primarily designed to describe a moving and/or changing shape in an image sequence. The method enables the structure of a moving shape to be described, together with any associated motion information and so they are termed spatio-temporal moments. They are formulated as a weighted sum of moments over a sequence of frames of length T, where the weight factor is a real-valued scalar function of the displacement of the COM between consecutive frames. Zernike velocity moments are based on orthogonal Zernike polynomials and so the moments are less correlated and require a lower precision for their calculation in comparison to their Cartesian counterparts. Zernike velocity moments are defined as:

$$A_{mn\alpha\gamma} = \frac{m+1}{\pi} \sum_{i=2}^{I} \sum_{x} \sum_{y} U(i,\alpha,\gamma) S(m,n) P_{i_{xy}} \tag{2.75}$$

where $P_{i_{xy}}$ is the pixel at location x,y of the ith image in the sequence. They are bounded so that $(x^2 + y^2) \leq 1$, while the shape's structure contributes through the orthogonal complex Zernike polynomials ([362]):

$$S(m,n) = [V_{mn}(r,\theta)]^{*} \tag{2.76}$$

and velocity is introduced using the COM ([362]):

$$U(i,\alpha,\gamma) = (\overline{x_i} - \overline{x_{i-1}})^{\alpha} (\overline{y_i} - \overline{y_{i-1}})^{\gamma} \tag{2.77}$$

where $\overline{x_i}$ is the current COM in the x direction, while $\overline{x_{i-1}}$ is the previous COM in the x direction, $\overline{y_i}$ and $\overline{y_{i-1}}$ are the equivalent values for the y direction. Velocity moments have been previously applied to analyze human motion ([340]). We can use the Zernike velocity moments to describe the breast MRI image sequences [136]. These spatio-temporal moments (or descriptors) allow us to capture the spatial variation of contrast enhancement over time within the image sequence (i.e., they capture the changes in signal intensity and shape between scans).

2.4.3.4 Writhe Number

The writhe number describes to which extent a surface is twisted and coiled in itself. It has been introduced by Fuller for the description of the writhing of curves in space [101]. Lauric et al. used this idea and transferred it to surfaces [205,206]. Their application to the analysis of intracranial aneurysms yielded promising results. The writhe number is computed for vertices along the surface of the tumor and represents geometrically the surface asymmetries. Additionally, it has an interesting physical interpretation: it describes the "twisting force" acting on an object. The writhe number quantifies the twisting force that acts at each point on the surface of a lesion. Given a surface S they defined a relationship w between two different points $\mathbf{p}, \mathbf{p}' \in S$ as:

$$w(\mathbf{p}, \mathbf{p}') = \frac{[\mathbf{n_p}, \mathbf{p}' - \mathbf{p}, \mathbf{n}'_\mathbf{p}]}{\|\mathbf{n_p}\| \cdot \|\mathbf{p}' - \mathbf{p}\| \cdot \|\mathbf{n}'_\mathbf{p}\|}$$

where $\mathbf{n_p}, \mathbf{n}'_\mathbf{p}$ denote the surface normals at the points $\mathbf{p}$ and $\mathbf{p}'$ and $[\mathbf{a}, \mathbf{b}, \mathbf{c}] := \mathbf{a}^\top(\mathbf{b} \times \mathbf{c})$. Thereby $\mathbf{b} \times \mathbf{c}$ denotes the cross product of the vectors $\mathbf{b}$ and $\mathbf{c}$. The writhe number of a point $\mathbf{p} \in S$ is then given as:

$$W(\mathbf{p}) = \int_{\mathbf{p}' \in S \setminus \{\mathbf{p}\}} \mathbf{w}(\mathbf{p}, \mathbf{p}') dS$$

The discrete counterpart is:

$$W(\mathbf{p}) = \sum_{\mathbf{p}' \in S \setminus \{\mathbf{p}\}} \mathbf{w}(\mathbf{p}, \mathbf{p}') \Delta \mathbf{p}'$$

The value $\Delta \mathbf{p}'$ is the area assigned to the point $\mathbf{p}'$.

We now want to compute the writhing number for the vertices in $S = V$. Let a face in the set F be given by the three vertices $\mathbf{v}_1, \mathbf{v}_2, \mathbf{v}_3 \in V$. We can then compute the area A and normal $\mathbf{n}$ for this face as:

$$n = (\mathbf{v}_2 - \mathbf{v}_1) \times (\mathbf{v}_3 - \mathbf{v}_1), \quad A = \frac{1}{2}\|n\|$$

The normal of each vertex is then computed by averaging the normals of the faces containing the vertex. Similarly, to obtain a measurement of the value $\Delta \mathbf{p}$ of a vertex $\mathbf{p}$, the areas of the faces containg $\mathbf{p}$ are averaged. Having these values we can compute the writhe number for every vertex. Finally, we compute the first order statistics of the values.

2.4.3.5 Minkowski Functionals

Another family of characteristics that capture morphologic attributes is generated by Minkowski functionals (MFs) [253]. Minkowski functionals (MFs) can characterize geometrical and topological space concepts such as shape, convexity, and connectivity and represent a simple yet precise tool for the analysis of geometrical structures in tumors. The most important MF for 3-D tumors is the Euler characteristic.

Since we consider three-dimensional objects, Hadwiger's theorem [252] states that in this case we have four MFs, the MFs are the volume V, the surface area S, the mean breadth B, and the Euler characteristic χ. The meaning of V and S is immediately clear and the Euler characteristic is the number of regions of connected white voxels plus the number of completely enclosed regions of black voxels minus the number of black tunnels through white regions, remembering that in our data the tumor is shown as a white object in a black area. The mean breadth is proportional to the integral mean curvature H of the tumor A which is defined as

$$H(A) = \frac{1}{2}\int_{\partial A}(\kappa_1 + \kappa_2)df \tag{2.78}$$

with κ_1, κ_2 the principal curvatures and df the area element of A.

In an MR image, for example, we are only computing an approximation of the MFs of the existing tumor. To calculate the MFs one can use an iterative technique that checks every voxel and updates the values if the voxel belongs to the tumor. The efficient algorithm that can be used is presented in [253]. For the classification it can also make sense to look at normed Minkowski functionals, which are

$$\tilde{V} = 1, \quad \tilde{S} = \frac{S}{N^{1/3}}, \quad \tilde{B} = \frac{2B}{N^{2/3}}\tilde{\chi} = \chi/N \tag{2.79}$$

The N denotes the number of voxels in the tumor when we assume a voxel size of one.

2.4.4 Structural Descriptors

The basic idea of this modeling technique is that the signal characteristics of interest are separable and deterministic, and together they describe the concept of interest. The feature information is not provided in numerical form but it is given in structural form. In other words, the signals can be accurately described in symbolic form. To make the description of structural information feasible, we need to define a suitable grammar. A grammar is defined as a tuplet $\mathbf{G} = (Z, V, S, P)$, with Z describing an alphabet of letters while V is an auxiliary alphabet of letter-valued variables. The start variable from V is denoted by S, while P describes a set of production rules [329].

2.4.5 Graph Descriptors

Graphs are a useful tool in structural pattern recognition in case the training set is too small to correctly infer pattern class grammars, or in case each pattern can be considered as class prototype. A directed graph is defined as $G_r = [V, E]$, where V is a set of nodes, and E is a subset of $N \times N$ edges in the graph [329].

The most important graph types are relational and attributed graphs.

- In general, a relational graph describes the structure of a pattern. Such graphs represent semantic networks where nodes describe pattern while edges describe the relations between nodes. A given input pattern first has to be converted into a relational graph, and then the correct class is determined by finding the matching between the given pattern and a relational graph belonging to that class. In other words, a library of relational graphs has to be designed providing for reference prototypes for every class.
- An attributed graph is a relational graph with attributes that represent measures of either symbolic or numeric patterns. These attributes are located at graph nodes. The attributed graph is defined as a 3-tuple $G_{ri} = [V_i, P_i, E_i]$, where V_i is the set of nodes, P_i the properties of nodes, and E_i the relations between nodes.

Graph descriptors form the basis for the model-based techniques in medical image processing. For instance, knowledge of the anatomic structure of certain body parts is essentially for designing models for these structures. On the other hand, a graph is very useful to describe these models. Its nodes determine the objects with associated attributes, while the edges define the relations between these objects. The model has to be patient insensitive and generalize very well. This modeling is frequently used in PET [311].

The extracted image features make it possible to establish a graph prototype of these properties. The model adaptation is accomplished by finding a corresponding subgraph.

2.4.6 Texture

Texture represents a very useful feature for medical imaging analysis by helping to segment images into regions of interest and to classify those regions.

Texture provides information about the spatial arrangement of the colors or intensities in an image. A more relevant definition is given in [52], saying that texture describes the spatial distribution of the gray levels of the pixels in a region. To measure texture is equivalent to quantifying the nature of the variation in gray level within an object. Texture is represented in an image if the gray levels vary significantly. Figure 2.7 shows examples of textured images taken from [36].

The literature distinguishes between random texture and pattern texture. Random texture induced by film grain noise exhibits no visible regularity while pattern texture such as cross-hatching does.

The problem of measuring texture leads us to search for methods to describe texture. A texture feature is derived from an image, and exploits the spatial relations underlying the gray-level distribution. It does not depend on the object's size, shape, orientation, and brightness.

Figure 2.7 Some examples of textures. *(Images courtesy of Dr. T. Randen.)*

A mathematical model for a texture pattern ρ is given in [214] as a function R of a small region S_k of pixels

$$\rho = R(S_k) \tag{2.80}$$

S_k is at the same time a function of the input image $I(m, n)$, which makes the above equation recursive.

Texture analysis based on eq. (2.80) is illustrated in Fig. 2.8.

A different way to look at texture analysis is based on region segmentation. This is always the case when an object differs from the rest of the image in texture but not in average brightness. This leads to image segmentation based on texture. Instead of

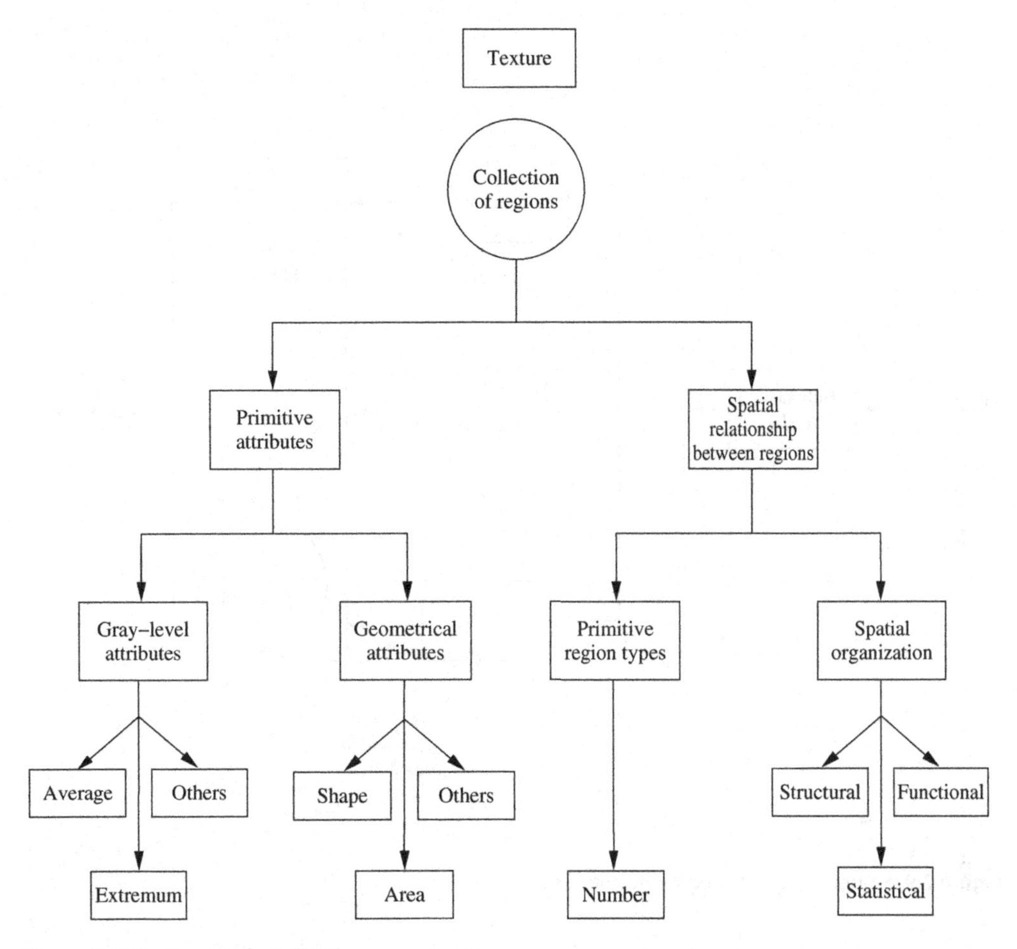

Figure 2.8 Texture analysis [214].

analyzing neighboring pixels, the focus is shifted toward neighboring regions. In other words, we have to define regions to be compared. Then the comparison is still performed based on similarity and proximity of local patterns. A model for texture segmentation is illustrated in Fig. 2.9.

To define exactly what texture is, we have to examine two main approaches: statistical and structural. In the statistical approach, texture represents a quantitative measure of the arrangement of intensities in a region. The structural approach considers texture as a set of primitive texels ordered in some regular or repeated relationship. These can be small objects that represent one unit of a repeated pattern. Then feature extraction is equivalent to locating the primitives and quantifying their spatial distribution. For example, cell nuclei in healthy liver tissue are uniformly distributed whereas unhealthy tissue has lost this ordered relationship. We can choose the nuclei as primitives, and

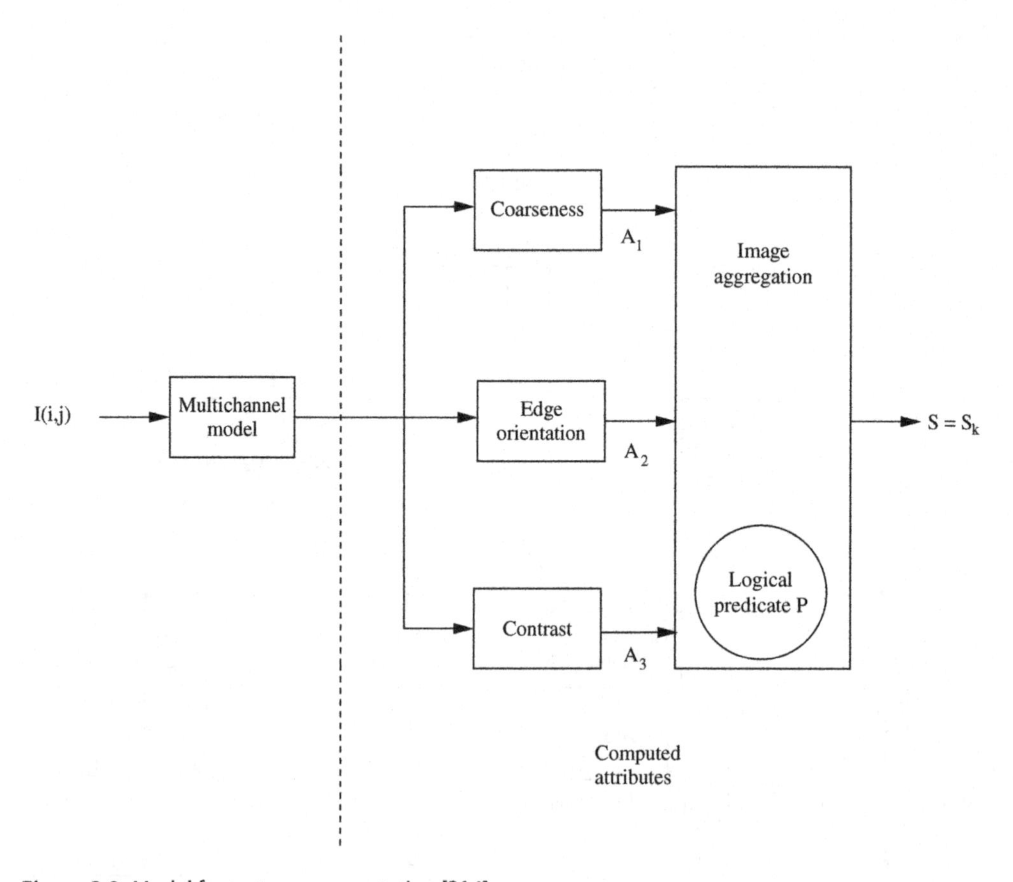

Figure 2.9 Model for texture segmentation [214].

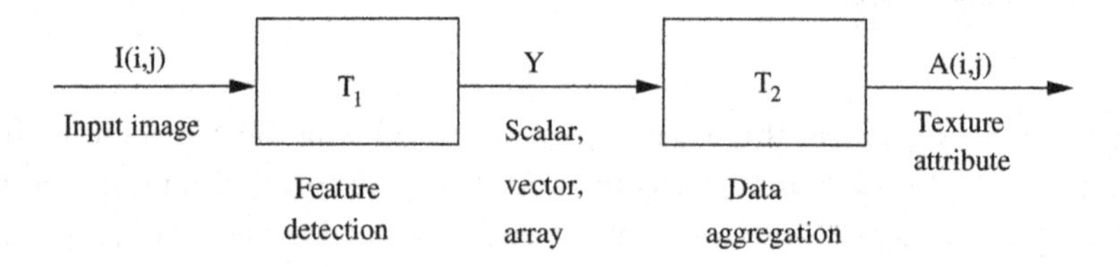

Figure 2.10 Statistical texture analysis.

the mean and standard deviations reflecting the spatial nuclei distribution are structural texture features.

While the structural approach works well for man-made, regular patterns, the statistical approach is more general and easy to compute, and therefore more appealing for practical problems. In the following, we will concentrate our discussion on statistical texture analysis. Figure 2.10 shows the block diagram for texture computation.

A transformation T_1 measures a statistical feature from the image region $I(i,j)$ and provides as an output a scalar, vector, or matrix. In case of a matrix or vector an additional transformation T_2 is needed in order to compress the output of transformation T_1. The goal is to achieve a texture attribute $A(i,j)$ that is a scalar and describes the texture in region $I(i,j)$.

The statistical approach is based on the Julesz conjecture [214]. It stated that the human visual system cannot distinguish between patterns with identical first- and second-order statistics. First-order statistics measure point properties while second-order statistics capture the spatial relationship in a texture pattern. Unfortunately, many counter examples to the conjecture have been found. But this approach has remained successful in machine vision.

Before defining these features based on first- and second-order statistics, we need to define first- and second-order statistics. In [214] it is done in terms of "dropping" a point and a line, respectively, at random on an image. First-order statistics is determined by the pixel gray level which was randomly hit and represents the probability distribution in terms of a histogram. Second-order statistics is determined by considering the gray levels at the two extremities of a randomly dropped line of varying length and orientation.

2.4.6.1 First-Order Statistics Features

Starting with Fig. 2.10, we see that in case of first-order statistics, Y is the image array and the transformation T_2 computes a histogram, from which the scalar A representing the mean value is easily found. Thus, the histogram (or first-order probability distribution) provides precise image information. Let us assume that $Y_k(i,j)$ represents the kth feature of a local region $W(i,j)$ of (i,j). Additionally it is assumed that Y_k takes only discrete values $\{y_1, \cdots, y_s, \ldots, y_t\}$.

Instead of considering the gray levels themselves, it is more useful to look at the gray-level difference between two pixels. Let's define a distance vector

$$\mathbf{d} = (\Delta_x, \Delta_y) \tag{2.81}$$

where Δ_x and Δ_y are both integers. The gray-level difference at the distance $\mathbf{d}$ is given by

$$Y(\mathbf{d}) = |I(i,j) - I(i+\Delta_x, j+\Delta_y)| \tag{2.82}$$

Let $H(y_s, \mathbf{d})$ be the probability that a given difference y_s derived from equation (2.82) occurs at a distance $\mathbf{d}$.

In [214], four distinct transformations T_2 were proposed for determining the texture attribute A:

1. Contrast:

$$A_1 = \sum_{y_s=y_1}^{y_t} y_s^2 H(y_s, \mathbf{d}) \tag{2.83}$$

This is the second moment of the histogram.

2. Angular second moment:

$$A_2 = \sum_{y_s=y_1}^{y_t} [H(y_s, \mathbf{d})]^2 \tag{2.84}$$

Flat histograms have a small A_2, whereas those concentrated around the origin yield a large A_2.

3. Entropy:

$$A_3 = -\sum_{y_s=y_1}^{y_t} H(y_s, \mathbf{d}) \log H(y_s, \mathbf{d}) \tag{2.85}$$

Entropy is maximal for uniform $H(y_s, \mathbf{d})$.

4. Mean:

$$A_4 = \left(\frac{1}{t}\right) \sum_{y_s=y_1}^{y_t} y_s H(y_s, \mathbf{d}) \tag{2.86}$$

A_4 is small for histograms concentrated around $y_s = y_l$ and larger otherwise.

2.4.6.2 Second-Order Statistics Features

While first-order statistics provide only point properties (or gray-level distribution of the image), second-order statistics describe the spatial relationships in a texture pattern. In other words, they provide information about the relative positions of various gray levels within the image.

For notational purposes we will introduce the co-occurrence matrix or spatial dependence matrix. It describes the probability of occurrence of particular pairs y_q and y_r of gray levels at the distance $\mathbf{d}$. It is interesting to point out the relationship between the first-order $H(y_s, \mathbf{d})$ and second-order difference statistics $H(y_q, y_r, \mathbf{d})$:

$$H(y_s, \mathbf{d}) = \sum_{|y_q - y_r| = y_s} H(y_q, y_r, \mathbf{d}) \tag{2.87}$$

Five transformations T_2 are mentioned in the literature [214] for computing texture attributes A:

1. Contrast:

$$A_1 = \sum_{y_r=y_1}^{y_t} \sum_{y_s=y_1}^{y_t} \delta(y_r, y_s) H(y_r, y_s, \mathbf{d}) \tag{2.88}$$

where $\delta(y_r, y_s)$ represents a feature-dependent dissimilarity measure.

2. Angular second moment:

$$A_2 = \sum_{y_r=y_1}^{y_t} \sum_{y_s=y_1}^{y_t} [H(y_r, y_s, \mathbf{d})]^2 \tag{2.89}$$

Describes uniformity or homogeneity.

3. Inverse difference moment:

$$A_3 = \sum_{y_r=y_1}^{y_t} \sum_{y_s=y_1}^{y_t} \left[\frac{H(y_r, y_s, \mathbf{d})}{1 + \delta(y_r, y_s)} \right] \tag{2.90}$$

with $\delta(y_r, y_s) = (y_r - y_s)^2$. Describes the local homogeneity of an image.

4. Entropy:

$$A_4 = - \sum_{y_r=y_1}^{y_t} \sum_{y_s=y_1}^{y_t} H(y_r, y_s, \mathbf{d}) \log H(y_r, y_s, \mathbf{d}) \tag{2.91}$$

A_4 is a measure of randomness. It takes low values for smooth images.

5. Correlation:

$$A_5 = \left[\frac{1}{\sigma_{y_r} \sigma_{y_s}} \right] \sum_{y_r=y_1}^{y_t} \sum_{y_s=y_1}^{y_t} (y_r - \mu_{y_r})(y_s - \mu_{y_s}) H(y_r, y_s, \mathbf{d}) \tag{2.92}$$

where σ_{y_r} and σ_{y_s} are the standard deviations, and μ_{y_r} and μ_{y_s} are the means of y_s and y_r. A_5 becomes larger if the elements of $H(y_q, y_r, \mathbf{d})$ become more similar.

2.4.6.3 Laws' Texture Energy Measures

A different approach to obtain texture features is to employ local masks to detect various types of texture. Each mask is designed to respond to a different local property. A set of texture energy measures (one for each convolved image) is computed for each pixel by taking the average absolute value of pixel values in a square 15×15 window. Since the texture energy measures are computed for each pixel, the feature value for a segmented object is computed by averaging the Laws' texture energy of the object pixels.

Laws [207,208] proposed a texture energy approach that determines the variation amount within a fixed-size window. A set of nine 5×5 convolution masks is applied to an image in order to determine the texture energy, which is then represented by a vector of nine numbers for each pixel of the image being analyzed [338].

Laws provides a set of five element, one-dimensional convolution kernels named $L5, E5, S5, W5$, and $R5$:

L5	(Level)	= [	1	4	6	4	1	]
E5	(Edge)	= [	−1	−2	0	2	1	]
S5	(Spot)	= [	1	0	2	0	−1	]
W5	(Wave)	= [	−1	2	0	−2	1	]
R5	(Ripple)	= [	1	−4	6	−4	1	]

The names of the kernels are mnemonics for level, edge, spot, wave, and ripple. Two-dimensional kernels are created by taking the outer product of the one-dimensional kernels.

After preprocessing which removes the effects of illumination, each of the sixteen 5×5 masks is applied to the preprocessed image, producing 16 filtered images.

With $F_k[i,j]$ we denote the result of the filtering with the kth mask at pixel $[i,j]$. The texture energy map E_k for filter k is given by

$$E_k[r,c] = \sum_{j=c-7}^{c+7} \sum_{i=r-7}^{r+7} |F_k[i,j]| \tag{2.93}$$

We immediately see that each texture energy map represents an image, describing the application of the kth mask to the input image.

From the 16 energy maps based on the combination of certain symmetric pairs, the following nine final maps are obtained:

L5E5/E5L5	L5S5/S5L5
L5R5/R5L5	E5E5
E5S5/S5E5	E5R5/R5E5
S5S5	S5R5/R5S5
R5R5	

$E5L5$ measures the horizontal edge content while $L5E5$ measures the vertical edge content, and the total edge content is given by the average of these two maps.

In summary, nine energy maps are obtained, or equivalently a single image is produced with a vector of nine texture attributes at each pixel.

Feature vectors derived from Laws masks provide a good spatial discrimination since the determined measure is well localized. However, the masks only operate at a single scale which reduces their ability to characterize texture. In theory a multiscale approach for masks could be developed, but only at the cost of impracticability.

Example 2.4.4 A problem with Laws' texture energy measures are the errors they introduce along the separating line between two texture regions [124] since the computed textural energy for a pixel over a given window contains more than one texture.

This poses a serious problem for microcalcification detection in mammography [393]. Microcalcifications are known for being very small, and the 15×15 window of each pixel most likely crosses these textural boundaries. A solution to this problem was given in

[392] by choosing a set of modified Laws' texture energy measures. The 5×5 convolution mask was preserved but instead 12 measures were determined: six measures were obtained by averaging the absolute values of the object pixels, while another six are obtained by averaging the absolute values of the border pixels.

This substantial improvement is based on the fact that pixels of different textural regions are not combined and thus it eliminates the smoothing effect at pixel level introduced by the original Laws energy measures.

2.5. GAUSSIAN MARKOV RANDOM FIELDS

The voxel grid in images is always arbitrarily laid on the object of interest. However, we can assume that in most cases voxel in one area covers the same tissue, and therefore correlation between voxels depends on the spatial structure. Here, we want to construct joint probability density functions (pdfs) with a correlation structure which naturally arises from the spatial structure. These approaches can be used to simply remove noise from images, but also for strengthening the analysis of single voxels by using information from correlated voxels.

2.5.1 Markov Random Field

In the following, we will consider an image as a realization of a stochastic process [8]. A stochastic process can be seen as a generalization of a scalar random variable on arbitrary dimensions. For our purposes, the dimension of the random field is the dimension of an image. We are only concerned with discrete random fields, as we have a countable number of voxels per image.

More general, a random field is a collection of random variables $\mathbf{X} = (X_i)_{i \in I}$, where I is the collection of sites (here: pixels or voxels), with same sample space Σ_X, given a probability space (Σ, P, p). Here Σ is the sample space of the random field, given as the tensor product of the sample spaces Σ_{X_i}. P is a σ-algebra and p a probability measure.

For images, we are interested in random fields which can describe neighborhood structures. One popular way to do so are Markov Random Fields (MRF) [178]. Here, the dependency structure of the univariate random variables X_i is described using the neighborhood structure of the sites I. A neighborhood structure or neighborhood system $\partial = \{\partial(s), s \in I\}$ is a set of sites, where

$$s \notin \partial(s), \nu \in \partial(s) \Leftrightarrow s \in \partial(\nu)$$

All sites $\nu \in \partial(s)$ are called *neighbors* of s, with the notation $s \sim \nu : s$ is neighbor of ν.

In images the sites comprise a regular grid. Usually, neighborhood is defined by direct contiguousness. That is, considering two-dimensional images, each site, that is, voxel, has four neighbors, one in each direction (with the exception of voxels lying at the image edges). More complex neighborhood structures have been considered, for example neighbors including diagonal voxel (eight neighbors in 2-D) [217]. In most applications, however, considering only the nearest four (in 2-D) or six (in 3-D) voxels as neighborhood is sufficient.

Given a neighborhood structure, a Markov Random Field is defined by the dependence of the conditional probability distribution of each site on its neighbors only. That is,

$$p(X_s|X_v;\, v \neq s) = p\,(X_s|X_v;\, v \in \partial(s)) \quad \text{for all } s \in I \tag{2.94}$$

With this, the random fields satisfy the following Markov properties:

- Local Markov property: Given its neighbors, each site is conditionally independent of all other sites
$$X_s \perp X_{I\setminus\{s\cup\partial(s)\}}|X_{\partial(s)}$$
- Pairwise Markov property: Two not neighboring sites are conditionally independent given all other sites
$$X_s \perp X_v|X_{I\setminus\{s,v\}} \quad \text{if } s \nsim v$$
- Global Markov property: Two subsets S and V of I are conditionally independent, if no site in S is neighbor of a site in V, that is, the subsets are separated by a subset $U = I \setminus (S \cup V)$
$$X_S \perp X_V|X_U$$

The probability measure of the Markov random field is defined using the local (conditional) probability density functions (pdf) of each site given its neighbors $p(X_s|X_{\partial(s)})$. For the computation of the joint pdf of all sites, Brook's Lemma can be used [37]. Given a fixed point $\mathbf{X}_0 = (X_{0,i})_{i\in I}$ in the sample space, the joint pdf is

$$\begin{aligned} p(X_1,\ldots,X_n) &= \frac{p(X_1\,|X_2,\ldots,X_n)}{p(X_{0,1}\,|X_2,\ldots,X_n)} \cdot \frac{p(X_2\,|X_{0,1},X_3,\ldots,X_n)}{p(X_{0,2}\,|X_{0,1},X_3,\ldots,X_n)} \\ &\cdots \frac{p(X_n\,|X_{0,1},\ldots,X_{0,n-1})}{p(X_{0,n}\,|X_{0,1},\ldots,X_{0,n-1})} p(X_{0,1},\ldots,X_{0,n}) \end{aligned} \tag{2.95}$$

where n is the number of sites in I. Please note, that the joint pdf can only be computed up to a constant, as $p(X_{0,1},\ldots,X_{0,n})$ is not known. However, this is not a problem, as the integral over a pdf is always one. Brook's Lemma does however not guarantee that for arbitrary definitions of the local conditional pdfs a joint distribution exists.

Therefore, another way to construct joint distribution of Markov random field is of interest: the Hammersley-Clifford theorem [64]. The Hammersley-Clifford theorem states that each Markov random field is a Gibbs random field, that is, a random field with joint distribution

$$p(\mathbf{X}) \propto \exp(-\Phi) = \exp\left(-\sum_{j\in C} \phi_j(X_j)\right) \tag{2.96}$$

Here, C is the set of all sites in I and all cliques. Cliques are all subsets in which each pair in the subset are neighbors. For neighborhood structures where neighborhood is defined by contiguousness, the cliques are just all pairs of neighboring sites. That is, (2.96) simplifies to

$$p(\mathbf{X}) \propto \exp\left(\sum_{i\in I}^{n} \phi_i(X_i) + \sum_{i\sim j} \phi_{i,j}(X_i, X_j)\right) \tag{2.97}$$

Again, the joint pdf can only be computed up to a normalizing constant.

The function Φ is called *Gibbs function* or *Gibbs energy*, and in statistics it is referred to as negative *log-likelihood*. The Hammersley-Clifford theorem defines the joint pdf for all possible Markov random fields.

2.5.2 Ising Model

Let's consider black and white images. We use the following notation: $X_i = +1$ is a white pixel, $X_i = -1$ is a black pixel. Now, we use the following Gibbs function

$$\Phi(X) = \beta \sum_{i\sim j} X_i X_j \tag{2.98}$$

Therefore, the Gibbs function is high if most of the neighboring pixels have the same color, that is, the image has large black and white areas. β is called the (global) inverse temperature of the system.

We now want to denoise an observed noisy black and white image. Let $\mathbf{Y}$ be the observed image and $\mathbf{X}$ be the estimated true image. We use the assumption that the true image is smoother than the noisy image, that is, it has larger black and white areas. On the other hand, the estimated true image should not be too far from the observed noisy image. We define the distance between images by the Hamming distance

$$D(X, Y) = \sum_{i\in I} X_i Y_i \tag{2.99}$$

that is, the number of pixels with noise (where the color changed from white to black or vice versa) minus the number of pixels unchanged. We combine this with the Gibbs function (2.98) and gain

$$H(X, Y) = \Phi(X) + D(X, Y) = \beta \sum_{i\sim j} X_i X_j + \sum_{i\in I} X_i Y_i \tag{2.100}$$

This is called the Ising model, named after Ernst Ising, who invented this mathematical model for ferromagnetics in statistical mechanics.

Minimizing $H(X, Y)$ in (2.100) with respect to X, we gain a denoised version of the noisy image Y. The result depends on the inverse "temperature" β in (2.100), which translates to the "smoothness" of the image. The lower the temperature, the smoother the resulting image, that is, we gain larger black and white areas. A high temperature corresponds to an image with more changes in color from pixel to pixel.

For images with more than two colors ($X_i \in \{1, \ldots, k\}$), the *Potts model* can be used. Here, the Gibbs function is

$$\left(\beta \sum_{i \sim j} I(X_i = X_j)\right) \tag{2.101}$$

with I the indicator function.

2.5.3 Gaussian Markov Random Fields

The Potts model is useful for low numbers of colors. For large numbers of gray scale colors, a Gibbs function based on the quadratic difference is more appropriate

$$\Phi = \sum_i m_i X_i + \sum_{i \sim j} w_{ij}(X_i - X_j)^2 \tag{2.102}$$

This Gibbs function leads to a joint normal distribution for all sites, that is, $\mathbf{X} \sim \mathrm{N}(\mu, P^{-1})$, with joint probability density function

$$p(\mathbf{X}) = (2\pi)^{-n/2}|P|^{1/2} \exp\left(-\frac{1}{2}(x - \mu)' P(x - \mu)\right) \tag{2.103}$$

where μ is the vector of expected values and $P = (p_{i,j})_{i,j \in I}$ is the precision matrix, that is, the inverse of the covariance matrix. This stochastic process is known as *Gaussian Markov Random Field* (GMRF). The precision matrix P has a certain structure:

$$\begin{aligned} p_{ij} &= 0 && \text{if } i \nsim j \\ p_{ij} &= -c_{ij}/\sigma 2_i && \text{if } i \sim j \\ p_{ii} &= 1/\sigma_i^2 \end{aligned} \tag{2.104}$$

From the joint normal distribution, the full conditional distribution of one site X_i given the rest of the sites X_{-i} can easily be derived:

$$X_i | X_{-i} \sim X_i | X_{\partial(i)} \sim \mathrm{N}\left(\mu_i + \sum_{j \sim i} c_{ij}(X_j - \mu_j), \sigma_i^2\right) \tag{2.105}$$

That is, given its neighbors each pixel has Gaussian distribution and the expected value of this pixel is a weighted mean of the neighboring pixels.

Typically, the expected vector μ is set to zero and all weights are set equally to one. This gives us the special case of the intrinsic GMRF

$$\mathbf{X} \sim N(0, \sigma^2 \mathbf{K}^{-1}) \tag{2.106}$$

with

$$\begin{aligned} k_{ij} &= 0 \quad \text{if } i \nsim j \\ k_{ij} &= -1 \quad \text{if } i \sim j \\ k_{ii} &= \text{number of neighbors of } i \end{aligned} \tag{2.107}$$

Please note that $\mathbf{K}$ does have full rank. The term $\mathbf{K}^{-1}$ therefore represents a generalized inverse of $\mathbf{K}$. This distribution of $\mathbf{X}$ is not a proper n-dimensional distribution (with n the number of pixels), as the n-dimensional integral over $p(\mathbf{X})$ is infinite. In (2.103) the term $|\mathbf{K}|^{1/2}$ is therefore replaced by the product of the non-zero eigenvalues of $\mathbf{K}$.

Using an intrinsic GMRF, the expected values in the full conditional (2.105) reduce to the mean of the neighboring pixels. The parameter τ^2 can be interpreted as smoothing parameter. A low value of τ^2, that is, a low variance between neighbors implies a smooth surface. A high τ^2 leads to a bumpy image.

2.5.4 Latent GMRF

The advantage of the intrinsic GMRF is the fact, that it does not impose heavy constraints on the random field and is easy to handle, in particular when used as a latent field. In a latent field approach, not the observed image itself is assumed to be a realization of a random field, but the observations per pixel are observations from conditionally independent distributions given a latent variable. The latent, i.e., unobserved variable has the same dimension and follows a Gaussian Markov Random Field. This allows to use arbitrary distributions for the image intensity and still using the convenient GMRF.

For example, in many applications the intensities in an image can be seen as counting variable. The standard distribution for counting variables is the Poisson distribution using the following *hierarchical* Log-Poisson approach, also known as Besag, York, Mollie (BYM) model [27]. The intensity per pixel X_i given a latent variable λ_i is independently Poisson-distributed

$$X_i | \lambda_i \sim Poisson(\lambda_i) \quad \text{for all } i \tag{2.108}$$

Now, we use an intrinsic GMRF on the natural logarithm of the latent field $\boldsymbol{\lambda}$,

$$\log(\boldsymbol{\lambda}) \sim N(0, \tau^2 K^{-1}) \tag{2.109}$$

The latent field, i.e., latent image, λ can be derived as the expectation of the posteriori distribution of λ given $\mathbf{X}$, which pdf can be computed using Bayes' formula.

2.5.5 Inferring from (Gaussian) Markov Random Fields

Several algorithms exist to estimate Markov random field $\mathbf{X}$ from an observed image $\mathbf{Y}$, that is to minimize the posterior energy or accordingly maximize the joint posterior probability density function $p(\mathbf{X}|\mathbf{Y})$. This is of course a high-dimensional problem, as the dimension of the joint posterior pdf is the number of voxels n. In general, the joint posterior pdf is given by

$$p(\mathbf{X}|\mathbf{Y}) \propto f(\mathbf{Y}|\mathbf{X}) \exp(-\Phi(\mathbf{X})) \tag{2.110}$$

where $f(\mathbf{Y}|\mathbf{X})$ is the pdf of the distribution of the observed image given the true image, i.e., the noise distribution, and Φ the Gibbs function of the Markov random field.

One possibility for inferring from GMRF models are Markov Chain Monte Carlo Methods, see Section 2.6. Another popular approach for optimization of the joint posterior pdf was proposed by Besag [25]: the *Iterated Conditional Modes* (ICM). Here, the joint pdf is maximized by iteratively maximizing the pdfs of each voxel conditioned on the observed image and all other voxels. Using the properties of the Markov random field, it can easily be seen that

$$p(X_i|X_{-i}, Y) = p(X_i|X_{\partial(i)}, Y_i) \propto f(Y_i|X_i) \exp\left(-\phi(X_i) - \sum_{j\sim i} \phi(X_i, X_j)\right) \tag{2.111}$$

This pdf can easily be computed and optimized.

The algorithm is not guaranteed to converge to the global maximum. However, in most cases a "good" local maximum can be found. Furthermore, the result depends on the order the voxels are optimized. It is suggested to use a random order in order to eliminate systematic biases introduced by the voxel order.

2.6. MARKOV CHAIN MONTE CARLO

The standard approach for evaluation of posterior probabilities in Bayesian models are Markov Chain Monte Carlo (MCMC) algorithms, see e.g., [38,103,106]. The idea behind Monte Carlo simulation is to produce random samples from the posterior pdf and to gain point estimates of unknown parameter from these samples. Typically, the posterior expected value of the unknown parameter is estimated using the mean of samples.

In Bayesian approaches, the joint posterior pdf is given by

$$p(\boldsymbol{\theta}|\mathbf{X}) = \frac{f(\mathbf{X}|\boldsymbol{\theta})p(\theta)}{\int f(\mathbf{X}|\boldsymbol{\theta})p(\theta)d\theta} \tag{2.112}$$

where $f(\mathbf{X}|\boldsymbol{\theta})$ is the pdf of the data distribution (or likelihood) and $p(\theta)$ is the prior assumption about the unknown parameter θ given as pdf. Sampling from the posterior pdf (2.112) is, in general, not trivial. In most cases, the constant integral $\int f(\mathbf{X}|\boldsymbol{\theta})p(\theta)d\theta$ cannot be computed analytically. For such cases, MCMC algorithms like the Metropolis-Hastings sampler and its special case, the Gibbs sampler, were proposed. For MCMC, the samples are not randomly drawn from the posterior, but form a Markov Chain. That is, each sample depends on the previous sample. However, given the previous sample the current sample is independent from all other samples before.

2.6.1 Metropolis-Hastings algorithms

Considering our aim to draw random sample of the random parameter $\boldsymbol{\theta}$ from the posterior $p(\boldsymbol{\theta}|\mathbf{X})$, the Metropolis-Hastings (MH) algorithm can be formulated as follows

(1) choose starting values $\boldsymbol{\theta}^{(0)}$,

(2) for each iteration $k = 1, \ldots, M$, for each voxel i,

  (a) draw a random number from a proposal distribution with pdf $q(\theta_i^*|\theta^{(k-1)})$,

  (b) calculate the *acceptance probability*

$$\alpha = \min\left\{1, \frac{p(\theta_i^*|\bullet)q(\theta^{(k-1)}|\theta_i^*)}{p(\theta^{(k-1)}|\bullet)q\left(\theta_i^*\right)|\theta^{(k-1)}}\right\}, \tag{2.113}$$

  (c) draw a random number u from a uniform distribution $U[0, 1]$,

  (d) if $u < \alpha$ accept θ_i^*, that is, $\theta_i^{(k)} = \theta_i^*$, otherwise $\theta_i^{(k)} = \theta_i^{(k-1)}$.

Here, $p(\theta_i^*|\bullet)$ is the pdf of θ_i^* given all other θ and the observed data X, the so-called *full condional* of θ_i.

It can be shown that independent of the starting values the resulting chain of samples converges to the distribution of the posterior pdf. That is, after b iterations, we actually draw random samples from the posterior pdf. b is called *burn-in*. As mentioned before, the samples are dependent on each other. In principle this is no problem, however, the more dependent the samples are, the more iterations we need to gain a good estimate. We want a good "mixing" of the chain, that is, the chain should not be to long in one state and should draw samples from the complete posterior in few iterations. Fig. 2.11 shows two MCMC chains with samples from the same posterior.

The mixing depends obviously on the proposal distribution pdf $q(\theta^*|\theta^{(k-1)})$. The choice of this proposal distribution depends on the problem on hand. The easiest way to construct a proposal distribution is the random walk, that is

$$\theta^* = \theta^{(k-1)} + \epsilon \tag{2.114}$$

with $E(\epsilon) = 0$. Typically, a Gaussian distribution with variance ν^2 is used for ϵ. ν^2 determines the mixing and has to be chosen carefully—this can be done by adjusting or "tuning" ν^2 after the burn-in, such that the actual acceptance rate of the proposed values is roughly 30% to 50%.

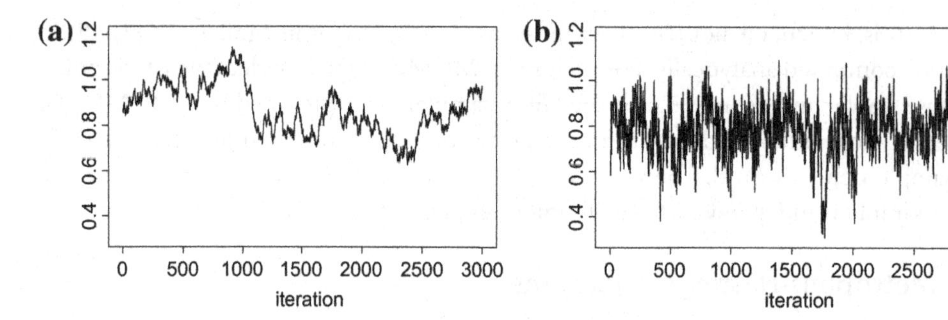

Figure 2.11 Two sample paths ("traces") from the same posterior pdf from Metropolis-Hastings algorithms with different proposal distributions: (a) bad mixing and (b) good mixing.

When using a random walk with symmetric distribution, i.e., $q(\theta_i^*)|\theta^{(k-1)} = q(\theta^{(k-1)}|\theta_i^*))$, the acceptance probabilty α reduces to

$$\alpha = \min\left\{1, \frac{p(\theta_i^*|\bullet)}{p(\theta^{(k-1)}|\bullet)}\right\} \tag{2.115}$$

This is also known as the *Metropolis algorithm* [244].

2.6.2 Gibbs Sampler

Another idea for a proposal distribution is to use the actual full conditional of each parameter θ_i—if we can draw random numbers from the full conditional. In this case (2.113) reduces to $\alpha = 1$ and all proposed values are accepted. This is the so-called *Gibbs sampler*. Using numerical algorithms, it is actually possible to draw random numbers in a broad range of problems, see the OpenBugs software project [227]. In other, more specific cases, the full conditional is a known standard distribution, where random number generators are available.

As an example, we use problem of denoising images using GMRFs. Let us assume we have an observed image $\mathbf{Y}$ with white noise, that is $Y_i \sim N(X_i, \sigma^2)$ independently for each $i = 1, \ldots, n$. We assume a smooth surface for the true image, that is, we assume an intrinsic Gaussian Markov random field $\mathbf{X} \sim \text{GMRF}(\tau^2)$, where τ^2 is a known smoothing parameter. As we do not know the observation error, we estimate the observation error jointly with $\mathbf{X}$. For this, we introduce a prior distribution $\sigma^2 \sim IG(a, b)$, an Inverse Gamma prior with parameters a and b.

The posterior pdf can then be written up to a constant as the product of the pdf of white noise and the prior distributions, that is, the pdf of the intrinsic GMRF and the inverse Gamma prior

$$p(\mathbf{X},\sigma^2) \propto f(\mathbf{Y}|\mathbf{X},\sigma^2)p(\mathbf{X})p(\sigma^2) \tag{2.116}$$

$$\begin{aligned} &\propto (\sigma^2)^{-n/2}\exp\left(\frac{1}{2\sigma^2}\sum_{i=1}^{n}(Y_i - X_i)^2\right) \\ &\cdot \exp\left(-\frac{1}{2\tau^2}\mathbf{X}^T\mathbf{K}\mathbf{X}\right) \\ &\cdot (\sigma^2)^{-(a+1)}\exp(-b/\sigma^2) \end{aligned} \tag{2.117}$$

Using the definition of conditional distributions

$$p(X_i|h, Y) = \frac{p(X_i, h|Y)}{p(h|Y)} \propto p(X_i, h|Y) \tag{2.118}$$

where h is the vector of all other unknown parameters, we see that the full conditional pdf of each parameter is proportional to the joint posterior pdf. Here, we get

$$p(X_i|\bullet) \propto p(X_i|X_{\partial(i)}, \sigma^2, Y) \tag{2.119}$$

$$\propto \exp\left(\frac{1}{2\sigma^2}(Y_i - X_i)^2\right)\exp\left(-\frac{1}{2\tau^2}\sum_{j\sim i}(X_i - X_j)^2\right) \tag{2.120}$$

$$\propto \exp\left(-\frac{1}{2s^2}(X_i - \mu)^2\right) \tag{2.121}$$

with $s^2 = (1/\sigma^2 + n_i/\tau^2)^{-1}$, n_i the number of neighbors of voxel i, and $\mu = s^2(Y_i/\sigma^2) + \sum_{j\sim i}(X_i/\tau^2)$. The last row is (up to the normalizing constant) the pdf of a Gaussian distribution with expected value μ and variance s^2. The full conditional of σ^2 turns also out to be a standard distribution, more specifically a $IG(a + n/2, b + \sum(Y_i - X_i)^2/2)$ distribution. Therefore, we can draw random samples from the posterior density using the following algorithm

(1) choose a starting value $(\sigma^2)^{(0)}$,
(2) for each voxel i draw $X_i^{(k)}$ from a $\mathrm{N}(\mu, s^2)$ distribution, where μ and s^2 depend on $(\sigma^2)^{(k-1)}$,
(3) draw $(\sigma^2)^{(k)}$ from $IG(a + n/2, b + \sum(Y_i - X_i)^2/2)$,
(4) iterate (2) and (3) for $k = 1, \ldots, m$,
(5) delete the first b samples (burn-in).

From the samples, we can finally estimate $\mathbf{X}$ using the voxel-wise mean of the samples. Additionally, we can gain estimates about the uncertainty of the estimate, for example using the 2.5% and the 97.5% quantiles to compute a 95% credible interval.

1	2	1	2	1	2
2	1	2	1	2	1
1	2	1	2	1	2
2	1	2	1	2	1
1	2	1	2	1	2
2	1	2	1	2	1
1	2	1	2	1	2

Figure 2.12 "Checker board" representation of voxels. Voxels in group 1 are independent from voxels in group 2.

2.6.3 Computational efficiency

A disadvantage of MCMC algorithms is the computational burden. Efficient computing is therefore essential. For example, in the example in Section 2.6.2 a GMRF is used. That is, the voxel is conditionally dependent on the rest of the image given their neighbors (see Section 2.5.3). Therefore, for 2-D images the voxels can be grouped into two groups of independent voxels, see Fig. 2.12. In each group the voxels are conditionally independent from each other given the voxels of the other groups. This is a direct consequence from the global Markov property. Therefore, the Gibbs sampler can be modified as follows

(1) choose a starting value $(\sigma^2)^{(0)}$,

(2a) for each voxel in group 1 i draw $X_i^{(k)}$ independently from a $N(\mu, s^2)$ distribution, where μ and s^2 depend on $(\sigma^2)^{(k-1)}$,

(2b) for each voxel in group 2 i draw $X_i^{(k)}$ independently from a $N(\mu, s^2)$ distribution, where μ and s^2 depend on $(\sigma^2)^{(k-1)}$,

(3) draw $(\sigma^2)^{(k)}$ from $IG(a + n/2, b + \sum(Y_i - X_i)^2/2)$,

(4) iterate (2) and (3) for $k = 1, \ldots, m$,

(5) delete the first b samples (burn-in).

Steps (2a) and (2b) can be parallelized due to independency. This can be a massive speed up of the MCMC algorithm.

An alternative is to use a block-wise approach. That is, instead of sampling single voxels, we sample a whole block of voxels. In the example above either the whole image or parts of the image. It can be shown that the full conditional of a subset C of voxels is a multivariate Gaussian distribution

$$\mathbf{X}_C|\mathbf{X}_{-C}, \sigma^2, \mathbf{Y} \sim N_{|C|}(\mathbf{Q}^{-1}\mathbf{b}, \mathbf{Q}^{-1}) \tag{2.122}$$

where X_{-C} is the rest of the voxels, $|C|$ is the number of voxels in the subset, $\mathbf{Q} = \mathbf{K}_C + (1/\sigma^2)\mathbf{I}$, $\mathbf{K}_C$ is the part of the precision matrix $\mathbf{K}$ of the GMRF corresponding to C, $\mathbf{I}$ is the identity matrix, $b = \frac{1}{\sigma^2}\mathbf{Y}_C + \mathbf{m}_C$, and for $i \in C$, $m_i = \sum j \sim i, j \notin CX_j/\tau^2$.

Similarly to $\mathbf{K}$, $\mathbf{Q}$ is a sparsely populated matrix. This allows a very efficient way of sampling from (2.122) proposed by Rue [316]

(1) compute the Cholesky decomposition $Q = LL'$,
(2) solve $Lw = b$,
(3) solve $L^T u = w$,
(4) draw $z \sim N(0, I_{|C|})$,
(5) solve $L^T v = z$,
(6) $\mathbf{X}_C = u + v$.

The sparsity of $\mathbf{Q}$ can be used in the Cholesky decomposition and when solving the linear system in (2), (3), and (5).

2.7. FEATURE SELECTION METHODS

There are several criteria to measure the classification performance of individual features and/or feature vectors. In this section we will review techniques to select a subset of l features out of m originally available.

The most important techniques described here are exhaustive search [329], branch and bound algorithm, max-min feature selection [364], and Fisher's linear discriminant [401].

2.7.1 Exhaustive Search

Let $\mathbf{y}$ with $\mathbf{y} = [y_1, \ldots, y_D]^T$ be a pattern vector. Exhaustive search selects the d best features out of the maximal available features D so as to minimize the classification error.

The resulting number of total combinations is

$$\binom{D}{d} = \frac{D!}{(D-d)!d!} \tag{2.123}$$

The main disadvantage of exhaustive search is that the total number of combinations increases exponentially with the dimension of the feature vector.

2.7.2 Branch and Bound Algorithm

The computational cost associated with the exhaustive search is very huge. Therefore, new techniques are necessary to determine the optimal feature set without explicit evaluation of all the possible combinations of d features. Such a technique is the branch and bound algorithm.

This technique is applicable when the separability criterion is monotonic, that is,

$$\text{If} \quad \chi_1 \subset \chi_2 \subset \cdots \chi_D, \quad \text{then} \quad J(\chi_1) \leq J(\chi_2) \leq \cdots J(\chi_D) \tag{2.124}$$

χ_i is one of the possible feature subsets with i features while $J(\chi_i)$ is the corresponding criterion function. This technique is visualized for an example in Fig. 2.13. The problem is to select out of the four features y_1, y_2, y_3, y_4 the two most relevant. At each stage, we

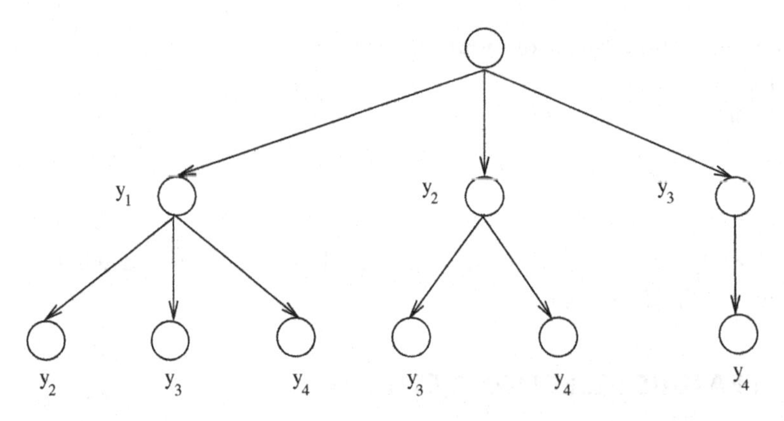

Figure 2.13 Feature selection based on branch and bound algorithm.

determine the maximal value of the criterion function and determine the correct path based on this. At each node then, we eliminate one feature from the initial feature set. The criterion function J is updated as well: if J is larger than its initial value J_0, then continue along the same path, else choose the next path at left.

In summary, the branch and bound method represents a computationally efficient way by formulating the problem as a combinatorial optimization task. The optimal solution is obtained without involving exhaustive enumeration of all possible combinations.

2.7.3 Max-Min Feature Selection

The main idea here is to perform feature selection based on the individual and pairwise merit of features [179]. Thus the criterion function has to be evaluated only in one- and two-dimensional spaces.

This technique determines, based on a given criterion function J, the individual and pairwise relevance of features. Let us assume that we have already determined k features and thus the remaining features are $Y - X_k$. By writing $Y = \{y_j | j = 1, \dots, D\}$, we denote the initial available feature set, while by X_k we denote the already chosen features from Y. Let $y_j \in Y - X_k$ and $x_l \in X_k$.

The max-min feature selection first determines

$$\Delta J(y_j, x_l) = J(y_j, x_l) - J(x_l) \tag{2.125}$$

and then it chooses

$$\Delta J(x_{k+1}, x_l) = \max_j \min_l \Delta J(y_j, x_l) \quad \text{with} \quad x_l \in X_k \tag{2.126}$$

2.7.4 Sequential Forward and Sequential Backward Selection

There are many situations in medical imaging when the determination of the optimal feature set based on the branch and bound algorithm is computationally too intensive.

Therefore the question of trade-off between the optimality and efficiency of algorithms leads us to seek suboptimal solutions.

The suboptimal search methods reviewed in this section represent an alternative to the above-mentioned problems and are shown to give very good results and to be computationally more effective than other techniques [401]. The simplest suboptimal techniques are the sequential forward and sequential backward selection algorithms [304]. We follow here the approach described in [179].

Sequential forward selection (SFS) is a bottom-up technique. We start with an empty feature set, and as the first feature we choose the individually best measurement. In the subsequent steps, we choose from the remaining feature set only the feature that, together with the previously selected ones, yields the maximum value of the criterion function. For a better understanding, let us assume that the feature set X_k consists of k features out of the original complete feature set $Y = \{y_j | j = 1, \dots, D\}$. We choose the $(k+1)$th feature from the set of the remaining features, $Y - X_k$, such that

$$J(X_{k+1}) = \max \quad J(X_k \cup y_j), \quad y_j \in Y - X_k \tag{2.127}$$

$$\text{Initialization}: \quad X_0 = \Phi \tag{2.128}$$

Sequential backward selection (SBS), on the other hand, is a top-down technique. Here, we start with the whole feature set Y, and at each step we eliminate only that feature which yields the smallest decrease of the criterion function.

We assume that k features are already eliminated from the initial feature set, $Y = \{y_j | j = 1, \dots, D\}$, such that the remaining feature set is given by X_{D-k}. The $(k+1)$th feature to be removed is selected from the set X_{D-k} such that

$$J(X_{D-k-1}) = \max J(X_{D-k} - y_j), \quad y_j \in X_{D-k} \tag{2.129}$$

$$\text{Initialization}: \quad X_D = Y \tag{2.130}$$

SFS is computationally less expensive than SBS since the criterion function is evaluated in d-dimensional spaces, while for the SBS it has to be evaluated in spaces of the dimensionality from d to D. The advantage of the SBS as a top-down technique is that at each stage the information loss can be supervised.

However, the disadvantage with both SBS and SFS lies in the so-called nesting effect. This means that once eliminated by SBS a feature cannot be reconsidered later, while a feature added by SFS cannot be removed later. To overcome these disadvantages associated with SBS and SFS, a new technique, the so-called floating search method, was proposed in [304]. Based on the traditional SBS and SFS, it has the flexibility to allow revision at later stages of added or discarded features in previous stages. The computational efficiency of the floating search technique makes it attractive for large-scale feature selection, when the number of features approaches 100. Besides avoiding the nesting of features, this technique compared to the branch and bound method is also tolerant to deviations from monotonic behavior of the feature selection criterion function.

2.7.5 Fisher's Linear Discriminant

The motivation of this technique is based on achieving a maximum class separability. This simple method is built upon information related to the way the sample vectors are scattered in the feature space.

Fisher's approach determines out of D given features the d best features based on a $d \times D$ linear transformation matrix $\mathbf{A}$. Let us assume that we are given a D-dimensional feature vector, and obtain a new d-dimensional feature vector $\mathbf{x}$ by

$$\mathbf{x} = \mathbf{A}\mathbf{y} \tag{2.131}$$

Let us now define the between-class scatter matrix $\mathbf{S_b}$ and the within-class scatter matrix $\mathbf{S_w}$ as

$$\mathbf{S_b} = \sum_{i}^{M} P(\omega_i) E[(\mu_i - \mu_0)(\mu_i - \mu_0)^T] \tag{2.132}$$

and

$$\mathbf{S_w} = \sum_{i}^{M} P(\omega_i) E[(\mathbf{y} - \mu_i)(\mathbf{y} - \mu_i)^T] \tag{2.133}$$

$P(\omega_i)$ is the a priori probability of class ω_i, μ_i is the mean vector of class ω_i, and μ_0 is the global mean vector over all classes ω_i with $i = 1, \ldots, M$.

Figure 2.14 illustrates three different cases of classes at different locations and within-class variances [364].

A reasonable measure of projected data separability is given by the following criterion function $J(\mathbf{A})$:

$$J(\mathbf{A}) = \frac{\mathbf{A^T S_b A}}{\mathbf{A^T S_w A}} \tag{2.134}$$

The goal is to determine an adequate $\mathbf{A}$ which maximizes the above equation. The solution yields a generalized eigenvector problem

$$\mathbf{S_b A} = \Lambda \mathbf{S_w A} \tag{2.135}$$

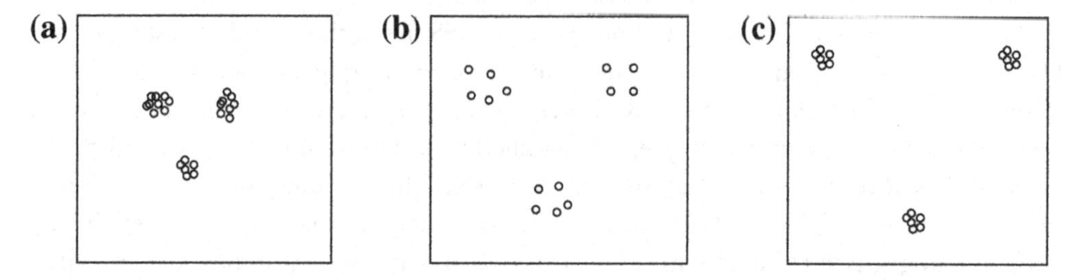

Figure 2.14 Classes with (a) small within-class variance and between-class distances, (b) large within-class variance and small between-class distances, and (c) small within-class variance and large between-class distances [364].

where Λ is the matrix of the significant eigenvalues, and $\mathbf{A}$ is at the same time also the matrix of the corresponding eigenvalues. Thus, $\mathbf{A}$ represents a projection matrix which transforms the feature vector $\mathbf{y}$ into a lower dimensional vector $\mathbf{x}$.

The algorithmic description of determining the relevant features is as follows:

1. Determine Fisher's covariance matrix [329]

$$J_{wb} = \mathbf{S_w^{-1} S_b} \tag{2.136}$$

2. Determine D eigenvalues of the matrix J_{wb} and the corresponding eigenvectors based on

$$J_{wb}\mathbf{Q} = \mathbf{Q}\Lambda \tag{2.137}$$

where $\Lambda = [\lambda_1, \ldots, \lambda_D]$ is the matrix of the eigenvalues and $\mathbf{Q}$ is the matrix of the corresponding eigenvectors. Let us assume that $\lambda_1 > \lambda_2 > \cdots > \lambda_D$ holds.

3. Determine the rank d of matrix J_{wb}.
4. Determine for each $i = 1, \ldots, D$ the term $U(i)$ which represents the sum of the absolute value of the first d eigenvectors:

$$U(i) = \sum_{j=1}^{d} |\mathbf{Q}(i,j)|, \quad j = 1, 2, \ldots, D \tag{2.138}$$

where D is the dimension of the feature vector $\mathbf{y}$.

5. The feature ranking is based on the value of U: the larger the value of U the better the selection.

Example 2.7.1 Acoustic neuromas are benign tumors which generally grow along the track of the acoustic nerve and are detected in MR images.

An automatic system for detection and classification of acoustic neuromas or candidate tumor regions (CTRs) works only as well as the relevance of the selected features describing such neuromas.

The following features were found to be relevant in characterizing acoustic neuromas [80]:

- Shape: These neuromas being roughly globular objects, a contour-based moment scheme can be employed requiring only four moment measurements.
- Compactness: This feature is concerned with shape and measures how closely the CTR combination approximates a circular disk.
- Position: Acoustic neuromas grow on and along the acoustic nerve and are therefore site specific. The position of a CTR combination was represented by the position of its centroid in relationship to the centroid of the head in the image. Two resulting values, the position and the angle, are treated as separate features.
- Symmetry: Acoustic neuromas induce a nonsymmetry in an MRI image as can be seen from Fig. 2.15. A descriptor consisting of four parameters describes the position of the two clusters F_1 and F_2.

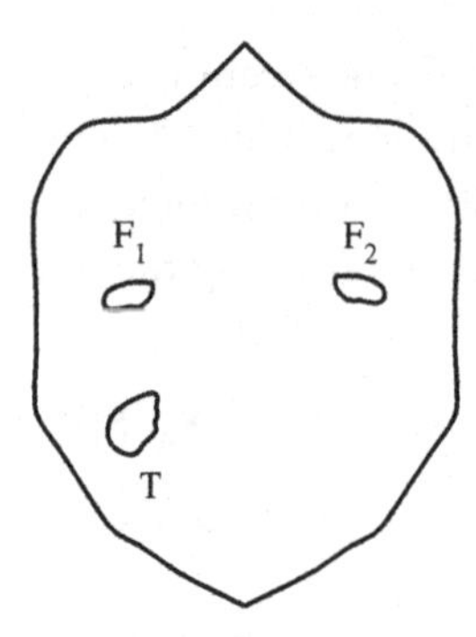

Figure 2.15 Typical position of fp-tumors F_1 and F_2. The rp-tumor is marked with T [80].

Table 2.3 Results of assessing the 127 subsets. The eight best feature subsets achieving class separability are presented. The higher the J_{wb} the better the class separability between tumor and nontumor classes. * shows that a feature was used in the feature subset, while – shows it was not considered [80].

Shape	Inter-clus. posit.	Global angle	Global dist.	Compactness	Mean intens.	Stand. deviat.	Dim. vector	J_{wb}
*	*	*	*	*	*	*	13	8.079
*	*	*	*	*	–	*	12	8.065
*	*	*	–	*	*	*	12	8.061
*	*	*	–	*	–	*	11	8.006
*	*	*	*	*	*	–	12	7.819
*	*	*	*	*	–	–	11	7.811
*	*	*	–	*	*	–	11	7.743
*	*	*	–	*	–	–	10	7.739

- Average pixel gray level: This encodes brightness and texture.
- Gray-level standard deviation: Also serves for representing brightness and texture.

The collection of these seven distinct features (position counts as two separable features) results in a vector of dimension of 13.

The data set consists of 105 pattern vectors with 70 describing tumors and 35 nontumors. The criterion function chosen as a measure for class separability was proposed by Fukunaga [100]

$$J_{wb} = \text{Trace}(\mathbf{S}_{\mathbf{w}}^{-1}\mathbf{S}_{\mathbf{b}}) \tag{2.139}$$

The best feature subset yields the highest value for J_{wb}. Table 2.3 gives the details of the feature subsets that give the eight best class separabilities.

2.8. EXERCISES

1. Show that the central moments shown in eq. (2.7) are translation invariant while the normalized moments shown in eq. (2.9) are both translation and scaling invariant.
2. Write a program to compute both the central and normalized moments described in eqs. (2.7) and (2.9) and apply it to any digitized image of your choice.
3. Write a program that determines the AR parameters for a noncausal prediction model. Then apply it to homogeneous and isotropic image with the following autocorrelation sequence

$$r(k, l) = \exp\left(-2\sqrt{k^2 + l^2}\right) \tag{2.140}$$

for a window W of order $p = q = 1$.
4. Write a program to determine the first- and second-order moments and apply it to any digitized image of your choice.
5. Given the following image array

$$\begin{bmatrix} 1 & 3 & 1 \\ 0 & 1 & 3 \\ 2 & 4 & 2 \end{bmatrix} \tag{2.141}$$

compute its two-dimensional DFT transform.
6. For the image given in Exercise 2.5, compute the two-dimensional discrete cosine transform (DCT) and discrete sine transform (DST).
7. Write a program to compute the two-dimensional DFT transform and apply it to any digitized image of your choice.
8. Write a program to compute the two-dimensional DCT and DST transform and apply it to any digitized image of your choice.
9. Write a program for sequential forward and sequential backward selection.
10. Apply Fishers's linear discriminant to separate the following data belonging to the two classes ω_1 and ω_2.

Class	ω_1	Class	ω_2
−0.25	1.83	1.18	−2.55
1.13	3.18	1.65	−2.60
−2.18	3.77	4.20	−0.28
0.55	−0.10	2.10	−0.94
−2.31	−0.41	2.21	−1.10

(2.142)

   a) Calculate the within- and between-class scatter matrices, $\mathbf{S_W}$ and $\mathbf{S_B}$.
   b) Write a program that computes the relevant features and apply it to the given data set.

CHAPTER THREE

Subband Coding and Wavelet Transform

Contents

3.1. INTRODUCTION

New transform techniques that specifically address the problems of image enhancement and compression, edge and feature extraction, and texture analysis received much attention in recent years especially in biomedical imaging. These techniques are often found under the names multiresolution analysis, time-frequency analysis, pyramid algorithms, and wavelet transforms. They became competitors to the traditional Fourier transform, whose basis functions are sinusoids. The wavelet transform is based on wavelets, which are small waves of varying frequency and limited duration. In addition to the traditional Fourier transform, they provide not only frequency but also temporal information on the signal.

In this chapter, we review the basics of subband coding and present in great detail the different types of wavelet transforms. By subband coding, we mean a transformation

Pattern Recognition and Signal Analysis in Medical Imaging
http://dx.doi.org/10.1016/B978-0-12-409545-8.00003-0

created by filtering and subsampling. The signal is separated approximately into frequency bands for efficient coding. The subband signals are quantized such that the objective measure is maximized. A wavelet is a basis function in continuous time. We know that a function $f(t)$ can be represented by a linear combination of basis functions, such as wavelets. The most important aspect of the wavelet basis is that all wavelet functions are constructed from a single mother wavelet. This wavelet is a small wave or a pulse.

This chapter will also reveal one of the most important aspects of signal processing: the connections among filters, filter banks, and wavelets. Historically, they were separately developed but they must be seen together. The lowpass filter coefficients of the filter bank determine the scaling functions; the highpass filter coefficients produce the wavelets.

3.2. THE THEORY OF SUBBAND CODING

In the following we will review the basics of subband decomposition which is strongly related to filter bank theory. A filter bank is a set of filters. The analysis bank splits the input signal into M subbands using a parallel set of bandpass filters. In other words, the input signal is separated into frequency bands. Those subsignals can be compressed more efficiently than the original signal. Then they can be transmitted or stored. This processing principle is called subband coding or subband decomposition. The synthesis bank reconstructs the original signal from the subband signals. The essential information is extracted from the subband signals in the processing block between the banks. In audio/image compression, the spectral contents are coded depending on their energies.

The analysis bank often has two filters, a lowpass and highpass filter. The lowpass filter takes "averages": it smooths out the high-frequency components. The response of this filter is small or near zero for the highest discrete-time frequency $\omega = \pi$. The highpass filter takes "differences": it smooths the low-frequency components. The response of this filter is small or zero for the frequencies near $\omega = 0$. As mentioned before, these two filters separate the signal into frequency bands. One difficulty arises: the signal length has doubled. This means twice as much storage is needed to keep two full-length outputs. Since this is unacceptable, the solution is to downsample or decimate the signal. Downsampling is an essential part of the filter bank. Only the even-numbered components of the two outputs from the lowpass and highpass filters are saved, while the others are removed. We see that the analysis part has two steps, filtering and downsampling. The synthesis part is the inverse of the analysis bank. The synthesis part also has two steps, upsampling and filtering. The order must be reversed as it is always for inverses. The first step is to recreate full-length vectors. Since downsampling is not invertible, upsampling is the solution. The odd-numbered components are replaced by zeros. The second step of the filter bank is the filtering. The responses of a lowpass and highpass filter are added together in order to reproduce the original input signal up to a delay.

When the synthesis bank recovers the input signal exactly (apart from a time delay), the filter bank is called a perfect reconstruction filter bank.

In the following we will describe mathematically the concepts of downsampling, upsampling, and perfect reconstruction filter banks.

3.2.1 Decimation and Interpolation

We previously have seen that in order to reduce the amount of data to be transmitted, the operation of downsampling or decimation is needed. The filters' outputs must be downsampled. Then in order to regain the original signal, upsampling or interpolation is needed. The removed components are replaced by zeros. This section will review the operation of decimation and interpolation in the time domain as well as in the frequency domain.

The interpolation and decimation techniques being a basic part of the theory of subband decomposition are described in Figs. 3.1 and 3.3. The decimation accomplishes a reduction of the sampling rate by a factor M of a given signal $\{x(n)\}$ after this signal passes through an antialiasing filter $h(n)$. The subsampler is represented in Fig. 3.1 by

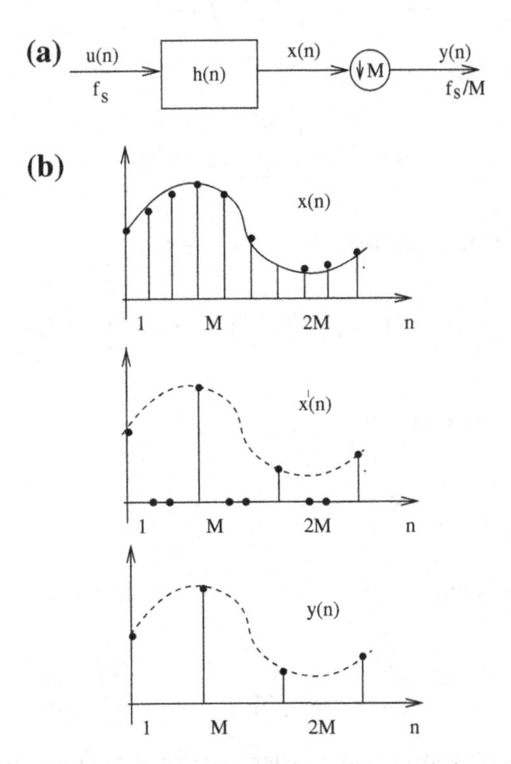

Figure 3.1 Representation of decimation technique: (a) filter and downsampler, (b) typical time sequences of the intermediate signals.

a circle containing the integer number M and a downward arrow. The subsampling considers only each Mth sample value and achieves such a new scaling of the samples on the x-axis.

The retained intermediate signal x' is shown in Fig. 3.1, as well as the new downsampled signal $y(n)$.

$$x' = \begin{cases} x(n), & n = 0, \pm M, \pm 2M, \dots \\ 0, & \text{else} \end{cases} \tag{3.1}$$

$$y(n) = x'(Mn) = x(Mn) \tag{3.2}$$

The intermediate signal $x'(n)$ has the same sampling rate as $x(n)$ and can be expressed mathematically as the product of $x(n)$ with a Dirac train

$$x' = \left[\sum_{r=-\infty}^{\infty} \delta(n - rM) \right] x(n) \tag{3.3}$$

The Dirac train can be written as a Fourier sum

$$\sum_{r} \delta(n - rM) = \frac{1}{M} \sum_{k=0}^{M-1} e^{j\frac{2\pi}{M}nk} \tag{3.4}$$

We see immediately that we can obtain

$$x'(n) = \frac{1}{M} \sum_{k=0}^{M-1} x(n) e^{j\frac{2\pi}{M}nk} \tag{3.5}$$

We now obtain as a Fourier transform $X'(z)$ of $x'(n)$:

$$X'(z) = \frac{1}{M} \sum_{k} Z\left\{x(n)\left(e^{j2\pi k/M}\right)^{n}\right\} = \frac{1}{M} \sum_{k=0}^{M-1} X(ze^{-j2\pi k/M}) \tag{3.6}$$

By choosing $W = e^{\frac{-j2\pi}{M}}$, we obtain

$$X'(z) = \frac{1}{M} \sum_{k=0}^{M-1} X(zW^{k}) \tag{3.7}$$

Then we get for the frequency response on the unit circle $z = e^{j\omega}$

$$X'(e^{j\omega}) = \frac{1}{M} \sum_{k=0}^{M-1} X\left(e^{j(\omega - \frac{2\pi k}{M})}\right) \tag{3.8}$$

The preceding expression shows the Fourier transform yields a sum of M delayed by $2\pi/M$ copies of the original signal. Next, we rename the time axis, such that we get a compression of the timescale with the factor M. We see immediately

$$Y(z) = \sum_{n=-\infty}^{\infty} x'(Mn)z^{-n} = \sum_{k=-\infty}^{\infty} x'(k)\left(z^{\frac{1}{M}}\right)^{-k} \tag{3.9}$$

or

$$Y(z) = X'\left(z^{\frac{1}{M}}\right) \tag{3.10}$$

and

$$Y(e^{j\omega}) = X'\left(e^{\frac{j\omega}{M}}\right) \tag{3.11}$$

For the M subsampler, we now obtain

$$Y(z) = \frac{1}{M}\sum_{k=0}^{M-1} X\left(z^{\frac{1}{M}} W^k\right) \tag{3.12}$$

or

$$Y(e^{j\omega}) = \frac{1}{M}\sum_{k=0}^{M-1} X\left(e^{j\left(\frac{\omega-2\pi k}{M}\right)}\right) \tag{3.13}$$

It is important to realize that a compression in the time domain leads to a dilation in the frequency domain in such a way that the interval 0 to $\frac{\pi}{M}$ now covers the band from 0 to π. The spectra of the intermediate signals occurring based on the subsampling are shown in Fig. 3.2.

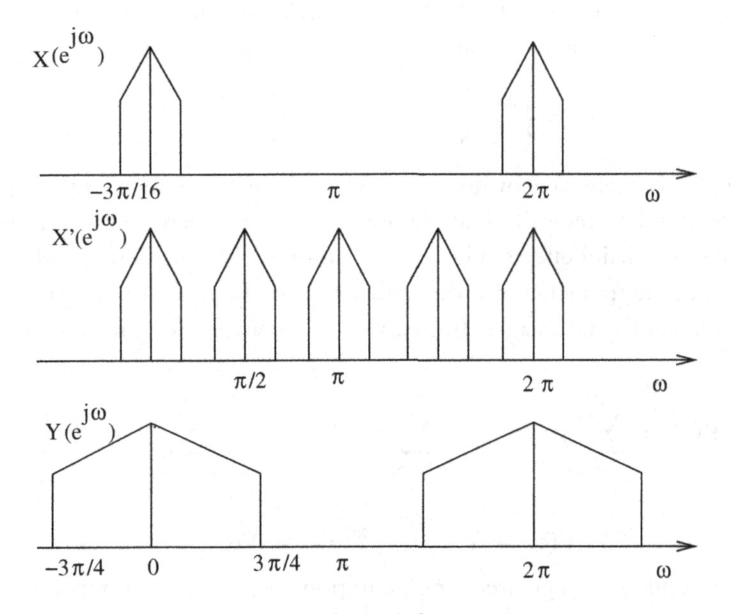

Figure 3.2 Frequency spectra of downsampled signals for $M = 4$.

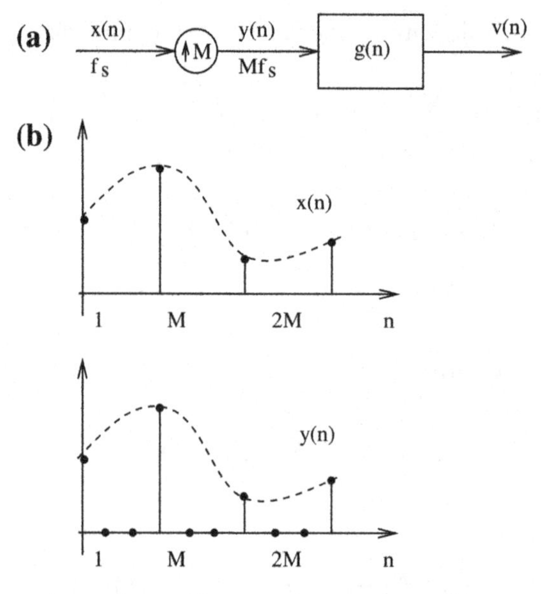

Figure 3.3 Upsampling or interpolation: (a) upsampling operation, (b) input and output waveforms for $M = 3$.

Interpolation, on the other hand, is a technique to increase the sampling rate of a given signal by a factor M. As shown in Fig. 3.3, this can be accomplished by a combination of an upsampler and a lowpass filter $g(n)$.

The upsampler is symbolically shown by a circle containing an upward arrow. The upsampler is described mathematically as

$$y(n) = \begin{cases} x(n/M), & n = 0, \pm M, \pm 2M, \ldots \\ 0, & \text{else} \end{cases} \tag{3.14}$$

Upsampling can be achieved by inserting $M - 1$ zeros between the sampled values and by reindexing the timescale. The sampling rate is thus increased by a factor M. The upsampling has two major effects: it leads to a dilation in the time domain while achieving a compression in the frequency domain, and based on the insertion of zeros it generates high-frequency signals and images. It is easy to prove these effects in the transformation domain:

$$Y(z) = \sum_{n=-\infty}^{\infty} y(n) z^{-n} = \sum_{n=-\infty}^{\infty} x\left(\frac{n}{M}\right) z^{-n} = \sum_{k=-\infty}^{\infty} x(k)(z^M)^{-k} \tag{3.15}$$

or

$$Y(z) = X(z^M), \quad Y(e^{j\omega}) = X(e^{j\omega M}) \tag{3.16}$$

To remove the undesired high-frequency components, a lowpass filter must follow the upsampler.

The interpolators and decimators are represented in the time domain as

$$y(n) = \sum_k h(Mn - k)u(k) \tag{3.17}$$

and

$$v(n) = \sum_k g(n - Mk)x(k) \tag{3.18}$$

Example 3.2.1 Consider the following alternating vector $\mathbf{x}(n) = (\ldots, 1, -1, 1, -1, 1, -1, \ldots)$. Downsampling produces the vector $(\downarrow)\mathbf{x} = (\ldots, 1, 1, 1, \ldots)$, while upsampling produces the vector $(\uparrow)\mathbf{x} = (\ldots, 1, 0, -1, 0, 1, 0, -1, 0, 1, 0, -1, \ldots)$.

We will finish this subsection with the following important remarks:

1. The result of a series circuit of a downsampler and an upsampler is shown in Fig. 3.4. The signal at the output of the interpolator is identical with the signal $x'(n)$ from Fig. 3.1. The spectra of the signals outputting both an upsampler and a downsampler are corrupted by aliasing.

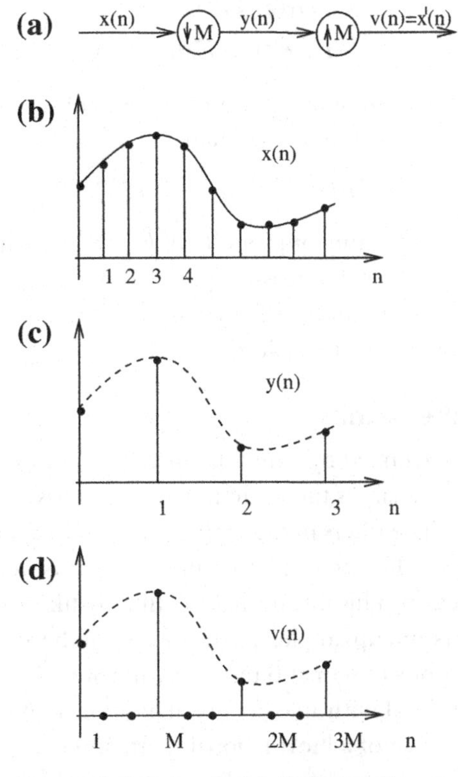

Figure 3.4 Decimation and interpolation signals: (a) downsampling and upsampling operation, (b) sampled input signal, (c) downsampled sequence, (d) upsampled sequence.

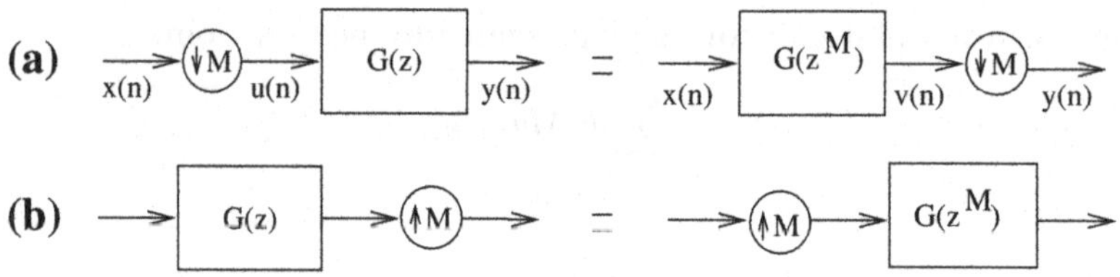

Figure 3.5 Equivalent structures: (a) equivalence for downsampling, (b) equivalence for upsampling.

2. There are equivalencies regarding transfer functions across down- and upsamplers as illustrated in Fig. 3.5. Filter and interpolator, or respectively decimator, can be interchanged, if the filter transfer functions are correspondingly changed.
3. A quadrature-mirror-filter [4] was first used in the elimination of aliasing in two-channel subband coders. Let $h_0(n)$ be a FIR lowpass filter having real coefficients. A mirror filter is described as

$$h_1(n) = (-1)^n h_0(n) \tag{3.19}$$

or, equivalently, in the transform domain as

$$\begin{aligned} H_1(z) &= H_0(-z) \\ H_1(e^{j\omega}) &= H_0(e^{j(\omega-\pi)}) \end{aligned} \tag{3.20}$$

In the following, we make the substitution $\omega \to \frac{\pi}{2} - \omega$, and note that the absolute value is an even function of ω. We thus obtain

$$|H_1(e^{j(\frac{\pi}{2}-\omega)})| = |H_0(e^{j(\frac{\pi}{2}+\omega)})| \tag{3.21}$$

The above equation shows the mirror property of H_0 and H_1 with respect to $\omega = \pi/2$, as illustrated by Fig. 3.6. The highpass response is a mirror image of the lowpass magnitude with respect to the middle frequency $\frac{\pi}{2}$—the quadrature frequency. This explains the term quadrature mirror filter.

3.2.2 Two-Channel Filter Banks

A filter bank is a set of filters, linked by sampling operators and sometimes by delays. In a two-channel filter bank, the analysis filters are normally lowpass and highpass. Those are the filters $H_0(z)$ and $H_1(z)$. The filters in the synthesis bank $G_0(z)$ and $G_1(z)$ must be specially adapted to the analysis filters in order to cancel the errors in the analysis bank or to achieve perfect reconstruction. This means that the filter bank is biorthogonal. The synthesis bank, including filters and upsamplers, is the inverse of the analysis bank. In the next chapter we will talk about biorthogonal filter banks and wavelets. If the synthesis is the transpose of the analysis, we deal with an orthogonal filter bank. A paraunitary (orthonormal) filter bank is a special case of a biorthogonal (perfect reconstruction) filter bank.

This section derives the requirements and properties for the perfect reconstruction by means of a two-channel filter bank. Such a filter bank is shown in Fig. 3.7. The

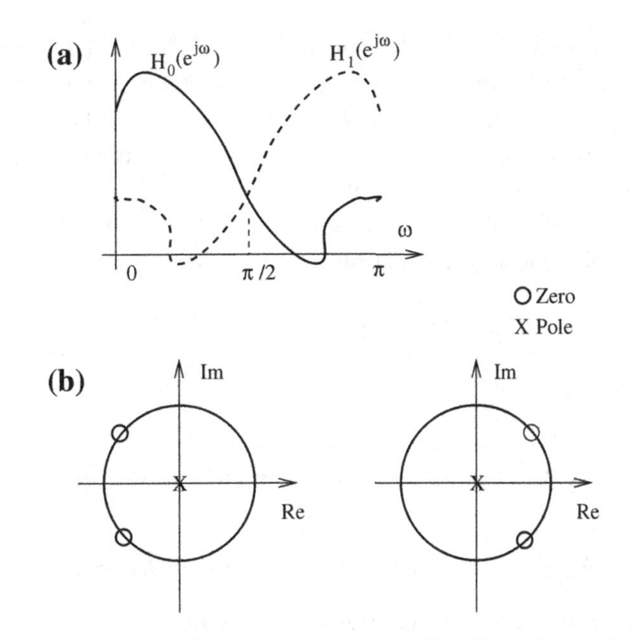

Figure 3.6 (a) Frequency response of the quadrature mirror filter, (b) pole-zero distribution.

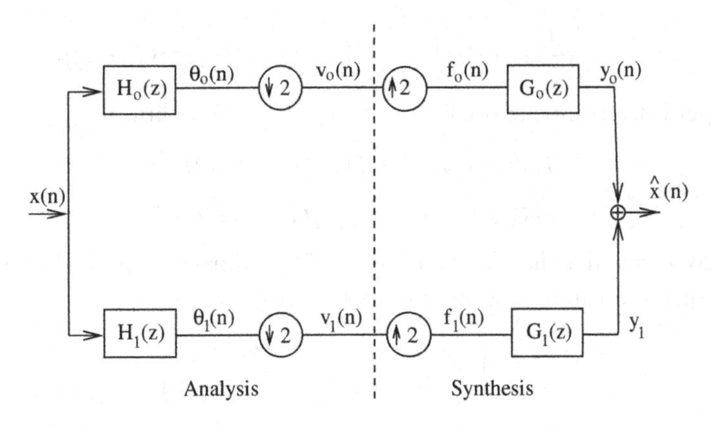

Figure 3.7 Two-channel filter bank.

input spectrum $X(e^{j\omega}), 0 \leq \omega\pi$ is going to be split in two subbands. The filter bank consists of two main parts: analysis and synthesis. The analysis part is performing the signal decomposition while the synthesis part is reconstructing the signal.

The analysis filters $H_0(z)$ and $H_1(z)$ separate the signal in two equal subbands. The filter output signals θ_0 and θ_1 are each downsampled by 2, and then we obtain, as output signals of the analysis part, the subband signals $v_0(n)$ and $v_1(n)$. These signals are quantized by employing a subband coder, coded, and sent to a receiver. Assuming ideal conditions, we expect to receive exactly $v_0(n)$ and $v_1(n)$ at the synthesis part. These signals are interpolated by 2, and the newly obtained signals are denoted by $f_0(n)$ and $f_1(n)$. These signals are processed by the interpolation filters $G_0(z)$ and $G_1(z)$. The reconstructed signal $\widehat{x}(n)$ is obtained by adding the signals $y_0(n)$ and $y_1(n)$. The requirements for a perfect reconstruction are described in great detail in [377].

Here, we derive these requirements for the time and frequency domain. The output signals for the interpolation and decimation filter are given by

$$V_0(z) = \frac{1}{2}[\theta_0(z^{1/2}) + \theta_0(-z^{1/2})]$$
$$F_0(z) = V_0(z) \tag{3.22}$$

and by

$$Y_0(z) = \frac{1}{2}G_0(z)[H_0(z)X(z) + H_0(-z)X(-z)] \tag{3.23}$$

and

$$Y_1(z) = \frac{1}{2}G_1(z)[H_1(z)X(z) + H_1(-z)X(-z)] \tag{3.24}$$

Next we z-transform the reconstructed signal and obtain

$$\begin{aligned}\widehat{X}(z) &= \frac{1}{2}X(z)[H_0(z)G_0(z) + H_1(z)G_1(z)] \\ &+ \frac{1}{2}X(-z)[H_0(-z)G_0(-z) + H_1(-z)G_1(-z)]\end{aligned} \tag{3.25}$$

To achieve a perfect reconstruction, we must impose two conditions:

$$H_0(-z)G_0(z) + H_1(z)G_1(z) = 0 \tag{3.26}$$

$$H_0(z)G_0(z) + H_1(z)G_1(z) = 2 \tag{3.27}$$

Equation (3.26) eliminates aliasing, while eq. (3.27) eliminates amplitude distortion. The two distinct equations can be rewritten in matrix form as

$$\begin{bmatrix} G_0(z) \\ G_1(z) \end{bmatrix} = \frac{2}{\det(\mathbf{H}_m(z))}\begin{bmatrix} H_1(-z) \\ -H_0(-z) \end{bmatrix} \tag{3.28}$$

where the nonsingular analysis modulation matrix $\mathbf{H}_m(z)$ is given by

$$\mathbf{H}_m = \begin{bmatrix} H_0(z) & H_0(-z) \\ H_1(z) & H_1(-z) \end{bmatrix} \tag{3.29}$$

Equation (3.28) describes the cross-modulation of the analysis and synthesis filters. This means the diagonally opposed filters in Fig. 3.7 are functionally related by $-z$ in the Z-domain: $G_1(z)$ is a function of $H_0(-z)$, while $G_0(z)$ is a function of $H_1(-z)$. For finite impulse response (FIR) filters, the determinate is a pure delay.

Three different solutions to eqs. (4.7) and (4.8) are given in Table 3.1 [110].

The second and third columns in Table 3.1 represent the quadrature mirror filter (QMF) solution and the conjugate quadrature mirror filters (CQFs) solution, while the

Table 3.1 Filter solutions for perfect reconstruction.

Filter	QMF	CQF	Orthonormal
$H_0(z)$	$H_0^2(z) - H_0^2(-z) = 2$	$H_0(z)H_0(z^{-1})+$ $H_0(-z)H_0(-z^{-1}) = 2$	$G_0(z^{-1})$
$H_1(z)$	$H_0(-z)$	$z^{-1}H_0(-z^{-1})$	$G_1(z^{-1})$
$G_0(z)$	$H_0(z)$	$H_0(z^{-1})$	$G_0(-z)G_0(z^{-1})+$ $G_0(-z)G_0(-z^{-1}) = 2$
$G_1(z)$	$-H_0(-z)$	$zH_0(-z)$	$-z^{-2K+1}G_0(-z^{-1})$

fourth column represents the orthonormal filters solution. $2K$ in Table 3.1 denotes the length or number of coefficients in each filter.

The orthonormality condition for perfect reconstruction filter banks is defined in the time domain as

$$\langle g_i(n), g_j(n+2m)\rangle = \delta(i-j)\delta(m), \quad i,j = \{0,1\} \tag{3.30}$$

This condition is different from the *biorthogonality* condition, which is given in the time domain as

$$\langle h_i(2n-k), g_j(k)\rangle = \delta(i-j)\delta(n), \quad i,j = \{0,1\} \tag{3.31}$$

It is useful to note that the biorthogonality condition holds for all two-band, real-coefficient perfect reconstruction filter banks.

Multirate techniques are considered among the most important tools for multiresolution spectral analysis. In this context, the perfect reconstruction quadrature-mirror-filter (PR QMF) bank is the most efficient signal decomposition block. The PR QMF bank splits the input spectrum into two equal subbands, a low (L) and high (H) band. This decomposition can be applied again to these (L) and (H) half bands to produce the quarter bands: (LL), (LH), (HL), and (HH). Two of these decomposition levels are illustrated in Fig. 3.8. The original signal at a data rate f_s is decomposed into four subband signals $v_0(n), \ldots, v_3(n)$, each operating at a rate $f_s/4$.

For perfect reconstruction, eqs. (3.26) and (3.27) must be fulfilled. This two-level structure is equivalent to the four-band bank shown in Fig. 3.9, if the following holds [4]:

$$H_0'(z) = H_0(z)H_0(z^2) \tag{3.32}$$

$$H_1'(z) = H_0(z)H_1(z^2) \tag{3.33}$$

$$H_2'(z) = H_1(z)H_0(z^2) \tag{3.34}$$

$$H_3'(z) = H_1(z)H_1(z^2) \tag{3.35}$$

The corresponding representation in the frequency domain given ideal bandpass filters is given in Fig. 3.9b.

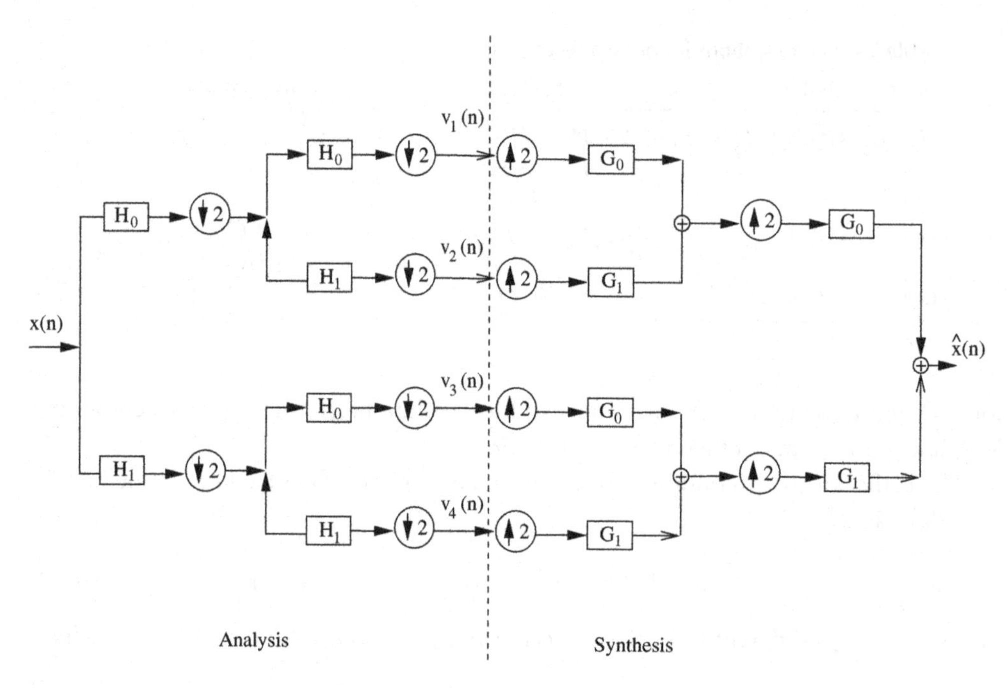

Figure 3.8 Four-band, analysis-synthesis tree structure.

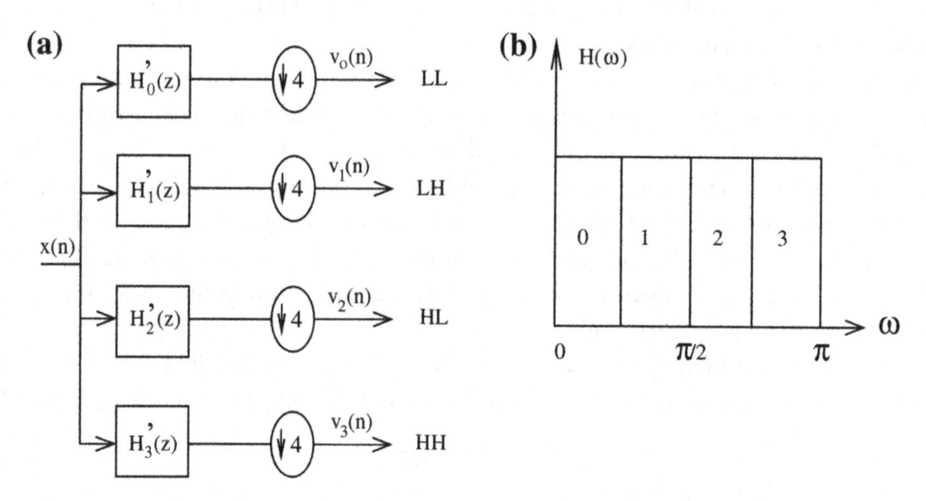

Figure 3.9 (a) Four-band equivalent to two-level regular binary tree, (b) frequency bands corresponding to the four-bands with ideal bandpass filters.

Almost all real-world signals such as speech concentrate most of their energy content in only subregions of their spectrum, indicating that only some spectral bands are significant, while the others are not. A logical consequence is that not all subbands of

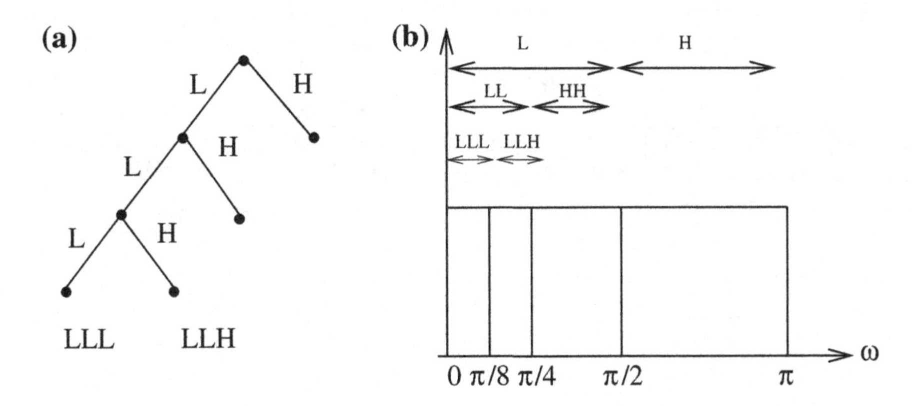

Figure 3.10 (a) A dyadic tree structure, (b) corresponding frequency band split.

the regular binary tree are needed, and therefore some of the fine frequency resolution subbands can be combined to achieve larger bandwidth frequency bands. This leads to a new irregular tree structure having unequal bandwidths. Figure 3.10 displays an arbitrary irregular subband tree with a maximum tree level $L = 3$ and the corresponding frequency bands. It represents a dyadic band tree, and it splits only the lower half of the spectrum at any level into two equal bands. The higher (detail) half-band component at any level is decomposed no further.

A closer look at the analysis-synthesis structure reveals that a half-resolution frequency step is used at any level. Based on this fact, this structure is also called the octave-band or constant-Q subband tree structure. As a first step, we need to obtain the low (L) and high (H) signal bands. It is relevant to note that band (L) gives a coarse representation of a signal while band (H) contains the detail information. Figure 3.11 shows the analysis-synthesis structure for a three-level dyadic tree. This structure is perfectly able to recover the original signal.

3.2.3 The Laplacian Pyramid for Signal Decomposition

The idea of multiscale signal decomposition (coarse to fine) goes back to Burt and Adelson [45] and was first applied to tasks related to image coding [45]. The basic motivation of multiresolution processing or image pyramids is seen in context with the examination of small and large objects or of low- and high-contrast objects in an image. Then it becomes evident that studying them at different resolution, either high or low, is of great relevance.

A pyramid represents a hierarchical data structure that contains successively compressed information, usually of an image [232]. Each level of such a pyramid produces a successively lower-resolution (more blurred) version of the given image. The detail information is given by the difference between the blurred representations of two neighboring levels.

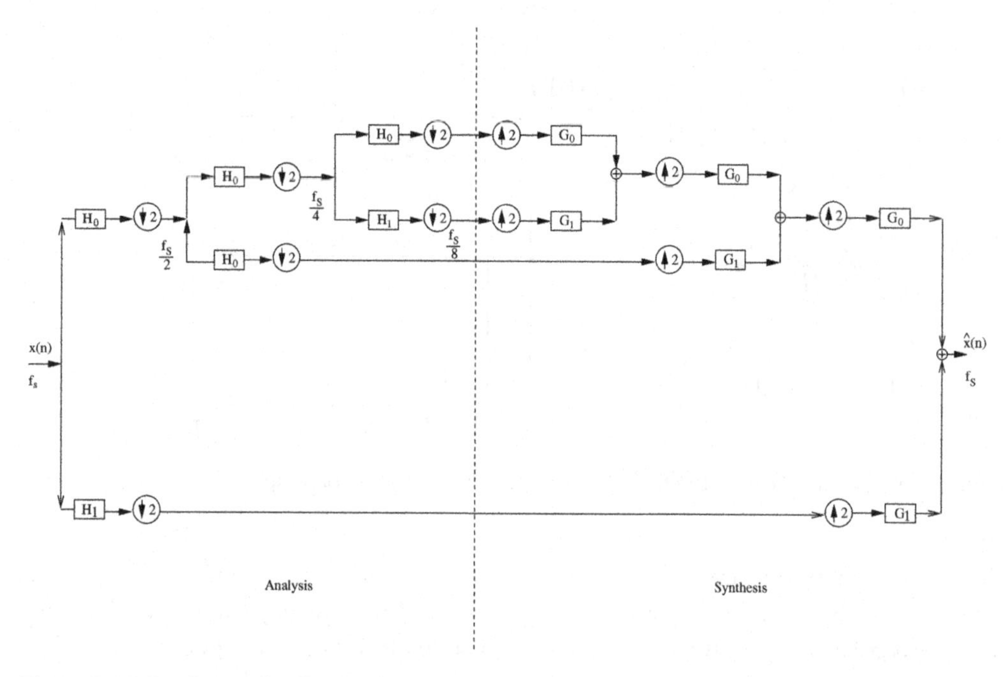

Figure 3.11 Dyadic wavelet filter bank.

The image decoding process based on a pyramid reconstruction is achieved in a progressive manner: It starts with the coarsest approximation of the image and adds successively finer details at each image level until the original image is obtained. The Laplacian pyramid is nothing other than a spectral or subband analysis accomplished by a dyadic tree [45].

The signal $x(n)$ in Fig. 3.12 is first lowpass filtered and then undersampled by 2. This newly obtained signal is denoted by $x_D^1(n)$. Next we upsample this signal by 2 and interpolate it, such that we get $x_I^1(n)$. The approximation error represents the high-resolution detail and is given by

$$x_L^1(n) = x(n) - x_I^1(n) \tag{3.36}$$

where L means Laplacian since the approximation error in most cases has a Laplacian-shaped pdf. For perfect reconstruction, we need to add the detail signal to the interpolated lowpass signal:

$$x(n) = x_I^1(n) + x_L^1(n) \tag{3.37}$$

Since $x_I^1(n)$ is obtained from $x_D^1(n)$, we achieve a perfect reconstruction of $x(n)$ based only on $x_L^1(n)$ and $x_D^1(n)$. The data rate of $x_D^1(n)$ is half of the data rate of $x(n)$. Therefore, we obtain a coarse-to-fine signal decomposition in terms of time and space. The downsampling and upsampling steps are repeated on the lowpass content of the signal until the desired level L is reached. Figure 3.12 illustrates a Laplacian pyramid and its frequency

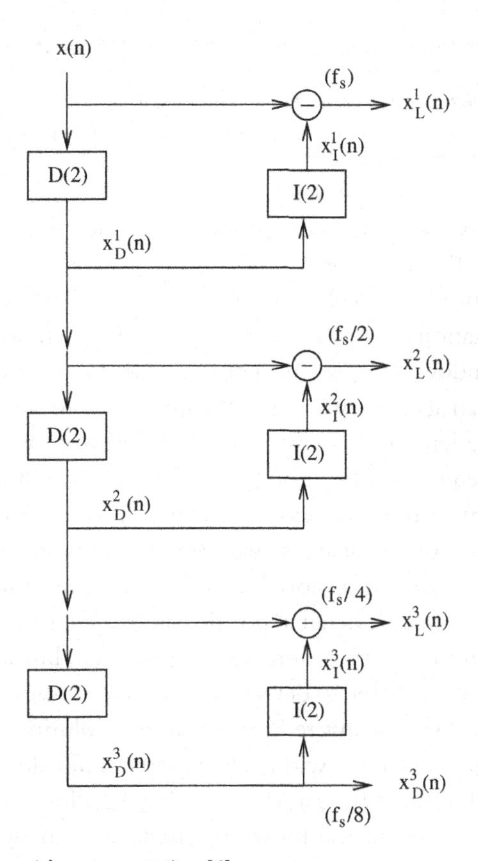

Figure 3.12 Laplacian pyramid representation [4].

resolution for $L = 3$ [4]. We can see that $x(n)$ is reconstructed from the lowpass signal and the detail signals $x_L^3(n)$, $x_L^2(n)$, and $x_L^1(n)$.

In general, the base of the pyramid contains the highest resolution, while the apex contains the lowest resolution. Moving upward, both size and resolution decrease. It is important to state that the Laplacian pyramid contains the prediction residuals needed to compute the Gaussian pyramid. The first-order statistics of the prediction residuals in the Laplacian pyramid are highly peaked around zero. This is different from their Gaussian counterparts, and therefore these images can be highly compressed by assigning fewer bits to the more probable values.

3.3. THE WAVELET TRANSFORM

3.3.1 Time-Frequency Representation

The discrete Fourier transform (DFT) analyzes a signal in terms of its frequency components by finding the signal's magnitude and phase spectra. Unfortunately, the DFT

Table 3.2 Time and frequency resolution by window width.

Narrow window	Good time resolution	Poor frequency resolution
Wide window	Poor time resolution	Good frequency resolution

cannot find the times at which various frequency components occur within the window. In other words, the DFT fails to distinguish signals whose characteristics change with time, known as nonstationary signals, from those whose characteristics do not change with time, known as stationary signals. Since the DFT is windowed, only the signal's behavior within the window is important. For signals that are stationary within the window, the DFT provides an accurate picture of the frequency content since the same signal behavior persists for the length of the window. For signals with changing behavior the DFT can report the frequencies but not when they occur. Choosing a variable window length is a solution to this problem. But while shorter windows can improve the DFT's ability to analyze nonstationary signals, there are resolution implications. If the sampling rate remains unchanged, a shorter window has fewer points, which means that the DFT cannot provide enough detail about the signal spectrum. So a shorter window means good time resolution by providing very local detail, but at the same time poor frequency resolution, since the time to detect signal characteristics is too short. A solution is to choose a larger window, but this reduces the time resolution since the DFT cannot pinpoint changes in signal behavior within its window. Thus good time and frequency resolution cannot be achieved at the same time. Table 3.2 illustrates this fact.

We can conclude that one of the most important tasks in signal analysis is to find a transform which represents the signal features simultaneously in time and frequency. Based on the standard Fourier analysis it is only possible to decompose a signal in its frequency components and to determine the corresponding amplitudes. Thus it is not possible to determine when a signal exhibits a particular frequency characteristics.

Therefore, the Fourier transform

$$F(\omega) = \int_{-\infty}^{\infty} f(t)e^{-j\omega t}\, dt \longleftrightarrow f(t) = \frac{1}{2\pi}\int_{-\infty}^{\infty} F(\omega)e^{j\omega t}\, d\omega \tag{3.38}$$

is not suitable for nonstationary signals. It sweeps over the whole time axis and does not detect local changes, such as high-frequency bursts. A solution is the short-time Fourier transform which was proposed by Gabor [102] and does not have the above-mentioned disadvantages. This transform works by sweeping a short-time window over the time signal, and thus determines the frequency content in each considered time interval. This technique has been successfully applied to speech processing [230,306,324].

The short-time Fourier transform works by positioning a window $g(t)$ at some point τ on the time axis and determining the Fourier transform of the signal within this window:

$$F(\omega, \tau) = \int_{-\infty}^{\infty} f(t)g^*(t-\tau)e^{-j\omega t}\, dt \tag{3.39}$$

The basis functions of this transformation are produced based on the modulation and translation of the window function $g(t)$. ω and τ represent in this context the modulation and translation parameters. It can be shown that this is equivalent to filtering the signal $f(t)$ using a bank of filters, each centered at a different frequency but all of them having the same bandwidth. This is its drawback, because low- and high-frequency signal components are analyzed through the same window in time, resulting in poor overall localization of events. The solution to this is to find a long window to analyze slowly time-varying low-frequency components and a narrow window to detect high-frequency short-time activities. As we saw, this is offered by a tree-structured octave-band filter bank associated with the discrete time wavelet transform.

In summary, the major problem with the short-time Fourier transform is that the window $g(t)$ has a fixed time duration, and at the same time it also has a fixed frequency resolution. The product of frequency interval and time interval is a stable quantity. The Heisenberg Uncertainty Principle [285] gives in this respect a precise definition and a lower bound for the product. The uncertainty principle states that for each transformation pair $g(t) \longleftrightarrow G(\omega)$ the relationship

$$\sigma_t \sigma_\omega \geq \frac{1}{2} \tag{3.40}$$

holds. The lower bound is given by the Gaussian function $f(t) = e^{-t^2}$. σ_T and σ_ω represent the squared variances of $g(t)$ and $G(\omega)$:

$$\sigma_T^2 = \frac{\int t^2 |g(t)|^2 \, dt}{\int |g(t)|^2 \, dt}$$
$$\sigma_\omega^2 = \frac{\int \omega^2 |G(\omega)|^2 \, d\omega}{\int |G(\omega)|^2 \, d\omega} \tag{3.41}$$

The window function $g(t)$ is defined as a prototype function. As τ increases, the prototype function is shifted on the time axis such that the window length remains unchanged. Figure 3.13 illustrates that each element σ_T and σ_ω of the resolution rectangle of the area $\sigma_T \sigma_\omega$ remains unchanged for each frequency ω and time-shift τ. The rectangles have the same form and area in the entire time-frequency plane.

Wavelet transforms are an alternative to the short-time Fourier transform. Their most important feature is that they analyze different frequency components of a signal with different resolutions. In other words, they address exactly the concern raised in connection with the short-time Fourier transform. To implement different resolutions at different frequencies requires the notion of functions at different scales. Like scales on a map, small scales show fine details while large scales show coarse features only. A scaled version of a function $\psi(t)$ is the function $\psi(t/a)$, for any scale a. When $a > 1$, a function of lower frequency is obtained that is able to describe slowly changing signals. When $a < 1$ a function of higher frequency is obtained that can detect fast signal changes. It is important to note that the scale is inversely proportional to the frequency.

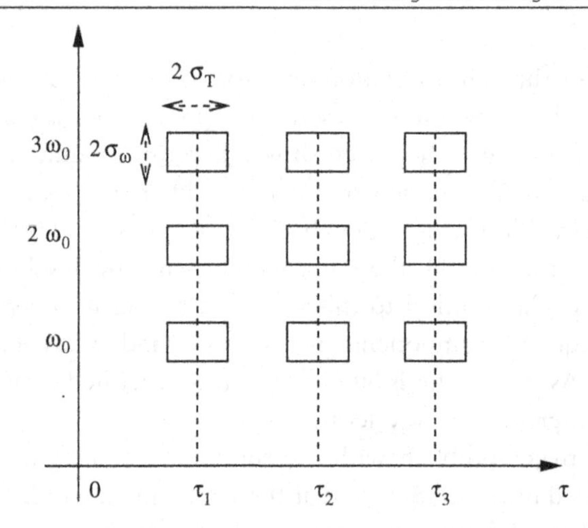

Figure 3.13 Short-time Fourier transform: time-frequency space and resolution cells.

Wavelet functions are localized in frequency in the same way sinusoids are, but they differ from sinusoids by being localized in time, as well. There are several wavelet families, each having a characteristic shape, and the basic scale for each family covers a known, fixed interval of time. The time spans of the other wavelets in the family widen for larger scales and narrow for smaller scales. Thus, wavelet functions can offer either good time resolution or good frequency resolution: Good time resolution is associated with narrow, small-scale windows, while good frequency resolution is associated with wide, large-scale windows.

To determine what frequencies are present in a signal and when they occur, the wavelet functions at each scale must be translated through the signal, to enable comparison with the signal in different time intervals. A scaled and translated version of the wavelet function $\psi(t)$ is the function $\psi\left(\frac{t-b}{a}\right)$, for any scale a and translation b. A wavelet function similar to the signal in frequency produces a large wavelet transform. If the wavelet function is dissimilar to the signal, a small transform will arise. A signal can be coded using these wavelets if it can be decomposed into scaled and translated copies of the basic wavelet function. The widest wavelet responds to the slowest signal variations and thus describes the coarsest features in the signal. Smaller-scale wavelets respond best to high frequencies in the signal and detect rapid signal changes, thus providing detailed information about this signal. In summary, smaller scales correspond to higher frequencies, and larger scales to lower frequencies. A signal is coded through the wavelet transform by comparing the signal to many scalings and translations of a wavelet function.

The wavelet transform (WT) is produced by a translation and dilation of a so-called prototype function ψ. Figure 3.14 illustrates a typical wavelet and its scalings. The bandpass characteristics of ψ and the time-frequency resolution of the WT can be easily detected.

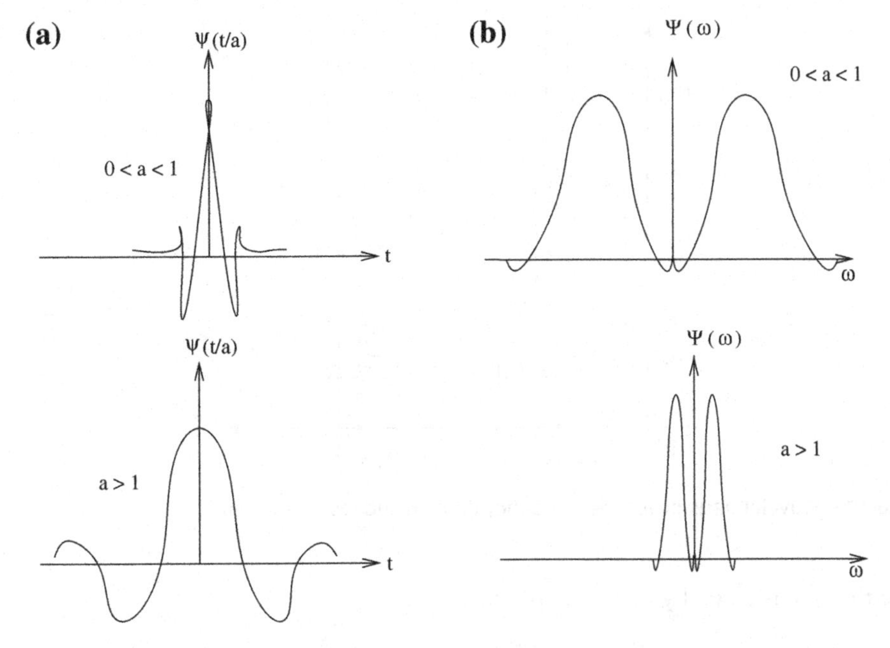

Figure 3.14 Wavelet in time and frequency domain: (a) scale parameter $0 < a < 1$, (b) $a > 1$.

The foundation of the WT is based on the scaling property of the Fourier transform. If

$$\psi(t) \longleftrightarrow \Psi(\omega) \tag{3.42}$$

represents a Fourier transform pair, then we have

$$\frac{1}{\sqrt{a}}\Psi\left(\frac{t}{a}\right) \longleftrightarrow \sqrt{a}\Psi(a\omega) \tag{3.43}$$

with $a > 0$ being a continuous variable. A contraction in the time domain produces an expansion in the frequency domain, and vice versa. Figure 3.15 illustrates the corresponding resolution cells in the time-frequency domain. The figure makes visual the underlying property of wavelets: they are localized in both time and frequency. While the functions $e^{j\omega t}$ are perfectly localized at ω they extend over all time: wavelets on the other hand, that are not at a single frequency, are limited to finite time. As we rescale, the frequency goes up by a certain quantity and at the same time the time interval goes down by the same quantity. Thus the uncertainty principle holds.

A wavelet can be defined by the scale and shift parameters a and b,

$$\psi_{ab}(t) = \frac{1}{\sqrt{a}}\psi\left(\frac{t-b}{a}\right) \tag{3.44}$$

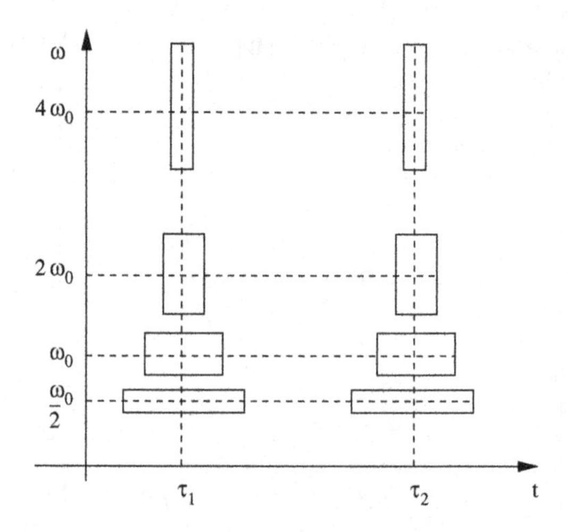

Figure 3.15 Wavelet transform: time-frequency domain and resolution cells.

while the WT is given by the inner product

$$W(a, b) = \int_{-\infty}^{\infty} \psi_{ab}(t) f^*(t) dt = \langle \psi_{ab}, f \rangle \tag{3.45}$$

with $a \in R^+, b \in R$.

The WT defines a $L^2(R) \rightarrow L^2(R^2)$ mapping which has a better time-frequency localization than the short-time Fourier transform.

In the following, we will describe the continuous wavelet transform (CWT) and show an admissibility condition which is necessary to ensure the inversion of the WT. Also, we will define the discrete wavelet transform (DWT) which is generated by sampling the wavelet parameters (a, b) on a grid or lattice. The quality of the reconstructed signals based on the transform values depends on the coarseness of the sampling grid. A finer sampling grid leads to more accurate signal reconstruction at the cost of redundancy, a coarse sampling grid is associated with loss of information. To address these important issues, the concept of frames is now presented.

3.3.2 The Continuous Wavelet Transform

The CWT transforms a continuous function into a highly redundant function of two continuous variables, translation and scale. The resulting transformation is important for time-frequency analysis and is easy to interpret.

The CWT is defined as the mapping of the function $f(t)$ on the timescale space by

$$W_f(a, b) = \int_{-\infty}^{\infty} \psi_{ab}(t) f(t) dt = \langle \psi_{ab}(t), f(t) \rangle \tag{3.46}$$

The CWT is invertible if and only if the resolution of identity holds:

$$f(t) = \underbrace{\frac{1}{C_\psi} \int_{-\infty}^{\infty} \int_{0}^{\infty} \frac{dadb}{a^2}}_{\text{summation}} \underbrace{W_f(a,b)}_{\text{Wavelet coefficients}} \underbrace{\psi_{ab}(t)}_{\text{Wavelet}} \tag{3.47}$$

where

$$C_\psi = \int_{o}^{\infty} \frac{|\Psi(\omega)|^2}{\omega} d\omega \tag{3.48}$$

assuming that a real-valued $\psi(t)$ fulfills the admissibility condition. If $C_\psi < \infty$, then the wavelet is called admissible. Then we get for the DC gain

$$\Psi(0) = \int_{-\infty}^{\infty} \psi(t) dt = 0 \tag{3.49}$$

We immediately see that $\psi(t)$ corresponds to the impulse response of a bandpass filter and has a decay rate of $|t|^{1-\epsilon}$. It is important to note that based on the admissibility condition, it can be shown that the CWT is complete if $W_f(a,b)$ is known for all a, b.

The Mexican-hat wavelet

$$\psi(t) = \left(\frac{2}{\sqrt{3}} \pi^{-\frac{1}{4}}\right) (1 - t^2) e^{-\frac{t^2}{2}} \tag{3.50}$$

is visualized in Fig. 3.16. It has a distinctive symmetric shape, and it has an average value of zero and dies out rapidly as $|t| \rightarrow \infty$. There is no scaling function associated with the Mexican-hat wavelet.

Figure 3.17 illustrates the multiscale coefficients describing a spiculated mass. Figure 3.17a shows the scanline through a mammographic image with a mass (8 mm) while Fig. 3.17b visualizes the multiscale coefficients at various levels.

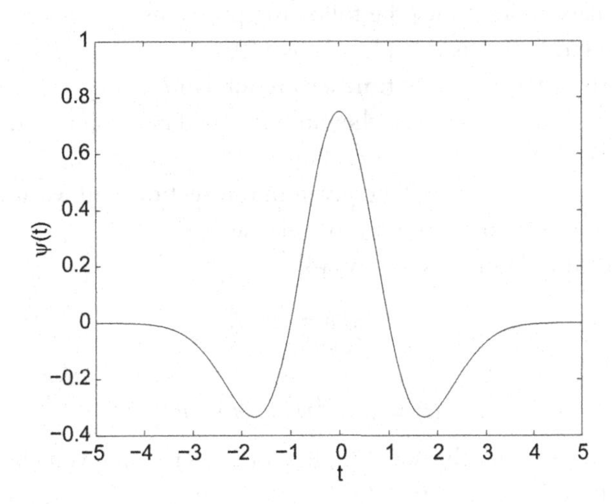

Figure 3.16 Mexican-hat wavelet.

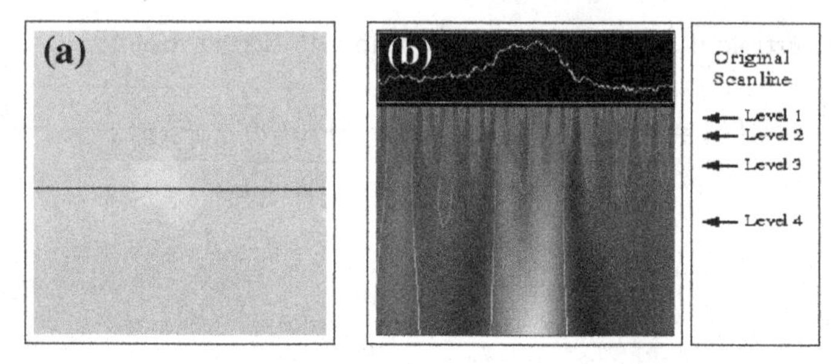

Figure 3.17 Continuous wavelet transform: (a) scan line, (b) multiscale coefficients. *Images courtesy of Dr. A. Laine, Columbia University.*

The short-time Fourier transform finds a decomposition of a signal into a set of equal-bandwidth functions across the frequency spectrum. The WT provides a decomposition of a signal based on a set of bandpass functions that are placed over the entire spectrum. The WT can be seen as a signal decomposition based on a set of constant-Q bandpasses. In other words, we have an octave decomposition, logarithmic decomposition, or constant-Q decomposition on the frequency scale. The bandwidth of each of the filters in the bank is the same in a logarithmic scale, or equivalently, the ratio of the filters' bandwidth to the respective central frequency is constant.

3.4. THE DISCRETE WAVELET TRANSFORMATION

The CWT has two major drawbacks: redundancy and lack of practical relevance. The first is based on the nature of the WT; the latter is because the transformation parameters are continuous. A solution to these problems can be achieved by sampling both parameters (a, b) such that a set of wavelet functions in the form of discrete parameters is obtained. We also have to look into the following problems:

1. Is the set of discrete wavelets complete in $L^2(R)$?
2. If complete, is the set at the same time also redundant?
3. If complete, then how coarse must the sampling grid be, such that the set is minimal or nonredundant?

A response to these questions will be given in this section, and we also will show that the most compact set is the orthonormal wavelet set.

The sampling grid is defined as follows [4]:

$$a = a_0^m b = nb_0 a_0^m \tag{3.51}$$

where

$$\psi_{mn}(t) = a^{-m/2}\psi(a_0^{-m}t - nb_0) \tag{3.52}$$

with $m, n \in Z$. If we consider this set complete in $L^2(R)$ for a given choice of $\psi(t), a, b$, then $\{\psi_{mn}\}$ is an affine wavelet. $f(t) \in L^2(R)$ represents a wavelet synthesis. It combines the components of a signal together again to reproduce the original signal $f(t)$. If we have a wavelet basis, we can determine a wavelet series expansion. Thus, any square-integrable (finite energy) function $f(t)$ can be expanded in wavelets:

$$f(t) = \sum_m \sum_n d_{m,n} \psi_{mn}(t) \tag{3.53}$$

The wavelet coefficient $d_{m,n}$ can be expressed as the inner product

$$d_{m,n} = \langle f(t), \psi_{mn}(t) \rangle = \frac{1}{a_0^{m/2}} \int f(t) \psi(a_0^{-m} t - nb_0) dt \tag{3.54}$$

These complete sets are called frames. An analysis frame is a set of vectors ψ_{mn} such that

$$A||f||^2 \leq \sum_m \sum_n |\langle f, \psi_{mn} \rangle|^2 \leq B||f||^2 \tag{3.55}$$

with

$$||f||^2 \triangleq \int |f(t)|^2 dt \tag{3.56}$$

$A, B > 0$ are the frame bounds. A tight, exact frame that has $A = B = 1$ represents an orthonormal basis for $L^2(R)$. A notable characteristic of orthonormal wavelets $\{\psi_{mn}(t)\}$ is

$$\int \psi_{mn}(t) \psi_{m'n'}(t) dt = \begin{cases} 1, & m = m', n = n' \\ 0, & \text{else} \end{cases} \tag{3.57}$$

They are additionally orthonormal in both indices. This means that for the same scale m they are orthonormal both in time and across the scales.

For the scaling functions the orthonormal condition holds only for a given scale

$$\int \phi_{mn}(t) \phi_{ml}(t) dt = \delta_{n-l} \tag{3.58}$$

The scaling function can be visualized as a lowpass filter. While scaling functions alone can code a signal to any desired degree of accuracy, efficiency can be gained by using the wavelet functions. Any signal $f \in L^2(R)$ at the scale m can be approximated by its projections on the scale space.

The similarity between ordinary convolution and the analysis equations suggests that the scaling function coefficients and the wavelet function coefficients may be viewed as impulse responses of filters, as shown in Fig. 3.18. The convolution of $f(t)$ with $\psi_m(t)$ is given by

$$y_m(t) = \int f(\tau) \psi_m(\tau - t) d\tau \tag{3.59}$$

where

$$\psi_m(t) = 2^{-m/2} \psi(2^{-m} t) \tag{3.60}$$

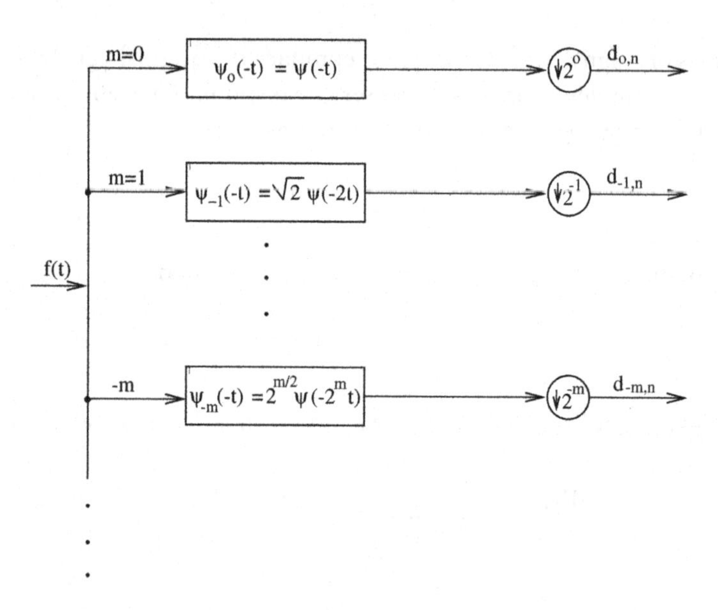

Figure 3.18 Filter bank representation of DWT.

Sampling $y_m(t)$ at $n2^m$ yields

$$y_m(n2^m) = 2^{-m/2} \int f(\tau)\psi(2^{-m}\tau - n)d\tau = d_{m,n} \tag{3.61}$$

Whereas in the filter bank representation of the short-time Fourier transform all subsamplers are identical, the subsamplers of the filter bank corresponding to the wavelet transform are dependent on position or scale.

The DWT dyadic sampling grid in Fig. 3.19 visualizes this aspect. Every single point represents a wavelet basis function $\psi_{mn}(t)$ at the scale 2^{-m} and shifted by $n2^{-m}$.

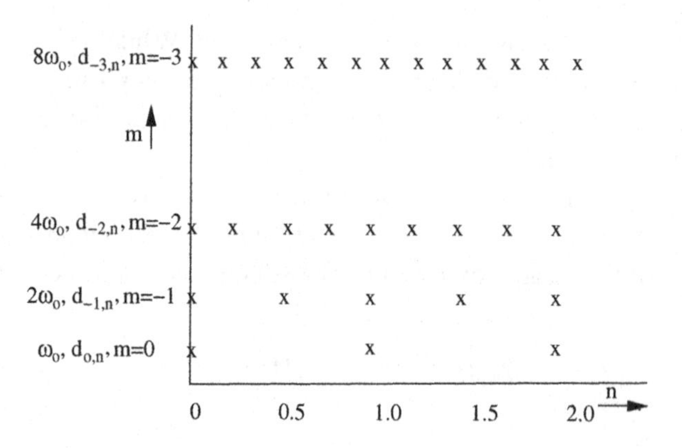

Figure 3.19 Dyadic sampling grid for the DWT.

3.5. MULTISCALE SIGNAL DECOMPOSITION

The goal of this section is to highlight an important aspect of the wavelet transform that accounts for its success as a method in pattern recognition: the decomposition of the whole function space into subspaces. That implies that there is a piece of the function $f(t)$ in each subspace. Those pieces (or projections) give finer and finer details of $f(t)$. For audio signals, these scales are essentially octaves. They represent higher and higher frequencies. For images and all other signals, the simultaneous appearance of multiple scales is known as multiresolution.

Mallat's and Meyer's method [232] for signal decomposition based on orthonormal wavelets with compact carrier is going to be reviewed here. We will establish a link between these wavelet families and the hierarchic filter banks. In the last part of this section, we will show that the FIR PR-QMF hold the regularization property, and produce orthonormal wavelet bases.

3.5.1 Multiscale-Analysis Spaces

Multiscale signal analysis provides the key to the link between wavelets and pyramidal dyadic trees. A wavelet family is used to decompose a signal into scaled and translated copies of a basic function. As stated before, the wavelet family consists of scaling and wavelet functions. Scaling functions $\phi(t)$ alone are adequate to code a signal completely, but a decomposition based on both scaling and wavelet functions is most efficient.

In mathematical terminology, a function $f(t)$ in the whole space has a piece in each subspace. Those pieces contain more and more of the full information in $f(t)$. These successive approximations converge to a limit which represents the function $f \in L^2$. At the same time they describe different resolution levels, as is known from the pyramidal representation.

A multiscale analysis is based on a sequence of subspaces $\{V_m | m \in Z\}$ in $L^2(R)$ satisfying the following requirements:

- *Inclusion:* Each subspace V_j is contained in the next subspace. A function $f \in L^2(R)$ in one subspace is in all the higher (finer) subspaces:

$$\begin{array}{c} \cdots V_2 \subset V_1 \subset V_0 \subset V_{-1} \subset V_{-2} \cdots \\ \leftarrow \text{coarser} \qquad\qquad \text{finer} \rightarrow \end{array} \tag{3.62}$$

- *Completeness:* A function in the whole space has a part in each subspace

$$\bigcap_{m \in Z} V_m = 0 \qquad \bigcup_{m \in Z} V_m = L^2(R) \tag{3.63}$$

- *Scale invariance:*

$$f(x) \in V_m \Longleftrightarrow f(2x) \in V_{m-1} \quad \text{for any function} \quad f \in L^2(R) \tag{3.64}$$

- *Basis-frame property:* This requirement for multiresolution concerns a basis for each space V_j. There is a scaling function $\phi(t) \in V_0$, such that $\forall m \in Z$, the set

$$\{\phi_{mn}(t) = 2^{-m/2}\phi(2^{-m}t - n)\} \tag{3.65}$$

forms an orthonormal basis for V_m:

$$\int \phi_{mn}(t)\phi_{mn'}(t)dt = \delta_{n-n'} \tag{3.66}$$

In the following, we will review mathematically the multiresolution concept based on scaling and wavelet functions, and thus define the approximation and detail operators.

Let $\phi_{mn}(t)$ with $m \in Z$ be defined as

$$\{\phi_{mn}(t) = 2^{-m/2}\phi(2^{-m}t - n)\} \tag{3.67}$$

Then the approximation operator P_m on functions $f(t) \in L^2(R)$ is defined by

$$P_m f(t) = \sum_n \langle f, \phi_{mn}\rangle \phi_{mn}(t) \tag{3.68}$$

and the detail operator Q_m on functions $f(t) \in L^2(R)$ is defined by

$$Q_m f(t) = P_{m-1}f(t) - P_m f(t) \tag{3.69}$$

It can be easily shown that $\forall m \in Z$, $\{\phi_{mn}(t)\}$ is an orthonormal basis for V_m [382], and that for all functions $f(t) \in L^2(R)$

$$\lim_{m\to-\infty} ||P_m f(t) - f(t)||_2 = 0 \tag{3.70}$$

and

$$\lim_{m\to\infty} ||P_m f(t)||_2 = 0 \tag{3.71}$$

An important feature of every scaling function $\phi(t)$ is that it can be built from translations of double-frequency copies of itself, $\phi(2t)$, according to

$$\phi(t) = 2\sum_n h_0(n)\phi(2t - n) \tag{3.72}$$

This equation is called a multiresolution analysis equation. Since $\phi(t) = \phi_{00}(t)$, both m, n can be set to 0 to obtain the above simpler expression. The equation expresses the fact that each scaling function in a wavelet family can be expressed as a weighted sum of scaling functions at the next finer scale. The set of coefficients $\{h_0(n)\}$ are called the scaling function coefficients and behave as a lowpass filter.

Wavelet functions can also be built from translations of $\phi(2t)$:

$$\psi(t) = 2\sum_{n} h_1(n)\phi(2t-n) \tag{3.73}$$

This equation is called the fundamental wavelet equation. The set of coefficients $\{h_1(n)\}$ are called the wavelet function coefficients and behave as a highpass filter. This equation expresses the fact that each wavelet function in a wavelet family can be written as a weighted sum of scaling functions at the next finer scale.

The following theorem provides an algorithm for constructing a wavelet orthonormal basis given a multiscale analysis:

Theorem 3.5.1 *Let $\{V_m\}$ be a multiscale analysis with scaling function $\phi(t)$ and scaling filter $h_0(n)$.*

Define the wavelet filter $h_1(n)$ by

$$h_1(n) = (-1)^{n+1} h_0(N-1-n) \tag{3.74}$$

and the wavelet $\psi(t)$ by eq. (3.73).

Then

$$\{\psi_{mn}(t)\} \tag{3.75}$$

is a wavelet orthonormal basis on R.

Alternatively, given any $L \in Z$,

$$\{\phi_{Ln}(t)\}_{n\in Z} \bigcup \{\psi_{mn}(t)\}_{m,n\in Z} \tag{3.76}$$

is an orthonormal basis on R.

The proof can be found in [382]. Some very important facts representing the key statements of multiresolution follow:

(a) $\{\psi_{mn}(t)\}$ is an orthonormal basis for W_m.

(b) If $m \neq m'$ then $W_m \perp W_{m'}$.

(c) $\forall m \in Z, V_m \perp W_m$ where W_m is the orthogonal complement of V_m in V_{m-1}.

(d) $\forall m \in Z, V_{m-1} = V_m \oplus W_m$. $\oplus$ stands for orthogonal sum. This means that the two subspaces are orthogonal and that every function in V_{m-1} is a sum of functions in V_m and W_m. Thus every function $f(t) \in V_{m-1}$ is composed of two subfunctions $f_1(t) \in V_m$ and $f_2(t) \in W_m$ such that $f(t) = f_1(t) + f_2(t)$ and $\langle f_1(t), f_2(t)\rangle = 0$.
The most important part of multiresolution is that the spaces W_m represent the differences between the spaces V_m, while the spaces V_m are the sums of the W_m.

(e) Every function $f(t) \in L^2(R)$ can be expressed as

$$f(t) = \sum_{m} f_m(t) \tag{3.77}$$

where $f_m(t) \in W_m$ and $\langle f_m(t), f_{m'} \rangle = 0$. This can be usually written as

$$\cdots \oplus W_j \oplus W_{j-1} \cdots \oplus W_0 \cdots \oplus W_{-j+1} \oplus W_{-j+2} \cdots = L^2(R) \tag{3.78}$$

Although scaling functions alone can code a signal to any desired degree of accuracy, efficiency can be gained by using the wavelet functions. This leads to the following new understanding of the concept of multiresolution. Multiresolution can be described based on wavelet W_j and scaling subspaces V_j. This means that the subspace formed by the wavelet functions covers the difference between the subspaces covered by the scaling functions at two adjacent scales.

3.5.2 A Very Simple Wavelet: The Haar Wavelet

The Haar wavelet is one of the simplest and oldest known orthonormal wavelets. However, it has didactic value because it helps to visualize the multiresolution concept.

Let V_m be the space of piecewise constant functions

$$V_m = \{f(t) \in L^2(R); \quad f \text{ is constant in } \quad [2^m n, 2^m(n+1)] \quad \forall n \in Z\} \tag{3.79}$$

Figure 3.20 illustrates such a function.

We can easily see that $\cdots V_1 \subset V_0 \subset V_{-1} \cdots$ and $f(t) \in V_0 \longleftrightarrow f(2t) \in V_{-1}$ and that the inclusion property is fulfilled. The function $f(2t)$ has the same shape as $f(t)$ but is compressed to half the width.

The scaling function of the Haar wavelet $\phi(t)$ is given by

$$\phi(t) = \begin{cases} 1, & 0 \le t \le 1 \\ 0, & \text{else} \end{cases} \tag{3.80}$$

and defines an orthonormal basis for V_0. Since for $n \neq m, \phi(t-n)$ and $\phi(t-m)$ do not overlap, we obtain

$$\int \phi(t-n)\phi(t-m)dt = \delta_{n-m} \tag{3.81}$$

The Fourier transform of the scaling function yields

$$\Phi(\omega) = e^{-j\frac{\omega}{2}} \frac{\sin \omega/2}{\omega/2} \tag{3.82}$$

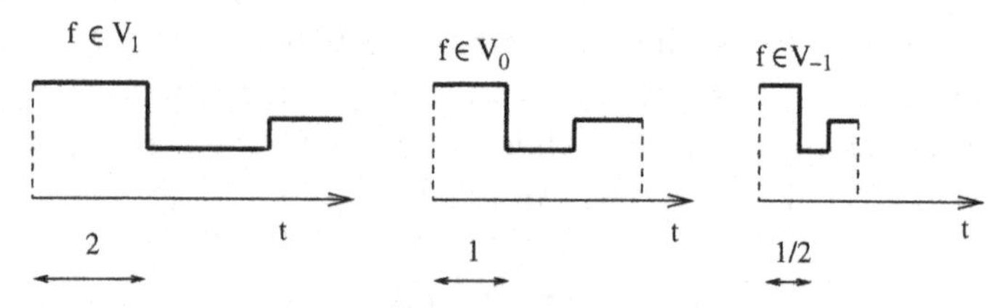

Figure 3.20 Piecewise constant functions in V_1, V_2, and V_{-1}.

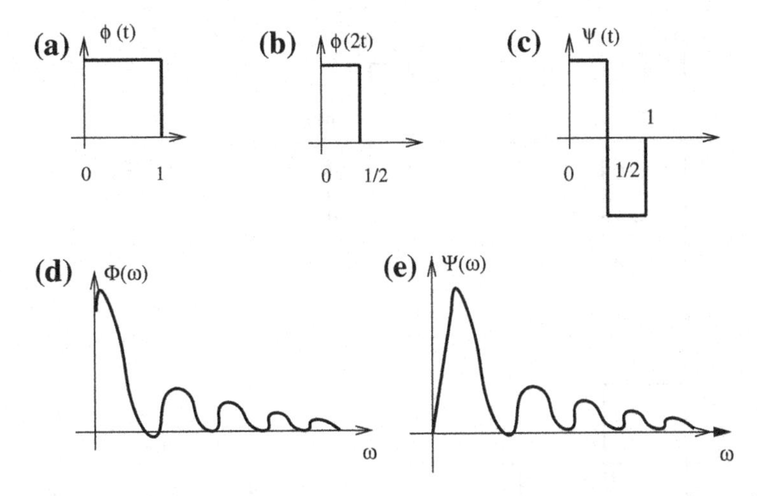

Figure 3.21 (a) and (b) Haar basis functions, (c) Haar wavelet, (d) Fourier transform of the scaling function, (e) Haar wavelet function.

Figure 3.21 shows that $\phi(t)$ can be written as the linear combination of even and odd translations of $\phi(2t)$

$$\phi(t) = \phi(2t) + \phi(2t-1) \tag{3.83}$$

Since $V_{-1} = V_0 \oplus W_0$ and $Q_0 f = (P_{-1}f - P_0 f) \in W_0$ represent the details from scale 0 to -1, it is easy to see that $\psi(t-n)$ spans W_0. The Haar mother wavelet function is given by

$$\psi(t) = \phi(2t) - \phi(2t-1) = \begin{cases} 1, & 0 \leq t < 1/2 \\ -1, & 1/2 \leq t < 1 \\ 0, & \text{else} \end{cases} \tag{3.84}$$

The Haar wavelet function is an up-down square wave and can be described by a half-box minus a shifted half-box. We also can see that the wavelet function can be computed directly from the scaling functions. In the Fourier domain it describes a bandpass, as can be easily seen from Fig. 3.21e. This is given by

$$\Psi(\omega) = je^{-j\frac{\omega}{2}} \frac{\sin^2 \omega/4}{\omega/4} \tag{3.85}$$

We can easily show that

$$\phi_{m+1,n} = \frac{1}{\sqrt{2}}[\phi_{m,2n} + \phi_{m,2n+1}] \tag{3.86}$$

and

$$\psi_{m+1,n} = \frac{1}{\sqrt{2}}[\phi_{m,2n} - \phi_{m,2n+1}] \tag{3.87}$$

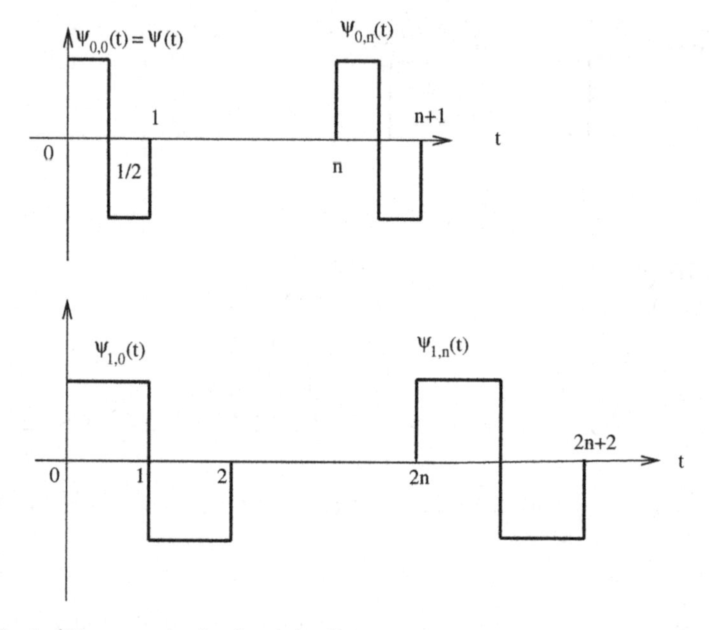

Figure 3.22 Typical Haar wavelet for the scales 0, 1.

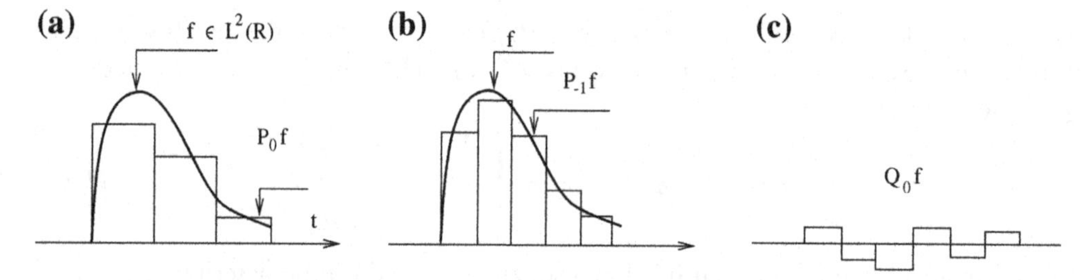

Figure 3.23 Approximation of (a) P_0f, (b) $P_{-1}f$, (c) the detail signal Q_0f, with $P_0f + Q_0f = P_{-1}f$.

Figure 3.22 illustrates a typical Haar wavelet for the scales 0 and 1. Figure 3.23 shows the approximations $P_0f, P_{-1}f$ and the detail Q_0f for a function f. As stated in context with multiresolution, the detail Q_0f is added to the coarser approximation P_0f in order to obtain the finer approximation $P_{-1}f$.

The scaling function coefficients for the Haar wavelet at scale m are given by

$$c_{m,n} = \langle f, \phi_{mn} \rangle = 2^{-m/2} \int_{2^m n}^{2^m(n+1)} f(t)dt \tag{3.88}$$

This yields an approximation of f at scale m

$$P_m f = \sum_n c_{m,n}\phi_{mn}(t) = \sum_n c_{m,n} 2^{-m/2}\phi(2^{-m}t - n) \tag{3.89}$$

In spite of their simplicity, the Haar wavelets exhibit some undesirable properties, which pose a difficulty in many practical applications. Other wavelet families such as Daubechies wavelets and Coiflet basis [4,382] are more attractive in practice. Daubechies wavelets are quite often used in image compression. The scaling function coefficients $h_0(n)$ and the wavelet function coefficients $h_1(n)$ for the Daubechies-4 family are nearly impossible to determine. They were obtained based on iterative methods [44].

3.5.3 Analogy Between Filter Banks and Wavelet Bases

In this section we will show the main approach to wavelets through two-channel filter banks. We will point out the connections between wavelets and filters. Historically, their development was separate; however, they are very closely related. We remember that the lowpass filter coefficients determine the scaling function while the highpass coefficients produce the wavelets.

The underlying problem here is that we again have an analysis and synthesis filter bank. The decimated lowpass and highpass filter can be expressed in one matrix. This matrix represents the whole analysis bank. It executes the lowpass and highpass channel (both decimated). There is also a synthesis bank. When the analysis and synthesis banks are transposes as well as inverses, we have an orthogonal filter bank. When they are inverses but not necessarily transposes, then we deal with a biorthogonal filter bank.

When the filter bank is orthonormal in discrete time, we hope for orthogonal basis functions in continuous time. All wavelets should be orthogonal to the scaling functions. Furthermore, the wavelets and the scaling functions should be mutually orthogonal.

Our starting point is the dilation equation given by (3.72). The Fourier transform of the dilation equation in (3.72) considering $h_0(n) \leftrightarrow H_1(e^{j\omega})$ and $\phi(t) \leftrightarrow \Phi(\omega)$ is given by

$$\Phi(\omega) = H\left(\frac{\omega}{2}\right)\Phi\left(\frac{\omega}{2}\right) = \Phi(0)\prod_{k=1}^{\infty} H_0(e^{j\omega/2^k}) \tag{3.90}$$

where, based on completeness and orthonormality property, we have

$$|\Phi(0)| = |\int_{-\infty}^{\infty}\phi(t)dt| = |H_0(e^{j\omega})\big|_{\omega=0} = 1 \tag{3.91}$$

Next we will establish the orthogonality conditions in the frequency domain. In wavelet terminology, this translates to finding for a given scaling function $\phi(t)$ and scale m the orthonormal set $\{\phi_{mn}\}$. If $\{\phi(t-n)\}$ spans V_0, then $\Phi(\omega)$ and H fulfill in the frequency domain

$$\sum_k |\Phi(\omega + 2\pi k)|^2 = 1 \tag{3.92}$$

and

$$|H_0(e^{j\omega})|^2 + |H_0(e^{j(\omega+\pi)}|^2 = 1 \tag{3.93}$$

The last equation gives the condition on the frequency response $H_0(e^{j\omega})$ to produce an orthonormal filter bank.

Similarly, we assume $h_1(n) \leftrightarrow H_1(e^{j\omega})$, and with eq. (3.90), we obtain for the Fourier transform of the wavelet equation (3.73)

$$\Psi(\omega) = H_1(e^{j\omega/2})\Phi\left(\frac{\omega}{2}\right) = H_1(e^{j\omega/2})\prod_{k=2}^{\infty} H_0(e^{j\omega/2^k}) \tag{3.94}$$

The orthonormal wavelet bases are complementary to the scaling bases. They fulfill the intra- and interscale orthonormalities

$$\langle \psi_{mn}(t), \psi_{kl}(t)\rangle = \delta_{m-k}\delta_{n-l} \tag{3.95}$$

with m and k being scale parameters, and n and l being translation parameters. Because the wavelets are orthonormal at all scales, this implies for the frequency domain

$$\sum_k |\Psi(\omega + 2\pi k)|^2 = 1 \tag{3.96}$$

Considering $\Psi(2\omega) = H_1(e^{j\omega})\Phi(\omega)$, we immediately obtain from the above equation

$$|H_1(e^{j\omega})|^2 + |H_1(e^{j(\omega+\pi)})|^2 = 1 \tag{3.97}$$

or

$$H_1(z)H_1(-z) + H_1(-z)H_1(-z^{-1}) = 1 \tag{3.98}$$

The scaling functions are orthogonal to the wavelets

$$\langle \phi_{mn}(t), \psi_{kl}(t)\rangle = 0 \tag{3.99}$$

This implies for the frequency domain

$$\sum_k \Phi(\omega - 2\pi k)\Psi^*(\omega - 2\pi k) = 0 \tag{3.100}$$

Considering eqs. (3.90), (3.94), and (3.98), we get for the frequency domain

$$H_0(e^{j\omega})H_1(e^{-j\omega}) + H_0(e^{j(\omega+\pi)})H_1(e^{-j(\omega+\pi)}) = 0 \tag{3.101}$$

or equivalently for the z-domain

$$H_0(z)H_1(-z) + H_0(-z)H_1(-z^{-1}) = 0 \tag{3.102}$$

For proving the analogy between filter banks and wavelet bases, we will introduce the paraunitary filter bank. The two-band paraunitary analysis filters must obey

$$\begin{bmatrix} H_0(z^{-1}) & H_0(-z^{-1}) \\ H_1(z^{-1}) & H_1(-z^{-1}) \end{bmatrix}\begin{bmatrix} H_0(z) & H_1(z) \\ H_0(-z) & H_1(-z) \end{bmatrix} = 2\begin{bmatrix} 1 & 0 \\ 0 & 1 \end{bmatrix} \tag{3.103}$$

Table 3.3 Properties of orthonormal wavelets.

$$\begin{aligned}
\Phi(\omega) &= \prod_{k=1}^{\infty} H_0(e^{j\omega/2^k}) \\
\Psi(\omega) &= H_1(e^{j\omega/2}) \prod_{k=2}^{\infty} H_0(e^{j\omega/2^k}) \\
H_1(z) &= z^{-(N-1)} H_0(-z^{-1}) \\
\phi(t) &= \sum h_0(n)\phi(2t-n) \\
\psi(t) &= \sum h_1(n)\psi(2t-n) \\
h_1(n) &= (-1)^{n+1} h_0(N-1-n) \\
\langle \psi_{mn}(t), \psi_{kl}(t) \rangle &= \delta_{m-k}\delta_{n-l} \\
\langle \phi_{mn}(t), \phi_{mn'}(t) \rangle &= \delta_{n-n'} \\
\langle \phi_{mn}(t), \psi_{kl}(t) \rangle &= 0
\end{aligned} \tag{3.106}$$

The conditions stated in eqs. (3.93), (3.98), and (3.101) are sufficient for a paraunitary two-band FIR perfect reconstruction QMF as given in Table 3.1.

Equation (3.101) holds if

$$H_1(z) = z^{-(N-1)} H_0(-z^{-1}), \quad N \text{ even} \tag{3.104}$$

or equivalently for the time domain we get

$$h_1(n) = (-1)^{n+1} h_0(N-1-n) \tag{3.105}$$

Thus, we have shown that compactly supported orthonormal wavelet bases imply FIR paraunitary QMF filter banks. In [4], it was shown that the converse is also true if $H_0(z)$ and $H_1(z)$ have at least one zero at $z = -1$ and $z = 1$.

We thus have shown the connection between the wavelet theory and filter banks.

The mathematical properties of orthonormal wavelets with compact carriers are summarized in Table 3.3 [4].

3.5.4 Multiscale Signal Decomposition and Reconstruction

In this section we will illustrate multiscale pyramid decomposition. Based on a wavelet family, a signal can be decomposed into scaled and translated copies of a basic function. As discussed in the last sections, a wavelet family consists of scaling functions, which are scalings and translations of a father wavelet, and wavelet functions, which are scalings and translations of a mother wavelet. We will show an efficient signal coding that uses scaling and wavelet functions at two successive scales. In other words, we give a recursive algorithm which supports the computation of wavelet coefficients of a function $f(t) \in L^2(R)$.

Assume we have a signal or a sequence of data $\{c_0(n)|n \in Z\}$, and $c_0(n)$ is the nth scaling coefficient for a given function $f(t)$:

$$c_{0,n} = \langle f, \phi_{0n} \rangle \tag{3.107}$$

for each $n \in Z$. This assumption makes the recursive algorithm work.

The decomposition and reconstruction algorithm is given by the following theorem [382]:

Theorem 3.5.2 *Let $\{V_k\}$ be a multiscale analysis with associated scaling function $\phi(t)$ and scaling filter $h_0(n)$. The wavelet filter $h_1(n)$ is defined by eq. (3.73), and the wavelet function is defined by eq. (3.74).*

Given a function $f(t) \in L^2(R)$, define for $n \in Z$

$$c_{0,n} = \langle f, \phi_{0n} \rangle \tag{3.108}$$

and for every $m \in N$ and $n \in Z$,

$$c_{m,n} = \langle f, \phi_{mn} \rangle \quad \text{and} \quad d_{m,n} = \langle f, \psi_{mn} \rangle \tag{3.109}$$

Then, the decomposition algorithm is given by

$$c_{m+1,n} = \sqrt{2} \sum_k c_{m,k} h_0(k - 2n), \quad d_{m+1,n} = \sqrt{2} \sum_k d_{m,k} h_1(k - 2n) \tag{3.110}$$

and the reconstruction algorithm is given by

$$c_{m,n} = \sqrt{2} \sum_k c_{m+1,n} h_0(n - 2k) + \sqrt{2} \sum_k d_{m+1,n} h_1(n - 2k) \tag{3.111}$$

From eq. (3.110) we obtain for $m = 1$ at resolution 1/2 the following wavelet $d_{1,n}$ and scaling coefficients $c_{1,n}$:

$$c_{1,n} = \sqrt{2} \sum h_0(k - 2n) c_{0,k} \tag{3.112}$$

and

$$d_{1,n} = \sqrt{2} \sum h_1(k - 2n) c_{0,k} \tag{3.113}$$

These last two so-called analysis equations relate the DWT coefficients at a finer scale to the DWT coefficients at a coarser scale. The analysis operations are similar to ordinary convolution. The similarity between ordinary convolution and the analysis equations suggests that the scaling function coefficients and wavelet function coefficients may be viewed as impulse responses of filters. In fact, the set $\{h_0(-n), h_1(-n)\}$ can be viewed as a paraunitary FIR filter pair. Figure 3.24 illustrates this.

The discrete signal $d_{1,n}$ is the WT coefficient at the resolution 1/2 and describes the detail signal or difference between the original signal $c_{0,n}$ and its smooth undersampled approximation $c_{1,n}$.

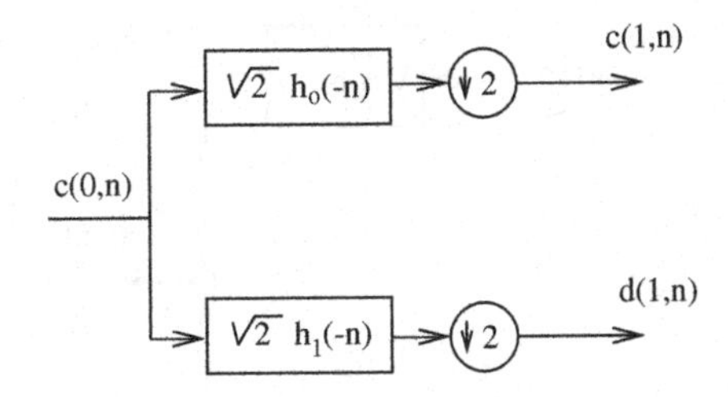

Figure 3.24 First level of the multiscale signal decomposition.

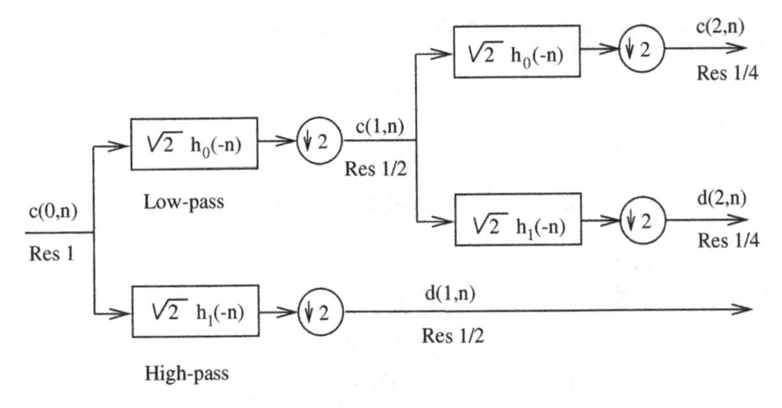

Figure 3.25 Multiscale pyramid decomposition.

For $m = 2$, we obtain at the resolution 1/4 the coefficients of the smoothed signal (approximation) and the detail signal (approximation error) as

$$c_{2,n} = \sqrt{2} \sum c_{1,k} h_0(k - 2n) \tag{3.114}$$

$$d_{2,n} = \sqrt{2} \sum c_{1,k} h_1(k - 2n) \tag{3.115}$$

These relationships are illustrated in the two-level multiscale pyramid in Fig. 3.25.

Wavelet synthesis is the process of recombining the components of a signal to reconstruct the original signal. The inverse discrete wavelet transformation, or IDWT, performs this operation. To obtain $c_{0,n}$, the terms $c_{1,n}$ and $d_{1,n}$ are upsampled and convoluted with the filters $h_0(n)$ and $h_1(n)$ as shown in Fig. 3.26.

The results of the multiscale decomposition and reconstruction of a dyadic subband tree are shown in Fig. 3.27 and describe the analysis and synthesis part of a two-band PR-QMF bank.

It is important to mention that the recursive algorithms for decomposition and reconstruction can be easily extended for a two-dimensional signal (image) [382] and play an important role in image compression.

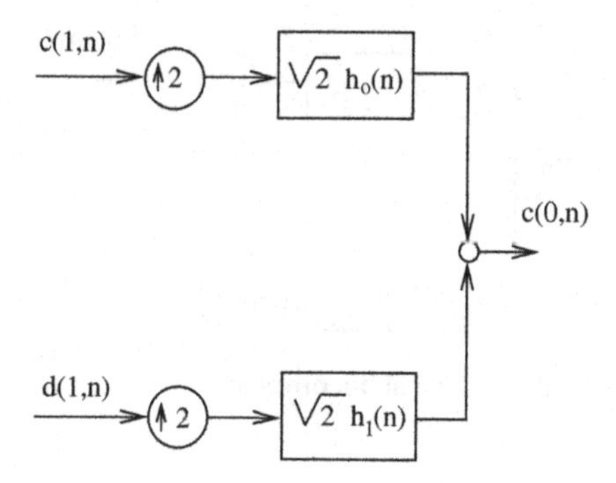

Figure 3.26 Reconstruction of a one-level multiscale signal decomposition.

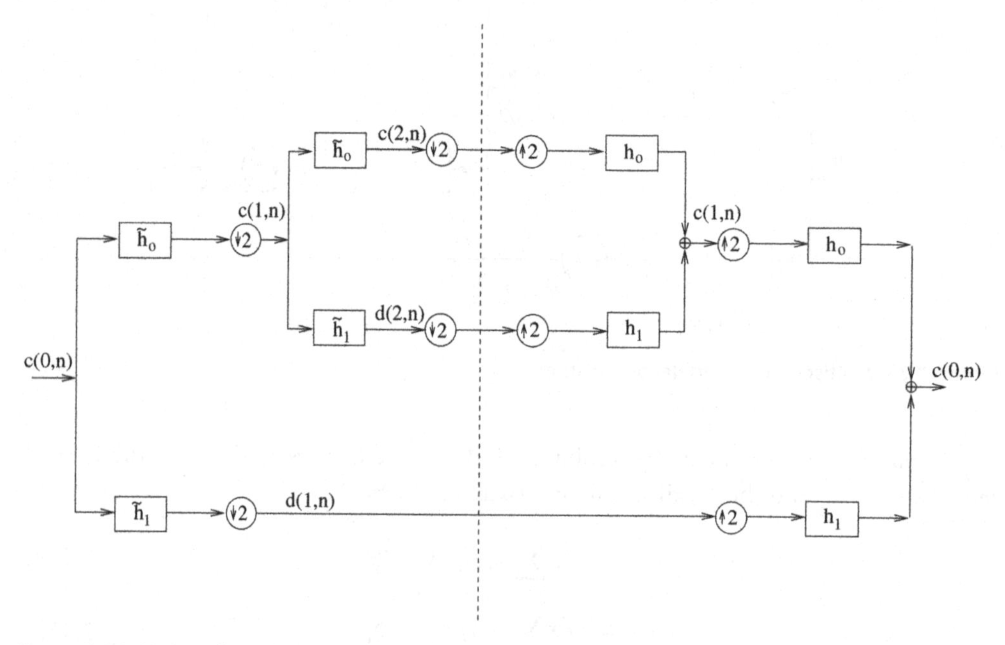

Figure 3.27 Multiscale analysis and synthesis.

3.5.5 Wavelet Transformation at a Finite Resolution

In this section we will show that a function can be approximated to a desired degree by summing together the scaling function and as many wavelet detail functions as necessary. Let $f \in V_0$ be defined as

$$f(t) = \sum c_{0,n}\phi(t-n) \tag{3.116}$$

As stated in previous sections, it also can be represented as a sum of signal at a coarser resolution (approximation) plus a detail signal (approximation error):

$$f(t) = f_v^1(t) + f_w^1(t) = \sum c_{1,n}2^{\frac{1}{2}}\phi\left(\frac{t}{2}-n\right) + \sum d_{1,n}2^{\frac{1}{2}}\psi\left(\frac{t}{2}-n\right) \tag{3.117}$$

The coarse approximation $f_v^1(t)$ can be rewritten as

$$f_v^1(t) = f_v^2(t) + f_w^2(t) \tag{3.118}$$

such that

$$f(t) = f_v^2(t) + f_w^2(t) + f_w^1(t) \tag{3.119}$$

Continuing with this procedure we have at scale J for $f_v^J(t)$

$$f(t) = f_v^J(t) + f_w^J(t) + f_w^{J-1}(t) + \cdots + f_w^1 \tag{3.120}$$

or

$$f(t) = \sum_{n=-\infty}^{\infty} c_{J,n}\phi_{J,n}(t) + \sum_{m=1}^{J}\sum_{n=-\infty}^{\infty} d_{m,n}\psi_{m,n}(t) \tag{3.121}$$

This equation describes a wavelet series expansion of function $f(t)$ in terms of the wavelet $\psi(t)$ and scaling function $\phi(t)$ for an arbitrary scale J. In comparison, the pure WT

$$f(t) = \sum_{m}\sum_{n} d_{m,n}\psi_{mn}(t) \tag{3.122}$$

requires an infinite number of resolutions for a complete signal representation.

From eq. (3.122) we can see that $f(t)$ is given by a coarse approximation at the scale L and a sum of L detail components (wavelet components) at different resolutions.

Example 3.5.1 Consider the simple function

$$y = \begin{cases} t^2, & 0 \le t \le 1 \\ 0, & \text{else} \end{cases} \tag{3.123}$$

Using Haar wavelets and the starting scale $J = 0$ we can easily determine the following expansion coefficients:

$$\begin{aligned} c_{0,0} &= \int_0^1 t^2\phi_{0,0}(t)dt = \frac{1}{3} \\ d_{0,0} &= \int_0^1 t^2\psi_{0,0}(t)dt = -\frac{1}{4} \\ d_{1,0} &= \int_0^1 t^2\psi_{1,0}(t)dt = -\frac{\sqrt{2}}{32} \\ d_{1,1} &= \int_0^1 t^2\psi_{1,1}(t)dt = -\frac{3\sqrt{2}}{32} \end{aligned} \tag{3.124}$$

Thus, we obtain the wavelet series expansion

$$y = \frac{1}{3}\phi_{0,0}(t) - \frac{1}{4}\psi_{0,0}(t) - \frac{\sqrt{2}}{32}\psi_{1,0}(t) - \frac{3\sqrt{2}}{32}\psi_{1,1}(t) + \cdots \tag{3.125}$$

3.6. OVERVIEW: TYPES OF WAVELET TRANSFORMS

The goal of this section is to provide an overview of the most frequently used wavelet types. Figure 3.28 illustrates the block diagram of the generalized time-discrete filter bank transform.

The mathematical representation of the direct and inverse generalized time-discrete filterbank transform is

$$v_k(n) = \sum_{m=-\infty}^{\infty} x(m)h_k(n_k n - m), \quad 0 \le k \le M-1 \tag{3.126}$$

and

$$\widehat{x}(n) = \sum_{k=0}^{M-1} \sum_{m=-\infty}^{\infty} v_k(m)g_k(n - n_k m) \tag{3.127}$$

Based on this representation, we can derive as functions of n_k, $h_k(n)$, and $g_k(n)$ the following special cases [91]:

1. *Orthonormal wavelets:* $n_k = 2^k$ with $0 \le k \le M-2$ and $n_{M-1} = n_{M-2}$. The basis function fulfills the orthonormality condition (3.57).
2. *Orthonormal wavelet packets:* They represent a generalization of the orthonormal wavelets because they make use of the recursive decomposition-reconstruction structure which is applied to all bands as shown in Fig. 3.8. The following holds: $n_k = 2^L$ with $0 \le k \le 2^L - 1$.

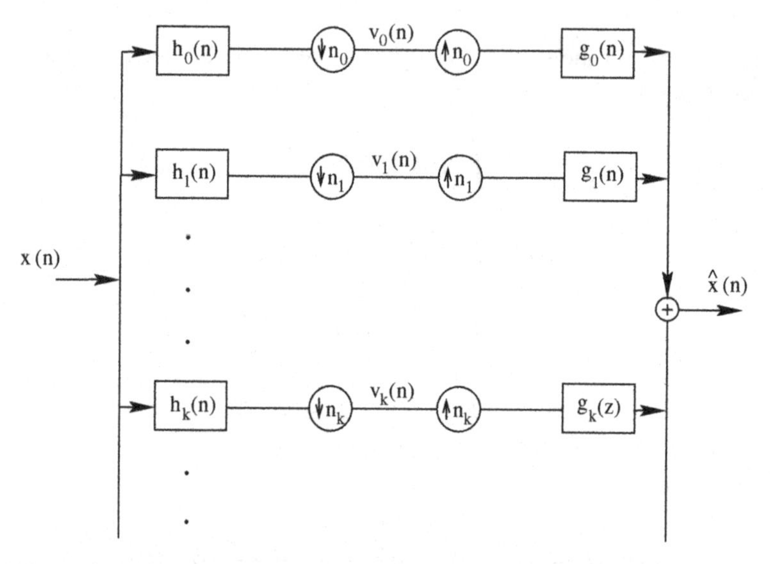

Figure 3.28 Generalized time-discrete filter bank transform.

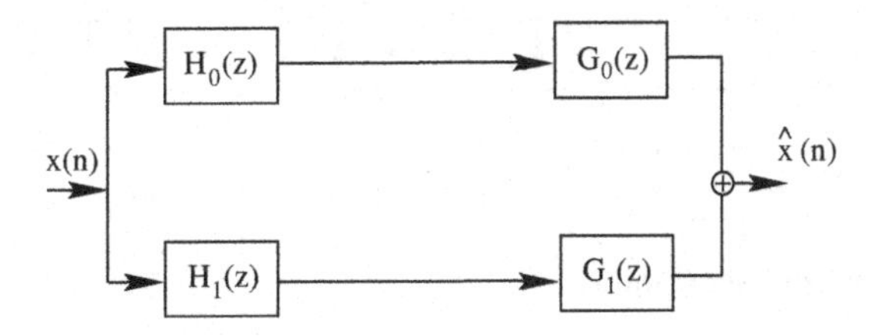

Figure 3.29 Oversampled wavelet transform.

3. *Biorthogonal wavelets:* They have properties similar to those of the orthogonal wavelets but are less restrictive. They will be explained in detail in Section 4.3.
4. *Generalized filter bank representations:* They represent a generalization of the (bi)orthogonal wavelet packets. Each band is split in two subbands. The basis functions fulfill the biorthonormality condition:

$$\sum_{m=-\infty}^{\infty} g_c(m - n_c l) h_k(n_k n - m) = \delta(c - k)\delta(l - n) \tag{3.128}$$

5. *Oversampled wavelets:* There is no downsampling or oversampling required, and $n_k = 1$ holds for all bands. An example of an oversampled WT is shown in Fig. 3.29.

The first four wavelet types are known as *nonredundant wavelet representations.* For the representation of oversampled wavelets more analysis functions ($\{u_k(n)\}$) than basis functions are required. The analysis and synthesis functions must fulfill

$$\sum_{k=0}^{M-1} \sum_{m=-\infty}^{\infty} g_k(m - l) h_k(n - m) = \delta(l - n), \quad \text{or} \quad \sum_{k=0}^{M-1} G_k(z) H_k(z) = 1 \tag{3.129}$$

This condition holds only in case of linear dependency. This means that some functions are represented as linear combinations of others.

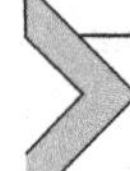

3.7. EXERCISES

1. Given

$$x(n) = 3\lambda^n u(n) \tag{3.130}$$

Let $y(n)$ be the downsampled signal of $x(n)$ for $M = 2$. Determine $X(z)$, $Y(z)$ and show the pole-zero plots. Determine and plot $|X(e^{j\omega})|$, and $|Y(e^{j\omega})|$. Next consider the upsampling of $x(n)$ and perform the same steps as for downsampling.

2. Consider the following alternating vector $\mathbf{x}(n) = \{\ldots, 1, -1, 1, -1, 1, -1, 1, -1, 1 \ldots\}$, which is delayed to $\mathbf{x}(n-2)$. Determine $(\downarrow 2)\mathbf{x}(n-2)$.
3. What are the components of $(\uparrow 3)(\downarrow 2)\mathbf{x}$ and $(\downarrow 2)(\uparrow 3)\mathbf{x}$ with $\mathbf{x}$ given as in Exercise 2?

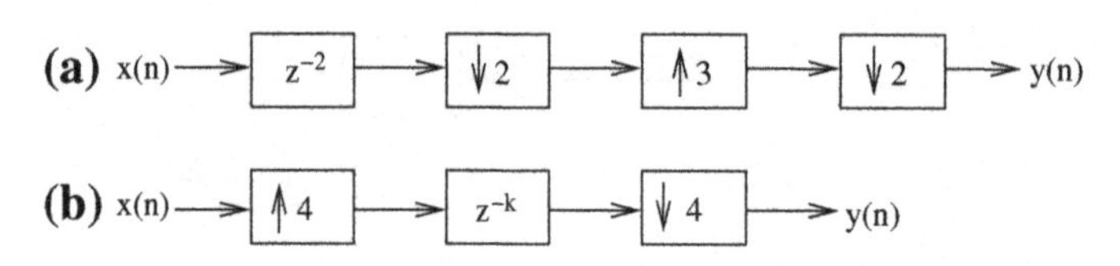

Figure 3.30 Systems for Exercises 6 and 7.

4. Show that the equivalence in Fig. 3.5a holds both in the z-domain and in the time domain.

5. Show that the equivalence in Fig. 3.5b holds both in the z-domain and in the time domain.

6. Simplify the following system given in Fig. 3.30a: Express $Y(z)$ in terms of $X(z)$. What is $\mathbf{y}(n)$ given
(a) $\mathbf{x}(n) = \delta(n)$ and
(b) $\mathbf{x}(n) = (\ldots, 1, 1, 1, 1, 1, 1, \ldots)$?

7. Simplify the following system given in Fig. 3.30b:
Express $Y(z)$ in terms of $X(z)$. What is $\mathbf{y}(n)$ in terms of $\mathbf{x}(n)$?

8. Given a $2^L \times 2^L$ image, does a $(L-1)$-level pyramid reduce or increase the amount of data required to represent the image? What compression or expansion ratio is thus achieved?

9. Show that the conjugate quadrature filter from Table 3.1 forms a perfect reconstruction filter bank.

10. Determine if the quadrature mirror filters are orthonormal filters.

11. Express the scaling space V_4 as a function of scaling function $\phi(t)$. Employ the Haar scaling function to plot the Haar V_4 scaling functions at translations $k = \{0, 1, 2\}$.

12. Plot the wavelet $\psi_{4,4}(t)$ for the Haar wavelet function. Express $\psi_{4,4}$ in terms of the Haar scaling function.

13. Discuss the differences between the CWT and DWT. When is the CWT and when is the DWT more useful for practical applications.

14. Verify the following for the Haar wavelet family:
(a) $\phi(4t) = \sum h_0(n)\phi(8t - k)$ and
(b) $\psi(4t) = \sum h_1(n)\psi(8t - k)$.

15. Split the function $f(t)$ from Fig. 3.31a into a scaling function and a wavelet.

16. Split the function $g(t)$ from Fig. 3.31b into its parts in the spaces V_0, W_0, and W_1.

17. The function $f(t)$ is given as:

$$f(t) = \begin{cases} 4, & 0 \leq t < 2 \\ 0, & \text{else} \end{cases} \tag{3.131}$$

Plot the following scaled and/or translated versions of $f(t)$:
(a) $f(t-1)$.

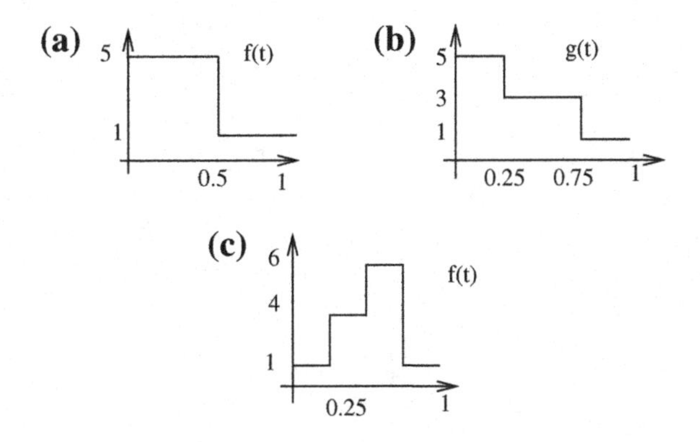

Figure 3.31 Examples of functions in Exercises 15, 16, and 18.

(b) $f(2t)$.
(c) $f(2t-1)$.
(d) $f(4t)$.
(e) $f(4t-5)$.
(f) $f(t/2)$.
(g) $f(t/2-1)$.

18. The Haar wavelet family must be used to code function $f(t)$ from Fig. 3.31c. Determine a coding using:
(a) $\phi(4t), \phi(4t-1), \phi(4t-2), \phi(4t-3)$.
(b) $\phi(2t), \phi(2t-1), \psi(2t), \psi(2t-1)$.
(c) $\phi(t), \psi(t), \psi(2t), \psi(2t-1)$.

CHAPTER FOUR

The Wavelet Transform in Medical Imaging

Contents

4.1. INTRODUCTION

Tremendous research efforts during the past 10 years have established the wavelet transform as the most important tool in biomedical imaging. This valuable technique specifically addresses for medical imaging the problems of image compression, edge and feature selection, denoising, contrast enhancement, and image fusion.

In this chapter, we present the theory of 2-D discrete wavelet transforms, and of biorthogonal wavelets, and we show several applications of the wavelet transform in medical imaging. The most remarkable applications are: (a) ability of the WT to make visible simple objects in a noisy background, which were previously considered to be invisible to a human viewer, (b) demonstrated superiority of the WT over existing techniques for unsharp mask enhancement and median filtering, and (c) enhancing the visibility of clinically important features.

4.2. THE TWO-DIMENSIONAL DISCRETE WAVELET TRANSFORM

For any wavelet orthonormal basis $\{\psi_{j,n}\}_{(j,n)\in Z^2}$ in $L^2(R)$, there exists a separable wavelet orthonormal basis also in $L^2(R)$:

$$\{\psi_{j,n}(x)\psi_{l,m}(y)\}_{(j,l,n,m)\in Z^4} \tag{4.1}$$

http://dx.doi.org/10.1016/B978-0-12-409545-8.00004-2

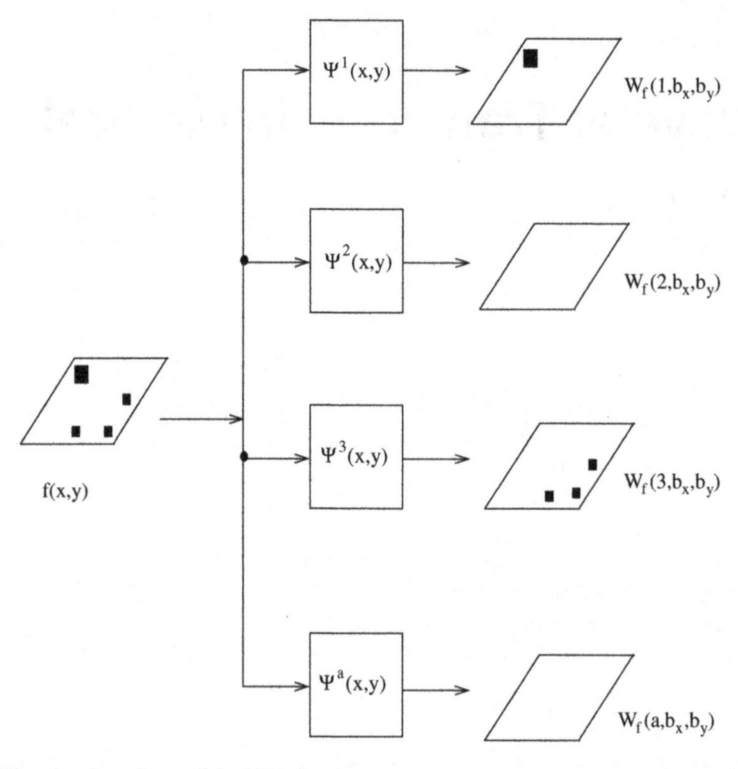

Figure 4.1 Filter-bank analogy of the WT of an image.

The functions $\psi_{j,n}(x)\psi_{l,m}(y)$ mix the information at two different scales 2^j and 2^l across x and y. This technique leads to a building procedure based on separable wavelets whose elements represent products of function dilation at the same scale. These multiscale approximations are mostly applied in image processing because they facilitate the processing of images at several detail levels. Low-resolution images can be represented using fewer pixels while preserving the features necessary for recognition tasks.

The theory presented for the one-dimensional WT can be easily extended to two-dimensional signals such as images. In two dimensions, a 2-D scaling function, $\phi(x, y)$ and three 2-D wavelets $\psi^1(x, y)$, $\psi^2(x, y)$, and $\psi^3(x, y)$ are required. Figure 4.1 shows a 2-D filter bank. Each filter $\psi_a(x, y)$ represents a 2-D impulse response and its output a bandpass filtered version of the original image. The set of the filtered images describes the WT.

In the following, we will assume that the 2-D scaling functions are separable, that is,

$$\phi(x, y) = \phi(x)\phi(y) \tag{4.2}$$

where $\phi(x)$ is a one-dimensional scaling function. If we define $\psi(x)$, the companion wavelet function, as shown in eq. (3.73), then based on the following three basis

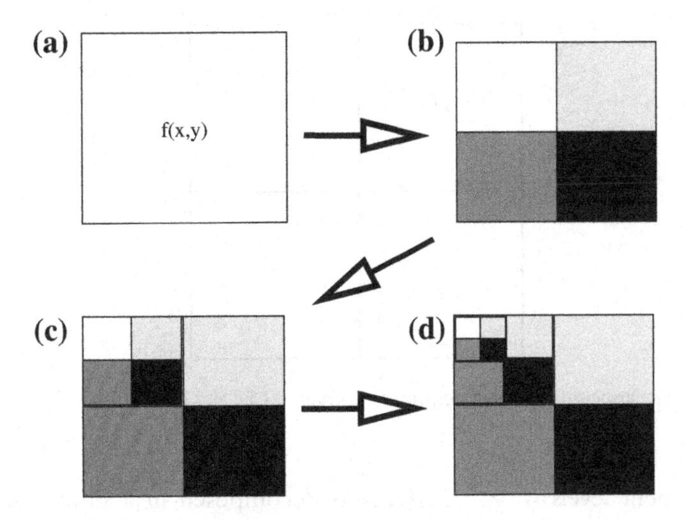

Figure 4.2 2-D discrete wavelet transform: (a) original image, (b) first, (c) second, and (d) third level.

functions,

$$\psi^1(x, y) = \phi(x)\psi(y) \quad \psi^2(x, y) = \psi(x)\phi(y) \quad \psi^3(x, y) = \psi(x)\psi(y) \tag{4.3}$$

we set up the foundation for the 2-D wavelet transform. Each of them is the product of a one-dimensional scaling function ϕ and a wavelet function ψ. They are "directionally sensitive" wavelets because they measure functional variations, either intensity or gray-level variations, along different directions: ψ^1 measures variations along the columns (horizontal edges), ψ^2 is sensitive to variations along rows (vertical edges), and ψ^3 corresponds to variations along diagonals. This directional sensitivity occurs as an implication of the separability condition.

To better understand the 2-D WT, let us consider $f_1(x, y)$, an $N \times N$ image, where the subscript describes the scale and N is a power of 2. For $j = 0$, the scale is given by $2^j = 2^0 = 1$ and corresponds to the original image. Allowing j to become larger doubles the scale and halves the resolution.

An image can be expanded in terms of the 2-D WT. At each decomposition level, the image can be decomposed into four subimages a quarter of the size of the original, as shown in Fig. 4.2. Each of these images stems from an inner product of the original image with the subsampled version in x and y by a factor 2. For the first level ($j = 1$), we obtain

$$\begin{aligned}
f_2^0(m, n) &= \langle f_1(x, y), \phi(x - 2m, y - 2n)\rangle \\
f_2^1(m, n) &= \langle f_1(x, y), \psi^1(x - 2m, y - 2n)\rangle \\
f_2^2(m, n) &= \langle f_1(x, y), \psi^2(x - 2m, y - 2n)\rangle \\
f_2^3(m, n) &= \langle f_1(x, y), \psi^3(x - 2m, y - 2n)\rangle
\end{aligned} \tag{4.4}$$

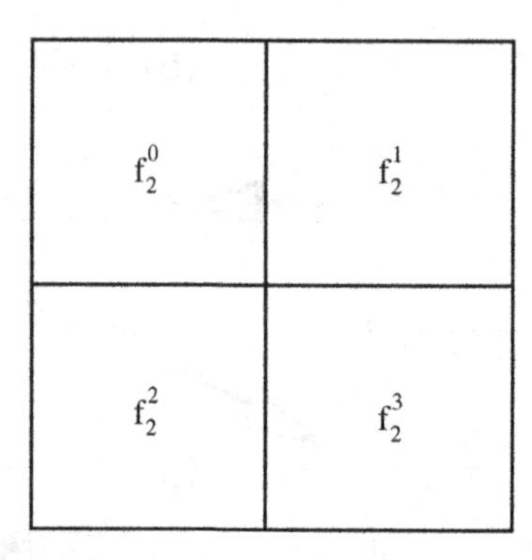

Figure 4.3 DWT decomposition in the frequency domain.

For the subsequent levels $(j > 1)$, $f_{2^j}^0(x, y)$ is decomposed in a similar way, and four quarter-size images at level 2^{j+1} are formed. This procedure is visualized in Fig. 4.2.

The inner products can also be written as a convolution:

$$\begin{aligned}
f_{2^{j+1}}^0(m, n) &= \left\{[f_{2^j}^0(x, y) * \phi(x, y)](2m, 2n)\right\} \\
f_{2^{j+1}}^1(m, n) &= \left\{[f_{2^j}^0(x, y) * \psi^1(x, y)](2m, 2n)\right\} \\
f_{2^{j+1}}^2(m, n) &= \left\{[f_{2^j}^0(x, y) * \psi^2(x, y)](2m, 2n)\right\} \\
f_{2^{j+1}}^3(m, n) &= \left\{[f_{2^j}^0(x, y) * \psi^3(x, y)](2m, 2n)\right\}
\end{aligned} \tag{4.5}$$

The scaling and the wavelet functions are separable, and therefore we can replace every convolution by a 1-D convolution on the rows and columns of $f_{2^j}^0$. Figure 4.4 illustrates this fact. At level 1, we convolve the rows of the image $f_1(x, y)$ with $h_0(x)$ and with $h_1(x)$ and then eliminate the odd-numbered columns (the leftmost is set to zero) of the two resulting arrays. The columns of each $N/2 \times N$ are then convolved with $h_0(x)$ and $h_1(x)$, and the odd-numbered rows are eliminated (the top row is set to zero). As an end result we obtain the four $N/2 \times N/2$-arrays required for that level of the WT. Figure 4.3 illustrates the localization of the four newly obtained images in the frequency domain. $f_{2^j}^0(x, y)$ describes the low-frequency information of the previous level, while $f_{2^j}^1(x, y)$, $f_{2^j}^2(x, y)$, and $f_{2^j}^3(x, y)$ represent the horizontal, vertical, and diagonal edge information.

The inverse WT is shown in Fig. 4.4. At each level, each of the arrays obtained on the previous level is upsampled by inserting a column of zeros to the left of each column. The rows are then convolved with either $h_0(x)$ or $h_1(x)$, and the resulting $N/2 \times N$-arrays are added together in pairs. As a result, we get two arrays which are oversampled to achieve an $N \times N$-array by inserting a row of zeros above each row. Next, the columns of the two new arrays are convolved with $h_0(x)$ and $h_1(x)$, and the two resulting arrays are added together. The result shows the reconstructed image for a given level.

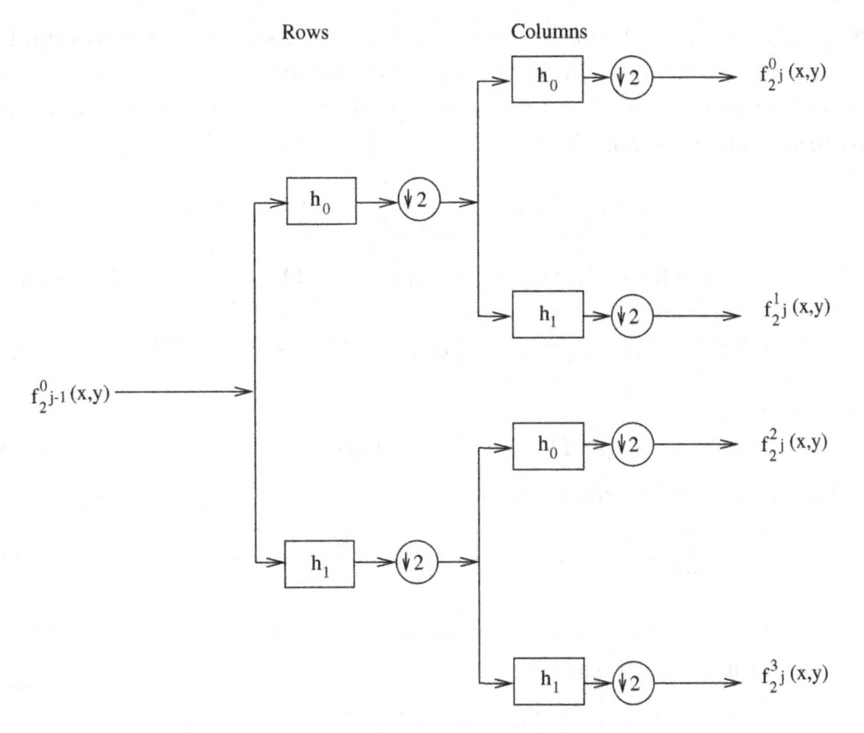

Figure 4.4 Image decomposition based on discrete WT.

4.3. BIORTHOGONAL WAVELETS AND FILTER BANKS

The functions that qualify as orthonormal wavelets with compact support lack a very important property, that of symmetry. By choosing two different wavelet bases, $\psi(x)$ and $\bar{\psi}(x)$, one for the analysis (decomposition) and the other for the synthesis (reconstruction), we get symmetric wavelets with compact support. The two wavelets are duals of each other, and the wavelet families $\{\psi_{jk}(x)\}$ and $\{\bar{\psi}_{jk}(x)\}$ are biorthogonal.

Let us look again at the standard two-band filter bank. The conditions for achieving perfect reconstructions are

$$\widehat{X}(z) = \frac{1}{2}X(z)[H_0(z)G_0(z) + H_1(z)G_1(z)] \\ + \frac{1}{2}X(-z)[H_0(-z)G_0(-z) + H_1(-z)G_1(-z)] \quad (4.6)$$

To achieve a perfect reconstruction, we must impose two conditions:

$$H_0(-z)G_0(z) + H_1(z)G_1(z) = 0 \quad (4.7)$$

$$H_0(z)G_0(z) + H_1(z)G_1(z) = 2 \quad (4.8)$$

Equation (4.7) eliminates aliasing, while eq. (4.8) eliminates amplitude distortion. Based on the perfect reconstruction conditions, we can demonstrate the biorthogonality of the analysis and synthesis filters. Let $P(z)$ be the product of the transfer functions of the lowpass analysis and synthesis filters. From eq. (3.28), we obtain

$$P(z) = G_0(z)H_0(z) = \frac{2}{\det(\mathbf{H}_m(z))} H_0(z)H_1(-z) \tag{4.9}$$

Similarly, we can write for $G_1(z)H_1(z)$, considering $\det(\mathbf{H}_m(z)) = -\det(\mathbf{H}_m(-z))$

$$P(-z) = G_1(z)H_1(z) = \frac{-2}{\det(\mathbf{H}_m(z))} H_0(-z)H_1(z) = G_0(-z)H_0(-z) \tag{4.10}$$

Thus, we obtain

$$G_0(z)H_0(z) + G_0(-z)H_0(-z) = 2 \tag{4.11}$$

Taking the inverse Z-transform, we obtain

$$\sum_k g_0(k)h_0(2n-k) = \langle g_0(k), h_0(2n-k)\rangle = \delta(n) \tag{4.12}$$

Similarly, by expressing G_0 and H_0 as a function of G_1 and H_1, we obtain the more general expression for biorthogonality:

$$\langle h_i(2n-k), g_j(k)\rangle = \delta(i-j)\delta(n), \quad i,j = \{0,1\} \tag{4.13}$$

We immediately see the main advantage with biorthogonal filters: linear phase and unequal filter length.

Let us now analyze this from the wavelet perspective. For this we must define two new hierarchies of subspaces:

$$\begin{gathered} \cdots V_2 \subset V_1 \subset V_0 \subset V_{-1} \subset V_{-2} \cdots \\ \cdots \bar{V}_2 \subset \bar{V}_1 \subset \bar{V}_0 \subset \bar{V}_{-1} \subset \bar{V}_{-2} \cdots \end{gathered} \tag{4.14}$$

The orthogonal complementary properties hold for $W_j \perp \bar{V}_j$ and $\bar{W}_j \perp V_j$.

This leads to

$$V_{j-1} = V_j \oplus \bar{W}_j \quad \text{and} \quad \bar{V}_{j\;1} = \bar{V}_j \oplus W_j \tag{4.15}$$

and to the following scaling and wavelet functions:

$$\begin{aligned} \phi(t) &= 2\sum h_0(n)\phi(2t-n) \\ \bar{\phi}(t) &= 2\sum \tilde{g}_0(n)\bar{\phi}(2t-n) \end{aligned} \tag{4.16}$$

and

$$\begin{aligned} \psi(t) &= 2\sum h_1(n)\phi(2t-n) \\ \bar{\psi}(t) &= 2\sum \tilde{g}_1(n)\bar{\phi}(2t-n) \end{aligned} \tag{4.17}$$

It is easy to show that the scaling and wavelet functions are interrelated for the biorthogonal case

$$\begin{aligned}\langle \bar{\phi}(t-k), \phi(t-l)\rangle &= \delta_{k-l}\\ \langle \bar{\psi}(t-k), \psi(t-l)\rangle &= \delta_{k-l}\end{aligned} \tag{4.18}$$

and

$$\begin{aligned}\langle \bar{\phi}(t-k), \psi(t-l)\rangle &= 0\\ \langle \bar{\psi}(t-k), \phi(t-l)\rangle &= 0\end{aligned} \tag{4.19}$$

Based on the foregoing, we see that any function $f \in L^2(R)$ can be written as

$$\begin{aligned}f(t) &= \sum_j \sum_k \langle f, \bar{\psi}_{jk}\rangle \psi_{jk}(t)\\ &= \sum_j \sum_k \langle f, \psi_{jk}\rangle \bar{\psi}_{jk}(t)\end{aligned} \tag{4.20}$$

The most important aspects achieved simultaneously based on biorthogonality are perfect reconstruction and linear phase filters. However, a drawback is the different filter lengths in the decomposition part, which means that the signal spectrum is unevenly distributed over the low-band and high-band segments. Figure 4.5 illustrates the image reconstruction based on the DWT.

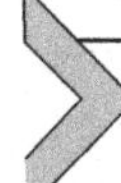

4.4. APPLICATIONS

4.4.1 Multiscale Edge Detection

The transient properties of signals and images are best described by points of sharp variations, which are mostly located at the boundaries of relevant image structures. To detect contours of small structures as well as boundaries of larger objects, a new wavelet-based concept, the so-called multiscale edge detection, was introduced [231,233]. The scale describes the size of the neighborhood where the signal variations are determined. The WT provides a better insight into these algorithms, because it is closely related to multiscale edge detection. For example, the Canny edge detector [47] corresponds to determining the local maxima of a wavelet transform modulus. Since the multiscale edges carry important information, it is possible to reconstruct images from multiscale edges. In [233] a compact image coding algorithm is proposed that keeps only the important edges. The image can be recovered from these main features, and although some small details are lost, it preserves its quality. In the following, we will describe how multiscale edge detection algorithms are related to the WT.

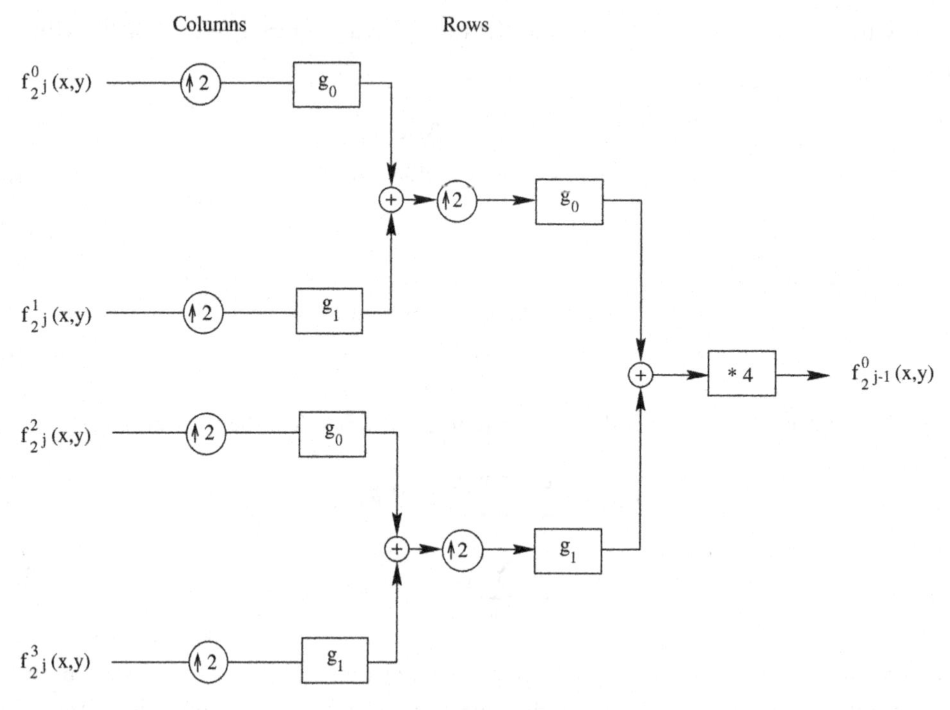

Figure 4.5 Image reconstruction based on the DWT.

Multiscale edge detectors typically smooth the signal at various scales and determine the sharp variation of points from their first- or second-order derivatives. The extrema of the first derivative describe the zero crossings of the second derivative and the inflection points of the smoothed signal.

A function $\theta(x)$ is defined as a smoothing function, if the following properties hold:

$$\int_{-\infty}^{\infty} \theta(x)dx = 1 \tag{4.21}$$

and

$$\lim_{x\to\infty} \theta(x) = 0 \tag{4.22}$$

The Gaussian function can be chosen as a smoothing function. For a smoothing function $\theta(x)$, we assume that it is twice differentiable and denote by $\psi^a(x)$ and $\psi^b(x)$ the first and second derivatives of $\theta(x)$

$$\psi^a(x) = \frac{d\theta(x)}{dx} \quad \text{and} \quad \psi^b(x) = \frac{d^2\theta(x)}{dx^2} \tag{4.23}$$

We immediately can see that $\psi^a(x)$ and $\psi^b(x)$ represent wavelets because the following holds:

$$\int_{-\infty}^{\infty} \psi^a(x)dx = 0 \quad \text{and} \quad \int_{-\infty}^{\infty} \psi^b(x)dx = 0 \tag{4.24}$$

Based on the new definition

$$\zeta_s(x) = \frac{1}{s}\zeta\left(\frac{x}{s}\right) \tag{4.25}$$

we can represent a dilation by a scaling factor s of the given function $\zeta(x)$.

The WT of a function $f(x)$ with respect to $\psi^a(x)$ and $\psi^b(x)$ is defined as

$$W_f(a,s) = \langle \psi_{as}(x), f(x)\rangle \tag{4.26}$$

and

$$W_f(b,s) = \langle \psi_{bs}(x), f(x)\rangle \tag{4.27}$$

From eqs. (4.23), (4.26), and (4.27) we obtain

$$W_f(a,s) = \left\langle f(x), \left(s\frac{d\theta_s(x)}{dx}\right)\right\rangle = s\frac{d\langle f(x), \theta_s(x)\rangle}{dx} \tag{4.28}$$

and

$$W_f(b,s) = \left\langle f(x), \left(s^2\frac{d^2\theta_s(x)}{dx^2}\right)\right\rangle = s^2\frac{d^2\langle f(x), \theta_s(x)\rangle}{dx^2} \tag{4.29}$$

We can see that the WTs $W_f(a,s)$ and $W_f(b,s)$ describe the first and second derivatives of the signal smoothed at scale s. The local extrema of $W_f(a,s)$ thus correspond to the zero crossings of $W_f(b,s)$ and to the inflection points of $\langle f(x), \theta_s(x)\rangle$. If we assume that the smoothing function $\theta(x)$ is Gaussian, then the zero-crossing detection corresponds to a Marr-Hildreth edge detection, and the extremum detection is equivalent to Canny edge detection [233]. The scale s plays an important role in edge detection: by choosing a large s and convolving it with the smoothing function corresponding to that scale, small signal fluctuations are removed, and only edges of large objects can be detected.

An inflection point of $\langle f(x), \theta_s(x)\rangle$ corresponds to either a maximum or minimum value of the first derivative. It is useful to point to the role of those values: the maximum value describes the fast fluctuations of $\langle f(x), \theta_s(x)\rangle$, while the minimum value describes the slow ones.

To apply the derived theoretical aspects to image processing, we need to extend the Canny edge detector in two dimensions. We will use the following notations:

$$\zeta_s(x,y) = \frac{1}{s^2}\zeta\left(\frac{x}{s},\frac{y}{s}\right) \tag{4.30}$$

describing the dilation by s of any 2-D function $\zeta(x,y)$.

We assume that the smoothing function has the properties

$$\int_{-\infty}^{\infty}\int_{-\infty}^{\infty}\theta(x,y)dx\,dy = 1 \tag{4.31}$$

and

$$\lim_{x,y\to\infty}\theta(x,y) = 0 \tag{4.32}$$

The image $f(x,y)$ is convolved with $\theta_s(x,y)$ and thus smoothed at different scales s. In the next step, we need to determine the gradient vector $\vec{\nabla}(\langle f(x,y),\theta_s(x,y)\rangle)$. The gradient vector points into the direction of the largest absolute value of the directional derivative of the given image. Edges are defined as points (x_0,y_0) where the gradient vector exhibits its greatest magnitude and are at the same time the inflection points of the surface $\langle f(x,y),\theta_s(x,y)\rangle$ [233].

As for the 1-D case, we will point out the equivalence between edge detection and the 2-D WT. Let $\psi_1(x,y)$ and $\psi_2(x,y)$ be two wavelet functions

$$\psi_1(x,y) = \frac{\partial\theta(x,y)}{\partial x} \quad \text{and} \quad \psi_2(x,y) = \frac{\partial\theta(x,y)}{\partial y} \tag{4.33}$$

and

$$\psi_{1s}(x,y) = \frac{1}{s^2}\psi_1\left(\frac{x}{s},\frac{y}{s}\right) \tag{4.34}$$

and, respectively,

$$\psi_{2s}(x,y) = \frac{1}{s^2}\psi_2\left(\frac{x}{s},\frac{y}{s}\right) \tag{4.35}$$

Let us assume that $f(x,y) \in L^2(R^2)$. The WT of $f(x,y)$ at the scale s is given by two distinct parts,

$$W_f(1,s) = \langle f(x,y),\psi_{1s}(x,y)\rangle \quad \text{and} \quad W_f(2,s) = \langle f(x,y),\psi_{2s}(x,y)\rangle \tag{4.36}$$

As seen in eqs. (4.26) and (4.27), we can also show here that the following relationship holds:

$$\begin{pmatrix} W_{1s}f(x,y) \\ W_{2s}f(x,y) \end{pmatrix} = s\begin{pmatrix} \frac{\partial\langle f(x,y),\theta_s(x,y)\rangle}{\partial x} \\ \frac{\partial\langle f(x,y),\theta_s(x,y)\rangle}{\partial y} \end{pmatrix} = s\vec{\nabla}\langle f(x,y),\theta_s(x,y)\rangle \tag{4.37}$$

We see here as well that edges can be determined based on the two components $W_{1s}f(x,y)$ and $W_{2s}f(x,y)$ of the WT.

Multiscale edge detection can be illustrated by means of a mammographic image. An overcomplete and continuous wavelet representation is used in the following for the extraction of features for mass detection. Overcomplete wavelet representations are more desirable for detection and enhancement because (1) they avoid the aliasing effects

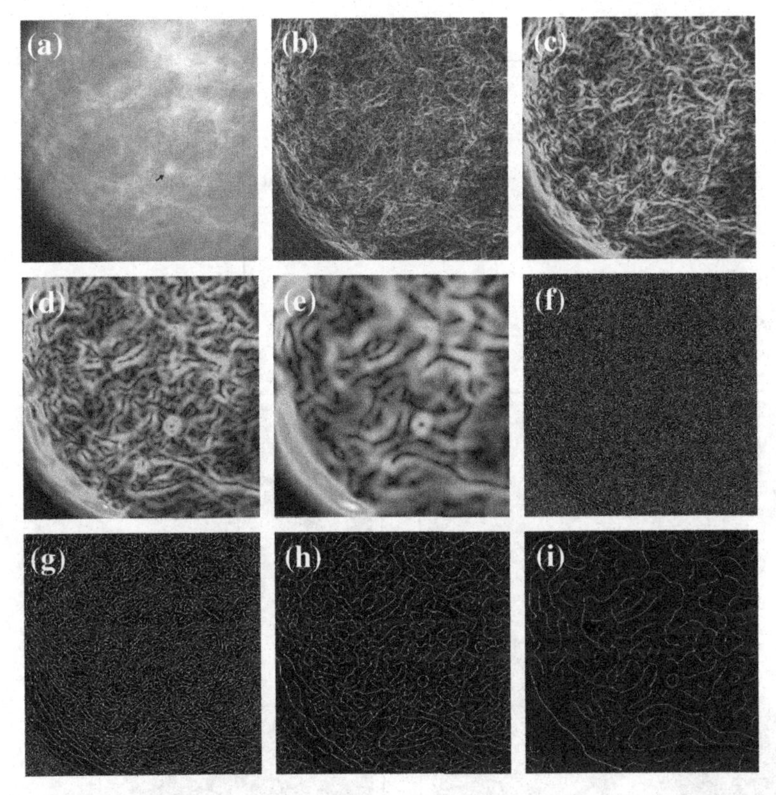

Figure 4.6 (a) Original mammogram containing a small spiculated mass, (b)–(e) multiscale coefficients, (f)–(i) multiscale edges. *(Images courtesy of Dr. A. Laine, Columbia University.)*

introduced by critically sampled representations, and (2) they yield a shift-invariant representation. Continuous wavelet representations are also useful for analysis because they yield wavelet representations at nondyadic scales.

Figure 4.6 shows a mammogram containing a small mass. The typical mammographic features are visualized at four different scales. It is important to note that across each scale, the spatial location of the mass is preserved because of the translation invariance property. Also, the transform coefficient matrix size at each scale remains the same as the spatial resolution of the original image since we do not down sample across scales.

4.4.2 Wavelet-Based Denoising and Contrast Enhancement

The DWT decomposes an image into components of different magnitudes, positions, and orientations. Similar to the linear filtering in the Fourier frequency domain, it is possible to modify the WT coefficients before we perform the inverse WT. This means to keep only the coefficients reflecting the relevant information and to omit those containing redundant information. Figure 4.7 elucidates this technique.

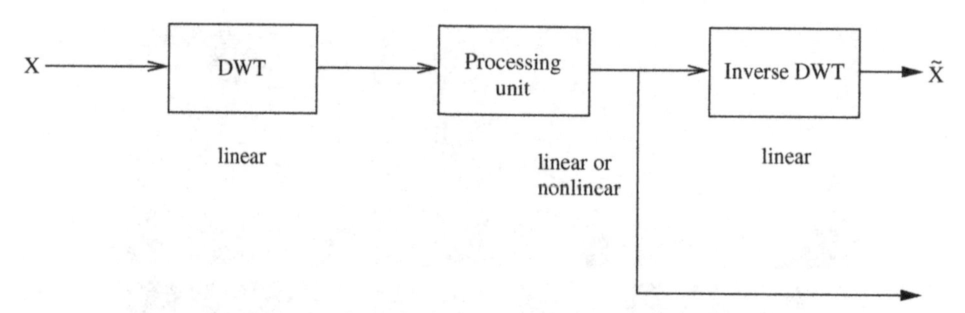

Figure 4.7 Wavelet-based signal processing.

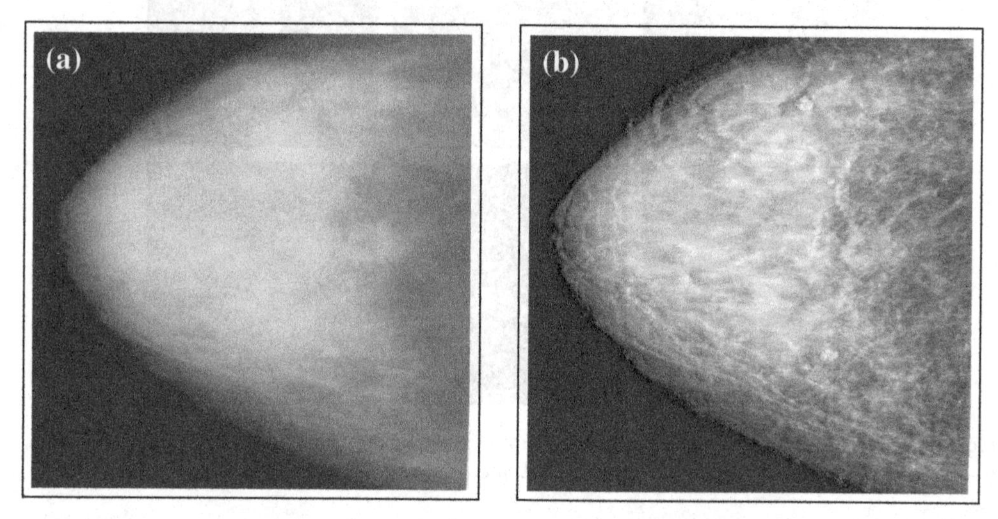

Figure 4.8 Global image enhancement via a sub-octave WT and adaptive nonlinear processing, (a) orginal mammogram, (b) enhanced mammogram. *(Images courtesy of Dr. A. Laine, Columbia University.)*

The middle part represents either a linear or nonlinear processing unit for modifying the WT coefficients.

Contrast enhancement can make more obvious unseen or barely seen features. In mammography, for example, early detection of breast cancer relies on the ability to distinguish between malignant and benign mammographic features. However, the detection of small malignancies and subtle lesions is in most cases very difficult. Contrast enhancement provides the physician with a better mammographic image without requiring additional radiation. Figure 4.8 illustrates a mammographic image containing a suspicious mass and the corresponding global enhanced image. In the new image we can observe that subtle features are made more clear, and the geometric shape of the spiculated mass is made more visible.

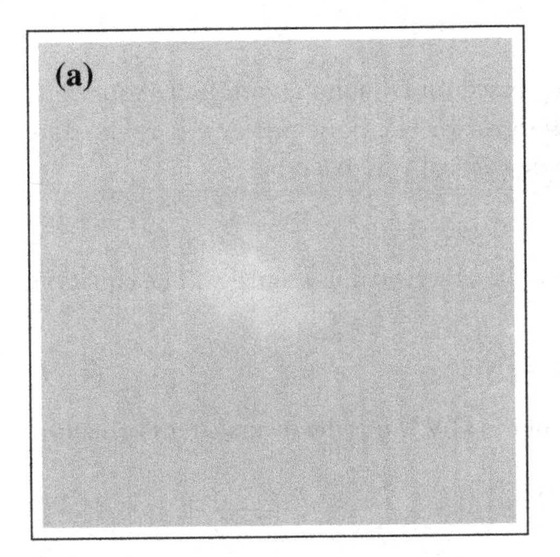

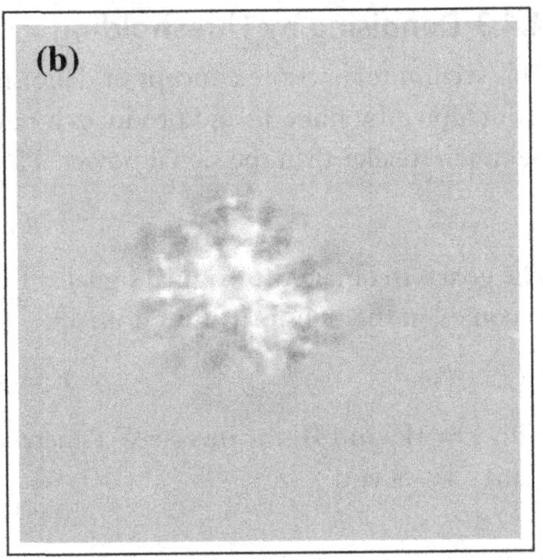

Figure 4.9 Local enhancement of mass via manual segmentation: original mammogram (a) and local multiscale enhanced image (b). *(Images courtesy of Dr. A. Laine, Columbia University.)*

A problem associated with image enhancement in digital mammography is the need to emphasize mammographic features while reducing the enhancement of noise. Local enhancement techniques apply only to some selected parts of an image, while global enhancement pertains to the whole image. A locally enhanced suspicious mass is shown in Fig. 4.9.

In the beginning of this section, we mentioned that we are interested in techniques for modifying the DWT before performing the inverse DWT. The crudest way is to remove layers of high-frequency detail from the signal by eliminating high-frequency coefficients from the DWT. In other words, the details can be dropped by simply zeroing all the high-frequency coefficients above a certain scale. At the same time, useful approximations to the original signal can be retained. However, scale reduction is not always the most effective way to compress a signal. A more selective zeroing of DWT coefficients can be achieved based on thresholding. Donoho and Johnstone [82,83] proposed a new technique for achieving high compression rates and denoising at the same time. Any DWT coefficient with a magnitude below a specified threshold is set to zero. Thresholding generally gives a lowpass version of the original signal and suppresses the noise amplitude. Based on the inverse WT, we then obtain an almost denoised version of the original signal. It is important to point out the difference between the linear filtering and wavelet shrinkage proposed by Donoho and Johnstone. The first technique requires that signal and noise spectra not overlap. The latter does not impose a separability condition; it requires only different amplitudes.

4.4.3 Denoising by Thresholding

This section reviews the concept of denoising based on Donoho's work [82]. A signal x_i is corrupted by noise n_i and produces a new signal y_i. The noise power ϵ is assumed to be much smaller than the signal power. The new signal y_i is given by

$$y_i = x_i + \epsilon n_i, \quad i = 1, \ldots, N \tag{4.38}$$

The goal is to obtain the original signal x_i from the observed noisy signal y_i. The equation becomes, in the transformation domain,

$$Y = X + N \tag{4.39}$$

with $y = Wy$ and W the inverse WT-matrix of the DWT. For the inverse transformation matrix we obtain

$$W^{-1}W = I \tag{4.40}$$

and

$$W^{-1} = W^T \tag{4.41}$$

$\widehat{X}$ describes the estimation of X based on the observed values Y. Next, we introduce the diagonal linear projection

$$\Delta = diag(\delta_1, \ldots, \delta_N), \quad \delta_i \in \{0, 1\}, \quad i = 1, \ldots, N \tag{4.42}$$

which gives the estimation

$$\hat{x} = W^{-1}\widehat{X} = W^{-1}\Delta Y = W^{-1}\Delta Wy \tag{4.43}$$

The estimated value $\widehat{X}$ is obtained by zeroing the wavelet coefficients. The resulting L_2-error becomes

$$R(\widehat{X}, X) = E\left[||\hat{x} - x||_2^2\right] = E\left[||W^{-1}(\widehat{X} - X)||_2^2\right] = E\left[||\widehat{X} - X||_2^2\right] \tag{4.44}$$

The last equal sign stems from the orthogonality of W. The optimal coefficients of the diagonal matrix Δ must be given by $\delta_i = 1$ for $x_i \geq \epsilon$ and $\delta_i = 0$ for $x_i < \epsilon$. Thus, only those values of Y are retained for which the corresponding X-value is larger than ϵ. The others are set to zero.

Donoho's algorithm for denoising is given below:

1. Compute the DWT $Y = Wy$.
2. Apply either hard or soft thresholding for the wavelet coefficients

$$\widehat{X} = T_h(Y, t) = \begin{cases} Y, & |Y| \geq t \\ 0, & |Y| < t \end{cases} \tag{4.45}$$

or

$$\widehat{X} = T_s(Y, t) = \begin{cases} \text{sgn}(Y)(|Y| - t), & |Y| \geq t \\ 0, & |Y| < t \end{cases} \tag{4.46}$$

3. Compute the inverse DWT $\hat{x} = W^{-1}\widehat{X}$.

Donoho's algorithm exhibits some interesting properties. The soft thresholding guarantees that $|\widehat{X}_i| < |X_i|$ holds and that $\widehat{X}$ has the same smoothness degree as X. Soft thresholding can be considered as an optimal estimation. On the other hand, the hard thresholding achieves the smallest L_2-error but lacks the required smoothness.

4.4.4 Nonlinear Contrast Enhancement

Image enhancement techniques have been widely applied in fields such as radiology, where the subjective quality of images is important for diagnosis. Low-contrast structures need to be resolved in all kinds of medical images. Obtaining high contrast in the raw image directly from the imaging device is almost always expensive in examination time or X-ray dose to the patient. Therefore, digital postprocessing can play a very important role.

From an image-processing point of view, the low contrast can be considered a result of "bad" distribution of pixel intensities over the dynamic range of the display device. This suggests the application of contrast enhancement methods in an attempt to modify the intensity distribution of the image. Many known enhancement algorithms have been proposed and applied to medical imaging. A comprehensive survey of the most relevant enhancement algorithms is given in [383]. The most commonly used algorithms are the histogram modification and the edge enhancement techniques.

Histogram modification is a simple and fast procedure and has achieved satisfactory results for some applications. This technique works as follows: a nonlinear transformation function is applied on a histogram of an input image in order to achieve a desired histogram. However, when applied to digital images information loss occurs as a result of quantization errors. For medical imaging this can lead to diagnostic losses because a subtle edge can be fused with the neighboring pixels and disappear.

A different approach is used for contrast enhancement. It is observed that any area in an image having highly visible edges has to be of high contrast. To achieve contrast enhancement, either the magnitude of edge gradients or the pixel intensities between both sides of edges must be amplified. Also, contrast enhancement appears to be spatially dependent, and an increase in contrast has to be perpendicular to edge contours. Standard techniques for edge enhancement are pointwise intensity transformation for edge pixels [21] or filter design for magnitude amplification of edge gradients [177]. The latter technique is identical to "unsharp masking" [52].

Recent advancement of wavelet theory has been the driving force in developing new methods for contrast enhancement [6,91,196,224]. The advantage of this new technique over traditional contrast enhancement techniques is illustrated in Fig. 4.10.

In the following, we will review simple nonlinear enhancement by point-wise functional mapping. Linear enhancement can be described as a mapping of wavelet coefficients by a linear function $E_m(x) = G_m x$ with m being the level and G_m the amplification.

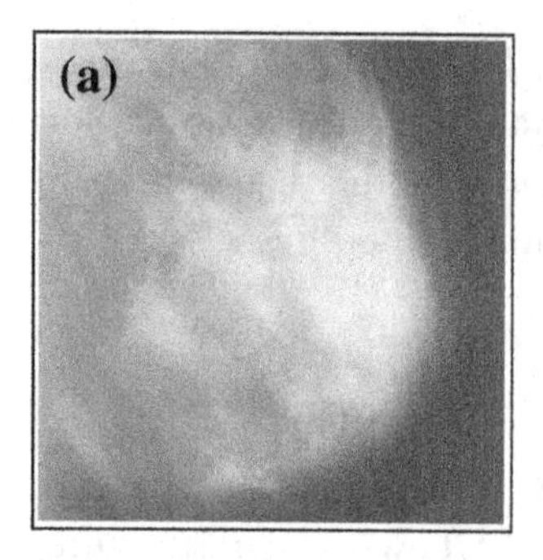

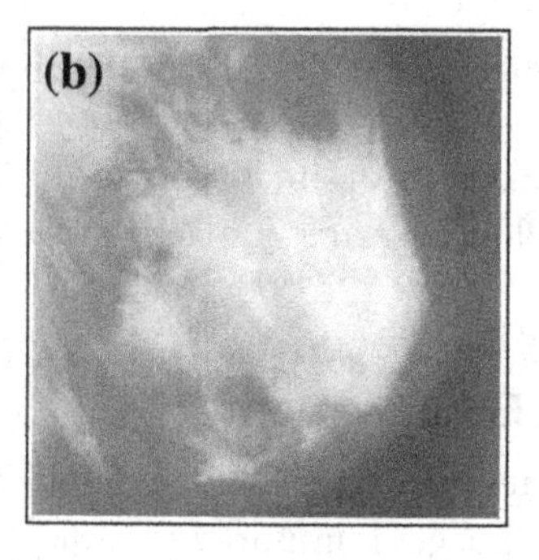

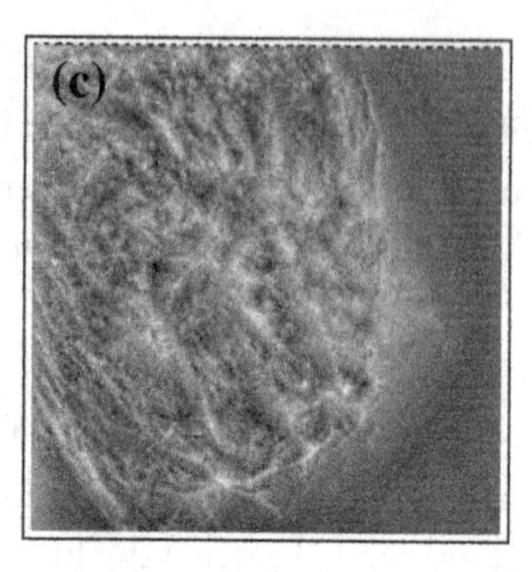

Figure 4.10 Multiscale contrast enhancement: original mammogram (a), histogram modified mammogram (b), and (c) multiscale contrast enhancement. *(Reprinted from [198] with permission from IEEE.)*

On the other hand, nonlinear enhancement represents a direct extension of the linear enhancement.

The idea of nonlinear contrast enhancement applied to digital mammography was first pioneered by Laine [5, 198]. The main difficulties are to choose a nonlinear function and to incorporate the dyadic wavelet information to accomplish contrast enhancement.

Laine [196] gives the following guidelines for designing a nonlinear enhancement function:

1. An area of low contrast should be enhanced more than an area of high contrast.
2. A sharp edge should not be blurred.
3. The enhancement function should be monotone in order to preserve the position of local extrema and not create new extrema.
4. The enhancement function should be antisymmetric in order to preserve phase polarity for "edge crispening."

A simple piecewise function that satisfies these requirements is shown in Fig. 4.11:

$$E(x) = \begin{cases} x - (K-1)T, & x < -T \\ Kx, & |x| \leq T \\ x + (K-1)T, & x > T \end{cases} \tag{4.47}$$

where $K > 1$.

For early detection of breast cancer, the ability to distinguish between malignant and mammographic features is crucial. However, the detection of small malignancies and subtle lesions encounters major difficulties. Global and local enhancement has the ability to make more visible unseen or barely seen features of a mammogram without requiring additional radiation. Figures 4.12–4.15 illustrate the necessary processing steps for global and local contrast enhancement in diagnostic radiology [197, 410]. The starting point is as usual a digitized radiograph. A unique component in the described technique is the "fusion" step shown in Figs. 4.12 and 4.13. The importance of this step lies in

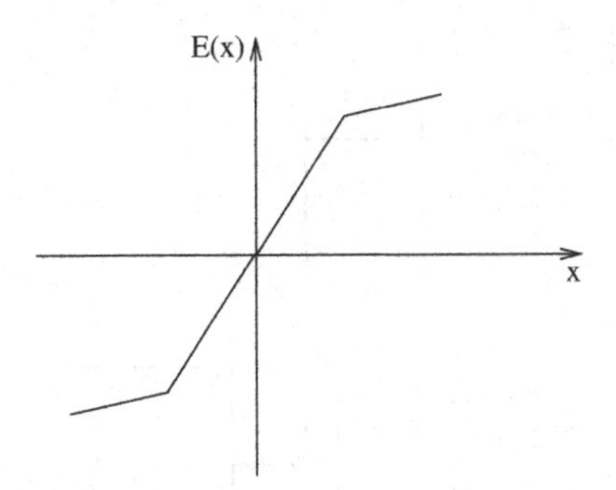

Figure 4.11 Nonlinear function for contrast enhancement.

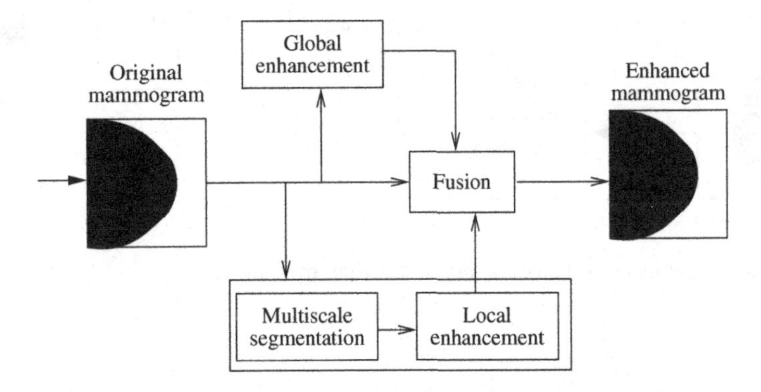

Figure 4.12 Processing overview for multiscale enhancement.

producing images that look more familiar to radiologists than do those produced using only multiscale global enhancement.

Therefore, by integrating the original mammogram with one or more enhanced versions, a more familiar image for radiologists is achieved. Image fusion combines two or more registered images of the same object into a single image that is more easily interpreted than any of the originals.

In [186] an image fusion method based on a steerable dyadic WT is described that enables multiscale image processing along arbitrary orientations. While steerable dyadic WTs when applied to image fusion share many properties with pyramids and WTs, they do not exhibit aliasing and translation noninvariance as sources for unwanted artifacts in a fused image. A steerable dyadic wavelet transform combines the properties of a discrete dyadic wavelet transform with the computational framework for analysis along arbitrary orientations. A steerable dyadic WT is implemented as a filter bank consisting of polar separable filters [195]. Input images are convolved with rotations, dilations, and translations of a mother wavelet. The coefficients corresponding to the greatest local oriented energy are included for reconstruction and thus produce the fused image.

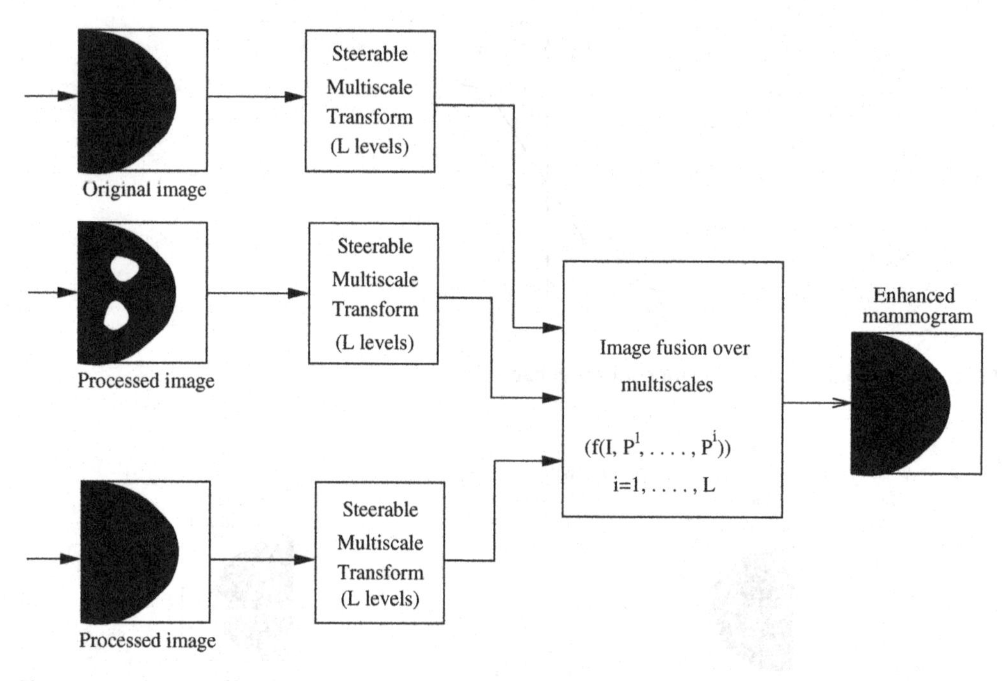

Figure 4.13 Fusion of locally and globally enhanced mammograms.

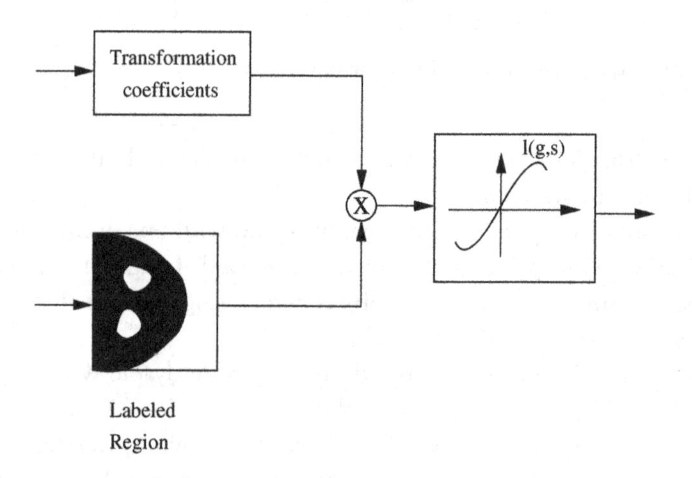

Figure 4.14 Local enhancement of segmented regions.

Figure 4.16 shows the results of enhancement via fusion.

In [196] it was shown that local enhancement techniques, focused on suspicious areas, provide a clear advantage over traditional global processing methods where enhancement is carried out within a specified ROI or within the entire mammogram.

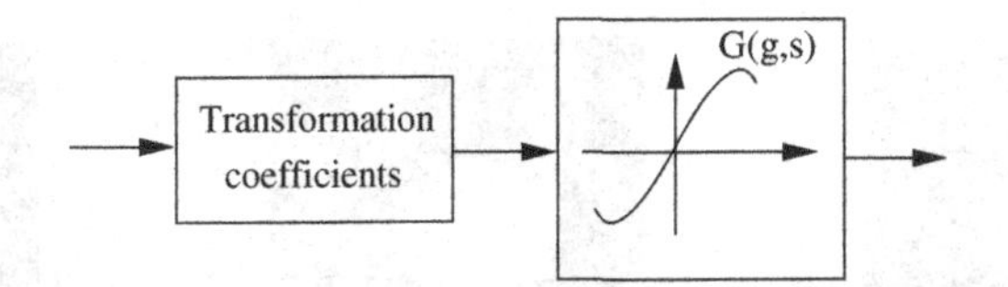

Figure 4.15 Global enhancement of multiscale coefficients.

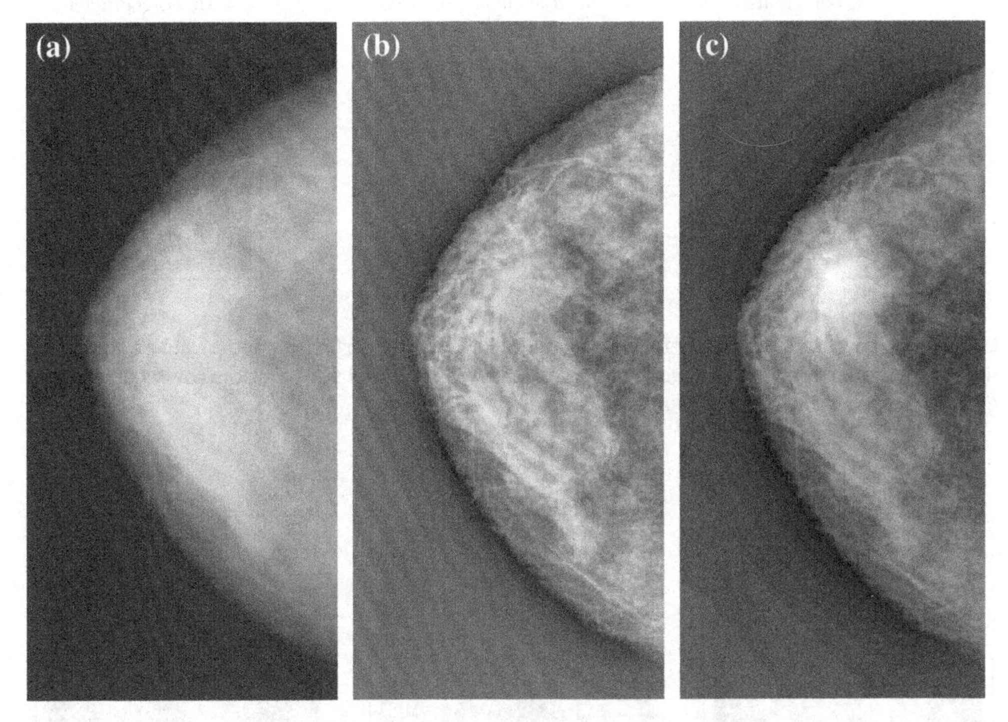

Figure 4.16 Enhancement via fusion: (a) original mammogram, (b) contrast enhancement by multiscale analysis, and (c) enhancement obtained by fusion of enhanced features. *(Reprinted from [187] with permission from IEEE.)*

An example of selection of localized transform coefficients is shown in Figs. 4.17 and 4.19.

4.4.5 Image Fusion

Image fusion combines two or more registered images of the same object into a single image that is more easily interpreted than any of the originals. The goal of image fusion, especially in medical imaging, is to create new images that are more suitable for the purposes of human visual perception. The simplest image fusion technique is to take the average of two input images. Applied directly, this leads to a feature contrast reduction.

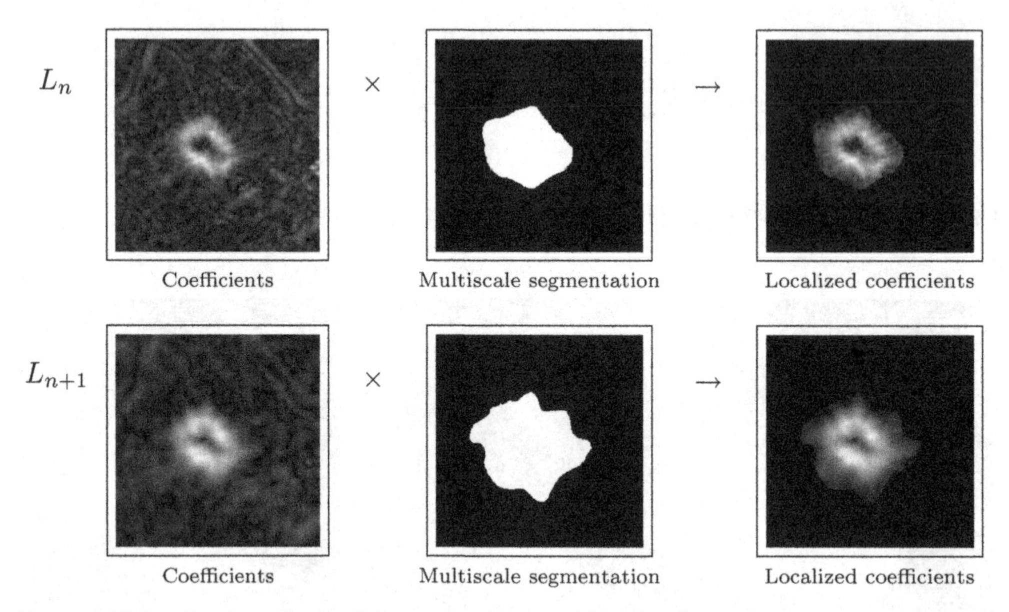

Figure 4.17 Localized wavelet coefficients of a suboctave band for a first-order derivative approximation of a smoothing function at two levels of analysis. Top row: finer scale. Bottom row: coarser scale. *(Images courtesy of Dr. A. Laine, Columbia University.)*

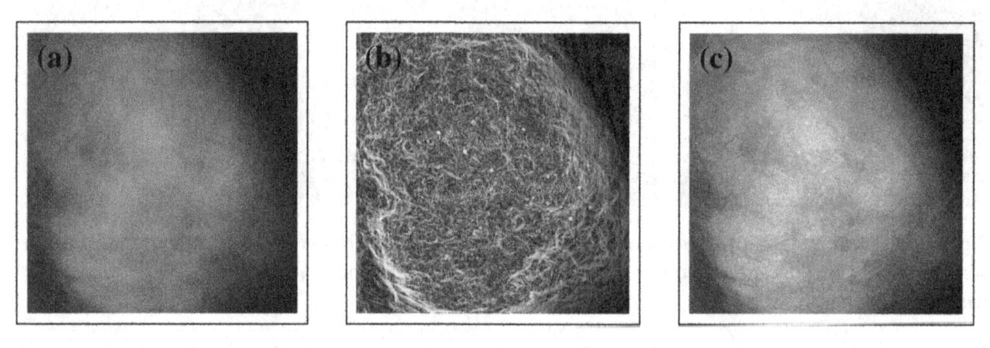

Figure 4.18 Fused image (c) of an original mammogram (a) with the selected modulus of dyadic wavelet coefficients (b). *(Images courtesy of Dr. A. Laine, Columbia University.)*

A solution to this technique offers a Laplacian pyramid-based image fusion [367], but at the cost of introducing blocking artifacts. Better fusion results were achieved based on the WT [215]. Figure 4.20 describes the fusion framework proposed in [215]. The steps are: (1) computation of WT for each image, (2) selection at every point of the coefficients having the highest absolute value, and (3) computing the inverse WT for the new image.

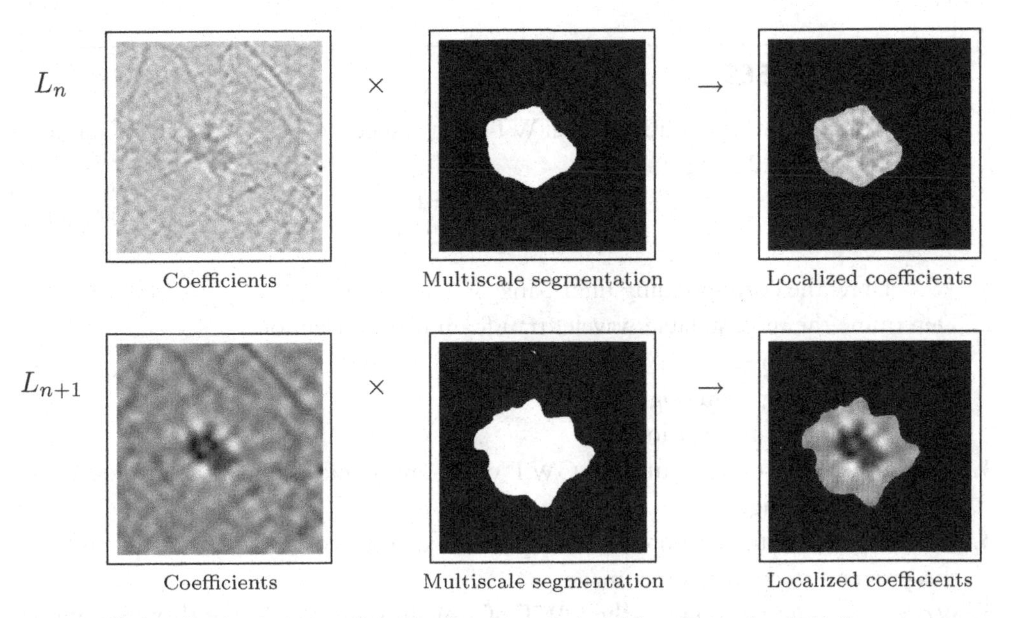

Figure 4.19 Localized wavelet coefficients of a suboctave band for a first-order derivative approximation of a smoothing function at two levels of analysis. Top row: finer scale. Bottom row: coarser scale. *(Images courtesy of Dr. A. Laine, Columbia University.)*

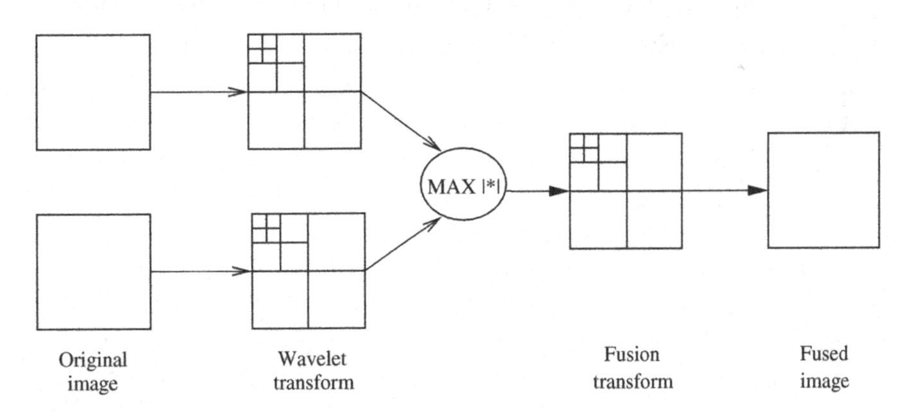

Figure 4.20 The block diagram of the image fusion scheme.

It is well known that the larger absolute value transform coefficients describe sharper brightness changes, and thus represent the salient features at different scales.

Figure 4.18 shows the image fusion results for an example from mammography. An original image is fused with a WT image and yields a better result than the original mammogram.

4.5. EXERCISES

1. Compute the two-dimensional WT with respect to Haar wavelets of the following 2×2 image:

$$\mathbf{F} = \begin{bmatrix} 4 & -1 \\ 8 & 2 \end{bmatrix} \tag{4.48}$$

   Draw the corresponding filter bank.
2. Determine the most suitable wavelet transform in an image for:
   (a) Detecting lines.
   (b) Compressing mammograms.
   (c) Performing image fusion.
3. Write a program to compute the CWT of an image and use it to determine small objects in the image.
4. Write a program to compute a wavelet series expansion of a medical image and use this program to compress the image.
5. Write a program to compute the DWT of a medical image and use this program to detect edges in the image.
6. Write a program to compute the DWT of a medical image and use this program to denoise the image by thresholding.
7. Write a program to compute the DWT of a medical image and use this program for nonlinear contrast enhancement of the image.

CHAPTER FIVE

Genetic Algorithms

Contents

5.1. INTRODUCTION

Genetic algorithms (GA) like neural networks are biologically inspired and represent a new computational model having its roots in evolutionary sciences. Usually GAs represent an optimization procedure in a binary search space, and unlike traditional hill climbers they do not evaluate and improve a single solution but a set of solutions or hypotheses, a so-called population. The GAs produce successor hypotheses by mutation and recombination of the best currently known hypotheses. Thus, at each iteration a part of the current population is replaced by offspring of the most fit hypotheses. In other words, a space of candidate hypotheses is searched in order to identify the best hypothesis, which is defined as the optimization of a given numerical measure, the so-called hypothesis fitness. Consider the case of function approximation based on given input-output samples: The fitness is the accuracy of the hypothesis (solution) over the training set.

The strength of this parallel process is enhanced by the mechanics of population modification, making GAs adequate candidates even for NP-hard problems. Mathematically, they are function optimizers and they encode a potential solution based on

Pattern Recognition and Signal Analysis in Medical Imaging
http://dx.doi.org/10.1016/B978-0-12-409545-8.00005-4

chromosome-like data structures. The critical information is preserved by applying recombination operators to these structures. Their most interesting properties are [251]:

- Efficiency.
- Simple programmability.
- Extraordinary robustness regarding the input data.

The most important property is robustness, and this represents an emulation of nature's adaptation algorithm of choice. Mathematically, it means that it is possible to find a solution even if the input data do not facilitate finding such a solution. GAs are mostly applied to nondifferentiable functions and functions with many local optima. Besides the above-mentioned advantages, there is one major disadvantage: GAs have to be carefully designed. An unfavorable choice of operators might affect the outcome of the application. Therefore, a precise knowledge of the basics and the context is crucial for any problem solution based on GAs.

Usually, the implementation of a GA starts with the random definition of a chromosome population. The structures are evaluated and then better reproductive opportunities are allocated for those chromosomes, leading to better solutions. The adequacy of a solution is typically related to the current population average, or the median population.

Let us look again at the definition of a GA. In a strict sense, the GA defines a model introduced and investigated by John Holland [137]. This computational paradigm—known also as the canonical genetic algorithm—is still a relevant research topic. In a more detailed sense, the GA represents a population-based model, which employs selection and recombination operators to generate new data points in a search space [387]. There are several GA models known in the literature, most of them designed as optimization tools for several applications.

GAs differ from classical optimization and search procedures: (1) direct manipulation of a coding, (2) search from a population of points and not a single solution, (3) search via sampling, a so-called blind search, and (4) search using stochastic operators, not deterministic rules.

In this section, we will review the basics of GAs, briefly describe the schema theorem and the building block hypothesis, and describe feature selection based on GAs, as one of the most important applications of GAs.

5.2. ENCODING AND OPTIMIZATION PROBLEMS

GAs have generally two basic parts that are problem-oriented: the problem encoding and the evaluation problem [95].

Let us consider the problem of minimizing (or maximizing) a function $F(x_1, x_2, \ldots, x_M)$. The goal is to optimize an output by finding the corresponding parameter combination. Since most of the problems are nonlinear, it is not possible to optimize

Table 5.1 Definition analogies.

Information theory	Biology/genetics
Vector, string	Chromosome
Feature, character	Gene
Feature value	Allele
Set of all vectors	Population

every single parameter, but the whole set of parameters revealing the interactions between them must be considered.

To solve this optimization problem, two assumptions have to be made. The first assumption concerns the representation of parameter variables as bit strings (concatenation of bits from the set $\{0,1\}$). Most test functions have a length of at least 30 bits. Considering the nonlinear nature of the optimization problem, we see immediately that the size of the search space is related to the number of bits used for the problem encoding. Considering a bit string of length L, we obtain a hypercube for the search space with a size of 2^L. In other words, the GA samples the corners of this L-dimensional hypercube. The success of the algorithm depends on the correct encoding. Binary encoding has a major disadvantage: the weighted representation. A bit has a larger weight if it is positioned more to the left in a bit string. This means that the same operator applied to a bit string achieves different results depending on the bit position in a string. This is not always acceptable, but applying the Gray code to the string corrects this.

The second assumption concerns the evaluation function, which is usually given as part of the problem description. It is desired that the evaluation function be easily computed. Although this is valid for all optimization problems, it is of special interest for GAs. GAs do not employ just one solution, but are based on a population of solutions being replaced for every iteration with their offspring. The offspring represent a new population that has to be evaluated anew. It is evident that the time factor plays an important role in that sense.

Most of the definitions used in context with GAs have their roots in genetics but have an equivalent in information theory. For a better visualization, we can find those correspondents in Table 5.1.

5.3. THE CANONICAL GENETIC ALGORITHM

Like all other optimization methods, GAs are searching iteratively for the optimal solution to a posed problem. Unlike the other methods, GAs are employing some very simple techniques for information processing. These usually are operations for string manipulations such as copying, inserting, deleting, and replacing of string parts and random number generation. Usually, the initial population is generated randomly. The simplicity of operation with the achieved computational power makes the genetic algorithm approach very appealing for a multitude of practical problems.

The most common operators are [387]:

- **Selection:**
  Based on selection, we can choose the best strings out of a set. The strings in the current population are copied in proportion to their fitness and placed in an intermediate generation. Through selection, we can ensure that only the fittest strings perpetuate. In a broader sense, we have a natural selection which enables that the strings of the new generation are closer to the desired final results. Mathematically, it means that the convergence of the population is given.
- **Crossover:**
  Crossover describes the swapping of fragments between two binary strings at a randomly choosen crossover point. In other words, it creates two new offspring from two parents. After recombination, two new strings are formed, and these are inserted into the next population. In summary, new sample points are generated by recombining two parent strings. Consider the following two strings 000111000 and 111000111. Using a single randomly chosen crossover point, recombination occurs as follows:

  $$000|111000$$
  $$111|000111$$

  The following offspring are produced by swapping the fragments between the two parents

  $$000000111 \quad \text{and} \quad 111111000 \tag{5.1}$$

  This operator also guarantees the convergence of the population.
- **Mutation:**
  This operator does not represent a critical operation for GAs. It produces new alleles on a random basis and is relatively simple in its concept. Mutation randomly generates a new bit or a new bit sequence in a given string, usually by flipping the bit or bits in the sequence. Mutation prevents an early convergence since it produces divergence and inhomogeneity within the population. On the other hand, it ensures that the population converges to a global maximum instead of a local maximum. What is much more, based on the random changes applied to the genetic material, new combinations are produced, which lead to better solutions. Mutation can be viewed as a way out of getting stuck in local minima and is often performed after crossover has been applied. Based on the arbitrary changes produced by this operator, one should employ mutation with care. At most one out of 1000 copied bits should undergo a mutation.

Apart from these very simple operations, many others have been proposed in the literature [251].

The implementation of a GA requires the generation of an initial population. Each member of this population represents a binary string of length L, called the chromosome, and thus determines the problem encoding. Usually, the initial population takes random values. After generating an initial population, every single string is evaluated and then its fitness value is determined. One should carefully distinguish between the fitness function and the evaluation function in the context of GAs. The evaluation function represents a performance measure for a particular set of parameters. The fitness function gives, based on the measured performance, the chance of reproductive opportunities. In other words, it defines the criterion for ranking potential hypotheses and for probabilistically selecting them for inclusion in the population of the next generation. In the case of learning classification rules, the fitness function determines the ranking of classification accuracy of the rule over a set of available training samples. The evaluation of a string describing a particular set of parameters is not related to any other string evaluation. However, the fitness of that string is related to the other chromosomes of the current population. Thus, the probability that a hypothesis is chosen is directly proportional to its own fitness, and inversely proportional to the rest of competing hypotheses for the given population.

For canonical GAs the definition of the fitness is given by $f_i/\bar{f}$, where f_i is the evaluation associated with chromosome i and $\bar{f}$ is the average evaluation of all strings in the population:

$$\bar{f} = \frac{1}{n}\sum_{i=1}^{n} f_i \tag{5.2}$$

The execution of a GA can be viewed as a two-level process. In the beginning, we have the current population. Based on selection we obtain an intermediate population, and afterward based on recombination and mutation we obtain the next population. The procedure of generating the next population from the current population represents one generation in the execution of a GA. Figure 5.1 shows this procedure [387].

The intermediate population is generated from the current population. In the beginning, the current population is given by the initial population. Next $f_i/\bar{f}$ is determined for all chromosomes of the current population, and then the selection operator is employed. The strings of the current population are copied or duplicated proportional to their fitness and then entered in the intermediate generation. If, for a string i, we obtain $f_i/\bar{f} > 1.0$, then the integer portion of fitness determines the number of copies of this string that directly enter the intermediate population. A string with a fitness of $f_i/\bar{f} = 0.74$ has a 0.74 chance of placing one string in the intermediate population while a string with a fitness of $f_i/\bar{f} = 1.53$ places one copy in the intermediate population. The selection process continues until the intermediate population is generated.

Next, the recombination operator is carried out. The recombination is a process of generating the next population from the intermediate population. Then crossover is applied to a pair of strings chosen in a random fashion to produce a pair of offspring

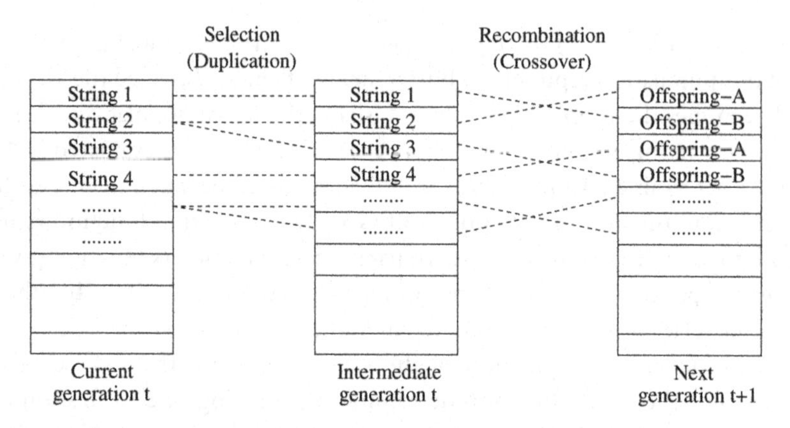

Figure 5.1 Splitting of a generation into a selection and recombination phase.

chromosomes which are syntheses of their parents. These strings are recombined with a probability of p_c and the newly generated strings are included in the next population. The last operation needed for producing the next population is the mutation operator. The mutation probability of a bit p_m is very small, usually $p_m \ll 1\%$. For practical applications, we normally choose p_m close to 0.01. Mutation changes the bit values and produces a near-identical copy with some components of the chromosome altered. Selection, recombination, and mutation are applied for each population. Combined, these operators form one generation in the execution of a GA.

The algorithmic description of a GA is given below:

Algorithm 1

1. Generate the initial population randomly for the strings a_i:
2. $\Pi = \{a_i\}, i = 1, \ldots, n$.
3. For $i \leftarrow 1$ to *Number of generations* do
4. **begin**
5. Initialize mating set $M \leftarrow \emptyset$ and Offspring O
6. For $j \leftarrow 1$ to n do
7. Add $f(a_i)/\bar{f}$ copies from a_i to M.
8. For $j \leftarrow 1$ to $n/2$ do
9. Choose two parents a_j and a_k from M and perform with the probability p_c $O = O \cup$ *Crossover* (a_j, a_k).
10. For $i \leftarrow 1$ to n do
11. For $j \leftarrow 1$ to d do
12. Mutate with the probability p_m the jth bit from $a_i \in O$
13. Update the population $\Pi \leftarrow$ combine (Π, O).
14. **end**

5.4. OPTIMIZATION OF A SIMPLE FUNCTION

A GA represents a general optimization method that searches a large space of candidates looking for a candidate that performs best according to the fitness function. It is not always guaranteed that an optimal candidate will be found; however, in most cases GAs succeed in finding a candidate with high fitness. Their most important application area is machine learning, where they have been successfully applied both to choosing the network topology for neural networks and to function approximation.

In this section, we will apply the most important basic operations of a GA to an example of function optimization [251].

The following function is considered:

$$g(x) = x \cdot \sin(10\pi \cdot x) + 1 \tag{5.3}$$

The goal is to find, based on a GA, the maximum of this function in the interval $[-1\ldots2]$:

$$g(x_0) \geq g(x), \quad \text{for all} \quad x \in [-1\ldots2] \tag{5.4}$$

To solve this optimization problem, some typical GA operators are employed.

5.4.1 Number Representation

The real-valued x have to be transformed into a binary vector (chromosome). The vector length is determined by the required precision, which for this example is six decimal places. The real-valued domain of the number x has the length of 3. The required precision means that the interval $[-1\ldots2]$ is split in 3,000,000 equal parts. This means that 22 bits are needed to represent a binary vector (chromosome):

$$2,097,152 = 2^{21} < 3,000,000 \leq 2^{22} = 4,194,304 \tag{5.5}$$

The transformation of a binary number $\langle b_{21}b_{20}\ldots b_0\rangle$ into a real number x is performed in two distinct steps:

- Transform the binary number $\langle b_{21}b_{20}\ldots b_0\rangle$ from basis 2 into basis 10:

$$(\langle b_{21}b_{20}\ldots b_0\rangle)_2 = \left(\sum_{i=0}^{21} b_i \cdot 2^i\right)_{10} = x' \tag{5.6}$$

- Determine the corresponding real number x:

$$x = -1 + x' \cdot \frac{3}{2^{22}-1} \tag{5.7}$$

where -1 is the left limit of the interval and 3 is the interval length.

5.4.2 Initial Population

The initial population is randomly generated. Each chromosome represents a 22-bit binary vector.

5.4.3 Evaluation Function

The evaluation function f of the binary vector $\mathbf{v}$ is equivalent to the function $g(x)$:

$$f(\mathbf{v}) = g(x) \tag{5.8}$$

The three given x-values $x_1 = 0.637197, x_2 = -0.958973$, and $x_3 = 1.627888$ correspond to the following three chromosomes:

$$\begin{aligned}
\mathbf{v}_1 &= (1000101110110101000111)\\
\mathbf{v}_2 &= (0000001110000000010000)\\
\mathbf{v}_3 &= (1110000000111111000101)
\end{aligned}$$

The evaluation function provides the following values:

$$\begin{aligned}
f(\mathbf{v}_1) &= g(x_1) = 1.586345\\
f(\mathbf{v}_2) &= g(x_2) = 0.0078878\\
f(\mathbf{v}_3) &= g(x_3) = 2.250650
\end{aligned}$$

We immediately see that $\mathbf{v}_3$ is the fittest chromosome since its evaluation function provides the largest value.

5.4.4 Genetic Operators

While the GA is executed, two distinct operators are employed to change the chromosomes: mutation and crossover.

We mutate first the fifth and then the 10th gene of the chromosome $\mathbf{v}_3$. We thus obtain

$$\mathbf{v}_3' = (1110\mathbf{1}00000111111000101) \tag{5.9}$$

and, respectively,

$$\mathbf{v}_3'' = (111000000\mathbf{1}111111000101) \tag{5.10}$$

This leads for $x_3'' = 1.630818$ and for $g(x_3'') = 2.343555$ to an increase in value to $g(x_3) = 2.250650$. Crossover is now applied to chromosomes $\mathbf{v}_2$ and $\mathbf{v}_3$. The fifth gene is chosen as a crossover point:

$$\begin{aligned}
\mathbf{v}_2 &= (00000|01110000000010000)\\
\mathbf{v}_3 &= (11100|00000111111000101)
\end{aligned}$$

Swapping the fragments between the two parents produces the following offspring:

$$\mathbf{v}_2' = (00000|00000111111000101)$$
$$\mathbf{v}_3' = (11100|01110000000010000)$$

Their evaluation functions are determined as

$$f(\mathbf{v}_2') = g(-0.998113) = 0.940865$$
$$f(\mathbf{v}_3') = g(1.666028) = 2.459245$$

The second offspring has a better evaluation function value than its parents.

5.4.5 Simulation Parameters

To determine the solution of the given optimization problem, we will choose the following parameters: the population consists of 50 distinct chromosomes, the crossover probability is $p_c = 0.25$, and the mutation probability is $p_m = 0.01$.

5.4.6 Simulation Results

Table 5.2 shows the generations where both the evaluation function and the value of the function have improved at the same time. The best chromosome after 150 generations is

$$\mathbf{v}_{max} = (1111001101000100000101) \tag{5.11}$$

Its value is $x_{max} = 1.850773$. We could demonstrate that the GA converges toward the maximum of the given function.

Table 5.2 Simulation results after 150 generations.

Generation	Evaluation function
1	1.441942
6	2.250003
8	2.250283
9	2.250284
10	2.250363
12	2.328077
39	2.344251
40	2.345087
51	2.738930
99	2.849246
137	2.850217
145	2.850227

5.5. THEORETICAL ASPECTS OF GENETIC ALGORITHMS

To obtain an understanding of GA performance, we realize that by exploiting important similarities in the coding we use for the raw available data, the search algorithm becomes more efficient. Thus, appropriate coding leads to good performance. This led to the development of the notion of similarity template, or schema, and further to a keystone of the genetic algorithm approach, the building block hypothesis.

The theory of GA relies on two concepts: representation of solutions in the form of binary strings, and the schemata concept for determination of similarity between chromosomes [251]. Strings that contain $*$ are referred to as schemata. They are more general than completely specified strings, since $*$ means don't care and stands for both 0 and 1. By specifying all the $*$ in a schema, we obtain a string named an instance.

The mathematical interpretation of this situation is very interesting. We recall that a GA can result in complex and robust search by implicitly sampling hyperplane partitions of a search space. A vector of length n defines a hyperspace M^n. Each schema corresponds to a hyperplane in the search space. For example, by using the data set $\{0, 1, *\}$ we have in M^n exactly 3^n schemata. Each element is an instance of 2^n schemata. A population of p chromosomes (strings of length n) has between 2^n and $p2^n$ schemata.

5.5.1 The Schema Theorem

The order of a schema $o(h)$ refers to the number of actual bit values contained in the schema. In other words, it is the number of fixed positions (in a binary alphabet, the number of 0's and 1's) present in the template. For a schema $1{*}{*}0{*}{*}1{*}$ we obtain $o(h) = 3$. The *defining length* of a schema is determined by the distance between the first and the last bit in the schema with value either 0 or 1. Thus, the defining length of the schema $1{*}{*}0{*}{*}1{*}$ is $l(h) = 6$, because the index of the leftmost occurrence of either 0 or 1 is 7 and the index of the rightmost occurrence is 1.

Schemata and their properties are important tools for discussing and classifying string similarities. Furthermore, they provide the basic means of analyzing reproduction and genetic operators on building blocks contained in the population.

There are some important aspects with respect to a schema:

- Schemata with a large defining length are very sensitive when carrying out crossover. It might happen that after crossover a string is no longer an instance of a schema.
- Schemata of a high order are very sensitive when mutation is carried out. However, schemata with many $*$ are less sensitive.
- The greater the fitness of a schema, the higher the probability that its instances are considered by the selection process. Therefore, schemata with a small fitness have a lower chance because their instances do not survive.

These observations can be summarized in the following theorem:

Theorem 5.5.1 *The best chances to reproduce in the next generations belong only to those schematas of GAs that have a large fitness value, a short defining length, and a low order.*

In other words, short, low-order above-average schemata will tend to grow in influence. Thus, the schema theorem roughly states that more fit schemas will gain influence, especially those containing a small number of defined bits preferably close to each other in the binary string.

Some very important conclusions can be drawn from the above theorem:

- Parts of a chromosome that encode similar information should be placed in neighboring positions on the string. This minimizes the probability of disruption during crossover.
- The segments of a string contributing decisively to the fitness value of that string should be efficiently coded so that the best chromosomes can perpetuate.

The schema theorem is a basic concept for the theory of GAs. However, in a way it is incomplete since it fails to consider the assumed positive contributions of crossover and mutation.

5.5.2 The Building Block Hypothesis

Schemata provide a better understanding of GAs. Short, low-order, and more fit schemata are sampled, recombined, and resampled to form strings of improved fitness. Using these schemata, the complexity of the optimization problem can be dramatically reduced: improved strings are formed continuously from the best partial solutions of past samplings.

Schemata with a high fitness value and at the same time of a low order are defined as building blocks [251]. They have several advantages compared with the regular schemata. They are hardly affected by crossover and mutation and have the best chances to be selected. In other words, they represent good partial solutions.

In this regard, we can state the following hypothesis:

Hypothesis: The fitness and thus the performance of the GA can be dramatically improved by combining building blocks.

This reduces the complexity of the posed problem. It is longer necessary to test all possible allele combinations, because highly fit strings can be formed based on the combination of building blocks. In other words, the experience of the last populations is included.

It is important to note that simple GAs depend upon the recombination of building blocks to seek the best points. In cases where the building blocks are misleading because of the coding used or the function itself, the problem may require long computational time to achieve near optimal solutions, and thus the performance of the GA is reduced.

As a closing remark to GAs, it is useful to emphasize the search method employed by evolutionary computing. GAs use a randomized beam search method to find a maximally fit hypothesis. This search method is quite different from the gradient descent method employed by most unsupervised statistical or cognitive methods. The gradient descent method moves smoothly from one hypothesis to a very similar new one, while GAs move quickly to a radically different hypothesis. Thus, the GA search is not as prone to getting stuck in local minima like gradient descent methods. However, there is one drawback with GAs: crowding. Crowding describes the fact that the fittest chromosomes reproduce quickly, and thus copies of those and similar chromosomes dominate the population. This means that the diversity of the population is reduced, and thus learning slows down.

5.6. FEATURE SELECTION BASED ON GENETIC ALGORITHMS

GAs are excellent candidates to reduce the dimension of a feature vector. They are among the few large-scale feature selection algorithms that can be efficiently applied to many classification problems [341].

The search space can be represented by an undirected graph, where the nodes correspond to points in the search space while an existing edge between two nodes shows that the feature set associated to one of these two connected nodes is either a subset or superset of the associated set of the second node. A four-dimensional feature selection lattice is shown in Fig. 5.2.

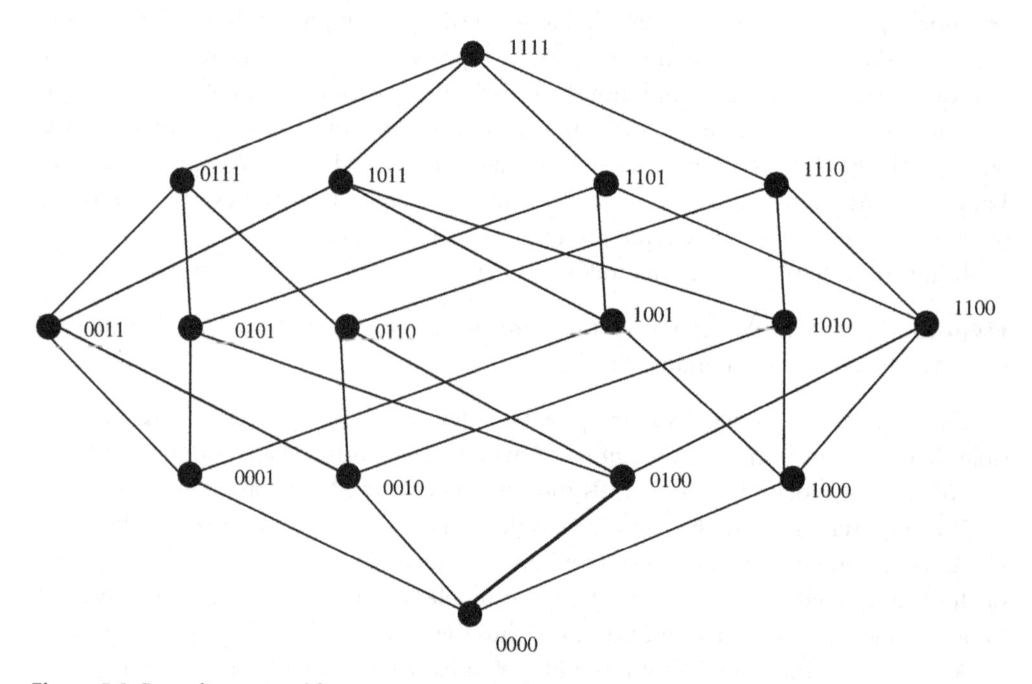

Figure 5.2 Four-dimensional feature selection graph [341].

The top node in the graph contains the whole feature subset while the bottom node contains the empty set. The topology of the graph in Fig. 5.2 reveals its level-structure with the bottom node being at level 0 and the top node being at level 4.

To find the subsets, we have to code each subset as a d-dimensional binary string, $a = (\alpha_1, \ldots, \alpha_d)$, the so-called feature selection vector. In this notation, d represents the initial number of features and α_i the feature selection variable with

$$\alpha_i = \begin{cases} 1, & \text{feature present in subset} \\ 0, & \text{else} \end{cases} \tag{5.12}$$

We immediately see that the search takes place in a binary d-dimensional space.

To better understand the concept of feature selection based on GAs, the monotonicity property for nodes was introduced in [341]. It is said that a function $e(\cdot)$ is monotonic, if for every pair of linked nodes in the lattice, a and b, the following holds:

$$\text{if} \quad l(a) < l(b) \quad \text{then} \quad e(a) > e(b) \tag{5.13}$$

where $l(\cdot)$ represents the level function

$$l(a) = \sum_{i=1}^{n} \alpha_i \tag{5.14}$$

In the following we will review the algorithm described in [341]. The approach determines the feasible subset of features for which the classifier's error rate is below the so-called feasibility threshold. In other words, it is searched for the smallest subset of features among all feasible subsets.

This yields to a constrained optimization problem, which is adapted for GAs by introducing a penalty function $p(e)$:

$$p(e) = \frac{\exp((e - t)/m) - 1}{\exp(1) - 1} \tag{5.15}$$

e is the error rate, t the feasibility threshold, and m a scale factor describing the tolerance margin. $p(e)$ is monotonic with respect to e. It can be shown that $p(e)$ approaches its minimum for $e < t$ and $e \to 0$:

$$p(0) = \frac{\exp((-t)/m) - 1}{\exp(1) - 1} > -\frac{1}{\exp(1) - 1} \tag{5.16}$$

Also, we obtain $p(t) = 0$ and $p(t + m) = 1$. There is a trade-off between penalty function and error rate: A large value for an error rate leads to a penalty function going toward infinity. The score $J(a)$ is obtained by adding the obtained penalty function to the level function

$$J(a) = l(a) + p(e(a)) \tag{5.17}$$

$l(a)$ is the level function representing the number of features in the evaluated binary subset a and represents the cost of extracting features.

Based on the properties of the penalty function described in eq. (5.15), it was shown in [341] that:

1. Feature subsets or chromosomes in GAs parlance producing an error rate below a feasibility threshold receive a negative penalty.
2. The adaptation of feature subsets is based on the associated error rates measured by the score from eq. (5.17).
3. It is possible to achieve a better subset at a certain level than that associated with the next higher level. This can be accomplished if feature subsets whose error rate e fulfills $t < e < t + m$ receive a small penalty.
4. On the other hand, if the error rate $e > t + m$, then the feature subsets receive a high penalty and cannot compete with subsets at the next higher level in the feature selection graph.

It remains now only to define the fitness function and the reproduction rule. Let $\Pi = \{a_1, \ldots, a_n\}$ be the population of feature selection vectors or chromosomes. The goal is to determine a minimum of the score $J(a)$ from eq. (5.17) and thus to find an adequate fitness function,

$$f(a_i) = (1 + \epsilon) \overset{max}{a_j \in \Pi} J(a_j) - J(a_i) \tag{5.18}$$

with ϵ a small positive constant such that $\min f(a_i) > 0$. In other words, even the least fit chromosomes get a chance to reproduce. The reproduction operator selects the best n chromosomes from $\Pi \cup O$.

Examples of how to use the GA in a large-scale feature selection are shown in Chapter 11.

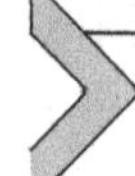

5.7. EXERCISES

1. Consider the function

$$g(x) = 21.5 + x \sin(4\pi x) \tag{5.19}$$

Find the maximum of this function in the interval $[-3 \ldots 12.1]$ by employing a GA.

2. Apply the GA to determine an appropriate set of weights for a $3 \times 2 \times 1$ multilayer perceptron. Encode the weights as bit strings and apply the required genetic operators. Discuss the advantages and disadvantages of using a GA for the weights' learning instead of the traditional backpropagation.

3. Consider the function

$$J = \sum_{i=1}^{N} d(\mathbf{x}, C_{\mathbf{v}_i}) \tag{5.20}$$

where $d(\mathbf{x}, C_{\mathbf{v}_i})$ describes the distance between an input vector $\mathbf{x}$ and a set using no representatives for the set. Propose a coding of the solutions for a GA that uses this function. Discuss the advantages and disadvantages of this coding.

CHAPTER SIX

Statistical and Syntactic Pattern Recognition

Contents

Pattern Recognition and Signal Analysis in Medical Imaging
http://dx.doi.org/10.1016/B978-0-12-409545-8.00006-6

6.1. INTRODUCTION

Modern classification paradigms such as neural networks, genetic algorithms, and neuro-fuzzy methods have gained an increasing importance over the past decade. However, the traditional and noncognitive classification methods, even though being challenged by these new approaches, are still in place, and provide sometimes a better solution than the modern ones.

Noncognitive classification paradigms encompass statistical and structural (syntactic) methods. If there is an underlying and quantifiable statistical basis for the generation of patterns, then we refer to statistical pattern recognition. Otherwise, the fundamental information is provided by the underlying pattern structure, and we refer to structural or syntactic pattern recognition. The latter classification paradigm is less popular than the statistical classification methods.

This chapter gives an overview about the most important approaches in statistical and syntactic pattern recognition and their application to biomedical imaging. Parametric and nonparametric estimation methods and binary decision trees form the basis for most classification problems related to bioimaging while grammatical inference and graphical methods are the basic classification paradigms in syntactic pattern recognition. The chapter also reviews the diagnostic accuracy of classification measured by ROC curves, and presents application examples based on statistical classification methods.

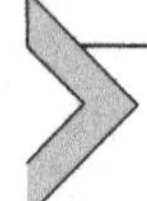

6.2. LEARNING PARADIGMS IN STATISTICAL PATTERN RECOGNITION

Pattern recognition techniques are widely used for a lot of miscellaneous practical problems. It can be either the statistical nature of pattern generation of special interest, or the relevant information which is captured by the pattern structure. This chapter deals with the design of a classifier in a pattern recognition system. There are several classification paradigms which lead to a reasonable solution: statistical, syntactic, or neural classification. Statistical and neural pattern recognition can be further differentiated in supervised and unsupervised.

6.2.1 Supervised Pattern Recognition

Here we assume that a set of training data is available, and the classifier is designed by exploiting this a priori known information. The most relevant techniques which belong to this group are [84]:

- *Maximum likelihood method:* Estimates unknown parameters using a set of known feature vectors in each class.
- *Bayes method:* It assumes a known a priori probability and minimizes the classification error probability. Also the class-conditional probability density functions describing the distribution of the feature vectors in each class need to be known. If not known, they can be estimated from the available training data.
- *Minimum distance classifier:* Classifies an input vector based on its distance to the learned prototypes.
- *Entropy criteria:* Classify an input vector based on the minimization of the randomness of this vector.

6.2.2 Unsupervised Pattern Recognition

Unsupervised pattern recognition concerns another type of pattern recognition tasks for which training data, of known class labels, are not available. Here, we deal with a given set of feature vectors and the goal is to find out the underlying similarities and cluster (group) "similar" vector together. Classifiers can be designed as either linear or nonlinear depending on the data separability.

The most relevant techniques which belong to this group are [84]:

- *Isodata or k-means or c-means clustering:* The goal is to achieve a close partitioning of the data space. The number of classes and the initial values of the class centers need to be a priori known. Learning is an iterative process which adapts the class centers according to the training data.
- *Vector quantization:* The input data are mapped onto a given number of code vectors (prototypes) which together form the code book. The goal is to achieve a code book with least possible distortion.
- *Cluster swapping:* Is applied mostly to a large data space and aims to avoid a suboptimal class partitioning. Based on a predefined criterion function, the classification as a whole is supervised during each iteration. This avoids that two feature vectors of different classes but erroneously assigned to the same class belong forever to this class.
- *Hierarchical clustering:* They produce instead of a single clustering a hierarchy of nested clustering. More specifically, these algorithms have as many steps as the number of data vectors. At each step a new clustering is obtained based on the clustering produced at the previous step.

  There are two main categories of hierarchical algorithms: agglomerative and divisive hierarchical algorithms. Agglomerative clustering is a bottom-up procedure which starts with as many clusters as the number of feature vectors. The cluster number is reduced by merging clusters describing the same classes. In this context, the dendogram is a useful tool for representing the sequence of clustering produced by an agglomerative algorithm. This procedure continues until the final clustering which contains as a single set the set of data. The divisive clustering is a top-down procedure and follows the inverse path of the agglomerative clustering. In the first step, the clustering consists of a single set. In the following they increase the number of clusters.

Statistical classification approaches are built upon probabilistic arguments stemming from the statistical nature of generated features. The goal is to design a classifier that classifies a pattern into the most probable of the classes.

Decision rules necessary for performing the classification task can be formulated in several ways:

- An a priori known class-related probability $P(\omega_i)$ is transformed into an a posteriori probability $P(\omega_i|\mathbf{x})$. $\mathbf{x}$ is an n-dimensional feature vector.
- Choice of a classification error rate and of a decision rule.

In the following, we will consider the standard classification problem. An unknown feature vector $\mathbf{x}$ has to be assigned to one of the M classes, $\omega_1, \omega_2, \ldots, \omega_M$. M conditional or a posteriori probabilities $P(\omega_i|\mathbf{x})$ with $i = 1, 2, \ldots, M$ can be defined. Each of them describes the probability that an unknown pattern belongs to a given class ω_i. The goal of the classification is to determine the maximum of these M probabilities, or equivalently, the maximum of an appropriately defined function of them.

6.3. PARAMETRIC ESTIMATION METHODS

Statistical pattern recognition assumes the knowledge of $p(\mathbf{x}_i|\omega_i)$ and $P(\omega_i)$ for each class. In addition, the number of classes needs to be known a priori.

Let's also assume that we have a set of training samples representative of the type of features and underlying classes, with each labeled as to its correct class. This yields a learning problem. When the form of the densities is known, we are faced with a parameter estimation problem.

This section reviews techniques for estimating probability density functions (pdfs) based on the available experimental data which are described by the feature vectors corresponding to the patterns of the training set.

Every pdf can be characterized by a specific parameter set. One of the most frequently used pdfs is the Gaussian or normal density function. This pdf is completely characterized by the mean value and covariance matrix which have to be estimated for each class ω_i from the training data set. This technique is known under the name of parametric estimation.

There are two approaches to parameter estimation known in the literature. These are:

- *Maximum likelihood estimation:* It is assumed the parameters are fixed, but unknown. This approach seeks the "best" parameter set estimate that maximizes the probability of obtaining the (given) training set.
- *Bayesian estimation:* It uses the training set to update the training-set conditioned density function of the unknown parameters. The training set acts as "observations" and allows the conversion of the a priori information into an a posteriori density.

Both techniques yield an uncertainty in the parameter estimation. While the Bayesian approach determines a density that approximates an impulse, the maximum likelihood approach determines parameter estimates that maximize a likelihood function.

6.3.1 Bayes Decision Theory

Bayes decision theory represents a fundamental statistical approach to the problem of pattern classification. This technique is based on the assumption that the decision problem is formulated in probabilistic terms, and that all relevant probability values are given. In this section, we develop the fundamentals of this theory.

A simple introduction to this approach can be given by an example which focuses on the two-class case ω_1, ω_2. The a priori probabilities $P(\omega_1)$ and $P(\omega_2)$ are assumed to be known since they can be easily determined from the available data set. Also known are the pdfs $p(\mathbf{x}_i|\omega_i)$, $i = 1, 2$. $p(\mathbf{x}_i|\omega_i)$ is also known under the name of the likelihood function of ω_i with respect to $\mathbf{x}$.

Recalling the Bayes rule, we have

$$P(\omega_i|\mathbf{x}) = \frac{p(\mathbf{x}|\omega_i)P(\omega_i)}{p(\mathbf{x})} \tag{6.1}$$

where $p(\mathbf{x})$ is the pdf of $\mathbf{x}$, and for which it holds

$$p(\mathbf{x}) = \sum_{i=1}^{2} p(\mathbf{x}|\omega_i)P(\omega_i) \tag{6.2}$$

The Bayes classification rule can now be stated for the two-class case ω_1, ω_2

$$\begin{array}{lll} \text{If} & P(\omega_1|\mathbf{x}) > P(\omega_2|\mathbf{x}), & \mathbf{x} \text{ is assigned to } \omega_1 \\ \text{If} & P(\omega_1|\mathbf{x}) < P(\omega_2|\mathbf{x}), & \mathbf{x} \text{ is assigned to } \omega_2 \end{array} \tag{6.3}$$

We immediately can conclude from above that a feature vector can be either assigned to one class or the other. Equivalently, we now can write

$$p(\mathbf{x}|\omega_1)P(\omega_1) \gtrless p(\mathbf{x}|\omega_2)P(\omega_2) \tag{6.4}$$

This corresponds to determining the maximum of the conditional pdfs evaluated at $\mathbf{x}$. Figure 6.1 visualizes two equiprobable classes and the conditional pdfs $p(x|\omega_i)$, $i = 1, 2$ as function of x. The dotted line at x_0 corresponds to a threshold splitting the one-dimensional feature space into two regions R_1 and R_2. Based on the Bayes classification rule, all values of $x \in R_1$ are assigned to class ω_1, while all values $x \in R_2$ are assigned to class ω_2.

The probability of the decision error is given by

$$P_e = \int_{-\infty}^{x_0} p(x|\omega_2)dx + \int_{x_0}^{+\infty} p(x|\omega_1)dx \tag{6.5}$$

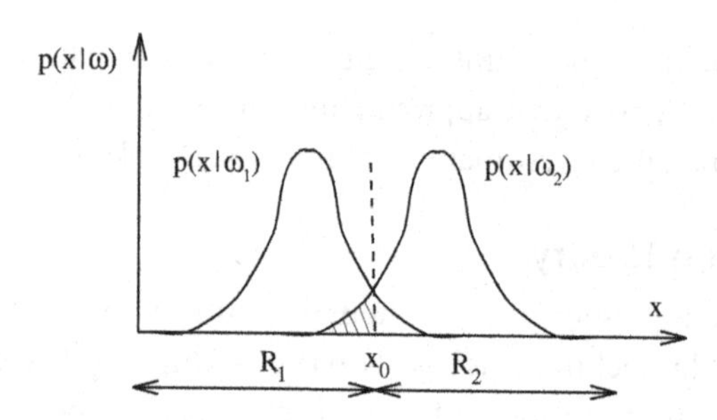

Figure 6.1 Two equiprobable classes and the corresponding regions R_1 and R_2.

The Bayes classification rule achieves a minimal error probability. In [84] it was shown that the classification error is minimal, if the partition of the feature set into the two regions R_1 and R_2 is chosen such that

$$\begin{aligned} R_1: &\quad P(\omega_1|\mathbf{x}) > P(\omega_2|\mathbf{x}) \\ R_2: &\quad P(\omega_2|\mathbf{x}) > P(\omega_1|\mathbf{x}) \end{aligned} \tag{6.6}$$

The generalization for M classes $\omega_1, \omega_2, \ldots, \omega_M$ is very simple. A feature vector $\mathbf{x}$ is assigned to class ω_i if

$$P(\omega_i|\mathbf{x}) > P(\omega_j|\mathbf{x}) \quad \forall j \neq i \tag{6.7}$$

Every time we assign an object to a class, we risk making an error. In multiclass problems, some misclassifications can have more serious repercussions than others. A quantitative way to measure this is given by a so-called cost function. Let $L(i,j)$ be the cost (or "loss") of assigning an object to class i when it really belongs to class j.

From the above, we see that a different classification possibility is achieved by defining a so-called cost term $L(i,j)$ with $i,j = 1, 2, \ldots, M$. The penalty term is equal to zero, $L(i,j) = 0$, if the feature vector $\mathbf{x}$ is correctly assigned to its class, and larger than zero, $L(i,j) > 0$, if assigned to class ω_j instead of the correct class ω_i. In other words, there is only loss if misclassification occurs.

The conditional loss term $R_i(\mathbf{x})$ with respect to the class assignment of $\mathbf{x}$ is

$$R_i(\mathbf{x}) = \sum_{j=1}^{M} L(i,j)P(\omega_j|\mathbf{x}) \tag{6.8}$$

or equivalently,

$$R_i(\mathbf{x}) = \sum_{j=1}^{M} L(i,j)p(\mathbf{x}|\omega_j)P(\omega_j) \tag{6.9}$$

For practical applications we choose $L(i,j) = 0$ for $i = j$, and $L(i,j) = 1$ for $i \neq j$.

Thus, given the feature vector, there is a certain risk involved in assigning the object to any group.

Based on the above definitions, we obtain a slightly changed Bayes classification rule: a feature vector $\mathbf{x}$ is assigned to a class ω_i for which $R_i(\mathbf{x})$ is minimal.

6.3.2 Discriminant Functions

There are many different possibilities to define pattern classifiers. One way, which can be considered as a canonical form for classifiers, are the so-called discriminant functions. In the M class case, they are used to partition the feature space.

Also, in many situations it is simpler instead to work directly with probabilities, to deal with an equivalent function of them, for example, $g_i(\mathbf{x}) = f(P(\omega_i|\mathbf{x}))$, where $f(\cdot)$ is a monotonically increasing function. $g_i(\mathbf{x})$ is called the discriminant function.

Based on the above relationships, we get for eq. (6.7) the following equivalent representation or simply, decision rule

$$\mathbf{x} \text{ belongs to class } \omega_i \quad \text{if} \quad g_i(\mathbf{x}) > g_j(\mathbf{x}) \quad \forall j \neq i \tag{6.10}$$

Assuming two regions R_1 and R_2 are neighboring, then they can be separated by a hyperplane in the multidimensional space. The equation defining the separation plane is given by

$$g_{ij}(\mathbf{x}) = g_i(\mathbf{x}) - g_j(\mathbf{x}) = 0, \quad i,j = 1,2,\ldots,M \quad i \neq j \tag{6.11}$$

The discriminant functions are extremely useful when dealing with Gaussian pdf. Let's consider the following normal density function

$$p(\mathbf{x}|\omega_i) = \frac{1}{(2\pi)^{n/2}|\Sigma_i|^{1/2}} \exp\left(-\frac{1}{2}(\mathbf{x}-\mu_i)^T\Sigma_i^{-1}(\mathbf{x}-\mu_i)\right), \quad i = 1,\ldots,M \tag{6.12}$$

where μ_i is the mean value and Σ_i the covariance matrix of the class ω_i. The covariance matrix Σ_i is given by

$$\Sigma_i = E\left[(\mathbf{x}-\mu_i)(\mathbf{x}-\mu_i)^T\right] \tag{6.13}$$

In this case, we choose a monotonic logarithmic discriminant function $\ln(\cdot)$

$$g_i(\mathbf{x}) = \ln\left(p(\mathbf{x}|\omega_i)P(\omega_i)\right) = \ln\ p(\mathbf{x}|\omega_i) + \ln\ P(\omega_i) \tag{6.14}$$

and we obtain such for the normal density function

$$g_i(\mathbf{x}) = -\frac{1}{2}(\mathbf{x}-\mu_i)^T\Sigma_i^{-1}(\mathbf{x}-\mu_i) + \ln\ P(\omega_i) + c_i \tag{6.15}$$

where $c_i = -(n/2)\ln 2\pi - (1/2)\ln|\Sigma_i|$ is a constant.

6.3.3 Bayesian Classification for Tumor Detection

When dealing with tumor detection, we are considering only two pattern classes: abnormal and normal. In other words, two discriminant functions have to be considered:

$g_{abnormal}(\mathbf{x})$ and $g_{normal}(\mathbf{x})$. A pattern vector is assigned to the abnormal class, if $g_{abnormal}(\mathbf{x}) > g_{normal}(\mathbf{x})$.

Assuming the a priori probabilities are equal for each class, we obtain from eq. (6.15) for tumor detection the following discriminant function

$$g(\mathbf{x}) = -\frac{1}{2}(\mathbf{x}-\mu)^T\Sigma^{-1}(\mathbf{x}-\mu) + \log|\Sigma| \tag{6.16}$$

If now each of the two classes has the same covariance matrix, they can be separated by a hyperplane, and hence the classifier is referred to as a linear classifier. If the covariance matrices are unequal, then the decision surfaces are hyperquadrics and the classifier is referred to as a quadratic classifier.

6.3.4 Minimum Distance Classifiers

Here we assume that two classes are equiprobable with the same covariance matrix $\Sigma_1 = \Sigma_2 = \Sigma$. This simplifies eq. (6.15) to

$$g_i(\mathbf{x}) = -\frac{1}{2}(\mathbf{x}-\mu_i)^T\Sigma^{-1}(\mathbf{x}-\mu_i) \tag{6.17}$$

considering the constant terms as being neglected.

Based on the covariance matrix shape, we can distinguish between two types of minimum distance classifiers:

- $\Sigma = \sigma^2\mathbf{I}$: In this case, the maximum of $g_i(\mathbf{x})$ corresponds to the minimum of

$$\text{Euclidean distance:} \quad d_e = ||\mathbf{x}-\mu_i|| \tag{6.18}$$

Thus, the feature vectors are assigned to classes according to the Euclidean distance between the class mean points and these vectors. Figure 6.2(a) visualizes the curves of equal distance $d_e = c$ to the stored class mean points. In this case, they are hyperspheres of radius c.

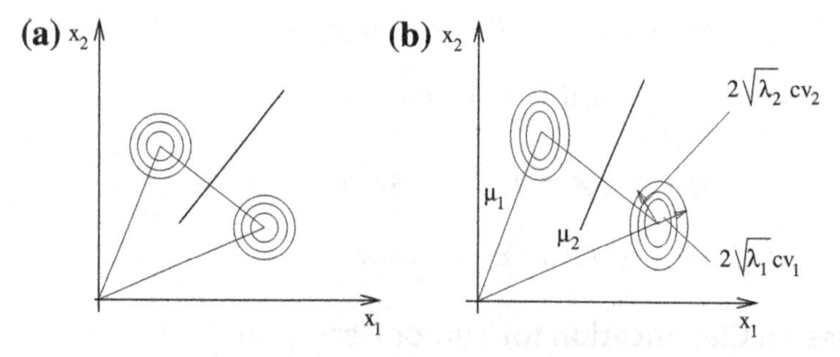

Figure 6.2 Curves of equal (a) Euclidean distance and (b) equal Mahalanobis distance from the mean values of each class.

- Nondiagonal covariance matrix: In this case, the maximum of $g_i(\mathbf{x})$ is equivalent to the minimum of the

$$\text{Mahalanobis distance:} \quad d_m = \left((\mathbf{x} - \mu_i)^T \Sigma^{-1} (\mathbf{x} - \mu_i)\right)^{1/2} \tag{6.19}$$

  The curves of constant distance $d_m = c$ represent hyperellipses.
  The covariance matrix Σ is a symmetric function. All points having the same distance from a class mean point are located on an ellipse. The center of mass of the ellipse is at μ_i, and the principal axes are aligned with the corresponding eigenvectors and have a length of $2\sqrt{\lambda_k c}$, as we can see from Fig. 6.2(b).

Example 6.3.1 Consider a two-class, two-dimensional classification problem. The feature vectors are generated by Gaussian pdfs with the same covariance matrix $\Sigma_1 = \Sigma_2 = \Sigma$

$$\Sigma = \begin{bmatrix} 1,1 & 0,3 \\ 0,3 & 1,9 \end{bmatrix} \tag{6.20}$$

and the mean vectors given by $\mu_1 = [1, 1]^T$ and respectively by $\mu_2 = [4, 4]^T$.

The vector $[2, 2]^T$ should be assigned to one of the two given classes according to the Bayesian classifier.

By determining the Mahalanobis distance of $[2, 2]^T$ from the two mean vectors, we obtain

$$\begin{aligned} d_m^2(\mu_1, \mathbf{x}) &= (\mathbf{x} - \mu_1)^T \Sigma^{-1} (\mathbf{x} - \mu_1) \\ &= [1, 1] \begin{bmatrix} 0.95 & -0.15 \\ -0.15 & 0.55 \end{bmatrix} \begin{bmatrix} 1 \\ 1 \end{bmatrix} = 1.5 \end{aligned}$$

and

$$\begin{aligned} d_m^2(\mu_2, \mathbf{x}) &= (\mathbf{x} - \mu_2)^T \Sigma^{-1} (\mathbf{x} - \mu_2) \\ &= [-2, -2] \begin{bmatrix} 0.95 & -0.15 \\ -0.15 & 0.55 \end{bmatrix} \begin{bmatrix} -2 \\ -2 \end{bmatrix} = 4.8 \end{aligned}$$

Thus the feature vector is assigned to the class ω_1 based on the minimal distance.

6.3.5 Estimation of Unknown Probability Density Functions

So far, we have assumed that the pdfs are known. However, in many practical problems this is not always the case, and they have to be estimated from the available data.

There are many elegant solutions known to this problem. Sometimes the type of the pdf is known (Gaussian, Laplacian), and we have only to estimate some parameters like mean value and variance. Or the opposite is the case, that we have knowledge of certain statistical parameters but we have no information about the pdf type. Depending on the available information, different approaches are known in the literature to determine the missing information. This section will give a review about the maximum likelihood parameter estimation and the maximum entropy estimation as a solution to the above-stated problem.

6.3.5.1 Maximum Likelihood Parameter Estimation

Let's consider the following classification task: feature vectors of a given distribution $p(\mathbf{x}|\omega_i), i = 1, 2, \ldots, M$ have to be assigned to the corresponding class out of M classes. Further, we are assuming that these likelihood functions are given in a parametric form and that the corresponding parameters form the unknown vectors θ. This dependence is reflected by writing $p(\mathbf{x}|\omega_i; \theta_i)$. The goal is to estimate the unknown parameters by using a set of known feature vectors for each class.

The joint pdf is given by $p(X; \theta)$ with $X = \{\mathbf{x_1}, \mathbf{x_2}, \ldots, \mathbf{x}_N\}$ being the set of the sample vectors.

$$p(X; \theta) = p(\mathbf{x_1}, \mathbf{x_2}, \ldots, \mathbf{x}_N; \theta) = \prod_{k=1}^{N} p(\mathbf{x_k}; \theta) \tag{6.21}$$

We are further assuming that the sample vectors are statistically independent.

The resulting function of θ is known under the name of likelihood function with respect to X. The estimation of θ is based on the maximum likelihood (ML) method and determines the maximum value of the likelihood function, that is,

$$\widehat{\theta}_{ML} = \arg \overset{max}{\theta} \prod_{k=1}^{N} p(\mathbf{x_k}; \theta) \tag{6.22}$$

The necessary condition that $\widehat{\theta}_{ML}$ is a maximum, is that the gradient of the likelihood function equals zero

$$\frac{\partial \prod_{k=1}^{N} p(\mathbf{x_k}; \theta)}{\partial \theta} = \mathbf{0} \tag{6.23}$$

A computational simplicity can be achieved by considering the monotonicity of the logarithmic function. A new function, the so-called loglikelihood function, can be defined

$$L(\theta) \equiv \ln \prod_{k=1}^{N} p(\mathbf{x_k}; \theta) \tag{6.24}$$

and based on this we obtain from eq. (6.23) the equivalent expression

$$\frac{\partial L(\theta)}{\partial \theta} = \sum_{k=1}^{N} \frac{\partial \ln p(\mathbf{x_k}; \theta)}{\partial \theta} = \sum_{k=1}^{N} \frac{1}{p(\mathbf{x_k}; \theta)} \frac{\partial p(\mathbf{x_k}; \theta)}{\partial \theta} = \mathbf{0} \tag{6.25}$$

In other words, we are determining now the maximum of the loglikelihood function.

Example 6.3.2 Let's assume that the sample vectors $\mathbf{x_1}, \mathbf{x_2}, \ldots, \mathbf{x_N}$ have a Gaussian pdf. Their covariance matrix is known, while the mean vector is unknown, and has to be estimated. The pdf takes the form

$$p(\mathbf{x_k};\mu) = \frac{1}{(2\pi)^{n/2}|\Sigma|^{1/2}} \exp\left(-\frac{1}{2}(\mathbf{x_k}-\mu)^T\Sigma^{-1}(\mathbf{x_k}-\mu)\right), \quad i=1,\ldots,M \tag{6.26}$$

For the N available samples, we have to determine the loglikelihood function

$$L(\mu) \equiv \ln \prod_{k=1}^{N} p(\mathbf{x_k};\mu) = -\frac{N}{2}\ln\left((2\pi)^n|\Sigma|\right) - \frac{1}{2}\sum_{k=1}^{N}(\mathbf{x_k}-\mu)^T\Sigma^{-1}(\mathbf{x_k}-\mu) \tag{6.27}$$

Taking the gradient of $L(\mu)$ with respect to μ, we get

$$\frac{\partial L(\mu)}{\partial \mu} \equiv \begin{bmatrix} \frac{\partial L}{\partial \mu_1} \\ \frac{\partial L}{\partial \mu_2} \\ \vdots \\ \frac{\partial L}{\partial \mu_l} \end{bmatrix} = \sum_{k=1}^{N}\Sigma^{-1}(\mathbf{x_k}-\mu) = \mathbf{0} \tag{6.28}$$

For the mean value we obtain

$$\widehat{\mu}_{ML} = \frac{1}{N}\sum_{k=1}^{N}\mathbf{x_k} \tag{6.29}$$

We immediately see that the ML estimate for normal densities is the mean of all sample vectors. In other words, the maximum likelihood estimate for the unknown population mean is just the arithmetic average of the samples—the sample mean.

6.3.6 Maximum Entropy Estimation

The definition of entropy stems from Shannon's information theory and gives in pattern recognition applications a measure of the randomness of the feature vectors. The entropy H for the density function $p(\mathbf{x})$ is given by

$$H = -\int_{\mathbf{X}} p(\mathbf{x}) \ln p(\mathbf{x}) d\mathbf{x} \tag{6.30}$$

Further, we assume that $p(\mathbf{x})$ is not known, and that some of the related constraints (mean value, variance, etc.) are known.

Example 6.3.3 Let's assume that the random variable x is nonzero for $a \leq x \leq b$ and is equal to zero outside of the interval. We want to determine the maximum entropy of its pdf.

In this sense, we have to determine the maximum of eq. (6.30) under the condition

$$\int_a^b p(x)dx = 1 \tag{6.31}$$

Employing Lagrange multipliers, we obtain an equivalent representation

$$H_L = -\int_a^b p(x)(\ln p(x) - \lambda)dx \tag{6.32}$$

The derivative of H_L with respect to $p(x)$ is given by

$$\frac{\partial H_L}{\partial p(x)} = -\int_a^b \{\ln p(x) - \lambda) + 1\}dx \tag{6.33}$$

Setting the above equation equal to zero, we have

$$\widehat{p}(x) = \exp(\lambda - 1) \tag{6.34}$$

From eq. (6.31) we obtain $\exp(\lambda - 1) = \frac{1}{b-a}$. The estimated value of $p(x)$ is thus

$$\widehat{p}(x) = \begin{cases} \frac{1}{b-a}, & a \leq x \leq b \\ 0, & \text{else} \end{cases} \tag{6.35}$$

In other words, the maximum entropy estimate of the unknown pdf $p(x)$ is the uniform distribution.

6.4. NONPARAMETRIC ESTIMATION METHODS

For classification tasks we need to estimate class-related pdfs since they determine the classifier's structure. So far, each pdf was characterized by a certain parameter set. For Gaussian distributions, the covariance and the mean value are needed, and they are estimated from the sample data. Techniques described in the last section fall under parametric estimation methods.

This section will deal with nonparametric estimation techniques. There is no information available about class-related pdfs, and so they have to be estimated directly from the data set. There are many types of nonparametric techniques for pattern recognition. One procedure is based on estimating the density functions $p(\mathbf{x}|\omega_i)$ from sample patterns. If the achieved results are good, they can be included in the optimal classifier. Another approach estimates directly the a posteriori probabilities $P(\omega_i|\mathbf{x})$, and is closely related to nonparametric decision procedures. They bypass probability estimation and go directly to decision functions.

The following nonparametric estimation techniques will be reviewed:

- Histogram method.
- Parzen windows.
- *k* nearest neighbor.
- Potential function.

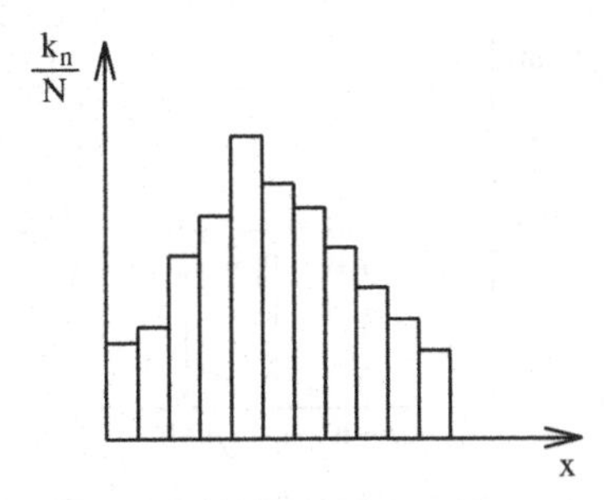

Figure 6.3 Histogram method applied to pdf estimation.

6.4.1 Histogram Method

This is one of the simplest and most popular pdf estimation methods. However, to achieve a robust estimation we need a relatively large data set. For the simple, one-dimensional case, the x-axis is divided into successive bins of length h, and then the probability of a sample x being located in a bin is estimated for each of these bins. Let N be the total number of samples and k_N of these samples are in a certain bin, then the probability of the corresponding class is given by $P \approx k_N/N$. Figure 6.3 illustrates this procedure.

6.4.2 Parzen Windows

One of the most important nonparametric methods for pdf estimation is "Parzen windows" [243,297]. For a better understanding, we will take the simple one-dimensional case. The goal is to estimate the pdf $p(x)$ at the point x. This requires to determine the number of the samples N_h within the interval $[x-h, x+h]$, and then to divide by the total number of all feature vectors M and by the interval length $2h$. Based on the described procedure, we will obtain an estimate for the pdf at x

$$\widehat{p}(\mathbf{x}) = \frac{N_h(x)}{2hM} \tag{6.36}$$

As a support function K_h, we will choose

$$K_h = \begin{cases} 0,5 & : \quad |m| \leq |1| \\ 0 & : \quad |m| > |1| \end{cases} \tag{6.37}$$

From eq. (6.36) we get

$$\widehat{p}(x) = \frac{1}{hM} \sum_{i=1}^{M} K\left(\frac{x - m_i}{h}\right) \tag{6.38}$$

with the ith component of the sum being equal to zero if m_i falls outside the interval $[x - h, x + h]$. This leads to

$$\gamma(x, m) = \frac{1}{h} K\left(\frac{x - m}{h}\right) \tag{6.39}$$

as it can be seen from Fig. 6.4.

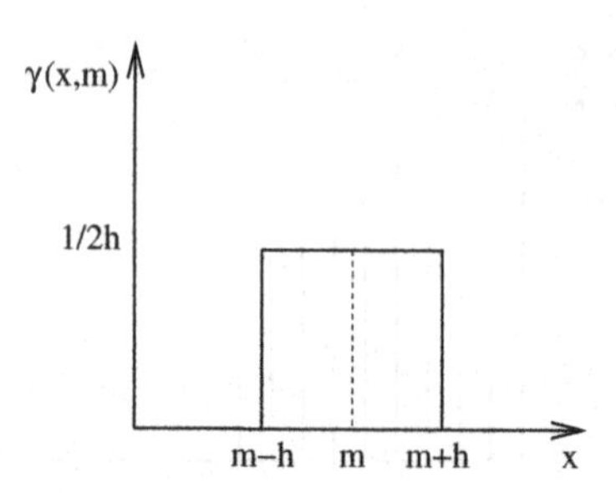

Figure 6.4 Clustering process of a two-dimensional vector table. The maximum number of vectors in one cluster was set to 7.

If $\widehat{p}(x)$ is considered to be a function corresponding to the number of samples, we obtain thus

$$\widehat{p}(x) = \widehat{p}(x, M) \tag{6.40}$$

Parzen showed that the estimate $\widehat{p}$ with $M \to \infty$ is bias free, if $h = h(M)$ and

$$\lim_{M \to \infty} h(M) = 0 \tag{6.41}$$

In practice, where only a finite number of samples is possible, a right compromise between M and h has to be made. The choice of h is crucial, and it is recommended to start with an initial estimate of h and then modify it iteratively to minimize the misclassification error. Theoretically, a large M is necessary for acceptable performance. But in praxis, a large number of data points increases the computational complexity unnecessarily.

Typical choices for the function $K(m)$ are

$$K(m) = (2\pi)^{\frac{-1}{2}} e^{-\frac{m^2}{2}} \tag{6.42}$$

$$K(m) = \frac{1}{\pi(1 + m^2)} \tag{6.43}$$

or

$$K(m) = \begin{cases} 1 - |m| & : \quad |m| \leq 1 \\ 0 & : \quad |m| > 1 \end{cases} \tag{6.44}$$

6.4.3 *k* Nearest Neighbor Density Estimation

In the Parzen windows estimation the length of the interval is fixed while the number of samples falling inside an interval varies from point to point. For the k nearest neighbor density estimation exactly the reverse holds: the number of samples k falling inside an interval is fixed while the interval length around x will be varied each time, to include the same number of samples k. We can generalize for the n-dimensional case: in low density areas the hypervolume $V(\mathbf{x})$ is large while in high density areas it is small.

The estimation rule can be given now as

$$\hat{p}(\mathbf{x}) = \frac{k}{NV(\mathbf{x})} \tag{6.45}$$

and reflects the dependence of the volume $V(\mathbf{x})$. N represents the total number of samples while k describes the number of points falling inside the volume $V(\mathbf{x})$.

This procedure can be very easily elucidated based on a two-class classification task: an unknown feature vector $\mathbf{x}$ should be assigned to one of the two classes ω_1 or ω_2. The decision is made by computing its Euclidean distance d from all the training vectors belonging to various classes. With r_1 we denote the radius of the hypersphere centered at $\mathbf{x}$ that contains k points from class ω_1 while r_2 is the corresponding radius of the hypersphere belonging to class ω_2. V_1 and V_2 are the two hypersphere volumes.

The k nearest neighbor classification rule in case of two classes ω_1 and respectively ω_2 can now be stated

$$\text{Assign } \mathbf{x} \text{ to class } \omega_1(\omega_2) \quad \text{if} \quad \frac{V_2}{V_1} > (<) \frac{N_1}{N_2} \frac{P(\omega_2)}{P(\omega_1)} \tag{6.46}$$

If we adopt the Mahalanobis distance instead of the Euclidean distance, then we will have hyperellipsoids instead of hyperspheres.

6.4.3.1 *k Nearest Neighbor for Tumor Detection*

The classification rule has to be slightly modified for abnormal and normal classes. A new pattern vector is assigned to a particular class, if at least l of the k nearest neighbors are in that particular class. This is especially attractive for applications where the penalty for misclassifying one class is much greater than that associated to the other class. Also, if there are more samples from one class than from the other class, in other words if we deal with an unbalanced training set, a majority vote might not be useful. For example, consider a training set with 50 sample vectors for class one, and 500 sample vectors for class two. In this case, we get more accurate classifications if $l \approx k/10$ for class one, because there are ten times more sample vectors in class two than one. Thus, a new pattern vector is assigned to class one if approximately 10% of the k nearest neighbors belong to class one.

6.4.4 Potential Functions

Potential functions represent a useful method for estimating an unknown pdf $\hat{p}(\mathbf{x})$ from the available feature vectors [13] and [243].

The estimated pdf is given by a superposition of potential functions $\gamma(\mathbf{x}, \mathbf{m})$

$$\hat{p}(\mathbf{x}) = \frac{1}{M} \sum_{j=1}^{M} \gamma(\mathbf{x}, \mathbf{m}_j) \tag{6.47}$$

Figure 6.5(a) and (b) illustrate an example of a potential function for the one-dimensional and two-dimensional case.

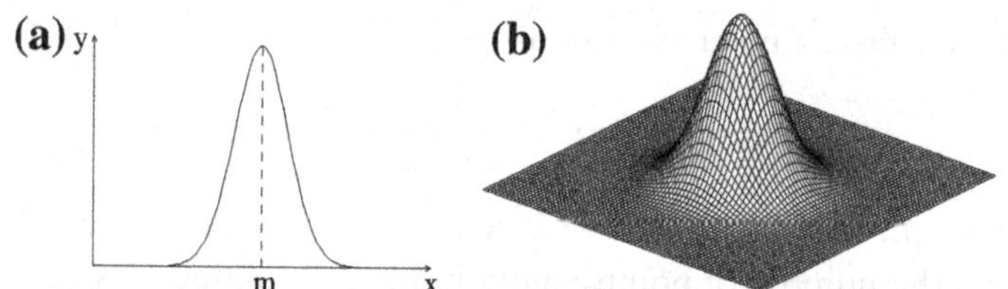

Figure 6.5 Potential functions for the (a) one-dimensional and (b) two-dimensional case.

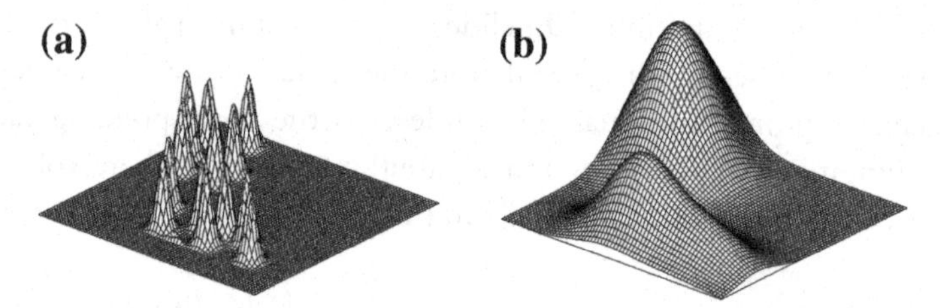

Figure 6.6 Importance of the width of the potential function (a) sharper surfaces and (b) smoother surfaces.

A possible potential function is

$$\gamma(\mathbf{x}, \mathbf{m}) = \frac{1}{(2\pi)^{\frac{n}{2}}\sigma^n} e^{\left(\frac{-||\mathbf{x}-\mathbf{m}||^2}{2\sigma^2}\right)} \tag{6.48}$$

where $||\mathbf{x}||$ is a norm in the n-dimensional space. The potential function defines a distance measure (Mahalanobis distance) between two feature vectors $\mathbf{x}$ and $\mathbf{m}$.

Equation (6.47) describes the complete algorithm for a prespecified potential function. $\widehat{p}(\mathbf{x})$ can be estimated for every $\mathbf{x}$ directly from eq. (6.47).

The choice of the potential function is not so trivial since the width of the potential function plays herein an important role. The smaller its width the more it peaks and the higher it is. This means it considers only feature vectors in its immediate neighborhood. Larger widths produce a potential function of a smoother shape, as it can be seen from Fig. 6.6.

The choice of a certain variance has a considerable importance on the overlapping degree of neighboring potential functions. This is especially critical when potential functions describing feature vectors belonging to different classes overlap.

There are several possible ways of determining σ [19]:

1. Let's assume that within the distance σ from a specific feature vector there are L other feature vectors. The average value describes the distance $D_L(\mathbf{m})$ from the feature vector to the Lth feature vector:

$$\sigma = \frac{1}{M} \sum_{i=1}^{M} D_L(\mathbf{m}_i) \tag{6.49}$$

To determine L, we have to take into account the distribution of the feature vectors. In many practical problems, $L = 10$ is a good choice.

2. σ can also be chosen as a multiple of the minimal distance between two feature vectors. This is necessary to achieve a certain overlapping degree. In [19] it is recommended to set the multiple equal to 4.

The potential function has the following general properties [243]:

1. $\gamma(\mathbf{x}, \mathbf{m})$ has its maximum at $\mathbf{x} = \mathbf{m}$.
2. $\gamma(\mathbf{x}, \mathbf{m})$ goes asymptotically toward zero if the distance between the two feature vectors is very large. This is of special importance for multimodal distributions.[1]
3. $\gamma(\mathbf{x}, \mathbf{m})$ is a continuous function decreasing monotonically on both sides from the maximum.
4. If $\gamma(\mathbf{x}_1, \mathbf{m}_1) = \gamma(\mathbf{x}_2, \mathbf{m}_1)$, then the feature vectors $\mathbf{x}_1$ and $\mathbf{x}_2$ have the same similarity degree with respect to $\mathbf{m}_1$.

There are several types of known potential functions: besides the above-mentioned unimodal or multimodal normal distributions, potential functions built from orthonormal functions are also of interest [243]. They have the following form

$$\gamma(\mathbf{x}, \mathbf{m}) = \sum_{i=1}^{R} \lambda_i^2 \Phi_i(\mathbf{x}) \Phi_i(\mathbf{m}) \tag{6.50}$$

where $\Phi_i(\mathbf{x})$ is an orthonormal function and λ is a constant. Orthonormal functions fulfill

$$\int \Phi_i(\mathbf{x}) \Phi_j(\mathbf{x}) dx = \begin{cases} 1 & i = j \\ 0 & \text{else} \end{cases} \tag{6.51}$$

with $\lambda_i = 1$.

It is easy to see that they are potential functions because

$$\sum_{i=1}^{\infty} \Phi_i(\mathbf{x}) \Phi_i(\mathbf{m}) = \delta(\mathbf{x} - \mathbf{m}) \tag{6.52}$$

where $\delta(\mathbf{x})$ is a Dirac function and $\{\Phi_i\}$ describes a set of orthonormal functions.

The resulting discriminant function can be determined from by eq. (6.51) and is given by

$$\widehat{p}(\mathbf{x}) = \frac{1}{M} \sum_{i=1}^{M} \gamma(\mathbf{x}, \mathbf{m}_i) = \frac{1}{M} \sum_{i=1}^{M} \sum_{k=1}^{R} \Phi_k(\mathbf{x}) \Phi_k(\mathbf{m}_i) \tag{6.53}$$

where $\widehat{p}(\mathbf{x})$ is the estimated pdf, that is,

$$\widehat{p}(\mathbf{x}) = \sum_{k=1}^{R} c_k \Phi_k(\mathbf{x}) \tag{6.54}$$

[1] There are more than one clusters for each class.

The estimated coefficients c_k are

$$c_k = \frac{1}{M} \sum_{i=1}^{M} \Phi_k(\mathbf{m}_i) \tag{6.55}$$

It is important to note that the potential function estimator is both unbiased, and asymptotically consistent. We also can easily see that the potential function method is related to Parzen windows. In fact, the smooth function used for estimation is known as either *kernels* or potential functions or Parzen windows.

6.5. BINARY DECISION TREES

Binary decision trees are nonlinear multistage classifiers. This classification system operates by rejecting sequentially classes until the correct class is found. In other words, the correct class corresponding to a feature vector is determined by searching a tree-based decision system. The feature space is divided into regions corresponding to the different classes.

This classification scheme is extremely useful when a large number of classes is given. Binary decision trees divide the search space into hyperrectangles with sides parallel to the axis. The tree is searched in a sequential manner and a decision of the form $x_i \leq \alpha$, with x_i being a feature and α a threshold value, is made at each node for individual features. This processing scheme is an essential part of many tree-based vector quantization algorithms.

Figure 6.7 illustrates an example of a binary decision tree. It shows the regions created by the successive sequential splitting of 23 reference vectors.

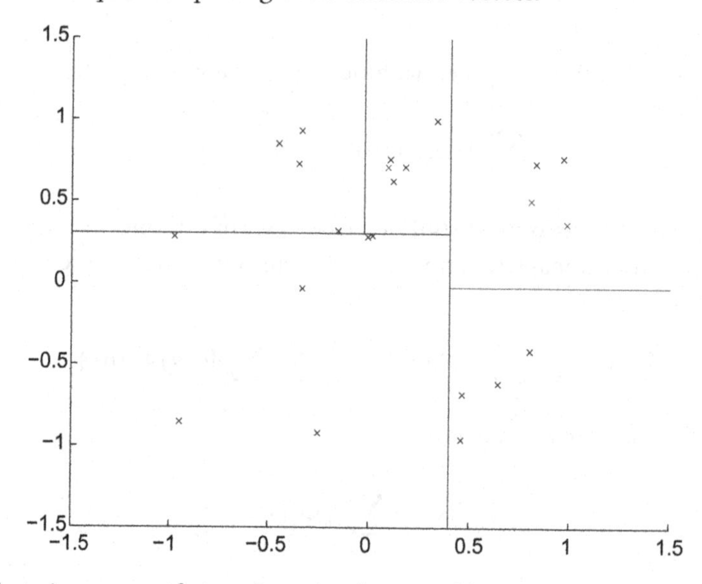

Figure 6.7 Clustering process of a two-dimensional vector table. The maximum number of vectors in one cluster was set to 7.

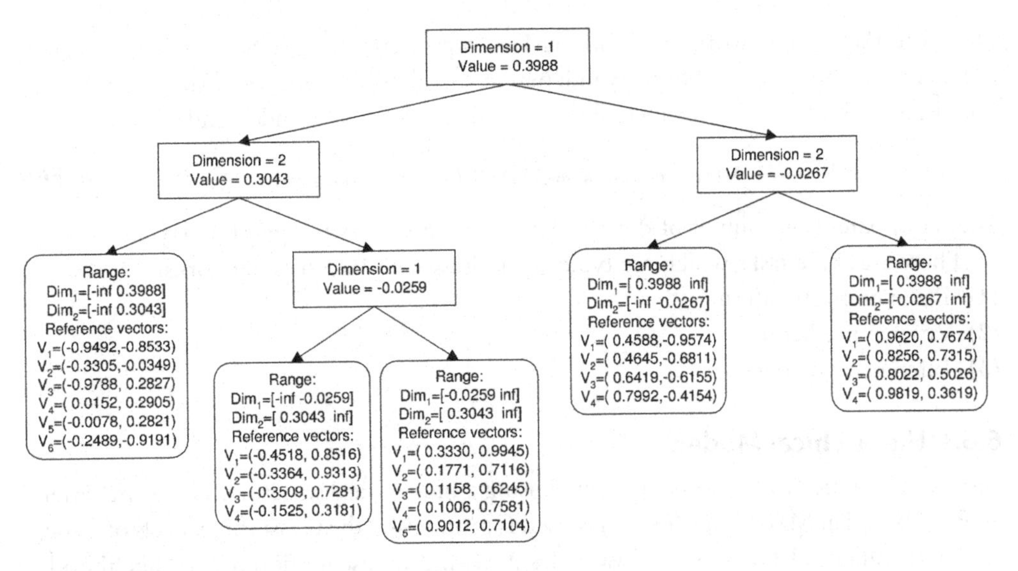

Figure 6.8 The internal tree structure after clustering.

Figure 6.8 shows the corresponding binary decision tree classification. A decision based on this classification scheme can be made without testing all available features.

6.5.1 Binary Decision Trees for Tumor Detection

Since binary decision trees have the tendency of poorly generalizing new sample vectors, overtraining is one of the unwanted side effects. In a pruning process which replaces subtrees by leaf nodes the new sample is assigned to the most frequent class. In other words, instead of a single class being associated with each leaf, there is a probability for each class that a new pattern vector belongs to that class. For example, if a leaf node contains seven training samples from the abnormal class, and three samples from the normal class, then a new pattern vector that reaches the leaf node is 70% abnormal.

6.6. BAYESIAN NETWORKS

Bayesian networks are frequently used as classification tools. A Bayesian network is a graphical model based on probabilistic beliefs. A Bayesian network consists of stochastic nodes, which either can be observed variables, unknown parameters, or latent, that is unobserved variables. In the first case, the distribution of the node is given by the distribution of the observation error or data distribution, in the other cases, prior distributions are specified for the nodes. Each node is connected with a subset of the other nodes, which influence the corresponding distribution.

The nodes in the Bayesian networks are determined by their relationships and the network is a directed acyclic graph. A node is only dependent on its "parent" nodes.

Therefore, Bayesian networks have a local Markov property (see also Section 2.5.3). Given its parents each node is conditionally independent of all other nondescendant nodes [318]. That is, the pdf for the random variable x_i in node i given all nondescendant nodes is

$$f(x_i|x_{j:j\in \text{ nondescendants of } i}) = f(x_i|x_{j:j\in \text{ parents of } i}) \tag{6.56}$$

The set of parents is a subset of the set of nondescendants as the graph is acyclic.

There are three tasks which are typically addressed in Bayesian networks:

(1) Parameter estimation.
(2) Variable prediction.
(3) Inference on edges.

6.6.1 Hierarchical Models

Most Bayesian networks can be described using a hierarchical structure. The term "Hierarchical Bayesian Modelling" (HBM) is used for models with at least three levels of nodes:

Level 1: data model on observed variables X with data distribution pdf or "likelihood": $f(X|\theta)$,
Level 2: prior models on the unknown parameters θ in level 1: $p(\theta|\nu)$,
Level 3: prior distributions on the unknown parameters ν in the priors (also known as hyper priors): $p(\nu)$.

The hierarchical structure can easily be extended, but three levels are usually sufficient.

The posterior pdf can be computed using Bayes' formula

$$p(\theta, \nu|X) = \frac{f(X|\theta)p(\theta|\nu)p(\nu)}{f(X)} \tag{6.57}$$

with $f(X)$ the marginal likelihood of the data

$$f(X) = \int f(X|\theta)p(\theta|\nu)p(\nu)d\theta\, d\nu \tag{6.58}$$

6.6.2 Parameter Estimation

Parameters can be estimated from the joint posterior probability density function. Using Bayes' theorem the posterior pdf can be computed from the data distribution pdf and the prior pdfs. Typical choices for the prior distribution are the Gaussian distribution, noninformative prior like Jeffreys' prior and priors based of maximum entropy methods, which make use of constraints on the nodes.

From the posteriori, parameters can be estimated using fully Bayes approaches with MCMC or with expectation-maximization (EM) algorithms, see also Section 2.6.

6.6.3 Variable Prediction

Given the Bayesian network, unobserved variables, for example new observations can be predicted using the so-called predictive posterior distribution. This can be computed as integral over the likelihood of the unobserved variable x^* and the posterior distribution of the unknown parameters,

$$p(x^*|\mathbf{X}) = \int f(x^*|\theta)p(\theta|\mathbf{X})d\theta \tag{6.59}$$

In simple cases integration (or marginalization) can be done analytically or numerically. Alternatively, simulation methods can be applied, for example importance sampling or MCMC.

6.6.4 Inference on Edges

In order to perform inference on the edges of a Bayesian network, an indicator variable is introduced for each possible edge, which is one if the edge exists and zero if it does not exist. Using Bayes' theorem, posterior probabilities of different Bayesian networks can be computed. Optimization should be performed by MCMC algorithms, to avoid local maxima.

6.7. SYNTACTIC PATTERN RECOGNITION

In some practical applications the pertinent information is not always available in the form of a numerical-valued feature vector as it is the case with the classification based on decision-theoretic methods. Therefore, it is not surprising that patterns contain structural information, that is difficult or impossible to quantify in feature vector form. Structure describes the way in which elements or components of a pattern are related. This structural relation can be either implicit in the model or be explicitly defined in detail.

This important aspect of structural information leads to a new classification paradigm, the so-called structural or syntactic classification concept. Syntactic pattern recognition is accomplished based on the following steps: (1) definition, (2) extraction of structural information, and (3) comparison based on measures of pattern structural similarity. Standard references for syntactic pattern recognition are [98,99,254].

Structural pattern recognition can be employed for both classification and description. For classification, metrics of pattern structural similarity have to be determined first. Classes can be defined according to a common structural description. Summarizing the above, we can state that the basics for syntactic pattern classification lies in the definition of the structure of an entity, which can be used afterwards for classification and description. The concept of syntactic pattern recognition is described in Fig. 6.9.

To make syntactic pattern recognition feasible, we have to require that "structure" is quantifiable. This can be accomplished based either on formal grammars or relational graphs.

The most important component of a grammar is the string. A string is a list of symbols being the elements of an alphabet, and represent the basic components of a pattern structure.

A string produced by a class-specific grammar can be easily classified based on the following procedures:

- *String-Matching:* Pattern structure is essential, but the available number of patterns is too small. Data structure is employed to represent both prototypes and unknown input patterns, and the classification is based on matching metrics between two structural representations.
- *Parsing:* The goal of this method is to determine, if an input string (pattern) can be generated by a given grammar. Therefore, parsers can be considered as syntax analyzers [329]. If an input string is successfully parsed, then a structural analysis of this pattern is also available since a production sequence has been found.

In many practical applications, the grammars are not explicitly given, but have to be determined based on the available set of feature vectors. This raises the issue of grammatical inference.

Statistical pattern recognition methods rely strongly on quantitative measurements and are not for all problems well qualified. Figure 6.10 shows an example where statistical method is not sufficient. Suppose that telocentric chromosomes of various sizes and

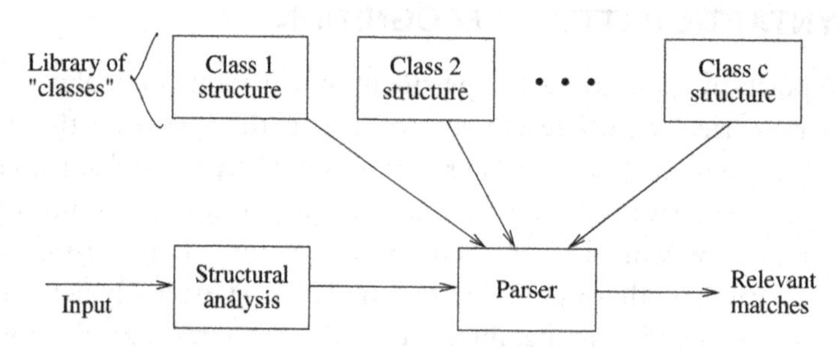

Figure 6.9 Syntactic pattern classification.

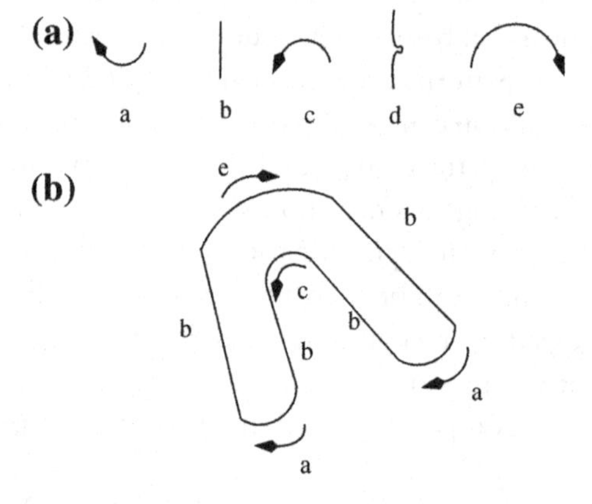

Figure 6.10 Boundary structure of telocentric chromosome [43] p. 5: (a) pattern primitives and (b) telocentric chromosome.

locations in the plane are to be classified. Only the boundary is to be considered, and because of the variable size of shapes, not the metrics but the structure proves to be critical. It is evident that the structure has to be extracted in some way, and then it has to be compared to that of other objects for the purpose of classification.

The following example of a telocentric chromosome [43] illustrates the concept of structural pattern recognition. Figure 6.10 shows the five curved line segments a, b, c, d, e which serve as primitives for the boundary description of the telocentric chromosome.

6.7.1 Elements of Formal Grammars and Languages

Syntactic pattern recognition is based on employing formal grammars for pattern class representation. The use of a grammar has some major advantages: it considers (1) recursively in the patterns and (2) common substructures shared by the members of a pattern class.

Formal grammars are based on words over finite sets of symbols. For a better understanding of this important concept, we will give some basic notations and descriptions.

An alphabet Z is defined as a finite, nonempty set of symbols such as

$$Z = \{a, b, \ldots, z\} \tag{6.60}$$

A *string* over Z is defined as either a single symbol from Z or as an ordered sequence of symbols of zero or more symbols from Z. In general, a string can be produced as follows: (1) letters from Z can be concatenated; (2) any letter in Z can be used several times; (3) strings can be concatenated; and (4) the empty string $\varnothing$ has no letters but is still considered as a valid string. Strings over Z define sentences. The set of all strings over Z is described by Z^*, and is also known as the closure set of Z. Two strings m and n can be concatenated to form the new string mn or nm. The string n is a substring of the string m whenever $m = inj$, where i and j are strings. i is the prefix string of and j is a suffix string of n with respect to m. The length of a string m is the number of letters it contains and is denoted by $|m|$.

Example 6.7.1 Let $Z = \{a, b\}$, $m = abaaabba$, and $n = aab$. Then $|m| = 8$, $|n| = 3$, n is a substring of m. The strings of all concatenations of m and n are: $mn = abaaabbaaab$ and $nm = aababaaabba$.

Formal linguistics can be applied to model and describe patterns in syntactic pattern recognition. A concept borrowed from linguistics is that of the language. Languages are generated by grammars.

Grammars are used with Z to impose some structure to a subset of strings $L \subseteq Z^*$. L is called a language. A language over the alphabet Z is defined as any subset of all strings over this given alphabet. Two sets of strings GH can also be elementwise concatenated via $GH = \{mn : m \in G, n \in H\}$. For any set of strings $M \subset Z^*$, we define: (a) $M^0 = \{\varnothing\}$, (b) $M^1 = M$, (c) $M^2 = MM$, (d) $M^n = MM^{n-1}$, (e) $M^* = \cup_{(n=0,\infty)} M^n$, and (f) $M^+ = \cup_{(n=1,\infty)} M^n$.

A grammar is defined as a four-tuple $\mathbf{G} = (Z, V, S, P)$, with Z describing an alphabet of letters while V is an auxiliary alphabet of letter-valued variables with $Z \cap V = \emptyset$. By S is the start variable from V denoted while P describes a set of production rules that determines the construction of sentences. This includes the intermediate "strings" composed of letters and letter-valued variables. The production rules serve as a basis for constructing strings from letters from Z and letter-valued variables from V. S is always used to begin any such string. We would like to add that the variables in V are called nonterminals while the letters from Z are called terminals. The set of productions P in a grammar G defines a structure or syntax on the language generated by G via an implicit relation on strings.

The language L generated by grammar $\mathbf{G}$, described as $L(\mathbf{G})$, forms the set of all strings that satisfy the following conditions: (1) each string contains only terminal symbols from Z, and (2) each string was produced from S using the production rules P from the given grammar G.

Example 6.7.2 The following grammar $G = \{Z, V, P, S\}$ is given with $V = \{S, A, B\}$, $Z = \{a, b, c\}$, and $P = \{S \rightarrow cAb, A \rightarrow aBa, B \rightarrow aBa, B \rightarrow cb\}$.

The language $L(G) = \{ca^n cba^n b | n \leq 1\}$ is generated based on the given grammar. In [43] is shown that the word *caacbaab* can be generated by the following sequence

$$S \rightarrow cAb \rightarrow caBab \rightarrow caaBaab \rightarrow caacbaab \tag{6.61}$$

It is evident that a grammar can generate a language of infinite words, even if all components of this grammar are finite.

6.7.2 Syntactic Recognition via Parsing and Grammars

Till now we considered the generation of syntactic or structural pattern description using formal grammars. Here, we will look at the "inverse" of this problem. Given the description of a pattern as a string produced by a class-specific grammar, the goal is to determine to which $L(\mathbf{G_i})$, $i = 1, 2, \ldots, c$ the string belongs.

It is evident that recognition of entities using syntactic descriptions can be considered as a matching procedure. Suppose that we are interested in a c-classification and therefore we develop class-specific grammars $\mathbf{G_1}, \mathbf{G_2}, \ldots, \mathbf{G_c}$. Every unknown description x can be classified by determining if $x \in L(\mathbf{G_i})$, for $i = 1, 2, \ldots, c$. Suppose that the language of each $\mathbf{G_i}$ could be generated and stored in a class-specific library of patterns. x has to be matched against each pattern in each library in order to determine its class membership. In other words, x is parsed to determine to which class it belongs. Parsing is basic concept for syntactic recognition and determines if an input pattern (string) is syntactically well formed in the context of one or more prespecified grammars. Parsers are producing the parsing process, and are known for being syntax analyzers. One possibility to parse the sentences is to use a finite state machine $L(\mathbf{G})$ for a decision. It was shown [98,99] that there exists a connection between grammars as generators of languages and finite state

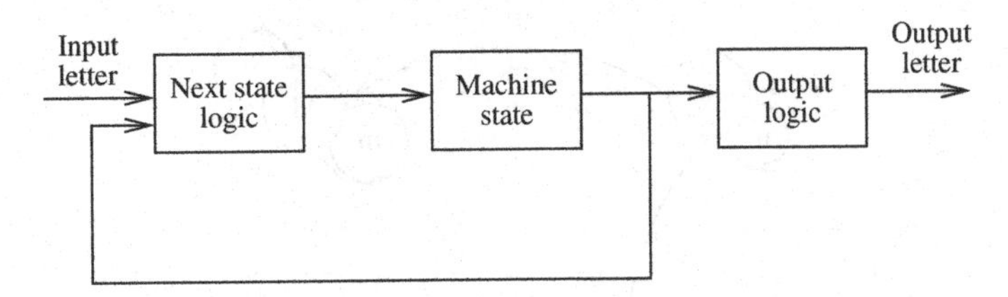

Figure 6.11 A finite state machine.

machines as recognizers of languages. An input string is processed by such a finite state machine, which then decides if this string is in a specific language L or not. Thus, the finite state machines are often used as the bases of language recognition algorithms.

A finite state machine is a memory-based machine of stored states. It then changes a state based on the current state, received input letter, and outputs a letter. The structure of a finite state machine is described in Fig. 6.11.

The finite state machine can be described abstractly by a tuple (Z, Q, g, h, q_0, Q_f), where Z is the alphabet of input/output letters, Q is the set of states, q_0 is the initial state, $Q_f \subset Q$ is the set of end states, g is the next state function $g : ZxQ \to Q$ that maps an input letter and current state into a new state, and h is the output function $h : QxZ \to Z$ that assigns an output letter in function of the current state and input letter.

Let's further define a finite state machine by β, and assume that it accepts a set of sentences defining a language $L(\beta)$. The existing correspondence between a regular grammar and a finite state machine can be expressed by the following relations: (a) $M \to aN$ describes a transition from state s_M to state s_N while being in state s_M and receiving a letter a as an input; and (b) $M \to a$ describes the transition from state s_M to a final state $s_F \in Q_F$ with Q_F being the set of final states.

Example 6.7.3 The following grammar is described by $\mathbf{G} = (Z, V, S, P)$ with $Z = \{m, n, o\}$, $V = \{S, M\}$, and $P = \{S \to mS, S \to nM, M \to nM, M \to mM, M \to o\}$. We immediately see, that $L(\mathbf{G})$ defines a regular language. Figure 6.12 describes the finite state machine. s_G is the rejected state while s_F is the final state.

6.7.3 Learning Classes via Grammatical Inference

In most practical applications, it is desired that the necessary grammars are directly inferred from a set of sample patterns. In this context, inference represents a procedure for constructing a structural description of items from a class. Grammatical inference is the process of learning grammars from a training set of sample sentences. The idea comes from Chomsky [60]. The most difficult part in syntactical pattern recognition is to learn classes from a sample of patterns. To describe the inference problem, we have first to consider a finite sample I of strings of letters stemming from an alphabet Z, and then to find a grammar $\mathbf{G}$ that accepts I, i.e., $I \subset L(\mathbf{G})$ as shown in [99,254].

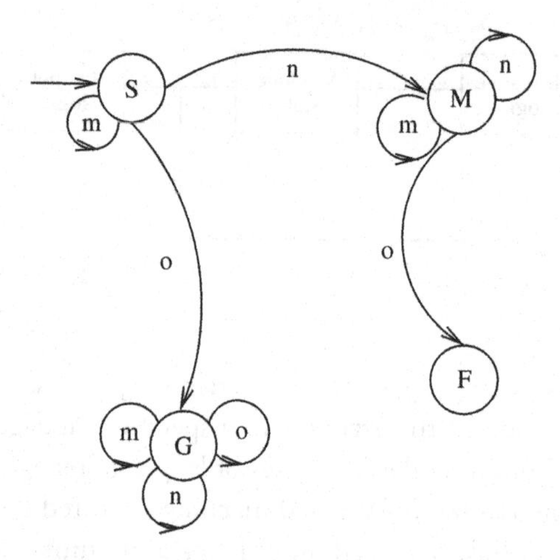

Figure 6.12 Finite state machine corresponding to grammar G.

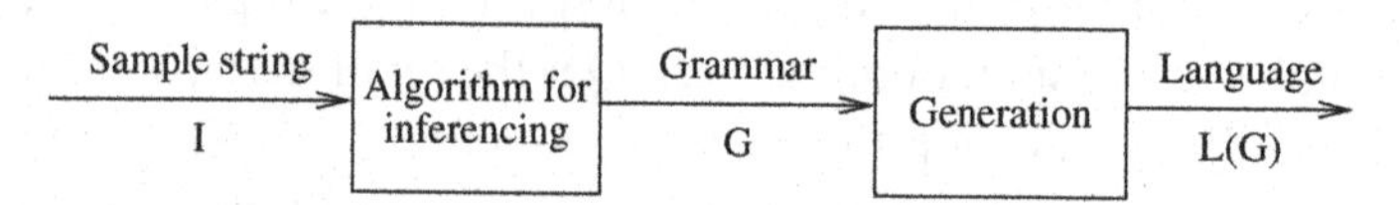

Figure 6.13 Grammatical inference.

The application of grammars in pattern recognition follows the following steps: (1) the alphabet Z is used to describe elementary patterns which can be extracted from an input pattern based on preprocessing or segmenting, (2) the auxiliary alphabet V is then used to describe subpatterns of greater complexity, which are formed by production rules from the primitive elements determined at the previous step, and finally (3) the language generated by the grammar is used to represent a whole class of patterns.

For syntactic pattern recognition, we need a grammar describing each pattern class. A given input string is parsed in order to determine its correspondence to one of the given classes. If none can be established, then the input pattern is rejected.

Since there are plenty of grammars that contain I, a major problem is caused by overlapping classes. The same sentence can for instance be accepted by several grammars. A solution to this problem is provided by introducing a distance function which determines the distance between a sentence and the regular grammars. As always in pattern recognition, the classification is then achieved by determining the minimal distance between the given sentence and the regular grammars [254]. The grammatical inference procedure is described in Fig. 6.13.

Example 6.7.4 For a given single sentence mn from $Z = \{m, n\}$, any of the following production rules is inferring a grammar whose language contains mn:

$$P_1 : \quad S \rightarrow mn$$

$$P_2 : \quad S \rightarrow AB, A \rightarrow m, B \rightarrow n$$

$$P_3: \quad S \to An, A \to m$$

$$P_4: \quad S \to mB, B \to n$$

We see immediately that a minimal grammar can be generated by the minimal set of productions $\{P_1\}$.

It is important to restate that there is not a unique relationship between a given language and some grammar. Thus, for a given string I there may be more than one grammar as solutions. Grammatical inference can be achieved by searching through the rule space of all context-free grammars for a grammar that can be inferred directly from a set of training patterns I.

A training pattern I is said to be structurally complete with respect to a grammar $\mathbf{G} = \{(Z, V, S, P)\}$ iff (a) $I \subset L(\mathbf{G})$, (b) I is defined over the alphabet Z, and (c) every production rule in P is employed in the generation of strings in I. Then the grammatical inference problem can be described as follows: (a) Find $\mathbf{G}$ such that $L(\mathbf{G}) \subset L(\mathbf{G})$; and (b) I is structurally complete with respect to $\mathbf{G}$ [223].

A methodological definition for grammatical inference was given in [43] as an "automatic learning of formal grammars from finite subsets of the language they generate".

Example 6.7.5 An example of how a grammar G can be employed for the description of a possibly infinite set of strings, i.e., the language $L(G)$, is given in [43].

Consider the previous example in Section 6.7.2 and its defined grammar. Figure 6.14 shows that the terminal symbols a, b, c are used to represent line segments of fixed length. A class of arrow-like shapes can be described based on these terminal symbols if it is proceeded in clockwise direction.

Thus, the class of patterns shown in Fig. 6.14 can be represented by the set of words

$$\{cacbab, caacbaab, caaacbaaab\} \tag{6.62}$$

which is identical with $L(G)$.

6.7.4 Graphical Approaches

Graph-based structural representations can be used as an approach in pattern recognition. Structural representations are replaced by graphical alternatives. They can be considered

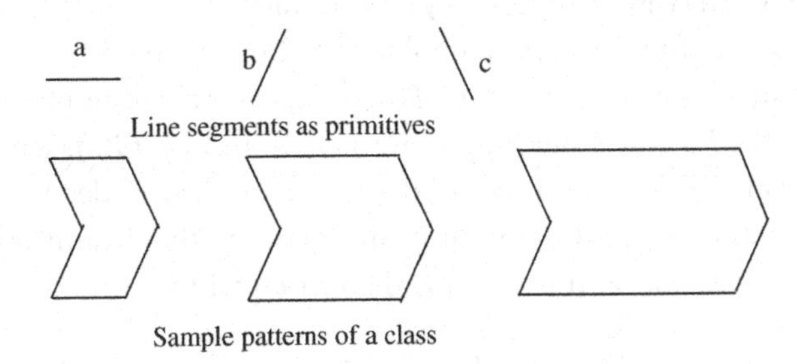

Figure 6.14 Inference of a symbolic data structure [43] p. 37.

as extensions of higher dimensional grammars, where instead of parsers a measure of graph similarity is employed.

In the following we will give some basic definitions from graph theory which is a fundamental area of discrete mathematics. A graph G is an ordered pair $G = (V, E)$, where $V = \{v_i, i = 1, \ldots, N\}$ is a set of vertices (nodes) and E is a set of edges (arcs) connecting some pairs of vertices. An edge which connects the vertices v_i and v_j is denoted either by e_{ij} or by (v_i, v_j). Often there is a significance attached to the direction of an edge, in the sense that an edge emanates from a vertice and is incident on another vertice. Therefore, $(a, b) \in R$ means there is an edge from node a to node b. It should be stated here that $(a, b) \in R$ does not imply $(b, a) \in R$. This directional significance characterizes a digraph, or directed graph. When the direction of edges is a graph, is not important, this means specification of either (a, b) or $(b, a) \in R$ is acceptable, an undirected graph results. A relational graph is an unlabeled digraph wherein nodes represent subpattern and primitives, and edges represent relationships between these subpatterns and primitives. To recognize structures using graphs requires that each pattern (structural) class is represented by a prototypical relational graph. An unknown input pattern has to be described by a graph, and this graph is then compared with the relational graphs for each class. We will illustrate this fact based on an example. Suppose, we have four classes, and that there are four prototypical relational graphs $G_{C1}, G_{C_2}, G_{C_3}, G_{C_4}$ available. For every new pattern data, a relational graph G_U has to be extracted from the data, and then compared with $G_{C1}, G_{C_2}, G_{C_3}, G_{C_4}$. To accomplish this, a discriminant function and a measure of similarity have to be established.

A directed graph G with p nodes can be converted to an adjacency matrix based on the following procedure:

1. Number each node and obtain a set of numbered graph nodes indexed by $[1, \ldots, p]$.
2. The adjacency matrix describes the existence or absence of an edge between two nodes in G:

$$ADJ[i,j] = \begin{cases} 1, & \text{for an existing edge from node } i \text{ to node } j \\ 0, & \text{otherwise} \end{cases} \tag{6.63}$$

In addition, if no cost is associated with the edges of a graph, the graph is called unweighted graph. Otherwise, a weighted graph results.

Some useful graph definitions [364] are illustrated in Fig. 6.15.

An undirected, unweighted graph with N nodes, each corresponding to a vector of the data set X, is called a threshold graph and is denoted by $G(a)$, where a represents the dissimilarity level. To determine if an edge between two nodes i and j with $i,j = 1, \ldots, N$ of the threshold graph $G(a)$ with N nodes exists, the dissimilarity between the corresponding vectors $\mathbf{x}_i$ and $\mathbf{x}_j$ must be less than or equal to a

$$(v_i, v_j) \in G(a), \quad \text{if} \quad d(\mathbf{x}_i, \mathbf{x}_j) \leq a, \quad i,j = 1, \ldots, N \tag{6.64}$$

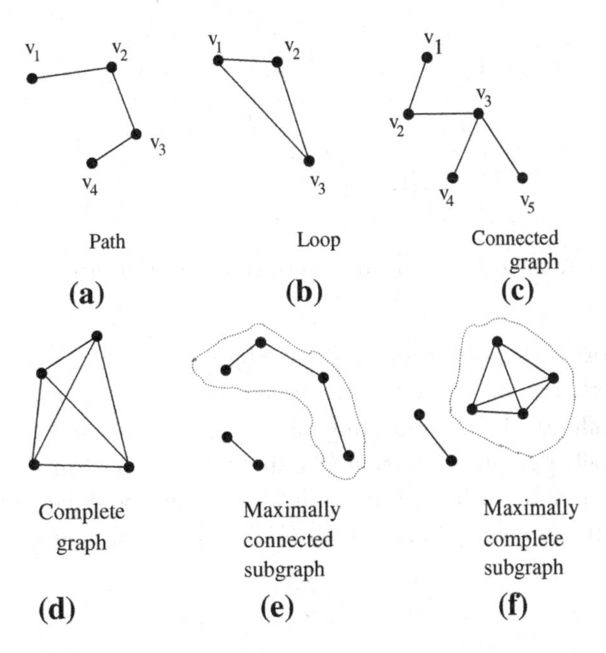

Figure 6.15 Basic graph types.

A threshold graph $G(a)$ whose edges (u_i, u_j) are weighted with the proximity measure between $\mathbf{x}_i$ and $\mathbf{x}_j$ is named a proximity graph $G_p(a)$. The proximity between two vectors can be defined based on a dissimilarity measure, and produces then a dissimilarity graph.

Example 6.7.6 Let $X = \{\mathbf{x}_i, i = 1, \dots, 5\}$, with $\mathbf{x}_1 = [1,1]^T, \mathbf{x}_2 = [2,1]^T, \mathbf{x}_3 = [5,4]^t, \mathbf{x}_4 = [6,5]^T$, and $\mathbf{x}_5 = [3,4]^T$. The pattern matrix of X is given by

$$D(X) = \begin{bmatrix} 1 & 1 \\ 2 & 1 \\ 5 & 4 \\ 6 & 5 \\ 3 & 4 \end{bmatrix} \tag{6.65}$$

This yields a corresponding dissimilarity matrix based on using the Euclidean distance

$$P(X) = \begin{bmatrix} 0 & 1 & 5 & 6.4 & 3.6 \\ 1 & 0 & 4.2 & 5.7 & 3.1 \\ 5 & 4.2 & 0 & 1.4 & 2 \\ 6.4 & 5.7 & 1.4 & 0 & 3.1 \\ 3.6 & 3.1 & 2 & 3.1 & 0 \end{bmatrix} \tag{6.66}$$

The resulting dissimilarity graph is shown in Fig. 6.16.

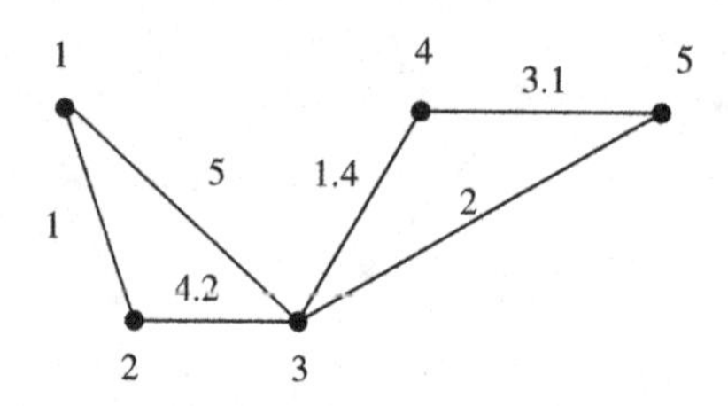

Figure 6.16 Dissimilarity graph defined by an Euclidean dissimilarity matrix.

The graph theory is a valuable tool for supporting agglomerative algorithms. In a graph G of N nodes, there is a correspondence between each node and vector of the data set X. Clusters can be built usually by connecting nodes together, and such forming connected subgraphs. A so-called graph property $h(k)$ imposes constraints on the subgraphs. In a very simple algorithm, the so-called single link algorithm [364], clusters C_r and C_s are merged based on the proximity function $g_{h(k)}(C_r, C_s)$ between their nodes

$$g_{h(k)}(C_i, C_j) = \overset{min}{\mathbf{x}_u \in C_r, \mathbf{x}_v \in C_s} \{d(\mathbf{x}_u, \mathbf{x}_v) = a\} \tag{6.67}$$

The subgraph $G(a)$ defined by $C_r \cup C_s$ is connected.

Example 6.7.7 Consider the following dissimilarity matrix

$$\mathbf{P} = \begin{bmatrix} 0 & 1.1 & 3 & 3.7 & 4.2 \\ 1.1 & 0 & 2.5 & 3.2 & 3.9 \\ 3 & 2.5 & 0 & 1.8 & 2.0 \\ 3.7 & 3.2 & 1.8 & 0 & 1.4 \\ 4.2 & 3.9 & 2.0 & 1.4 & 0 \end{bmatrix} \tag{6.68}$$

In the first step, $\mathcal{R}_0$, each pattern vector from X forms a single cluster. The next step, $\mathcal{R}_1$, requires the computation of $g_{h(k)}(C_r, C_s)$ for all cluster pairs. Based on the fact that $g_{h(k)}(\{\mathbf{x}_1\}, \{\mathbf{x}_2\}) = 1.1$ is the minimum value, the pattern vectors $\{\mathbf{x}_1\}$ and $\{\mathbf{x}_2\}$ are merged and we obtain

$$\mathcal{R}_1 = \{\{\mathbf{x}_1, \mathbf{x}_2\}, \{\mathbf{x}_3\}, \{\mathbf{x}_4\}, \{\mathbf{x}_5\}\}\}$$

The next minimum we find for $g_{h(k)}(\{\mathbf{x}_4\}, \{\mathbf{x}_5\}) = 1.4$, and so we obtain

$$\mathcal{R}_2 = \{\{\mathbf{x}_1, \mathbf{x}_2\}, \{\mathbf{x}_3\}, \{\mathbf{x}_4, \mathbf{x}_5\}\}$$

For $\mathcal{R}_3$, we get

$$\mathcal{R}_3 = \{\{\mathbf{x}_1, \mathbf{x}_2\}, \{\mathbf{x}_3, \mathbf{x}_4, \mathbf{x}_5\}\}$$

In the end, we obtain $g_{h(k)}(\{\mathbf{x}_1, \mathbf{x}_2\}, \{\mathbf{x}_3, \mathbf{x}_4, \mathbf{x}_5\}) = 2.5$, and this leads to $\mathcal{R}_4$.

6.8. DIAGNOSTIC ACCURACY OF CLASSIFICATION MEASURED BY ROC CURVES

Often a clinical researcher has to determine how accurate a particular laboratory test is in identifying diseased cases. The ability of a test to discriminate diseased cases from normal cases is based on receiver operating characteristic (ROC) analysis. The ROC represents an analytical technique derived from biometrics and is used for providing both the desired accuracy index and utility in terms of cost and benefit [245].

At the same time, a method to specify the performance of a classifier is in terms of its ROC curve. The ROC curve represents a plot describing the classifier's true positive detection rate versus its false positive rate. The false positive (FP) rate is the probability of incorrectly classifying a nontarget object (normal tissue region) as a target object (tumor region). On the other hand, the true positive (TP) detection rate is the probability of correctly classifying a target object as being indeed a target object. Both the TP and FP rates are specified in the interval from 0.0 to 1.0, inclusive. In medical imaging the TP rate is commonly referred to as sensitivity, and (1.0-FP rate) is called specificity.

The schematic outcome of a particular test in two populations, one population with a disease, the other population without the disease, is summarized in Table 6.1.

In general, the sensitivity S_e and the specificity S_p of a particular test can be mathematically determined.

Sensitivity S_e is the probability that a test result will be positive when the disease is present (true positive rate, expressed as a percentage).

$$S_e = \frac{TP}{FN + TP} \tag{6.69}$$

Specificity S_p is the probability that a test result will be negative when the disease is not present (true negative rate, expressed as a percentage).

$$S_p = \frac{TN}{TN + FP} \tag{6.70}$$

Sensitivity and specificity are a function of each other and counterrelated.

Table 6.1 Results of a particular test in two populations, one population with a disease.

	Disease present	Disease absent	Sum
Test positive	true positive (TP)	false positive (FP)	(TP + FP)
Test negative	false negative (FN)	true positive (TN)	(FN + TN)
Sum	(TP + FN)	(FP + RN)	

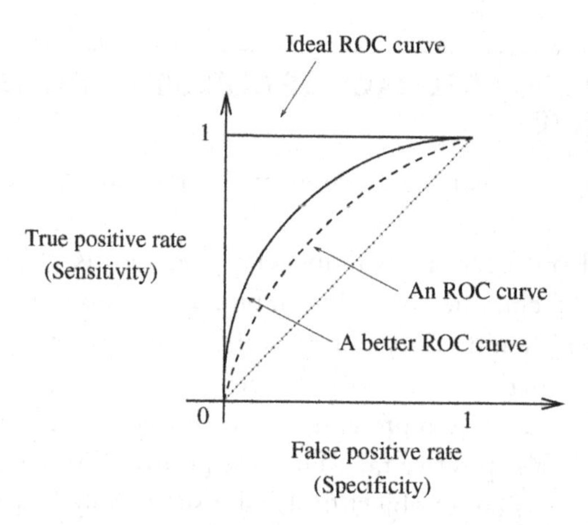

Figure 6.17 Typical ROC curves.

Parameters associated with statistical classifiers can be varied in order to change the TP and FP rates. Each set of parameter values may result in a different (TP,FP) pair, or operating point. It is possible to trade a lower (higher) FP rate for a higher (lower) TP detection rate by choosing appropriate values for the parameters in question.

A typical ROC curve is shown in Fig. 6.17. The area under the curve (AUC) is an accepted modality of comparing classifier performance. A perfect classifier has a TP rate of 1.0 and a FP rate of 0.0, resulting in an AUC of 1.0. Random guessing would result in an AUC of 0.5.

An ROC curve demonstrates several things:

- It shows the trade-off between sensitivity and specificity (any increase in sensitivity will be accompanied by a decrease in specificity).
- The closer the curve approaches the left-hand border and then the top border of the ROC space, the more accurate the test.
- The closer the curve comes to the 45° diagonal of the ROC space, the less accurate the test.
- The area under the curve is a measure of comparing classifier performance. An ideal classifier has an area of 1.

In the following it is shown how ROC curves are generated for statistical classifiers [393].

6.8.1 Bayesian Classifier

A Bayesian classifier can be trained by determining the mean vector and the covariance matrices of the discriminant functions for the abnormal and normal classes from the training data. Instead of computing the maximum of the two discriminant functions $g_{abnormal}(x)$ and $g_{normal}(x)$, the decision was based in [393] on the ratio $g_{abnorm}(x)/_{normal}(x)$.

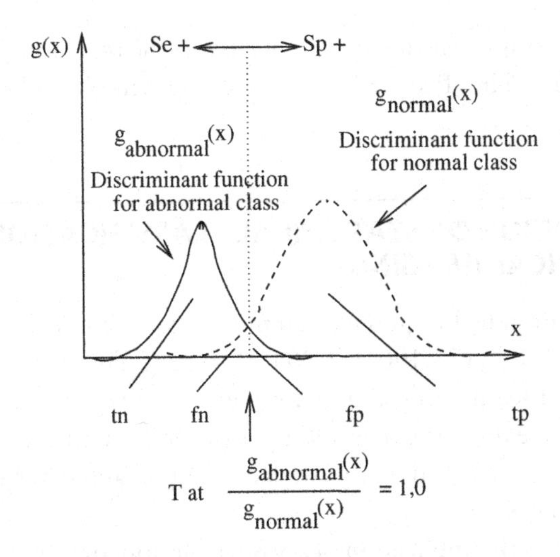

Figure 6.18 Discriminant functions for two population, one with a disease, the other population without the disease. A perfect separation between the two groups is rarely given while an overlap is mostly observed. The FN, FP, TP, TN areas are denoted.

A decision threshold T was set, such that if the ratio is larger than T the unknown pattern vector is classified as abnormal, else as normal. By changing T, the sensitivity/specificity trade-off of the Bayes classifier can be altered. A larger T will result in lower TP and FP rates, while a smaller T will result in higher TP and FP rates. The procedure described in [393] is illustrated in Fig. 6.18.

6.8.2 *k* Nearest Neighbor Classifier

The k nearest neighbor classifier described in [393] is based on two different parameters: k is the number of nearest neighbors to base the decision on, and a threshold T which specifies the minimum number of nearest neighbors that represent the disease before a new pattern vector is classified as an abnormality. The ROC points for a given k are determined by varying T from 1 to k, and monitoring the resulting TP and FP rates. It is immediately evident that as T is increased, the TP and FP rates will decrease. In [393] k is varied from 1 to 250, and for each k an ROC curve is produced by altering T from 1 to 40. Then the AUC is computed. For a subsequent classification the k producing the maximum AUC is selected. The operating point of this classifier is selected by choosing an adequate threshold T.

6.8.3 Binary Decision Trees

In [393] it was shown that for obtaining ROC points the threshold for the abnormal class has to be varied over some range on the leaf nodes of the tree. Usually, a new pattern vector is assigned to the class with the highest probability. Dealing only with patterns being either abnormal or normal, the threshold T is at 50%. Increasing T for the abnormal class will lead to a decrease in TP and FP rates, while lowering T will have a reverse effect.

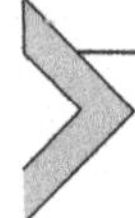

6.9. APPLICATION OF STATISTICAL CLASSIFICATION METHODS IN BIOMEDICAL IMAGING

Pattern classification techniques play an important role in the detection and classification of tumors [167,171,173,190,263,394]. The procedure is simple and follows the following scheme. After segmenting a suspicious region, a feature extraction and selection is performed in order to extract the relevant information from this region. A classification paradigm is chosen such that based on the available features and determined tumor classes the best results are achieved.

This section describes the application of statistical or noncognitive classification methods to breast cancer detection.

A standard mammogram screening case consists of four images, two views of each breast. These are a cranio-caudal (top-bottom) view, and a medio-lateral (middle-to-outside) view. Sometimes, an abnormality or suspicious region can be missed by a radiologist in one view, but easily detected in the other view.

A radiologist is looking for distinctive cues and characteristics indicative of cancer when evaluating a mammogram. Among these signs are the presence of clustered microcalcifications, spiculated or stellate lesions, circumscribed or well-defined masses, ill-defined or irregular masses, and architectural distortions. Sometimes asymmetry and developing densities are also considered.

A microcalcification is a tiny calcium deposit that has accumulated in the breast tissue, and represents a small bright spot on the mammogram. A cluster is typically defined to include at least three to five microcalcifications within a 1 cm^2 region [361]. Up to 50% of malignant masses demonstrate clustered microcalcifications, and sometimes the clusters represent the only sign of malignancy. They can be embedded in dense parenchymal tissue. The size of calcifications ranges from smaller than 0.1 mm to 5 mm in diameter. Their detection is normally very difficult for a radiologist who is usually locating them with a magnifier. When analyzing individual calcifications, two aspects are usually considered: (1) size, shape, and radiographic density for individual calcifications and (2) number and distribution of calcifications within a cluster.

There are a number of different types of masses which may be malignant. A circumscribed mass with a distinct border and circular in shape is easy to identify. High density radiopaque circumscribed masses are mostly malignant, while radiolucent masses are almost always benign. A detection difficulty is posed by ill-defined masses which have an irregular shape and less distinctive borders. Contour, density, shape, orientation, and size of the mass are important features for classifying a visible mass.

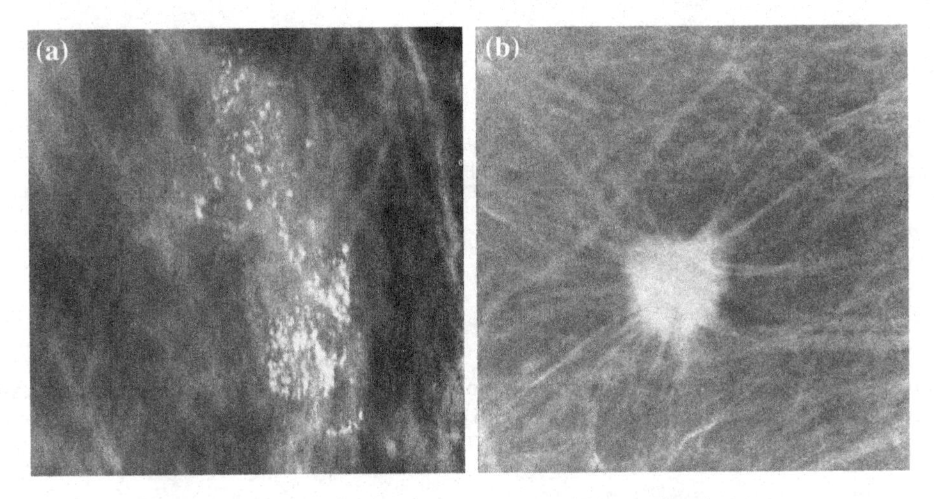

Figure 6.19 Example of (a) a clustered microcalcification and (b) a stellate lesion.

Spiculated lesions with a star-shaped appearance and blurred borders are far more difficult to detect. Like circumscribed and ill-defined masses, the spiculated lesions have a radiographically dense tumor compared to the surrounding glandular tissue. The distinct star shape is the result of radially oriented spicules extending from the tumor center into the surrounding breast tissue. They are always malignant, and the density of the tumor center and the radiating structure are the most important features [361].

Architectural distortions in the connective tissue of the breast represent one of the most difficult abnormalities to detect. The breast structures are loosely arranged along the duct lines which have the tendency to radiate in curvilinear arcs outward from the nipple to the chest wall. Both malignant and benign cancer forms are responsible for a disruption in this symmetrical pattern.

The most frequent breast cancer forms are clustered microcalcifications and stellate lesions as shown in Fig. 6.19.

In this section, we will show how clustered microcalcifications and stellate lesions are detected based on statistical classification methods. Chapter 13 describes the computer-aided diagnosis of breast cancer based on emerging biologically inspired techniques.

6.9.1 Microcalcification Detection

The ground truth location of each individual microcalcification is described by the center and radius of a circle. Since both microcalcifications nor spiculated masses are perfectly round, each ground truth circle always includes some pixels belonging to normal breast tissue.

To validate a positive calcification in an image, we have first to check if the centroid of the detected object lies inside a circle denoting the location of a known calcification

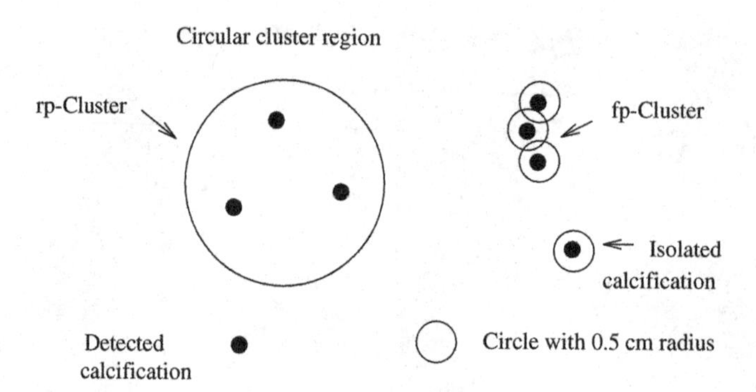

Figure 6.20 Examples of TP and FP microcalcification cluster detection. A TP is detected since at least two calcifications are found inside the ground truth circle.

in an image. When the centroid of the detected object lies within a ground truth circle, then a TP is detected otherwise a FP. To measure TP and FP cluster detections, detected microcalcifications have to be grouped together. A detected calcification is grouped with another detection if their distance is of less than one centimeter. We have a TP detection if two or more calcifications are detected within the location of the cluster described by a circular area, else we have a FP detection. This situation is visualized in Fig. 6.20.

In [393] are described three categories of features, in all 42 features, extracted from a segmented suspicious region corresponding to a clustered microcalcification from the MIAS data set. These categories are:

1. Object size, contrast, and gray-level distribution within the object which in case of calcification detection refers to all pixels that are part of the segmented object.
2. Shape description of the segmented object.
3. Texture information from the image. Among these are the Laws' texture energy measures, modified Laws' measures, and fractal-based texture. They describe the local intensity distribution properties such as smoothness and coarseness, and the structural properties such as regular patterns of image primitives.

The relevant features are determined based on an exhaustive search method. Figure 6.21 shows that a feature vector with seven components is optimal for both linear and quadratic Bayesian classifier while a feature vector with six components is optimal for the k nearest neighbor classifier.

Table 6.2 shows the selected features for each classifier. A contrast measurement and various Laws' texture energy measures are common to all three vectors. This makes sense because these features must aid in the detection of small, bright spots.

The AUC of four classifiers for microcalcification detection is shown in Table 6.3. The k nearest neighbor classifier is statistically significantly better than the others [393].

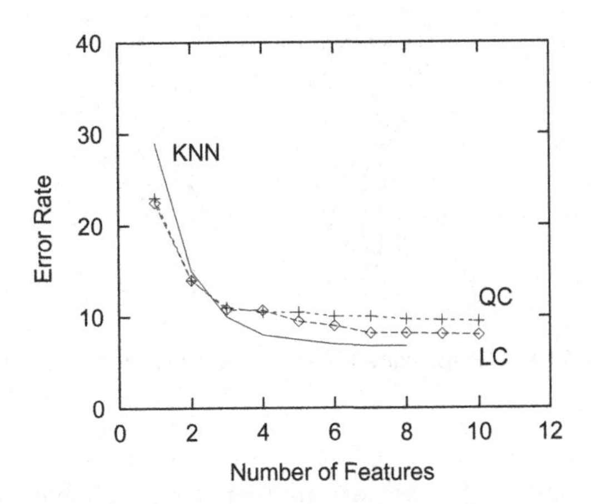

Figure 6.21 Feature selection results for microcalcification detection. Classification error rate versus number of features is shown for k nearest neighbor ($k = 1$), linear and quadratic Bayesian classifier [393].

Table 6.2 Feature vectors selected for microcalcification detection [393].

Classifier	Feature vector selected
Quadratic	absolute contrast, Laws' $E5{*}S5$, Laws' $E5{*}E5$, modified Laws' $E5{*}E5$, shape moment 3, fractal texture 1, fractal texture 2, fractal texture 2
Linear	average gray level, absolute contrast, Laws' $E5{*}S5$, Laws' $S5{*}S5$, modified Laws' $L5{*}E5$, average border gradient, fractal texture 2
k nearest	average gray level, object size, contrast, neighbor gray-level moment 1, Laws' $L5{*}E5$, Laws' $S5{*}S5$

Table 6.3 AUC for individual classifiers applied to microcalcification detection [393].

Classifier	AUC
k nearest neighbor	93.0
quadratic	91.7
linear	90.0
binary decision tree	85.0

6.9.2 Spiculated Lesion Detection

Breast cancer is mostly observed in the form of a mass (lesion) with well-distinguished margins from a surrounding breast parenchyma. Cancerous lesions have a tumor center

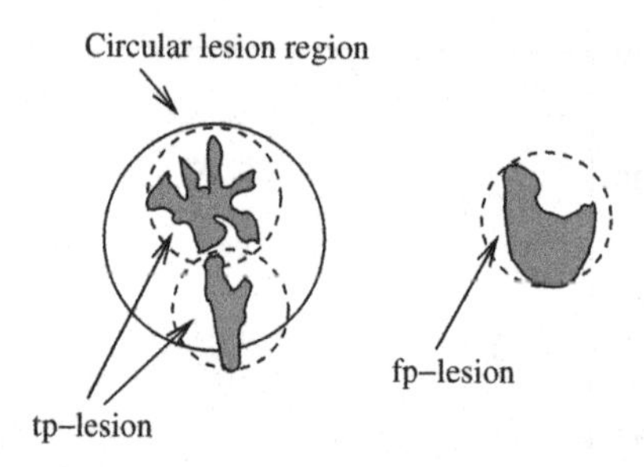

Figure 6.22 Examples of TP and FP spiculated lesion detections. The actual tumor is segmented into two TP detection objects.

being radiologically denser than the surrounding tissue, and have spiculated (stellate) margins. A central tumor mass being either very small or embedded in dense parenchymal tissue is difficult to perceive. Therefore all efforts are focused in detecting and diagnosing lesions when they are under one centimeter in diameter.

The ground truth location of each individual spiculated lesion is described by the center and radius of a circle. The center of the circle is at the centroid of the object, and the radius equals the distance from the centroid to the furthest point of the object's perimeter. If there exists an overlapping of a ground truth circle with the area of a circle surrounding a detected object by at least 50%, then the detection is considered a TP else is a FP. Figure 6.22 visualizes this aspect.

Segmentation of spiculated lesions is usually performed in two steps: (1) extraction of locally bright spots of appropriate size, and a (2) growing region procedure to improve the shape estimation of the regions initially extracted.

Tumor classification is usually either pixel-based or region-based. Texture is in general computed at the pixel level. In fact, the human vision system responds to texture at the "pixel" level. For example, features based on Laws' texture energy measures discriminate very well between various textured regions in an image when pixel-based classification is required.

A typical spiculated lesion has a distinct tumor center with spicules radiating outward. In [393] the discriminating power of each texture measurement when the feature value is computed only for pixels located in various regions in and around the segmented objects was therefore examined. This is because the texture within and around different regions of a segmented object is different, and therefore the discriminating power of the texture features is lost if the computation is performed across the various regions. An object and its background are divided into three disjoint regions: middle, border, and the surrounding. This is shown in Fig. 6.23.

As for microcalcifications, a total of 42 features were extracted from the segmented lesion. The three categories of features mentioned in [393] are:

1. Object size, contrast, and gray-level distribution within the object.

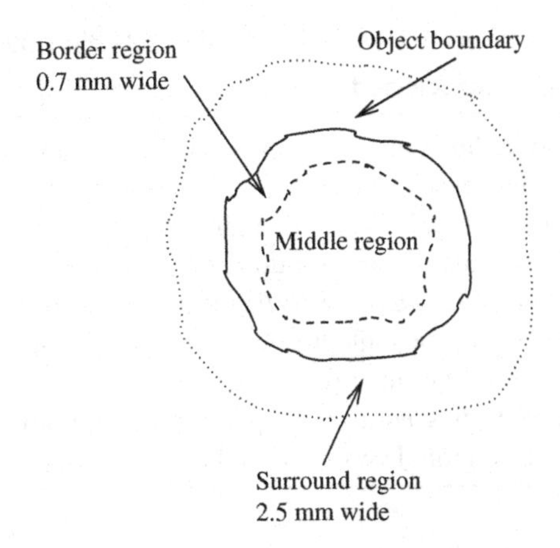

Figure 6.23 Regions of an object and its local background for which texture features are computed [393].

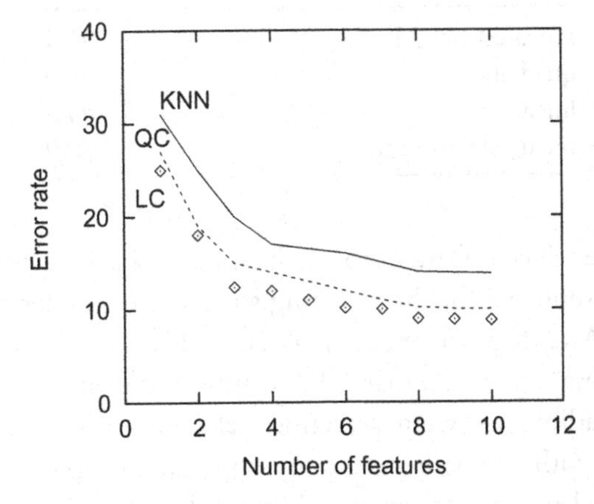

Figure 6.24 Feature selection results for spiculated lesion detection. Classification error rate versus number of features is shown for k nearest neighbor ($k = 1$), linear and quadratic Bayesian classifier [393].

2. Local edge features such as the ALOE features [171][2].
3. Texture information from the image. Among these are the Laws' texture energy measures, and fractal-based texture.

The relevant features are determined based on an exhaustive search method. Figure 6.24 shows that a feature vector with seven components is optimal for both linear Bayesian and k nearest neighbor classifier while a feature vector with eight components is optimal for the quadratic Bayesian classifier.

[2] Analysis of Local Edge Orientation (ALOE).

Table 6.4 Feature vectors selected for spiculated lesion detection [393].

Classifier	Feature vector selected
Quadratic	ALOE of radius, contrast, shape moment 2, shape moment 3, relative extrema density of outer region, Laws' $R5*R5$ of object, fractal texture 3, Laws' $L5*E5$ of angle image 2
Linear	relative extrema density of outer region, shape moment 1, compactness 1, Laws' $R5*R5$ of object, relative extrema density of outer region of angle image 2, average edge strength, ALOE of the object and the surround
k nearest neighbor	ALOE of Radius, contrast, shape moment 3, relative extrema density of outerregion, Laws' $E5*E5$, Laws' $S5*S5$, compactness 1

Table 6.5 AUC for individual classifiers applied to microcalcification detection [393].

Classifier	AUC
k nearest neighbor	95.9
quadratic	95.0
linear	94.6
binary decision tree	78.9

Table 6.4 shows the selected features for each classifier. Each vector contains an ALOE feature which is representative for the radiating structure of the lesion spicules, and two shape measurements. Also the relative extrema density feature present in all vectors determines if spicules are present. In summary, the feature selection corresponds to the technique followed by a radiologist when detecting stellate lesions: the search for an oval or circular tumor center with a radiating structure consisting of spicules.

The AUC of four classifiers for spiculated lesion detection is shown in Table 6.5. As in the microcalcifications case, the k nearest neighbor classifier offers the best performance [393].

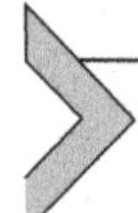

6.10. APPLICATION OF SYNTACTIC PATTERN RECOGNITION TO BIOMEDICAL IMAGING

Application of syntactic image recognition methods enables a more in-depth analysis by describing the semantic content of the image than just the simple recognition of pathological lesions. The main importance of these methods lies in the design of intelligent visual and information diagnostic system. These systems employ syntactic reasoning based on grammar use.

Syntactic pattern recognition based on grammatical reasoning has been widely used in medical image analysis in several application areas: (1) performing of simple semantic

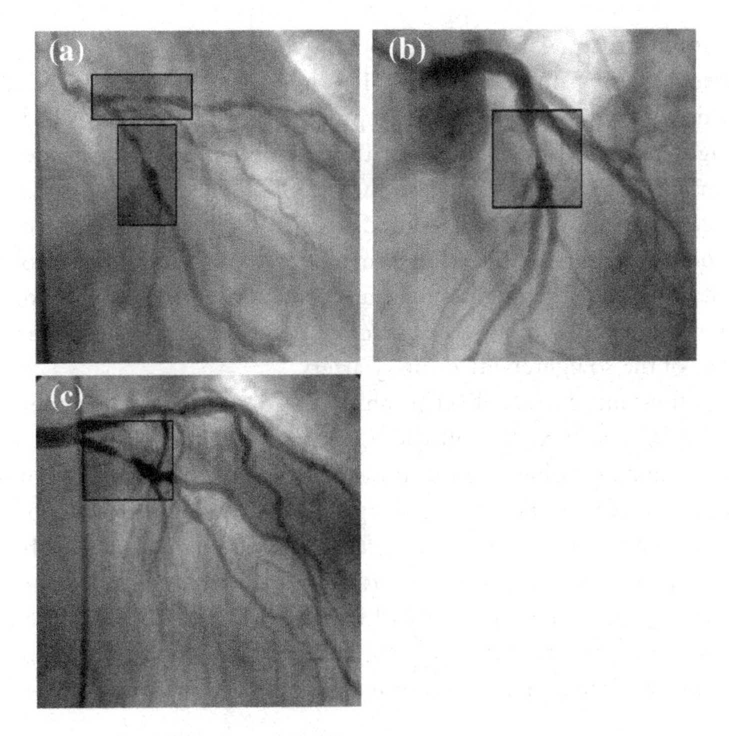

Figure 6.25 Examples of cardiac stenoses [268].

reasoning related to their content [209], (2) morphology description of important shapes such as ECG course analysis [373], and (3) recognition of local irregularities in ureter lumens, and examination of the morphology of the renal pelvis and calyxes [269,270].

Syntactic pattern recognition is applied to produce a perceptual description of the analyzed structures and pathological signs by defining first simple shape elements, and then by building the general grammar description of the considered organ. These descriptions based on sequences or terminal symbols belong to the languages generated by the introduced grammar. Thus, the analysis and recognition of pathological signs is performed by parsing algorithms which analyze input sequences and reduce them to one of the known categories.

This section describes the application of syntactic pattern recognition to support the diagnosis of the ischaemic heart disease based on coronography images [268]. The goal is to analyze and diagnose significant stenoses of the coronary lumen. Using context-free grammars enables the diagnosis of profiles of the examined artery width both of concentric stenoses as well as eccentric stenoses. Concentric stenoses appear on a cross-section as a uniform stricture of the whole lumen, and present a characteristics of a stable disturbance of heart rhythm, while eccentric stenoses appear only on one vascular wall, and characterize an unstable angina. Examples of cardiac stenoses are visualized in Fig. 6.25.

The goal of the structural analysis is to obtain width graphs which show the pathological changes occurring in these arteries. It is important to preserve the sequence of operations which are part of the required preprocessing. To obtain these width diagrams, a specific straightening transformation algorithm is employed which provides besides the width diagrams of the analyzed vessels also the morphological lesions occurring in them and the correct ramification of the diagnosed arteries. The required steps for the analysis of cardiac stenose images are [268]: (1) segmentation and skeletonization of the examined coronary artery, and (2) straightening transformation of the external contour of the examined artery from a two-dimensional space to a one-dimensional diagram visualizing thus the profile of the straightened coronary artery.

The recognition and correct description of the degree of advancement of lesions is done in [268] by a context-free attributed grammar of the look-ahead $LR(1)$-type. The attributed grammar enables the diagnosis of coronographic X-ray examinations by defining all potential shapes of expected morphological lesions. Automate of the $LR(1)$-type analyzes the input sequence from the left-hand to the right-hand side by processing its successive elements. The output represents the opposite of the right-hand derivation. At each analysis step, only one input symbol is examined. The detailed definition of the attributed grammar and of the $LR(1)$-type automate can be found in [268].

The following attributed grammar is proposed in [268] for various shape recognition of stenoses:

$$G_{CA} = (V_N, V_T, SP, STS) \tag{6.71}$$

where V_N is the set of nonterminal symbols, V_T the set of terminal symbols, SP the production set, and STS the grammar start symbol.

$$V_N = \{SYMPTOM, STENOSIS, H, V, NV\}$$

$$V_T = \{h, v, nv\} \quad \text{for} \quad h \in (-10^o, 10^o), v \in (11^o, 90^o), nv \in (-11^o, -90^o)$$

STS=SYMPTOM

SP:

The first production rule (Symtom: = Stenosis) describes that a vessel stenosis was discovered in the input sequence. The second and third production rules define potential shapes of cardiac stenoses while successive rules define the upward and downward ramifications of the analyzed stenosis. Thus, the semantic variables h_e and w_e specify the altitude and length of the terminal section labeled e, and make the diagnosis more specific by defining in percentage the degree of the cardiac stenosis.

1. SYMPTOM $\rightarrow$ STENOSIS	SYMPTOM:=STENOSIS	
2. STENOSIS $\rightarrow$ NVHV		
3. STENOSIS $\rightarrow$ NVv|NHV		
4. V $\rightarrow v\|Vv$	$w_{sym} := w_{sym} + w_{nv};$	$h_{sym} := h_{sym} + h_{nv}$
5. NV $\rightarrow nv\|NVnv$	$w_{sym} := w_{sym} + w_v;$	$h_{sym} := h_{sym} + h_v$
6. H $\rightarrow h\|Hh$	$w_{sym} := w_{sym} + w_h;$	$h_{sym} := h_{sym} + h_h$

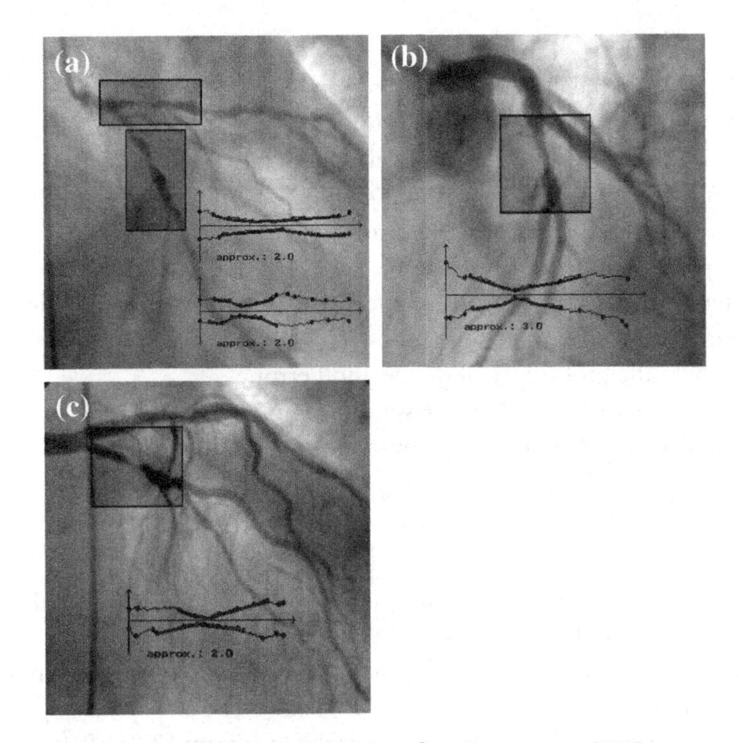

Figure 6.26 Results of structural pattern recognition of cardiac stenoses [268].

A recognition rate of 93% was reported in [268] based on the proposed structural analysis. The recognition rate describes the percentage of accurately recognized and measured vessel stenoses versus the total number of tested images. The recognition pertains to determining the correct location of a stenosis and its type (concentric or eccentric).

In summary, the objective of the study performed in [268] was to determine the percentage of the correct recognition of artery stenosis and its size. Figure 6.26 shows examples from [268] of detected cardiac stenosis. The recognized symptoms were marked with a bold line.

6.11. EXERCISES

1. The following class-dependent pdfs are given

$$p(\mathbf{x}|\omega_1) = \begin{cases} \frac{1}{16}, & \text{for} \quad 1 \leq x_1 \leq 5 \quad \text{and} \quad 1 \leq x_2 \leq 5 \\ 0, & \text{else} \end{cases} \tag{6.72}$$

$$p(\mathbf{x}|\omega_2) = \begin{cases} 1, & \text{for} \quad 2 \leq x_1 \leq 3 \quad \text{and} \quad 2 \leq x_2 \leq 3 \\ 0, & \text{else} \end{cases} \tag{6.73}$$

where $\mathbf{x}$ is a two-dimensional vector. Assume $P(\omega_1) = p(\omega_2) = \frac{1}{2}$.

(a) Draw the decision regions.
(b) Develop the classification strategy.
(c) Compute the classification error.

2. Consider the following one-dimensional, two-class case where $p(x|\omega_1)$ is a Gaussian pdf with mean $\mu = 0$ and $\sigma_i^2 = 1$ and $p(x|\omega_2)$ is a uniform pdf with mean $\mu = 2$ and $\sigma_1^2 = \frac{1}{3}$. We also assume equal a priori probabilities.
   (a) Plot $p(x|\omega_1)$ and $p(x|\omega_2)$.
   (b) Determine the classification method that minimizes the classification error. What are the discriminant functions $g_1(x)$ and $g_2(x)$?
   (c) Compute the probability of the decision error.
3. In a three-class two-dimensional classification problem the feature vectors in each class are normally distributed with covariance matrix

$$\Sigma = \begin{bmatrix} 1.2 & 0.5 \\ 0.5 & 1.8 \end{bmatrix} \tag{6.74}$$

   The mean vectors for each class are $\mu_1 = [0.1, 0.2]^T, \mu_2 = [2.2, 1.9]^T$, $\mu_3 = [-1.4, 2.1]$. Assume that the classes are equiprobable.
   (a) Classify the feature vector $[1.8, 1.6]^T$ according to the Bayes minimum error probability classifier.
   (b) Draw the curves of equal Mahalanobis distance from $[2.2, 1.8]^T$.
4. Consider the multivariate Gaussian pdf with $\sigma_{ij} = 0$ and $\sigma_{ii} = \sigma_i^2$.
   (a) Show that this pdf is given by

$$p(\mathbf{x}) = \frac{1}{\prod_{i=1}^{d} \sqrt{2\pi}\sigma_i} \exp\left[-\frac{1}{2}\sum_{i=1}^{d}\left(\frac{x_i - \mu_i}{\sigma_i}\right)^2\right] \tag{6.75}$$

   (b) describe the contours of constant density, and write an expression for the Mahalanobis distance from $\mathbf{x}$ to $\boldsymbol{\mu}$.
5. Show that if the likelihood function is Gaussian, we get as estimates for the mean $\boldsymbol{\mu}$ and the covariance matrix Σ the following values

$$\widehat{\boldsymbol{\mu}} = \frac{1}{N}\sum_{i=1}^{N} \mathbf{x}_k \tag{6.76}$$

   and

$$\widehat{\Sigma} = \frac{1}{N}\sum_{k=1}^{N} (\mathbf{x}_k - \widehat{\boldsymbol{\mu}})(\mathbf{x}_k - \widehat{\boldsymbol{\mu}})^T \tag{6.77}$$

6. The pdf of a random variable x is given by

$$p(x;\theta) = \theta^2 x \exp(-\theta x) u(x) \tag{6.78}$$

with $u(x)$ being the unit step function. Determine the maximum likelihood estimate for θ under the condition N measurements, $x_1, \ldots, x_N$, are known.

7. The pdf of a random variable has the following exponential distribution

$$p(x|\alpha) = \begin{cases} \theta \; \exp \; -\alpha x, & x \geq 0 \\ 0, & \text{else} \end{cases} \tag{6.79}$$

(a) Plot $p(x|\alpha)$ versus x for a fixed value of the parameter θ.

(b) Plot $p(x|\alpha)$ versus $\theta > 0$ for a fixed value of x.

(c) Suppose that n samples are drawn independently according to $p(x|\alpha)$. Show that the maximum likelihood estimate for α is given by

$$\widehat{\alpha} = \frac{1}{\frac{1}{n}\sum_{k=1}^{n} x_k} \tag{6.80}$$

8. The pdf of a random variable x is given by

$$p(x) = \begin{cases} \frac{1}{3}, & 1 < x < 3 \\ 0, & \text{else} \end{cases} \tag{6.81}$$

Employ for its approximation the Parzen window method and use a Gaussian kernel with $\mu = 0$ and $\sigma = 1$. Let the smoothing parameters be

(a) $h = 0.5$

(b) $h = 3$. For (a) and (b) plot the approximation based on $N = 32, N = 256$, and $N = 5000$ points, which are produced by a pseudorandom generator according to $p(x)$.

9. The pdf of a random variable x is given by

$$p(x) = \begin{cases} 1, & -2.5 < x < -2 \\ 0.125, & 0 < x < 4 \\ 0, & \text{else} \end{cases} \tag{6.82}$$

Employ for its approximation the Parzen window method and use a Gaussian kernel with $\mu = 0$ and $\sigma = 1$. Let the smoothing parameters be

(a) $h = 0.25$

(b) $h = 1$.

For (a) and (b) plot the approximation based on $N = 32, N = 256$, and $N = 5000$ points, which are produced by a pseudorandom generator according to $p(x)$.

10. Determine $L(\mathbf{G})$ for each of the following grammars [329], p. 142:

(a) Context-Free:

$$\begin{aligned} S &\to aAa \\ A &\to a \\ A &\to b \end{aligned} \tag{6.83}$$

(b) Context-Sensitive:

$$\begin{aligned} S &\rightarrow SC \\ CB &\rightarrow Cb \\ aB &\rightarrow aa \\ bB &\rightarrow bb \end{aligned} \tag{6.84}$$

(c) Finite-State:

$$\begin{aligned} S &\rightarrow aA_1 \\ S &\rightarrow bA_1 \\ A_1 &\rightarrow a \\ A_1 &\rightarrow b \end{aligned} \tag{6.85}$$

11. Define a grammar that produces the language

$$L = \{a^n b^{n+2} | n > 0\} \tag{6.86}$$

12. For the Finite-State Grammar productions shown in Example 6.10c
(a) Determine $L(\mathbf{G})$.
(b) How does $L(\mathbf{G})$ change if the production rules are changed to

$$\begin{aligned} S &\rightarrow bA_1 \\ S &\rightarrow aA_1 \\ A_1 &\rightarrow b \\ A_1 &\rightarrow a \end{aligned} \tag{6.87}$$

13. Let $X = \{\mathbf{x}_i, i = 1, \ldots, 5\}$, with $\mathbf{x}_1 = [1,1]^T, \mathbf{x}_2 = [2,1]^T, \mathbf{x}_3 = [4,6]^T, \mathbf{x}_4 = [2,7]^T$, and $\mathbf{x}_5 = [3,4]^T$.
(a) Determine the dissimilarity matrix of the data set when the Euclidean distance is in use.
(b) Perform on the data set agglomerative clustering based on the single link algorithm [364] described in eq. (6.67).

CHAPTER SEVEN

Foundations of Neural Networks

Contents

Pattern Recognition and Signal Analysis in Medical Imaging
http://dx.doi.org/10.1016/B978-0-12-409545-8.00007-8

7.1. INTRODUCTION

Advances in clinical medical imaging have brought about the routine production of vast numbers of medical images that need to be analyzed. As a result, an enormous amount of computer vision research effort has been targeted at achieving automated medical image analysis. This has proved to be an elusive goal in many cases. The complexity of the problems encountered in analyzing these images has prompted considerable interest in the use of neural networks for such applications.

Artificial neural networks are an attempt to emulate the processing capabilities of biological neural systems. The basic idea is to realize systems capable of performing complex processing tasks by interconnecting a high number of very simple processing elements which might even work in parallel. They solve cumbersome and intractable problems by learning directly from data. An artificial neural network usually consists of a large amount of simple processing units, i.e., neurons, with mutual interconnections. It learns to solve problems by adequately adjusting the strength of the interconnections according to input data. Moreover, it can be easily adapted to new environments by learning. At the same time, it can deal with information that is noisy, inconsistent, vague, or probabilistic. These features motivate extensive research and developments in artificial neural networks.

The main features of artificial neural networks are their massive parallel processing architectures and the capabilities of learning from the presented inputs. They can be utilized to perform a specific task only by means of adequately adjusting the connection weights, i.e., by training them with the presented data. For each type of artificial neural network, there exists a corresponding learning algorithm by which we can train the network in an iterative updating manner. Those learning algorithms fit into two main categories: supervised learning and unsupervised learning.

For supervised learning, not only the input data but also the corresponding target answers are presented to the network. Learning is done by the direct comparison of the actual output of the network with known correct answers. This is also referred to as learning with a teacher. In contrast, if only input data without the corresponding target answers are presented to the network for learning, we have unsupervised learning. In fact, the learning goal is not defined at all in terms of specific correct examples. The available information is in the correlations of the input data. The network is expected to create categories from these correlations and to produce output signals corresponding to the input category.

Neural networks have been successfully employed to solve a variety of computer vision problems. They are systems of interconnected, simple processing elements. Generally speaking, a neural network aims to learn the nonlinear mapping from an input space describing the sensor information onto an output space describing the classes to which the inputs belong.

There are three basic mapping neural networks known in the literature [129]:

1. *Recurrent networks:* They are nonlinear, fully interconnected systems and form an autoassociative memory. Stored patterns correspond to stable states of a nonlinear system. The neural network is trained via a storage prescription that forces stable states to correspond to local minima of a network "energy" function. The memory capacity, or allowed number of stored patterns, is related to the network size. The Hopfield neural network [139] is the most popular recurrent neural network. A similar model is the bidirectional associative memory (BAM) [188].
2. *Universal feedforward neural networks:* They implement a nonlinear mapping between an input and output space and are nonlinear function approximators. Also, the network is able to extract higher-order statistics. The multilayer perceptron (MLP) [219], the backpropagation-type neural network [71], the radial basis neural network [255], and the support vector machine (SVM) [335] are the best-known universal feedforward neural networks.
3. *Local interaction-based neural networks:* These networks are based on competitive learning; the output neurons of the network compete among themselves to be activated. The output that wins the competition is called a winning neuron. One way of introducing competition among the output neurons is to use lateral inhibitory connections between them [314]. The local interaction is found in Kohonen maps [183], ART maps [117,118], and in the von der Malsburg model [76,390].

The classification in terms of architecture and learning paradigm of the neural networks described in this chapter is shown in Fig. 7.1.

There exist many types of neural networks that solve a wide range of problems in the area of image processing. There are also many types of neural networks, determined by

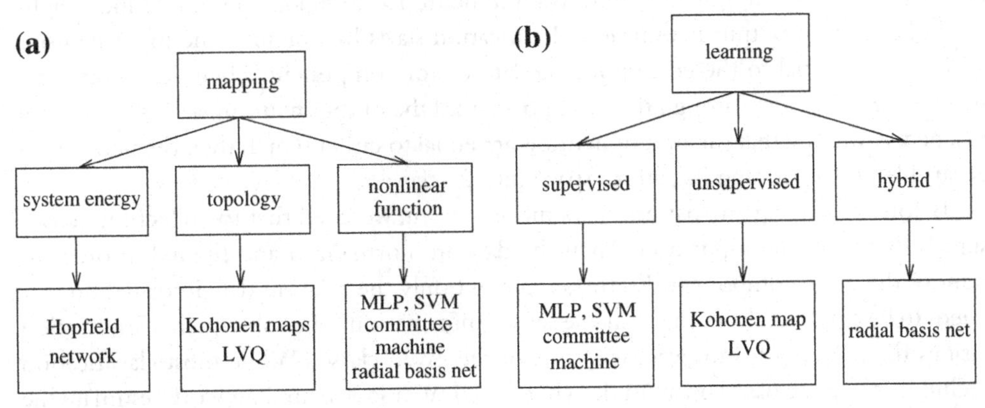

Figure 7.1 Classification of neural networks based on (a) architecture type and (b) learning algorithm.

the type of connectivity between the processing elements, the weights (synapses) of the connecting links, the processing elements' characteristics, and training or learning rules. These rules specify an initial set of weights and indicate how weights should be modified during the learning process to improve network performance.

The theory and representation of the various network types is motivated by the functionality and representation of biological neural networks. In this sense, processing units are usually referred to as neurons, while interconnections are called synaptic connections.

Although different neural models are known, all have the following basic components in common:

1. A finite set of neurons $x(1), x(2), \ldots, x(n)$ with each neuron having a specific activity at time t, which is described by $x_t(i)$.
2. A finite set of neural connections $\mathbf{W} = (w_{ij})$, where w_{ij} describes the strength of the connection of neuron $x(i)$ with neuron $x(j)$.
3. A propagation rule $\tau_t(i) = \sum_{j=1}^{n} x_t(j)w_{ij}$.
4. An activation function f, which has τ as an input value and produces the next state of the neuron $x_{t+1}(i) = f(\tau_t(i) - \theta)$, where θ is a threshold and f is a nonlinear function such as a hard limiter, threshold logic, or sigmoidal function.

7.2. MULTILAYER PERCEPTRON (MLP)

Multilayer perceptrons are one of the most important types of neural nets because many applications are successful implementations of MLPs. The architecture of the MLP is completely defined by an input layer, one or more hidden layers, and an output layer. Each layer consists of at least one neuron. The input vector is processed by the MLP in a forward direction, passing through each single layer. Figure 7.2 illustrates the configuration of the MLP.

A neuron in a hidden layer is connected to every neuron in the layers above and below it. In Fig. 7.2 weight w_{ij} connects input neuron x_i to hidden neuron h_j and weight v_{jk} connects h_j to output neuron o_k. Classification starts by assigning the input neurons $x_i, 1 \leq i \leq l$, equal to the corresponding data vector component. Then data propagates in a forward direction through the perceptron until the output neurons $o_k, 1 \leq k \leq n$, are reached. Assuming that the output neurons are equal to either 0 or 1, then the perceptron is capable of partitioning its pattern space into 2^n classes.

Before employing any pattern recognition system, we need first to collect the sensor signals that form the input data. Then the data are normalized and filtered in order to remove the noise component. Features represent only the relevant data information that needs to be extracted from the available sensor information. The subsequent classification assigns the resulting feature vector to one of the given classes. While most classification techniques operate based on an underlying statistical model, neural networks estimate the necessary discriminant functions. Mathematically, the multilayer perceptron performs a nonlinear approximation using sigmoidal kernel functions as hidden units and linear

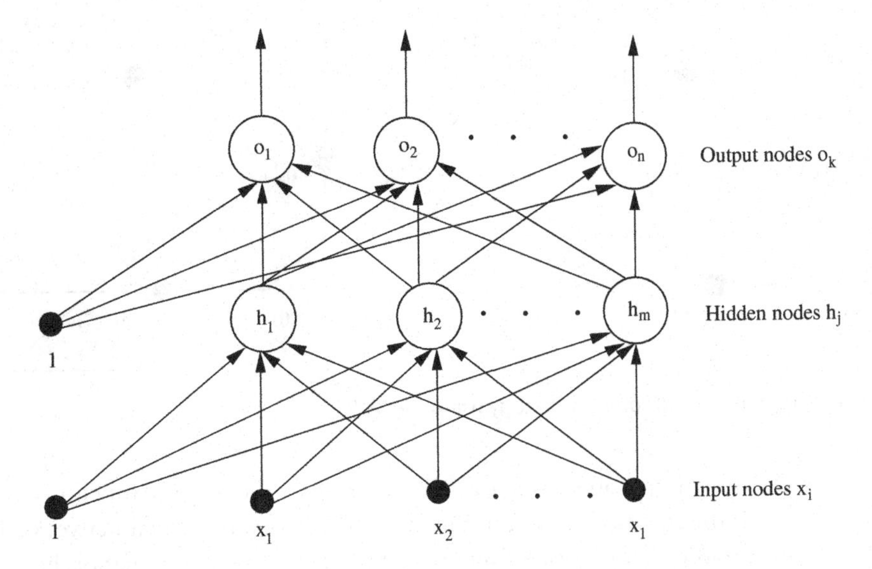

Figure 7.2 Two-layer perceptron.

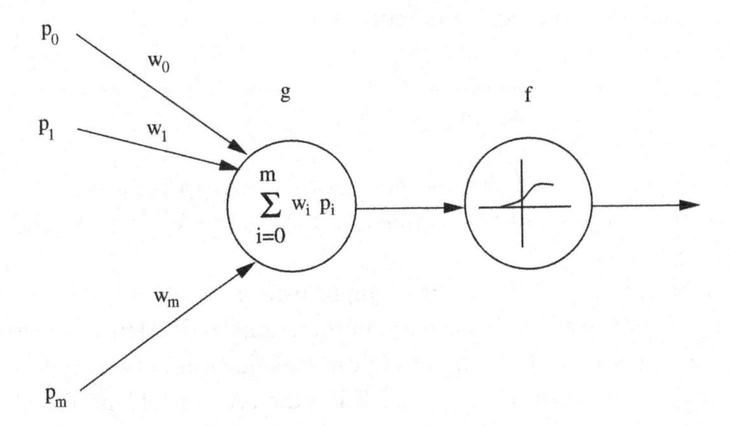

Figure 7.3 Propagation rule and activation function for the MLP network.

weights. During the learning of feature vectors, the weights are continuously adapted by minimizing the error between desired outputs and the computed network's outputs.

The steps that govern the data flow through the perceptron during classification are [310]:

1. Present the pattern $\mathbf{p} = [p_1, p_2, \ldots, p_l] \in \mathbf{R}^{\mathbf{l}}$ to the perceptron, that is, set $x_i = p_i$ for $1 \leq i \leq l$.
2. Compute the values of the hidden-layer neurons as illustrated in Fig. 7.3.

$$h_j = \frac{1}{1 + \exp\left[-\left(w_{0j} + \sum_{i=1}^{l} w_{ij}x_i\right)\right]}; \quad 1 \leq j \leq m \tag{7.1}$$

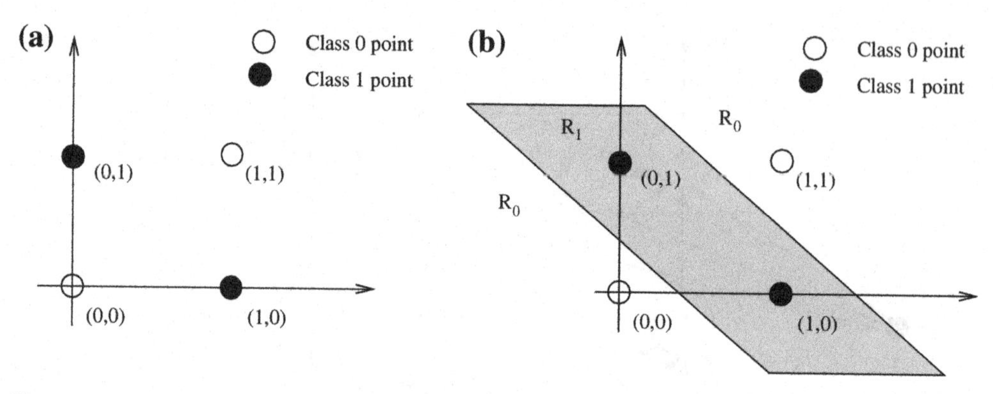

Figure 7.4 XOR problem and solution strategy using the MLP.

The activation function of all units in the MLP is given by the sigmoid function $f(x) = \frac{1}{1+\exp(-x)}$ and is the standard activation function in feedforward neural networks. It is defined as a monotonic increasing function representing an approximation between nonlinear and linear behavior.

3. Calculate the values of the output neurons based on

$$o_k = \frac{1}{1 + \exp\left(v_{0k} + \sum_{j=1}^{m} v_{jk} h_j\right)}; \quad 1 \leq k \leq n \tag{7.2}$$

4. The class $\mathbf{c} = [c_1, c_2, \ldots, c_n]$ that the perceptron assigns $\mathbf{p}$ must be a binary vector. So o_k must be the threshold of a certain class at some level τ and depends on the application.

5. Repeat steps 1,2,3, and 4 for each given input pattern.

Multilayer perceptrons are highly nonlinear interconnected systems and serve for both nonlinear function approximation and nonlinear classification tasks. A typical classification problem that can be solved only by the MLP is the XOR problem. Based on a linear classification rule, R^m can be partitioned into regions separated by a hyperplane. On the other hand, the MLP is able to construct very complex decision boundaries as depicted in Fig. 7.4.

Applied to image processing the MLP has as input features extracted from images or from regions of these images. Such features can be shape, size, or texture measures and attempt to capture the key aspects of an image.

Multilayer perceptrons have been applied to a variety of problems in image processing, including optical character recognition [320] and medical diagnosis [77,78].

7.2.1 Backpropagation-Type Neural Networks

Multilayer perceptrons (MLPs) have been applied successfully to sophisticated classification problems. The training of the network is accomplished based on a supervised learning technique that requires given input-output data pairs. The training technique, known as the error backpropagation algorithm, is bidirectional consisting of a forward and backward direction. During the forward direction a training vector is presented to the network and classified. The backward direction consists of a recursive updating of the weights in all layers based on the computed errors. The error terms of the output layer are a function of $\mathbf{c}^t$ and output of the perceptron $(o_1, o_2, \ldots, o_n)$.

The algorithmic description of the backpropagation is as follows [71]:

1. **Initialization:** Initialize the weights of the perceptron randomly with numbers between -0.1 and 0.1; that is,

$$\begin{aligned} w_{ij} &= \text{random}([-0.1, 0.1]), \quad 0 \leq i \leq l, \quad 1 \leq j \leq m \\ v_{jk} &= \text{random}([-0.1, 0.1]), \quad 0 \leq j \leq m, \quad 1 \leq k \leq n \end{aligned} \tag{7.3}$$

2. **Presentation of training patterns:** Present $\mathbf{p}^t = [p_1^t, p_2^t, \ldots, p_l^t]$ from the training pair $(\mathbf{p}^t, \mathbf{c}^t)$ to the perceptron and apply steps 1, 2, and 3 from the perceptron classification algorithm previously described.
3. **Forward computation (output layer):** Calculate the errors δ_{ok}, $1 \leq k \leq n$, in the output layer using

$$\delta_{ok} = o_k(1 - o_k)\left(c_k^t - o_k\right) \tag{7.4}$$

where $\mathbf{c}^t = [c_1^t, c_2^t, \ldots, c_n^t]$ represents the correct class of $\mathbf{p}^t$. The vector $(o_1, o_2, \ldots, o_n)$ describes the output of the perceptron.

4. **Forward computation (hidden layer):** Calculate the errors δ_{hj}, $1 \leq j \leq m$, in the hidden-layer neurons based on

$$\delta_{hj} = h_j(1 - h_j) \sum_{k=1}^{n} \delta_{ok} v_{jk} \tag{7.5}$$

5. **Backward computation (output layer):** Let v_{jk} denote the value of weight v_{jk} after the tth training pattern has been presented to the perceptron. Adjust the weights between the output layer and the hidden layer based on

$$v_{jk}(t) = v_{jk}(t-1) + \eta \delta_{ok} h_j \tag{7.6}$$

The parameter $0 \leq \eta \leq 1$ represents the learning rate.

6. **Backward computation (hidden layer):** Adjust the weights between the hidden layer and the input layer using

$$w_{ij}(t) = w_{ij}(t-1) + \eta \delta_{hj} p_i^t \tag{7.7}$$

7. **Iteration:** Repeat steps 2–6 for each pattern vector of the training data. One cycle through the training set defines an iteration.

7.2.2 Design Considerations

MLPs are global approximators and can be trained to implement any given nonlinear input-output mapping. In a subsequent testing phase, they prove their interpolation ability by generalizing even in sparse data space regions.

When designing a neural network, specifically deciding for a fixed architecture, performance and computational complexity considerations play a crucial role. Mathematically, it has been proved [126] that even one hidden-layer MLP is able to approximate the mapping of any continuous function.

As with all neural networks, the dimension of the input vector dictates the number of neurons in the input layer, while the number of classes to be learned dictates the number of neurons in the output layer. The number of chosen hidden layers and the number of neurons in each layer have to be empirically determined. As a rule of thumb, the neurons in the hidden layers are chosen as a fraction of those in the input layer. However, there is a trade-off regarding the number of neurons: Too many produce overtraining; too few affect generalization capabilities. Usually, after satisfactory learning is achieved, the neurons' number can be gradually reduced. Then subsequent retraining of a reduced-size network exhibits much better performance than the initial training of the more complex network.

Great care must be taken when selecting the training data set. It has to be representative of all data classes to allow a good generalization. During training, input vectors are presented in random order so that the network achieves a global learning and not class-selective learning.

7.2.3 Complexity of the Multilayer Perceptron

The complexity of a classifier is solely described by its adjustable parameters. Thus, the MLP's adaptable parameters are the number of weights and biases. Theoretically, there is always an optimal complexity for a given classification problem. If this optimal complexity for an MLP is exceeded, then the classification outcome becomes prone to the intrinsic data noise. Although the MLP learns all training examples very well, it fails to correctly recognize new feature vectors it did not learn. This effect is called overfitting and always occurs when the number of weights exceeds the number of input patterns. Underfitting occurs when we have a lower complexity than the optimum, and thus an insufficient discrimination capability. In other words, each classification problem requires a certain optimum MLP complexity, and thus complexity control becomes an issue.

Complexity control can be obtained by the following techniques:

- *Data preprocessing:* Feature extraction, feature selection, feature dimensionality reduction, as shown in Chapter 2.
- *Training scheme:* Early stopping, cross-validation.
- *Network structure:* Committee machines.

7.2.3.1 Cross-Validation

The simplest method to avoid overfitting is to provide more training samples so that the MLP cannot model the noise of the training data set. In practical applications, it may happen that there is a shortage of labeled data available.

A solution to this dilemma poses a technique from statistics known as cross-validation [352]. The MLP is trained on a disjoint subset of the input data, and then a cross-validation subset of the data is used. Thus, the MLP is trained on the training set and then the classification error is checked on the validation set. This is repeated, and then the training is stopped when the cross-validation measure given by the classification errors stops improving. This means that continued training results in overtraining on the training data subset of input data, and this implies that the MLP loses its generalization ability on the cross-validation set.

This procedure proves to be extremely useful when we want to achieve a good generalization (optimal number of hidden neurons) and when we want to decide when the training of the MLP has to be stopped.

7.2.3.2 The Process of Training and Validation

The goal of training a neural network is to achieve a good generalization. However, it is difficult to decide when it is best to stop training without ending up with overfitting. Overfitting can be easily detected by using cross-validation.

In general, the process of achieving good generalization with an MLP consists of three parts: (1) training, (2) validation, and (3) verification. This means, that we need three disjoint sets of input data: a training set that is the largest set, a validation set, and a verification set or test set. Validation has to be performed after cycles of several iterations during training to check when the classification error stops decreasing. The classification error is an ideal sensitivity measure in this respect since it deteriorates when the training of the MLP is too specialized on the training set. The advantage of this method is that we can stop the training before the classification error starts degradating and choose the respective MLP configuration as its final. Hence the name "early stopping." Figure 7.5

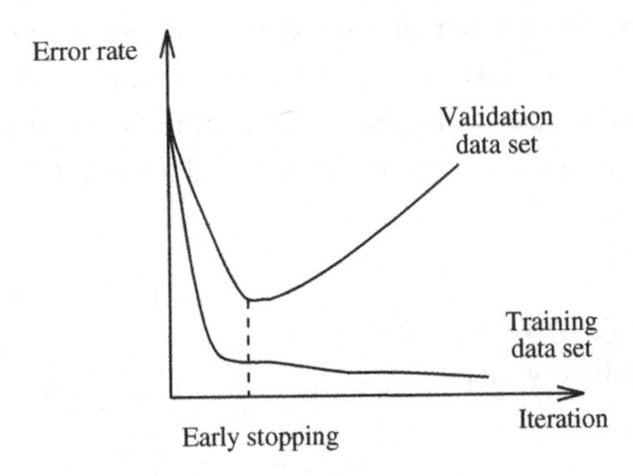

Figure 7.5 Early stopping method of training.

visualizes this technique. The validation subset is small compared to the training set. The final step represents the testing phase when the trained MLP is verified to demonstrate that it is indeed implementing a correct classification mechanism. Usually, the test set is half the size of the training set.

7.2.3.3 Committee Machines

Complex classification problems in practice require the contribution of several neural networks for achieving an optimal solution. Thus a complex task is split into smaller ones whose contributions are necessary to provide the global complex solution. In neural network terminology, it means to allocate the learning task among a number of experts, which in turn split the input data into a set of subspaces. This idea leads to the concept of the committee machine, which is based on a combination of concurrent experts, and thus implements the famous principle of divide and conquer. In this way, it combines the knowledge of every single expert and achieves a global decision that is superior to that achieved by a single expert. The idea of the committee machine was first introduced by Nilsson [262].

Committee machines fall into the category of universal approximators and can have either a static or a dynamic structure. In the dynamic structures, the input signal directly influences the mechanism that fuses the output decisions of the individual experts into an overall output decision. There are two groups of dynamic structures: mixture of experts and hierarchical mixture of experts.

- *Mixture of experts:* A single gating network represents a nonlinear function of the individual responses of the experts, and thus the divide-and-conquer algorithm is applied only once.
- *Hierarchical mixture of experts:* A hierarchy of several gating networks forms a nonlinear function of the individual responses of the experts, and thus the divide and conquer algorithm is applied several times at the corresponding hierarchy levels.

Figure 7.6 shows a hierarchical mixture of experts [157]. The architecture is a tree in which the gating networks sit at the nonterminals of the tree. These networks receive the vector $\mathbf{x}$ as input and produce scalar outputs that are a partition of unity at each point in the input space. The expert networks sit at the leaves of the tree. They are linear with the exception of a single output nonlinearity. Expert network (i, j) produces its output μ_{ij} as a generalized function of the input vector $\mathbf{x}$ and a weight matrix $\mathbf{U}_{ij}$

$$\mu_{ij} = \mathbf{U}_{ij}\mathbf{x} \tag{7.8}$$

The neurons of the gating networks are nonlinear.

Let ξ_i be an intermediate variable

$$\xi_i = \mathbf{v}_i^T \mathbf{x} \tag{7.9}$$

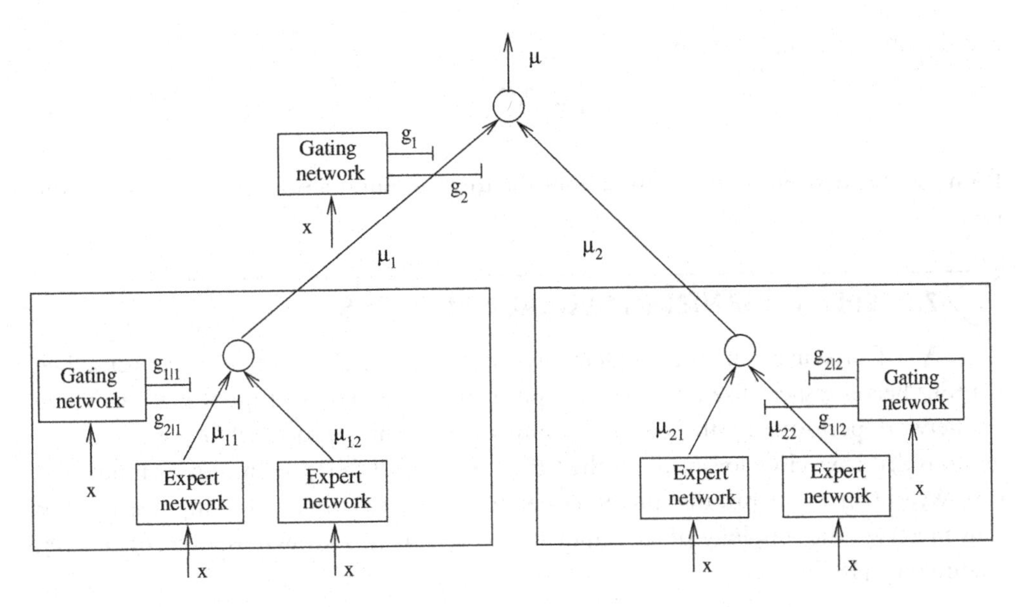

Figure 7.6 Hierarchical mixture of experts.

$\mathbf{v}_i$ is a weight vector. Then the ith output of the top-level gating network is the "softmax" function of ξ_i given as

$$g_i = \frac{\exp(\xi_i)}{\sum_k \exp(\xi_k)} \tag{7.10}$$

Note that $g_i > 0$ and $\sum_i g_i = 1$. The g_i values can be interpreted as providing a "soft" partitioning of the input space.

Similarly, the neurons of the gating networks at lower levels are also nonlinear. Define ξ_{ij} as

$$\xi_{ij} = \mathbf{v}_{ij}^T \mathbf{x} \tag{7.11}$$

Then

$$g_{i|j} = \frac{\exp(\xi_{ij})}{\sum_k \exp(\xi_{ik})} \tag{7.12}$$

is the output of the jth unit in the ith gating network at the second level of the architecture.

The output vector at each nonterminal of the tree is the weighted output of the experts below the nonterminal. That is, the output at the ith nonterminal in the second layer is

$$\mu_i = \sum_j g_{j|i}\, \mu_{ij} \tag{7.13}$$

and the output at the top level is

$$\mu = \sum_i g_i \mu_i \tag{7.14}$$

Both g and μ depend on the input $\mathbf{x}$; thus the total output is a nonlinear function of the input.

7.3. SELF-ORGANIZING NEURAL NETWORKS

A self-organizing map is implemented by a 1-D or 2-D lattice of neurons. The neurons become specialized to various input patterns or classes of input patterns while the network performs an unsupervised competitive learning process. The weights of the neurons that have close locations on the lattice are trained to represent similar input vectors. We obtain a network that preserves neighborhood relations in the input space, and thus preserves the topology of the input space. This map is known as a self-organizing feature map [184].

7.3.1 Self-Organizing Feature Map

Mathematically, the self-organizing map (SOM) determines a transformation from a high-dimensional input space onto a one- or two-dimensional discrete map. The transformation takes place as an adaptive learning process such that when it converges the lattice represents a topographic map of the input patterns. The training of the SOM is based on a random presentation of several input vectors, one at a time. Typically, each input vector produces the firing of one selected neighboring group of neurons whose weights are close to this input vector.

The most important components of such a network are:

1. A 1-D or 2-D lattice of neurons that encodes an input pattern of an arbitrary dimension into a specific location on it, as shown in Fig. 7.7a.
2. A method that selects a "winning neuron" based on the similarity between the weight vector and the input vector.
3. An interactive network that activates the winner and its neighbors simultaneously. A neighborhood $\Lambda_{i(\mathbf{x})}(n)$ which is centered on the winning neuron is a function of the discrete time n. Figure 7.7b illustrates such a neighborhood, which first includes the whole array and then shrinks gradually to only one "winning neuron," represented by the black circle.
4. An adaptive learning process that reinforces all neurons in the close neighborhood of the winning neuron, and inhibits all those that are farther away from the winner.

The learning algorithm of the self-organized map is simple and is described below:

1. **Initialization:** Choose random values for the initial weight vectors $\mathbf{w}_j(0)$ to be different for $j = 1, 2, \ldots, N$, where N is the number of neurons in the lattice. The magnitude of the weights should be small.

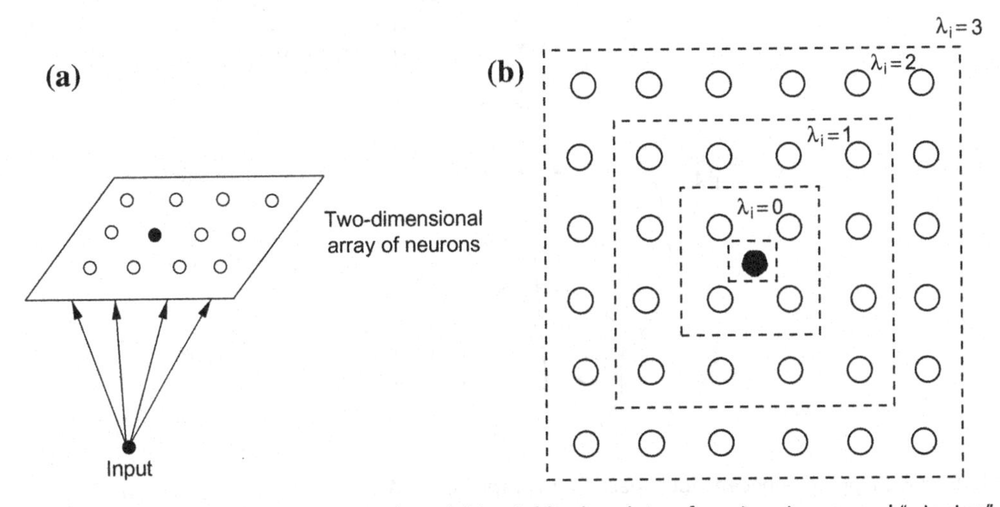

Figure 7.7 (a) Kohonen neural network; and (b) neighborhood Λ_i, of varying size, around "winning" neuron i, identified as a black circle.

2. **Sampling:** Draw a sample $\mathbf{x}$ from the input data; the vector $\mathbf{x}$ represents the new pattern that is presented to the lattice.
3. **Similarity matching:** Find the "winner neuron" $i(\mathbf{x})$ at time n based on the minimum distance Euclidean criterion:

$$i(\mathbf{x}) = \arg\min_j ||\mathbf{x}(n) - \mathbf{w}_j(n)||, \quad j = 1, 2, \ldots, N \tag{7.15}$$

4. **Adaptation:** Adjust the synaptic weight vectors of all neurons (winners or not), using the update equation

$$\mathbf{w}_j(n+1) = \begin{cases} \mathbf{w}_j(n) + \eta(n)[\mathbf{x}(n) - \mathbf{w}_j(n)], & j \in \Lambda_{i(\mathbf{x})}(n) \\ \mathbf{w}_j(n), & \text{else} \end{cases} \tag{7.16}$$

   where $\eta(n)$ is the learning-rate parameter, and $\Lambda_{i(\mathbf{x})}(n)$ is the *neighborhood function* centered around the winning neuron $i(\mathbf{x})$; both $\eta(n)$ and $\Lambda_{i(\mathbf{x})}$ are a function of the discrete time n and thus continuously adapted for optimal learning.
5. **Continuation:** Go to step 2 until there are no noticeable changes in the feature map.

The presented learning algorithm has some interesting properties, which are described based on Fig. 7.8.

Mathematically, the feature map represents a nonlinear transformation Φ from a continuous input space X to a spatially discrete output space A:

$$\Phi : X \to A \tag{7.17}$$

The map preserves the topological relationship that exists in the input space, if the input space is of lower or equal dimensionality compared to the output space. In all other cases,

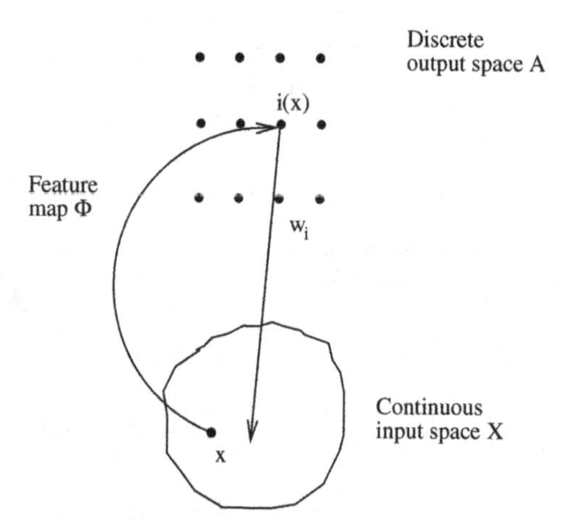

Figure 7.8 Mapping between input space X and output space A.

the map is said to be neighborhood preserving, in the sense that neighboring regions of the input space activate neurons that are adjacent on the lattice. Thus, in cases when the topological structure of the input space is not known a priori or is too complicated to be specified, Kohonen's algorithm necessarily fails in providing perfectly topology-preserving maps.

In the following, we will give some basic properties of self-organizing maps. The first property pertains to an approximation of the input space. The self-organizing feature map Φ, completely determined by the neural lattice, learns the input data distribution by adjusting its synaptic weight vectors $\{\mathbf{w}_j | j = 1, 2, \ldots, N\}$ to provide a good approximation to the input space X.

The second property refers to the topological ordering achieved by the nonlinear feature map: There is a correspondence between the location of a neuron on the lattice and a certain domain or distinctive feature of the input space.

Kohonen maps have been applied to a variety of problems in image processing, including texture segmentation [273] and medical diagnosis [190].

7.3.1.1 Design Considerations

A successful application of the Kohonen neural networks is highly dependent on the main parameters of the algorithm, namely, the learning-rate parameter η and the neighborhood function Λ_i. Since there is no theoretical foundation for the choice of these parameters, we have to rely on practical hints [183]: The learning-rate parameter $\eta(n)$ employed for adaptation of the synaptic vector $\mathbf{w}_j(n)$ should be time-varying. For the first 100 iterations $\eta(n)$ it should stay close to unity and decrease slowly thereafter, but remain above 0.1.

The neighborhood function Λ_i has to be wisely selected in order to ensure topological ordering of the weight vectors $\mathbf{w}_j$. Λ_i can have any geometrical form but should always include the winning neuron in the middle. In the beginning, the neighborhood function Λ_i includes all neurons in the network and then slowly shrinks with time. It takes in most cases about 1000 iterations for the radius of the neighborhood function Λ_i to shrink linearly with time n to a small value of only a couple of neighboring neurons.

7.3.2 Learning Vector Quantization

Vector quantization (VQ) provides an efficient technique for data compression. Compression is achieved by transmitting the index of the codeword instead of the vector itself.

VQ can be defined as a mapping that assigns each vector $\mathbf{x} = (x_0, x_1, \ldots, x_{n-1})^T$ in the n-dimensional space R^n to a codeword from a finite subset of R^n. The subset $\mathbf{Y} = \{\mathbf{y}_i : i = 1, 2, \ldots, M\}$ representing the set of possible reconstruction vectors is called a codebook of size M. Its members are called the codewords. In the encoding process, a distance measure is evaluated to locate the closest codeword for each input vector $\mathbf{x}$. Then, the address corresponding to the codeword is assigned to $\mathbf{x}$ and transmitted. The distortion between the input vector and its corresponding codeword $\mathbf{y}$ is defined by the distance, $d(\mathbf{x}, \mathbf{y}) = ||\mathbf{x} - \mathbf{y}||$, where $||\mathbf{x}||$ represents the norm of $\mathbf{x}$.

A vector quantizer achieving a minimum encoding error is referred to as a Voronoi quantizer. Figure 7.9 shows an input data space partitioned into four different regions, called Voronoi cells, and the corresponding Voronoi vectors. These regions describe the collection of only those input vectors that are very close to the respective Voronoi vector.

Recent developments in neural network architectures have led to a new VQ concept, the so-called learning vector quantization (LVQ). It represents an unsupervised learning algorithm associated with a competitive neural network consisting of one input and one output layer. The algorithm permits only the update of the winning prototype, that is, the

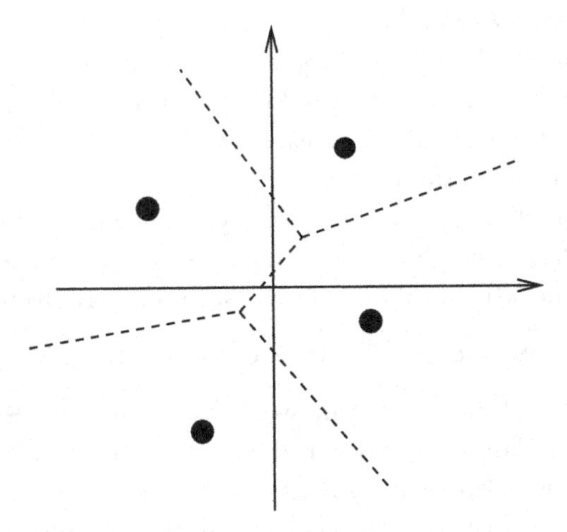

Figure 7.9 Voronoi diagram involving four cells. The small circles indicate the Voronoi vectors and are the different region (class) representatives.

closest prototype (Voronoi vector) of the LVQ network. If the class of the input vector and the Voronoi vector match, then the Voronoi vector is moved in the direction of the input vector $\mathbf{x}$. Otherwise the Voronoi vector $\mathbf{w}$ is moved away from this vector $\mathbf{x}$.

The LVQ algorithm is simple:

1. **Initialization:** Initialize the weight vectors $\{\mathbf{w}_j(0)|j = 1, 2, \ldots, N\}$ by setting them equal to the first N exemplar input feature vectors $\{\mathbf{x}_i|i = 1, 2, \ldots, L\}$.
2. **Sampling:** Draw a sample $\mathbf{x}$ from the input data; the vector $\mathbf{x}$ represents the new pattern that is presented to the LVQ.
3. **Similarity Matching:** Find the best matching codeword (Voronoi vector) $\mathbf{w}_j$ at time n based on the minimum distance Euclidean criterion:

$$\arg\min_j ||\mathbf{x}(n) - \mathbf{w}_j(n)||, \quad j = 1, 2, \ldots, N \tag{7.18}$$

4. **Adaptation:** Adjust only the best matching Voronoi vector, while the others remain unchanged. Assume that a Voronoi vector $\mathbf{w}_c$ is the closest to the input vector $\mathbf{x}_i$. By $C_{\mathbf{w}_c}$ we define the class associated with the Voronoi vector $\mathbf{w}_c$, and by $C_{\mathbf{x}_i}$ the class label associated to the input vector $\mathbf{x}_i$. The Voronoi vector $\mathbf{w}_c$ is adapted as follows:

$$\mathbf{w}_c(n+1) = \begin{cases} \mathbf{w}_c(n) + \alpha_n[\mathbf{x}_i - \mathbf{w}_c(n)], & C_{\mathbf{w}_c} = C_{\mathbf{x}_i} \\ \mathbf{w}_c(n) - \alpha_n[\mathbf{x}_i - \mathbf{w}_c(n)], & \text{otherwise} \end{cases} \tag{7.19}$$

   where $0 < \alpha_n < 1$.
5. **Continuation:** Go to step 2 until there are no noticeable changes in the feature map.

The learning constant α_n is chosen as a function of the discrete time parameter n and decreases monotonically. The relative simplicity of the LVQ and its ability to work in unsupervised mode have made it a useful tool for image segmentation problems [190].

7.3.2.1 The "Neural-Gas" Algorithm

The "neural-gas" algorithm [236] is an efficient approach which, applied to the task of vector quantization, (1) converges quickly to low distortion errors, (2) reaches a distortion error E lower than that from Kohonen's feature map, and (3) at the same time obeys a gradient descent on an energy surface.

Instead of using the distance $||\mathbf{x} - \mathbf{w}_j||$ or the arrangement of the $||\mathbf{w}_j||$ within an external lattice, it utilizes a neighborhood ranking of the reference vectors $\mathbf{w}_i$ for the given data vector $\mathbf{x}$. The adaptation of the reference vectors is given by

$$\Delta\mathbf{w}_i = \epsilon e^{-k_i(\mathbf{x},\mathbf{w}_i/\lambda)}(\mathbf{x} - \mathbf{w}_i) \quad i = 1, \ldots, N \tag{7.20}$$

where N is the number of units in the network. The step size $\epsilon \in [0, 1]$ describes the overall extent of the modification, and k_i is the number of the closest neighbors of the reference vector $\mathbf{w}_i$. λ is a characteristic decay constant.

In [236], it was shown that the average change of the reference vectors can be interpreted as an overdamped motion of particles in a potential that is given by the negative data point density. Added to the gradient of this potential is a "force" in the direction of the space where the particle density is low. This "force" is based on a repulsive coupling between the particles (reference vectors). In form it is similar to an entropic force and tends to uniformly distribute the particles (reference vectors) over the input space, as is the case with a diffusing gas. Hence the name "neural-gas" algorithm. It is also interesting to mention that the reference vectors are slowly adapted, and therefore pointers that are spatially close at an early stage of the adaptation procedure might not be spatially close later. Connections that have not been updated for a while die out and are removed.

Another important feature of the algorithm compared to the Kohonen algorithm is that it does not require a prespecified graph (network). In addition, it can produce topology-preserving maps, which is only possible if the topological structure of the graph matches the topological structure of the data manifold. In cases, however, where an appropriate graph cannot be determined from the beginning, such as where the topological structure of the data manifold is not known in advance or is too complex to be specified, Kohonen's algorithm always fails in providing perfectly topology-preserving maps.

To obtain perfectly topology-preserving maps we employ a powerful structure from computational geometry: the Delaunay triangulation, which is the dual of the already mentioned Voronoi diagram [302]. In a plane, the Delaunay triangulation is obtained if we connect all pairs $\mathbf{w}_j$ by an edge if their Voronoi polyhedra are adjacent. Figure 7.10 shows an example of a Delaunay triangulation. The Delaunay triangulation arises as a graph matching to the given pattern manifold.

The neural gas algorithm is simple:

1. **Initialization:** Randomly initialize the weight vectors $\{\mathbf{w}_j | j = 1, 2, \ldots, N\}$ and the training parameters $(\lambda_i, \lambda_f, \epsilon_i, \epsilon_f)$, where λ_i, ϵ_i are initial values of $\lambda(t), \epsilon(t)$ and λ_f, ϵ_f are the corresponding final values.
2. **Sampling:** Draw a sample $\mathbf{x}$ from the input data; the vector $\mathbf{x}$ represents the new pattern that is presented to the neural gas network.

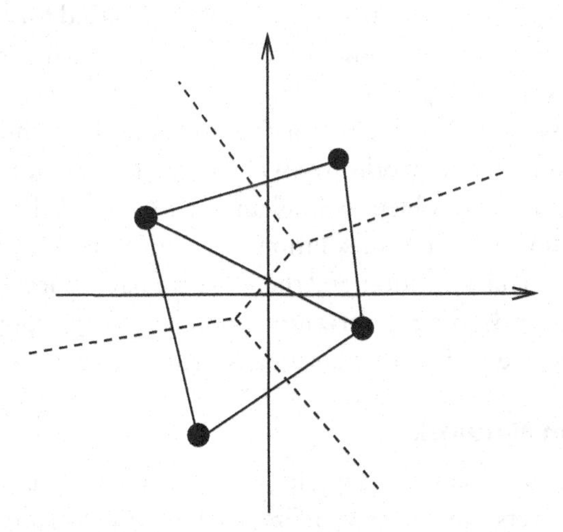

Figure 7.10 Delaunay triangulation.

3. **Distortion:** Determine the distortion set $D_{\mathbf{x}}$ between the input vector $\mathbf{x}$ and the weights $\mathbf{w}_j$ at time n based on the minimum distance Euclidean criterion:

$$D_{\mathbf{x}} = ||\mathbf{x}(n) - \mathbf{w}_j(n)||, \quad j = 1, 2, \ldots, N \tag{7.21}$$

Then order the distortion set in ascending order.
4. **Adaptation:** Adjust the weight vectors according to

$$\Delta \mathbf{w}_i = \epsilon e^{-k_i(\mathbf{x}, \mathbf{w}_i/\lambda)}(\mathbf{x} - \mathbf{w}_i) \quad i = 1, \ldots, N \tag{7.22}$$

where $i = 1, \ldots, N$. The parameters have the following time dependencies: $\lambda(t) = \lambda_i(\lambda_f/\lambda_i)^{\frac{t}{t_{max}}}$ and $\epsilon(t) = \epsilon_i(\epsilon_f/\epsilon_i)^{\frac{t}{t_{max}}}$.
Increment the time parameter t by 1.
5. **Continuation:** Go to step 2 until the maximum iteration number t_{max} is reached.

7.4. RADIAL BASIS NEURAL NETWORKS (RBNN)

Radial basis neural networks describe a new concept of classification, a so-called hybrid learning mechanism. Their architecture is quite simple, based on a three-layer model having only one hidden layer. The hybrid learning mechanism combines a self-organized learning of the hidden layers' neurons with a supervised learning applied to the output layer's weight adaptation.

The design of a neural network can be viewed as a curve-fitting (nonlinear approximation) problem in a high-dimensional space. Thus, learning is equivalent to finding the surface in a multidimensional space that provides the best match to the training data. To be specific, let us consider a system with n inputs and m outputs and let $\{x_1, \ldots, x_n\}$ be an input vector and $\{y_1, \ldots, y_m\}$ the corresponding output vector describing the system's answer to that specific input. After the training process is completed, the system has learned the input and output data distribution and is able to find for any input the correct output. In other words, we look for the "best" approximation function $\widehat{f}(x_1, \ldots, x_n)$ of the actual input-output mapping function [84,296].

In the following, we will describe the mathematical background that is necessary to understand radial basis neural networks. We will review the concept of interpolation and show how the interpolation problem is implemented by a radial basis neural network. Since there are an infinite number of solutions for a given approximation problem, we want to choose one particular solution, and therefore we have to impose some restrictions on the solutions. This leads to a regularization approach to the approximation problem and can be easily implemented by the radial basis neural network.

7.4.1 Interpolation Networks

The interpolation problem can be very elegantly solved by a three-layer feedforward neural network. The layers play entirely different roles in the learning process. The input layer is made up of source nodes that receive sensor information. The second layer, the only hidden layer, applies a nonlinear transformation from the input space to the hidden space. The nonlinear transformation is followed by a linear transformation from the hidden layer to the output layer. This leads to the theory of multivariable interpolation in the high-dimensional feature space.

The mathematical formulation of the interpolation problem is given below.

The Interpolation Prolem 7.4.1 *Assume that to N different points* $\{\mathbf{m}_i \in \mathcal{R}^n | i = 1, \ldots N\}$ *correspond N real numbers* $\{d_i \in \mathcal{R} | i = 1, \ldots, N\}$. *Then find a function* $F : \mathcal{R}^n \rightarrow \mathcal{R}$ *that satisfies the interpolation condition:*

$$F(\mathbf{m}_i) = d_i \quad \textit{for } i = 1, \ldots, N \tag{7.23}$$

The radial basis function technique consists of choosing a function F that has the following form [296]:

$$F(\mathbf{x}) = \sum_{i=1}^{N} c_i h(||\mathbf{x} - \mathbf{m}_i||) + \sum_{i=1}^{M} d_i p_i(\mathbf{x}), \quad M \leq N \tag{7.24}$$

h is a smooth function, known as a radial basis function; $||.||$ is the Euclidian norm in $\mathcal{R}^n$; $\{p_i | i = 1, \ldots, m\} : \mathcal{R}^n \rightarrow \mathcal{R}$ represents a basis of algebraic polynomials in the linear space $\pi_{k-1}(\mathcal{R}^n)$ of maximum order $k - 1$ for a given k.

Let the radial basis function $h(r)$ be continuous on $[0, \infty)$ and its derivatives on $[0, \infty)$ strict monotone. Then we get a simplified form of the preceding equation:

$$F(\mathbf{x}) = \sum_{i=1}^{N} c_i h(||\mathbf{x} - \mathbf{m}_i||) \tag{7.25}$$

It now becomes evident that we can implement the approximation problem with a three-layer neural network as shown in Fig. 7.11. The number of input neurons corresponds to the dimension of the feature space; the number of hidden neurons is problem-specific and is determined during the clustering process. The output neurons perform the superposition of the weighted basis functions and thus solve the approximation problem. Figure 7.11 depicts the architecture of the network for a single output. Clearly, such an architecture can be readily extended to accommodate any desired number of network outputs.

The network architecture is solely determined by the input-output mapping to be achieved. Moreover, from the viewpoint of approximation theory, the network has three desirable properties:

1. The network is a universal approximator that approximates arbitrarily well any multivariate continuous function, given a sufficiently large number of hidden units. This can be shown based on the Stone-Weierstrass theorem [297].

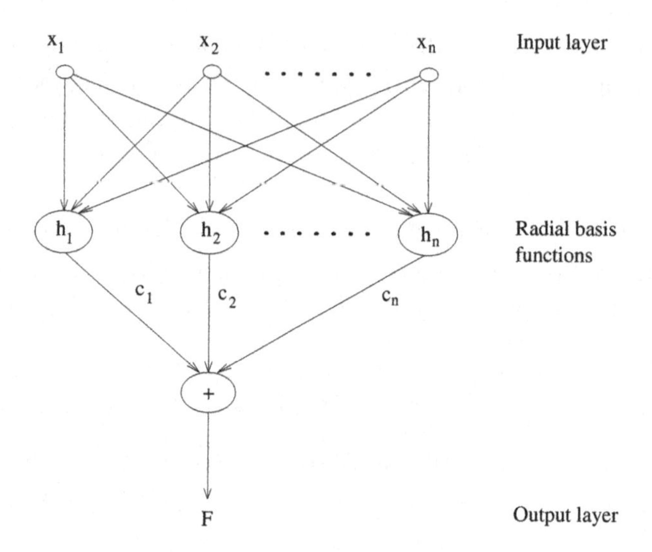

Figure 7.11 Approximation network.

2. The network has the best-approximation property. This means that given an unknown nonlinear function f, there always exists a choice of coefficients that approximate f better than other possible choices. This property holds for all other approximation schemes based on Green's functions as basis functions. It is not the case for a three-layer perceptron having sigmoid functions as activation functions. Sigmoid functions are not Green's functions, being additionally translation and rotation invariant.
3. By taking a closer look at the approximation approach,

$$F(\mathbf{x}) = \sum_{i=1}^{N} c_i h(||\mathbf{x} - \mathbf{m}_i||) \tag{7.26}$$

we can obtain a simplified form if we specify for each of the N radial basis functions a width σ_i,

$$F(\mathbf{x}) = \sum_{i=1}^{N} c_i\, g\left(\frac{||\mathbf{x} - \mathbf{m}_i||}{\sigma_i}\right) \tag{7.27}$$

In [286] it was shown that even then we can solve the approximation problem optimally, if all radial basis functions have the same width $\sigma_i = \sigma$. In neural network terminology, it means that neurons having the same Gaussian function as an activation function for the hidden nodes can approximate any given function.

7.4.2 Regularization Networks

The previous section dealt with the interpolation problem, and we showed how a function $F(\mathbf{x})$ can be determined to satisfy the interpolation condition. Assuming that the data are noise-corrupted, we immediately see that the problem is ill posed since it has an infinite number of solutions. To be able to choose one particular solution, we need to include some a priori knowledge of the function $F(\mathbf{x})$. This can be some prior smoothness information about the function, in the sense that similar inputs yield similar outputs. In [299], smoothness is defined as a measure describing the "oscillatory" behavior of a function. Thus, a smooth function exhibits few oscillations. Translated into the frequency domain, this means that a smooth function has less energy in the high-frequency range and therefore has a smaller bandwidth.

The regularization theory arose as a new approach for solving ill-posed problems [366]. The main idea of this method is to stabilize the solution by means of some auxiliary non-negative functional that includes a priori information about the solution. Thus, the solution to this ill-posed problem can be obtained from a variational principle, which includes both the data and prior smoothness information. Smoothness is considered by introducing a smoothness functional $\phi[F]$. A small value of $\phi[F]$ corresponds to smoother functions.

Let $\{\mathbf{m}_i \in \mathcal{R}^n | i = 1, \ldots N\}$ be N input vectors and $\{d_i \in \mathcal{R} | i = 1, \ldots, N\}$ be N real numbers and represent the desired outputs. The solution to the regularization problem is denoted by $F(\mathbf{x})$, and is required to approximate the data well and at the same time to be smooth. Thus, $F[\mathbf{x}]$ should minimize the following functional:

$$H[F] = \sum_{i=1}^{N} [d_i - F(\mathbf{m}_i)]^2 + \lambda \phi[F] \tag{7.28}$$

where λ is a positive number, called the regularization term. The first term in the equation enforces closeness to the data and measures the distance between the data and the desired solution F. The second term enforces smoothness and measures the cost associated with the deviation from smoothness. The regularization parameter $\lambda > 0$ controls the trade-off between these two terms.

The Principle of Regularization 7.4.1 *Determine the function $F(\mathbf{x})$ that minimizes the Tikhonov functional defined by*

$$H[F] = \sum_{i=1}^{N} [d_i - F(\mathbf{m}_i)]^2 + \lambda \phi[F] \tag{7.29}$$

with λ being the regularization parameter.

The solution for the regularization problem is derived in [299] and is given here in its simplified form

$$F(\mathbf{x}) = \frac{1}{\lambda} \sum_{i=1}^{N} [d_i - F(\mathbf{m}_i)]\, G(\mathbf{x}, \mathbf{m}_i) \tag{7.30}$$

The minimizing solution $F(\mathbf{x})$ to the regularization problem is thus given by a linear superposition of N Green's functions. The $\mathbf{m}_i$s represent the centers of the expansion and the weights $w_i = [d_i - F(\mathbf{m}_i)]\lambda$ represent the coefficients of the expansion.

Equation (7.30) can be similarly expressed as

$$F(\mathbf{x}) = \sum_{i=1}^{N} w_i G(\mathbf{x}_j, \mathbf{m}_i) \tag{7.31}$$

The Green's function is a symmetric function with $G(\mathbf{x}, \mathbf{m}_i) = G(\mathbf{m}_i, \mathbf{x})$. If the smoothness functional $\phi[F]$ is both translationally and rotationally invariant, then the Green's function $G(\mathbf{x}, \mathbf{m}_i)$ will depend only on $||\mathbf{x} - \mathbf{m}_i||$. Under these conditions, the Green's function has the same invariance properties, and eq. (7.31) takes a special form:

$$F_\lambda(\mathbf{x}) = \sum_{i=1}^{N} w_i G(||\mathbf{x} - \mathbf{m}_i||) \tag{7.32}$$

This solution is a strict interpolation, because all training data points are used to generate the interpolating function. The approximation scheme of eq. (7.31) can be easily interpreted as a network with only one hidden layer of units, which is referred to as a regularization network as shown in Fig. 7.11.

Let's take a closer look at the regularization parameter. It can be interpreted as an indicator of the sufficiency of the available data to specify the solution $F(\mathbf{x})$. If $\lambda \to 0$, the solution $F(\mathbf{x})$ is sufficiently determined from the available examples; if $\lambda = 0$, then this corresponds to pure interpolation. If $\lambda \to \infty$, this means the examples are unreliable. In practice, λ is between these two limits in order to enable both the sample data and the prior information to contribute to the solution $F(\mathbf{x})$.

The regularization problem represents an alleviation to a well-known and frequent problem in training neural networks, the so-called overtraining. If noise is added to input data, the neural network learns all the details of the input-output pairs but not their general form. This phenomenon is overtraining. A smart choice of the regularization parameter can prevent this.

Example 7.4.1 Figure 7.12 shows the effect of overtraining based on the example of learning an input-output mapping.

In practical applications, we have two choices for the Tikhonov functional in order to avoid overtraining:

- Zero-order regularization pertains to the weights of the network. The minimizer of the Tikhonov functional has a penalty term for the weights and in that way avoids overtraining:

$$\mathcal{E}(F) = \frac{1}{2} \sum_{p} \sum_{i} (d_i - F(\mathbf{m}_i))^2 + \lambda \sum_{i} \sum_{j} w_{ij}^2 \tag{7.33}$$

- Accelerated regularization pertains to the neurons of the hidden layer. This is a fast regularization network, since it tries to determine the centers and the widths of the radial basis functions.

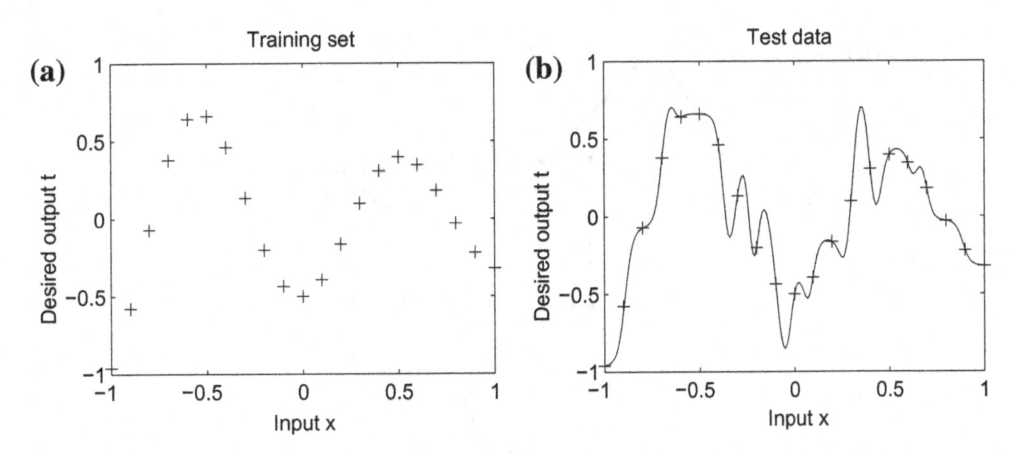

Figure 7.12 Overtraining: (a) learning data; and (b) testing data.

7.4.3 Data Processing in Radial Basis Function Networks

Data processing in a radial basis function neural network is quite different from standard supervised or unsupervised learning techniques. The MLP's backpropagation algorithm represented an optimization method known in statistics as stochastic approximation, while the "winner takes all" concept is neurobiologically inspired strategy used for training self-organizing feature maps or in LVQ.

In this section we present a different method by looking into the design of a neural network as an approximation problem in a high-dimensional space. In neural network parlance, the hidden units are kernel functions that constitute an arbitrary "basis" for the input patterns (vectors) when they are mapped onto the hidden-unit space; these functions are known as radial basis functions. However, there is a neurobiological equivalent to the mathematical concept of kernel or radial basis functions: the receptive field. It describes the local interaction between neurons in a cortical layer.

Fundamental contributions to the theory, design, and application of radial basis function networks are presented in an examplary manner in [255,298].

The construction of a radial basis function (RBF) network in its most basic form involves three different layers. For a network with N hidden neurons, the output of the ith output node $f_i(\mathbf{x})$ when the n-dimensional input vector $\mathbf{x}$ is presented is given by

$$f_i(\mathbf{x}) = \sum_{j=1}^{N} w_{ij}\Psi_j(\mathbf{x}) \tag{7.34}$$

where $\Psi_j(\mathbf{x}) = \Psi(||\mathbf{x} - \mathbf{m}_j||/\sigma_j)$ is a suitable radially symmetric function that defines the output of the jth hidden node. Often $\Psi(.)$ is chosen to be the Gaussian function where the width parameter σ_j is the standard deviation. $\mathbf{m}_j$ is the location of the jth centroid, where each centroid is represented by a kernel/hidden node, and w_{ij} is the

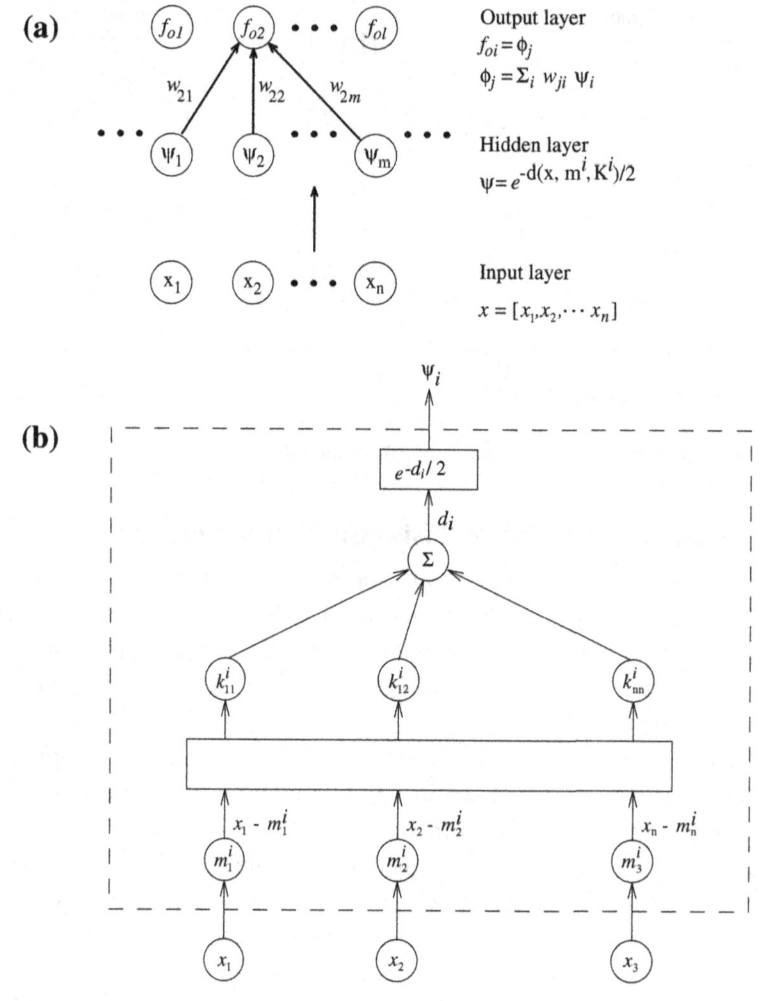

Figure 7.13 RBF network: (a) three-layer model; and (b) the connection between the input layer and the hidden layer neuron.

weight connecting the jth kernel/hidden node to the ith output node. Figure 7.13a illustrates the configuration of the network.

The steps that govern the data flow through the radial basis function network during classification and represent a typical RBF network learning algorithm are described below:

1. **Initialization:** Choose random values for the initial weights of the RBF network. The magnitude of the weights should be small. Choose the centers $\mathbf{m}_i$ and the shape matrices $\mathbf{K}_i$ of the N given radial basis functions.
2. **Sampling:** Draw randomly a pattern $\mathbf{x}$ from the input data. This pattern represents the input to the neural network.

3. **Forward computation of hidden layer's activations:** Compute the values of the hidden layer nodes as illustrated in Fig. 7.13b:

$$\psi_i = \exp\left(-d(\mathbf{x}, \mathbf{m}_i, \mathbf{K}_i)/2\right) \tag{7.35}$$

$d(\mathbf{x}, \mathbf{m}_i) = (\mathbf{x} - \mathbf{m}_i)^T \mathbf{K}_i (\mathbf{x} - \mathbf{m}_i)$ is a metric norm and is known as the Mahalanobis distance. The *shape matrix* $\mathbf{K}_i$ is positive definite and its elements k^i_{jk};

$$\mathbf{K}_i jk = \frac{h_{jk}}{\sigma_j * \sigma_k} \tag{7.36}$$

represent the correlation coefficients h_{jk} and σ_j the standard deviation of the ith shape matrix.
We choose for $h_{jk} : h_{jk} = 1$ for $j = k$ and $|h_{jk}| \leq 1$ otherwise.
4. **Forward computation of output layer's activations:** Calculate the values of the output nodes according to

$$f_{oj} = \phi_j = \sum_i w_{ji} \psi_i \tag{7.37}$$

5. **Updating:** Adjust weights of all neurons in the output layer based on a steepest descent rule.
6. **Continuation:** Continue with step 2 until no noticeable changes in the error function are observed.

The algorithm is based on the a priori knowledge of the radial basis centers and its shape matrices. In the following section, we will review some design strategies for RBF networks and show how we can determine the centers and widths of the radial basis functions.

RBF networks have been applied to a variety of problems in image processing, such as image coding and analysis [321], and in medical diagnosis [410].

7.4.3.1 *Design Considerations*

RBF networks represent, in contrast to the MLP, local approximators to nonlinear input-output mapping. Their main advantages are a short training phase and a reduced sensitivity to the order of presentation of training data. In many cases, however, we find that a smooth mapping is only achieved if the number of radial basis functions required to span the input space becomes very large. This hampers the feasibility of many practical applications.

The RBF network has only one hidden layer, and the number of basis functions and their shape is problem-oriented and can be determined online during the learning process [211,295]. The number of neurons in the input layer equals the dimension of the feature vector. Likewise, the number of nodes in the output layer corresponds to the number of classes.

7.4.4 Training Techniques for RBF Networks

As stated in the beginning, the learning process of the RBF networks is considered hybrid. We distinguish two different dynamics: the adjustment of the linear weights of the output neurons and the nonlinear activation functions of the hidden neurons. The weight adjustment of the output layer's neurons represents a linear process compared to the nonlinear parameter adjustment of the hidden layer neurons. Since the hidden and output layers of an RBF network perform different tasks, we must separate the optimization of the hidden and output layers, and thus employ different learning techniques. The output layer's synapses are updated according to a simple delta rule as shown in the MLP case.

We have some degrees of freedom in choosing the kernel functions of an RBF network: in the simple case they are fixed, or they can be adjusted during the training phase.

There are several techniques to design an RBF network, depending on how the centers of the radial basis functions of the network are determined. The best-known strategies chosen for practical applications are reviewed in the following.

7.4.4.1 Fixed Centers of the RBF Selected at Random

The simplest case is to assume fixed kernel functions with a center location chosen at random from the training set.

Specifically, a radial basis function centered at $\mathbf{m}_i$ is defined as

$$G(||\mathbf{x} - \mathbf{m}_i||^2) = \exp\left(-\frac{M}{d_{max}^2}||\mathbf{x} - \mathbf{m}_i||^2\right), \quad i = 1, \ldots, M \tag{7.38}$$

M represents the number of centers, and d_{max}^2 is the maximum distance between the selected centers. In fact, the standard deviation of all neurons is fixed at

$$\sigma = \frac{d_{max}}{\sqrt{2M}} \tag{7.39}$$

This guarantees that they are neither too spiky nor too flat. The only parameters that have to be adjusted here are the linear weights in the output layer of the network. They can be determined based on the pseudoinverse method [39]. Specifically, we have

$$\mathbf{w} = \mathbf{G}^+\mathbf{y} \tag{7.40}$$

where $\mathbf{y}$ is the desired response vector in the training set.

$\mathbf{G}^+$ is the pseudoinverse of the matrix $\mathbf{G} = \{g_{ji}\}$, which is itself defined as $\mathbf{G} = \{g_{ji}\}$, where

$$g_{ji} = \exp\left(-\frac{M}{d_{max}^2}||\mathbf{x_j} - \mathbf{m}_i||^2\right), \quad j = 1, \ldots, N; \quad i = 1, \ldots, M \tag{7.41}$$

$\mathbf{x}_j$ is the jth input vector of the training sample.

7.4.4.2 Self-Organized Selection of the Centers of the RBFs

The locations of the centers of the kernel functions (hidden units) are adapted based on a self-organizing learning strategy, whereas the linear weights associated with the output neurons are adapted based on a simple supervised delta rule. It's important to mention that this procedure allocates hidden units only in preponderantly input data regions.

The self-organized learning process is based on a clustering algorithm that partitions the given data set into homogeneous subsets. The k-means clustering algorithm [84] sets the centers of the radial basis functions only in those regions where significant input data are present. The optimal number M of radial basis functions is determined empirically. Let $\{\mathbf{m}_i(n)\}_{i=1}^{M}$ be the centers of the radial basis functions at iteration n of the algorithm.

Then, the k-means algorithm applied to center selection in RBF neural networks proceeds as follows:

1. **Initialization:** Choose randomly the initial centers $\{\mathbf{m}_i(0)\}_{i=1}^{M}$ of the radial basis functions.
2. **Sampling:** Draw a sample $\mathbf{x}$ from the data space with a certain probability.
3. **Similarity matching:** Determine the best-matching (winning) neuron $k(\mathbf{x})$ at iteration n using the Euclidean distance:

$$k(\mathbf{x}) = \overset{\text{argmin}}{k} \quad ||\mathbf{x}(n) - \mathbf{m}_k(n)||, \quad k = 1, 2, \dots M \tag{7.42}$$

   $\mathbf{m}_k(n)$ is the center of the kth radial basis function at iteration n.

4. **Updating:** Adjust the centers of the radial basis functions, using the update equation

$$\mathbf{x}_k(n+1) = \begin{cases} \mathbf{x}_k(n) + \eta(n)[\mathbf{x}(n) - \mathbf{x}_k(n)], & k = k(\mathbf{x}) \\ \mathbf{x}_k(n), & \text{else} \end{cases} \tag{7.43}$$

   η is the learning rate that lies in the range $0 < \eta < 1$.

5. **Continuation:** Go back to step 2 and continue the procedure until no noticeable changes are observed in the radial basis centers.

After determining the centers of the radial basis functions, the next and final stage of the hybrid learning process is to estimate the weights of the output layer based, for example, on the least-mean-square (LMS) algorithm [285].

7.4.4.3 Supervised Selection of Centers

All free parameters of the kernel functions including centers are adapted based on a supervised learning strategy. In such a case, we deal with an RBF network in its most generalized form. This can be realized by an error-correction learning, which requests an implementation based on a gradient-descent procedure.

In the first step, we define a cost function

$$E = \frac{1}{2} \sum_{j=1}^{P} e_j^2 \tag{7.44}$$

Table 7.1 Adaptation formulas for the linear weights and the position and widths of centers for an RBF network [129].

1.	**Linear weights of the output layer** $\frac{\partial \mathcal{E}(n)}{\partial w_i(n)} = \sum_{j=1}^{N} e_j(n) G(\|\|\mathbf{x}_j - \mathbf{m}_i(n)\|\|)$ $w_i(n+1) = w_i(n) - \eta_1 \frac{\partial \mathcal{E}(n)}{\partial w_i(n)}, \quad i = 1, \ldots, M$
2.	**Position of the centers of the hidden layer** $\frac{\partial \mathcal{E}(n)}{\partial \mathbf{m}_i(n)} = 2w_i(n) \sum_{j=1}^{N} e_j(n) G'(\|\|\mathbf{x}_j - \mathbf{m}_i(n)\|\|) \mathbf{K}^{\mathbf{i}}[\mathbf{x}_j - \mathbf{m}_i(n)]$ $\mathbf{m}_i(n+1) = \mathbf{m}_i(n) - \eta_2 \frac{\partial \mathcal{E}(n)}{\partial \mathbf{m}_i(n)}, \quad i = 1, \ldots, M$
3.	**Widths of the centers of the hidden layer** $\frac{\partial \mathcal{E}(n)}{\partial \mathbf{k}^{\mathbf{i}}(n)} = -w_i(n) \sum_{j=1}^{N} e_j(n) G'(\|\|\mathbf{x}_j - \mathbf{m}_i(n)\|\|) \mathbf{Q}_{ji}(n)$ $\mathbf{Q}_{ji}(n) = [\mathbf{x}_j - \mathbf{m}_i(n)][\mathbf{x}_j - \mathbf{m}_i(n)]^T$ $\mathbf{K}^{\mathbf{i}}(n+1) = \mathbf{K}^{\mathbf{i}}(n) - \eta_3 \frac{\partial \mathcal{E}(n)}{\partial \mathbf{K}^{\mathbf{i}}(n)}$

where P is the size of the training sample and e_j is the error defined by

$$e_j = d_j - \sum_{i=1}^{M} w_i G(||\mathbf{x}_j - \mathbf{m}_i||_{C_i}) \quad (7.45)$$

The requirement is to find the free parameters $w_i, \mathbf{m}_i$, and $\mathbf{\Sigma}_i^{-1}$ such that the error E is minimized.

The results of this minimization [129] are summarized in Table 7.1. From the table we can see that the update equations for $w_i, \mathbf{x}_i$, and $\mathbf{\Sigma}_i^{-1}$ have different learning rates, thus showing the different timescales. Unlike the backpropagation algorithm, the gradient-descent procedure described in Table 7.1 for an RBF network does not employ error backpropagation.

7.5. TRANSFORMATION RADIAL BASIS NEURAL NETWORKS

The selection of appropriate features is an important precursor to most statistical pattern recognition methods. A good feature selection mechanism helps to facilitate classification by eliminating noisy or nonrepresentative features that can impede recognition. Even features that provide some useful information can reduce the accuracy of a classifier when the amount of training data is limited. This so-called "curse of dimensionality," along with the expense of measuring and including features, demonstrates the utility of obtaining a minimum-sized set of features that allow a classifier to discern pattern classes well. Well-known methods in the literature applied to feature selection are the floating search methods [304] and genetic algorithms [341].

Radial basis neural networks are excellent candidates for feature selection. It is necessary to add an additional layer to the traditional architecture to obtain a representation

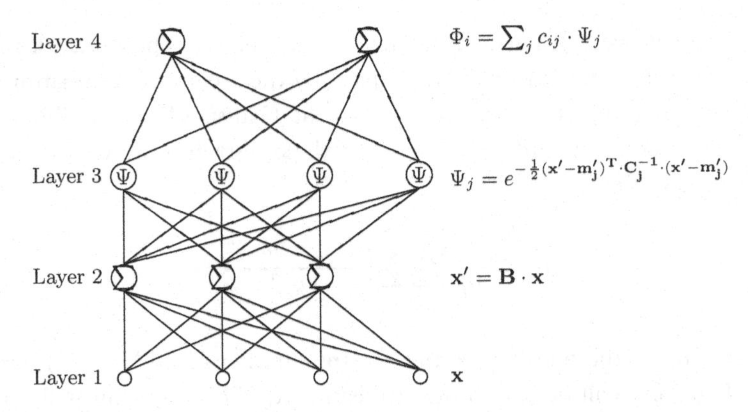

Figure 7.14 Linear transformation radial basis neural network.

of relevant features. The new paradigm is based on an explicit definition of the relevance of a feature and realizes a linear transformation of the feature space.

Figure 7.14 shows the structure of a radial basis neural network with the additional layer 2, which transforms the feature space linearly by multiplying the input vector and the center of the nodes by the matrix $\mathbf{B}$. The covariance matrices of the input vector remain unmodified:

$$\mathbf{x}' = \mathbf{Bx}, \quad \mathbf{m}' = \mathbf{Bm}, \quad \mathbf{C}' = \mathbf{C} \tag{7.46}$$

The neurons in layer 3 evaluate a kernel function for the incoming input while the neurons in the output layer perform a weighted linear summation of the kernel functions:

$$\mathbf{y}(\mathbf{x}) = \sum_{i=1}^{N} w_i \exp\left(-d(\mathbf{x}', \mathbf{m}_i')/2\right) \tag{7.47}$$

with

$$d(\mathbf{x}', \mathbf{m}_i') = (\mathbf{x}' - \mathbf{m}_i')^T \mathbf{C}_i^{-1} (\mathbf{x}' - \mathbf{m}_i') \tag{7.48}$$

Here, N is the number of neurons in the second hidden layer, $\mathbf{x}$ is the n-dimensional input pattern vector, $\mathbf{x}'$ is the transformed input pattern vector, $\mathbf{m}_i'$ is the center of a node, w_i are the output weights, and $\mathbf{y}$ represents the m-dimensional output of the network. The $n \times n$ covariance matrix $\mathbf{C}_i$ is of the form

$$C_{jk}^i = \begin{cases} \frac{1}{\sigma_{jk}^2} & \text{if} \quad m = n \\ 0 & \text{otherwise} \end{cases} \tag{7.49}$$

where σ^{jk} is the standard deviation. Because the centers of the Gaussian potential function units (GPFUs) are defined in the feature space, they will be subject to transformation by $\mathbf{B}$, as well. Therefore, the exponent of a GPFU can be rewritten as

$$d(\mathbf{x}, \mathbf{m}_i') = (\mathbf{x} - \mathbf{m}_i)^T \mathbf{B}^T \mathbf{C}_i^{-1} \mathbf{B} (\mathbf{x} - \mathbf{m}_i) \tag{7.50}$$

and is in this form similar to eq. (7.48).

For the moment, we will regard **B** as the identity matrix. The network models the distribution of input vectors in the feature space by the weighted summation of Gaussian normal distributions, which are provided by the Gaussian Potential Function Units (GPFU) Ψ_j. To measure the difference between these distributions, we define the relevance ρ_n for each feature x_n:

$$\rho_n = \frac{1}{PJ} \sum_p \sum_j \frac{(x_{pn} - m_{jn})^2}{2\sigma_{jn}^2} \tag{7.51}$$

where P is the size of the training set and J is the number of GPFUs. If ρ_n falls below the threshold ρ_{th}, one will decide to discard feature x_n. This criterion will not identify every irrelevant feature: If two features are correlated, one of them will be irrelevant, but this cannot be indicated by the criterion.

7.5.1 Learning Paradigm for the Transformation Radial Basis Neural Network

We follow the idea of [211] for the implementation of the neuron allocation and learning rules for the TRBNN. The network generation process starts initially without any neuron.

The mutual dependency of correlated features can often be approximated by a linear function, which means that a linear transformation of the input space can render features irrelevant.

First we assume that layers 3 and 4 have been trained so that they comprise a model of the pattern-generating process while **B** is the identity matrix. Then the coefficients B_{nr} can be adapted by gradient descent with the relevance ρ'_n of the transformed feature x'_n as the target function. Modifying B_{nr} means changing the relevance of x_n by adding x_r to it with some weight B_{nr}. This can be done online, that is, for every training vector $\mathbf{x_p}$ without storing the whole training set. The diagonal elements B_{nn} are constrained to be constant 1, because a feature must not be rendered irrelevant by scaling itself. This in turn guarantees that no information will be lost. B_{nr} will only be adapted under the condition that $\rho_n < \rho_p$, so that the relevance of a feature can be decreased only by some more relevant feature. The coefficients are adapted by the learning rule

$$B_{nr}^{new} = B_{nr}^{old} - \mu \frac{\partial \rho_n}{\partial B_{nr}} \tag{7.52}$$

with the learning rate μ and the partial derivative

$$\frac{\partial \rho_n}{\partial B_{nr}} = \frac{1}{PJ} \sum_p \sum_j \frac{(x'_{pn} - m'_{jn})}{\sigma_{jn}^2} (x'_{pr} - m'_{jr}) \tag{7.53}$$

In the learning procedure, which is based on, for example, [211], we minimize according to the LMS criterion the target function:

$$E = \frac{1}{2} \sum_{p=0}^{P} |y(\mathbf{x}) - \Phi(\mathbf{x})|^2 \tag{7.54}$$

where P is the size of the training set. The neural network has some useful features, such as automatic allocation of neurons, discarding of degenerated and inactive neurons, and variation of the learning rate depending on the number of allocated neurons.

The relevance of a feature is optimized by gradient descent:

$$\rho_i^{new} = \rho_i^{old} - \eta \frac{\partial E}{\partial \rho_i} \tag{7.55}$$

Based on the newly introduced relevance measure and the change in the architecture we get the following correction equations for the neural network:

$$\begin{aligned}
\frac{\partial E}{\partial w_{ij}} &= -(y_i - \Phi_i)\Psi_j \\
\frac{\partial E}{\partial m_{jn}} &= -\sum_i (y_i - \Phi_i) w_{ij} \Psi_j \sum_k (x'_k - m'_{jk}) \frac{B_{kn}}{\sigma_{jk}^2} \\
\frac{\partial E}{\partial \sigma_{jn}} &= -\sum_i (y_i - \Phi_i) w_{ij} \Psi_j \frac{(x'_n - m'_{jn})^2}{\sigma_{jn}^3}
\end{aligned} \tag{7.56}$$

In the transformed space the hyperellipses have the same orientation as in the original feature space. Hence they do not represent the same distribution as before. To overcome this problem, layers 3 and 4 will be adapted at the same time as **B**. If these layers converge fast enough, they can be adapted to represent the transformed training data, providing a model on which the adaptation of **B** can be based. The adaptation with two different target functions (E and ρ) may become unstable if **B** is adapted too fast, because layers 3 and 4 must follow the transformation of the input space. Thus μ must be chosen $\ll \eta$. A large gradient has been observed, causing instability when a feature of extreme high relevance is added to another. This effect can be avoided by dividing the learning rate by the relevance, that is, $\mu = \mu_0 / \rho_r$.

7.6. SUPPORT VECTOR MACHINES

The support vector machine [69,335,379] is an elegant universal feedforward network with a single hidden layer of nonlinear processing units and can be applied to a wide range of classification problems.

The basic idea of the SVM algorithm is to construct a hyperplane

$$f_{\mathbf{w}}, b(\mathbf{x}) = \mathbf{w} \cdot \mathbf{x} + b \tag{7.57}$$

with normal vector $\mathbf{w} \in \mathbf{R}^n$ and bias b, which separates the labeled training data into two classes with maximum-margin: given a set of N labeled training examples $\{(\mathbf{x}, y)_i\}, i = 1, \ldots, N$, $\mathbf{x}_i \in R^n$ belonging to two different classes $y_i \in \{-1, 1\}$, a maximum-margin hyperplane is determined which separates the training examples of the two different categories so that the distance between the hyperplane and the closest examples (the margin γ) is maximized. This hyperplane is fully specified by a subset of the training examples representing those points that lie closest to the decision surface and pose the biggest challenge in terms of classification.

Formally speaking, the margin $\gamma = \frac{1}{|\mathbf{w}|}$ of the hyperplane (7.57) is maximized by solving the following constrained optimization problem:

$$\min_{\mathbf{w},b} \quad \mathbf{w} \cdot \mathbf{w} \tag{7.58}$$

$$\text{subject to:} \quad y_i(\mathbf{w} \cdot \mathbf{x}_i + b) \geq 1, \quad \forall i = 1, \ldots, N \tag{7.59}$$

This optimization problem is solved by employing the Lagrange theory, leading to the maximum-margin hyperplane normal vector

$$\mathbf{w} = \sum_{i=1}^{N} \alpha_i y_i \mathbf{x}_i \tag{7.60}$$

with α_i being the Lagrange coefficients. In practice, the decision function

$$f_{\mathbf{w},b}(\mathbf{x}) = \sum_{i=1}^{N} \alpha_i y_i \mathbf{x}_i \cdot \mathbf{x} + b \tag{7.61}$$

is frequently determined by only a small subset of training examples—the so-called support vectors—with $\alpha_i > 0$, while the remaining examples with $\alpha_i = 0$ can be neglected.

In case of nonseparable data, we substitute in the above equation the inner product $\mathbf{x}_i \cdot \mathbf{x}_j$ by a nonlinear kernel function $K(\mathbf{x}_i, \mathbf{x}_j) = \Phi(\mathbf{x}_i) \cdot \Phi(\mathbf{x}_j)$. This kernel function evaluates the inner product between two examples after their transformation by a nonlinear function $\Phi(\mathbf{x})$. By doing the same in the Lagrangian form of the constrained quadratic optimization problem, the hyperplane is optimized in a new feature space and corresponds to a nonlinear decision function in the original data space. A frequently used nonlinear kernel function is the Gaussian kernel

$$K(\mathbf{x}_i, \mathbf{x}_j) = \exp\left(\frac{\|\mathbf{x}_i - \mathbf{x}_j\|^2}{2\sigma^2}\right). \tag{7.62}$$

Another frequently used kernel is the polynomial kernel

$$K(\mathbf{x}_i, \mathbf{x}_j) = \langle \mathbf{x}_i, \mathbf{x}_j \rangle^p \tag{7.63}$$

which corresponds to a mapping Φ in the space of all monomials of degree p and becomes for $p = 1$ a linear kernel. Less employed is sigmoid kernel

$$K(\mathbf{x}_i, \mathbf{x}_j) = \tan h(\mathbf{x}_i, \mathbf{x}_j) \tag{7.64}$$

Solving multiclass problems with the SVM algorithm requires a suitable decomposition of the classification task into a sequence of binary subtasks, which each can be handled by employing the standard SVM algorithm. The outputs of the binary classifiers are then recombined to the final multiclass prediction of the multiclass SVM (MSVM).

As an example, let's assume that three different tissue classes have to be distinguished. Each tissue class is considered in one of the binary subtasks as the target class to be distinguished from the union of the remaining classes (one-vs-all decomposition scheme). The MSVM then returns three-dimensional vectors with components reflecting the outcomes $f^c_{\mathbf{w},b}(\mathbf{x}),\ c = 1, \ldots, 3$ of the three binary SVMs. In order to increase the interpretability of the classification outcome, it is transformed into posteriori probabilities by postprocessing with a parameterized softmax function

$$P(\text{class}_k|\mathbf{x}) = \frac{\exp(a^k_1 f^k_{\mathbf{w},b}(\mathbf{x}) + a^k_0)}{\sum_{c=1}^{3} \exp(a^c_1 f^c_{\mathbf{w},b}(\mathbf{x}) + a^c_0)} \tag{7.65}$$

The parameters a^c_0, a^c_1 are estimated by minimizing the cross-entropy error on a subset of the training data.

7.7. HOPFIELD NEURAL NETWORKS

Hopfield neural networks represent a new neural computational paradigm by implementing an autoassociative memory. They are recurrent or fully interconnected neural networks. There are two versions of Hopfield neural networks: in the binary version all neurons are connected to each other but there is no connection from a neuron to itself, and in the continuous case all connections including self-connections are allowed.

A pattern, in N-node Hopfield neural network parlance, is an N-dimensional vector $\mathbf{p} = [p_1, p_2, \ldots, p_N]$ from the space $\mathbf{P} = \{-1, 1\}^N$. A special subset of $\mathbf{P}$ represents the set of stored or reference patterns $\mathbf{E} = \{\mathbf{e^k} : 1 \leq k \leq K\}$, where $\mathbf{e^k} = [e^k_1, e^k_2, \ldots, e^k_N]$. The Hopfield net associates a vector from $\mathbf{P}$ with a certain stored (reference) pattern in $\mathbf{E}$. The neural net splits the binary space $\mathbf{P}$ into classes whose members bear in some way similarity to the reference pattern that represents the class. The Hopfield network finds a broad application area in image restoration and segmentation.

As already stated in the Introduction, neural networks have four common components. For the Hopfield net we have the following:

Neurons: The Hopfield network has a finite set of neurons $\mathbf{x}(i), 1 \leq i \leq N$, which serve as processing units. Each neuron has a value (or state) at time t described by $\mathbf{x}_t(i)$. A neuron in the Hopfield net has one of the two states, either -1 or $+1$; that is, $\mathbf{x}_t(i) \in \{-1, +1\}$.

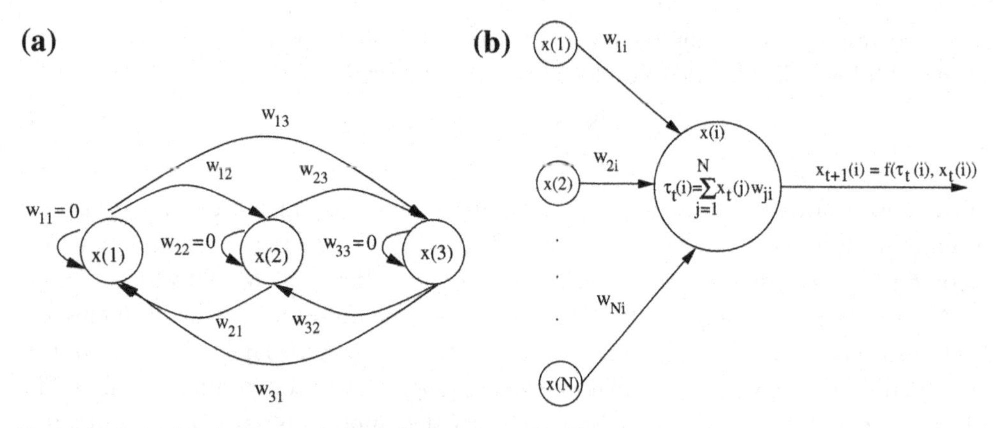

Figure 7.15 (a) Hopfield neural network and (b) propagation rule and activation function for the Hopfield network.

Synaptic connections: The learned information of a neural net resides within the interconnections between its neurons. For each pair of neurons, $\mathbf{x}(i)$ and $\mathbf{x}(j)$, there is a connection w_{ij} called the synapse between $\mathbf{x}(i)$ and $\mathbf{x}(j)$. The design of the Hopfield net requires that $w_{ij} = w_{ji}$ and $w_{ii} = 0$. Figure 7.15a illustrates a three-node network.

Propagation rule: This defines how states and synapses influence the input of a neuron. The propagation rule $\tau_t(i)$ is defined by

$$\tau_t(i) = \sum_{j=1}^{N} \mathbf{x}_t(j) w_{ij} + b_i \tag{7.66}$$

b_i is the externally applied bias to the neuron.

Activation function: The activation function f determines the next state of the neuron $\mathbf{x}_{t+1}(i)$ based on the value $\tau_t(i)$ computed by the propagation rule and the current value $\mathbf{x}_t(i)$. Figure 7.15b illustrates this fact. The activation function for the Hopfield net is the hard limiter defined here:

$$\mathbf{x}_{t+1}(i) = f(\tau_t(i), \mathbf{x}_t(i)) = \begin{cases} 1, & \text{if} \quad \tau_t(i) > 0 \\ -1, & \text{if} \quad \tau_t(i) < 0 \end{cases} \tag{7.67}$$

The network learns patterns that are N-dimensional vectors from the space $\mathbf{P} = \{-1, 1\}^N$. Let $\mathbf{e^k} = [e_1^k, e_2^k, \ldots, e_n^k]$ define the kth exemplar pattern where $1 \leq k \leq K$. The dimensionality of the pattern space is reflected in the number of nodes in the net, such that the net will have N nodes $\mathbf{x}(1), \mathbf{x}(2), \ldots, \mathbf{x}(N)$.

The training algorithm of the Hopfield neural network is simple and is outlined below:

1. **Learning:** Assign weights w_{ij} to the synaptic connections:

$$w_{ij} = \begin{cases} \sum_{k=1}^{K} e_i^k e_j^k, & \text{if} \quad i \neq j \\ 0, & \text{if} \quad i = j \end{cases} \tag{7.68}$$

Keep in mind that $w_{ij} = w_{ji}$, so it is necessary to perform the preceding computation only for $i < j$.

2. **Initialization:** Draw an unknown pattern. The pattern to be learned is now presented to the net. If $\mathbf{p} = [p_1, p_2, \ldots, p_N]$ is the unknown pattern, set

$$\mathbf{x}_0(i) = p_i, \quad 1 \leq i \leq N \tag{7.69}$$

3. **Adaptation:** Iterate until convergence. Using the propagation rule and the activation function we get for the next state,

$$\mathbf{x}_{t+1}(i) = f\left(\sum_{j=1}^{N} \mathbf{x}_t(j) w_{ij}, \mathbf{x}_t(i)\right) \tag{7.70}$$

This process should be continued until any further iteration will produce no state change at any node.

4. **Continuation:** For learning a new pattern, repeat steps 2 and 3.

In case of the continuous version of the Hopfield neural network, we have to consider neural self-connections $w_{ij} \neq 0$ and choose as an activation function a sigmoid function. With these new adjustments, the training algorithm operates in the same way.

The convergence property of Hopfield's network depends on the structure of $\mathbf{W}$ (the matrix with elements w_{ij}) and the updating mode. An important property of the Hopfield model is that if it operates in a sequential mode and $\mathbf{W}$ is symmetric with nonnegative diagonal elements, then the energy function

$$\begin{aligned} E_{hs}(t) &= \frac{1}{2} \sum_{i=1}^{n} \sum_{j=1}^{n} w_{ij} x_i(t) x_j(t) - \sum_{i=1}^{n} b_i x_i(t) \\ &= -\frac{1}{2} \mathbf{x}^T(t) \mathbf{W} \mathbf{x}(t) - \mathbf{b}^T \mathbf{x}(t) \end{aligned} \tag{7.71}$$

is nonincreasing [138]. The network always converges to a fixed point.

Hopfield neural networks are applied to solve many optimization problems. In medical image processing, they are applied in the continuous mode to image restoration, and in the binary mode to image segmentation and boundary detection.

The continuous version will be extensively described in Chapter 8 as a subclass of additive activation dynamics.

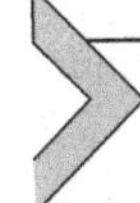

7.8. COMPARING STATISTICAL, SYNTACTIC, AND NEURAL PATTERN RECOGNITION METHODS

The delimitations between statistical, syntactic, and neural pattern recognition approaches are not necessarily clear. All these approaches share common features and have a correct classification result as a common goal. The decision to choose a particular approach over another is based on analysis of underlying statistical components, or grammatical structure or on the suitability of neural network solution.

Table 7.2 Comparing statistical, syntactic, and neural pattern recognition approaches.

	Statistical	**Syntactic**	**Neural**
Pattern generation basis	Probabilistic models	Formal grammars	Stable state or weight matrix
Pattern classification basis	Estimation or decision theory	Parsing	Neural network properties
Feature organization	Input vector	Structural relations	Input vector
Training mechanism			
Supervised	Density estimation	Forming grammars	Determining neural network parameters
Unsupervised	Clustering	Clustering	Clustering
Limitations	Structural information	Learning structural rules	Semantic information

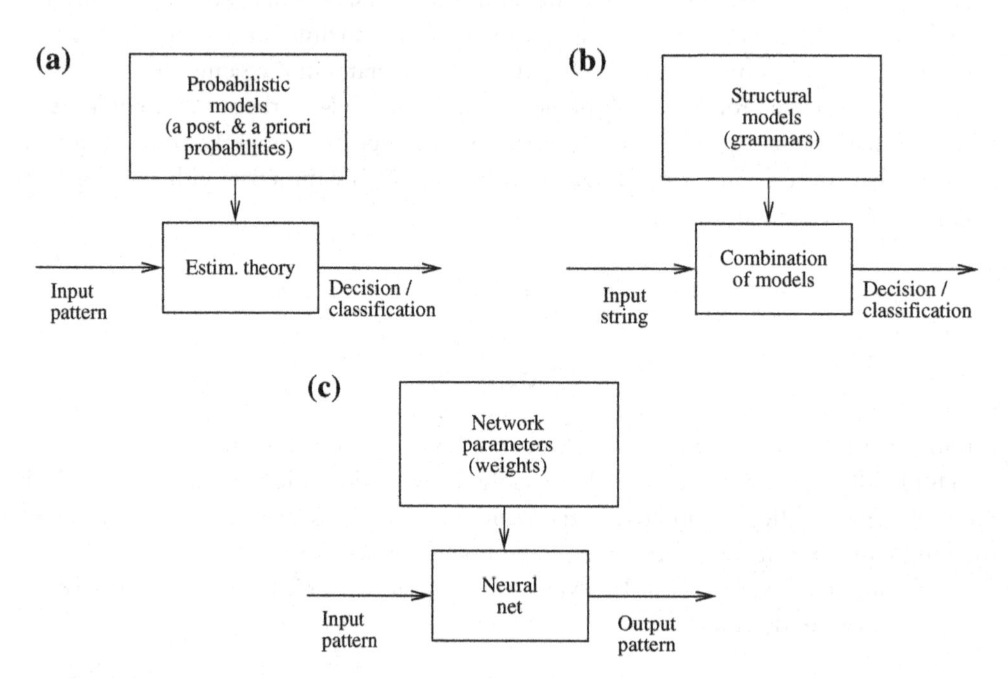

Figure 7.16 Pattern recognition approaches: (a) statistical approach; (b) syntactic approach; and (c) neural approach.

Table 7.2 and Fig. 7.16 summarize the differences between the three pattern recognition approaches [329].

For statistical pattern recognition, the pattern structure is considered insignificant. If structural information from pattern is available, we choose the syntactic approach. The neural network approach, sometimes considered an alternative technique, provides useful methods for processing patterns derived from biological systems and can emulate the computational paradigm of the human brain.

A more sophisticated approach is the newly introduced concept of biologically oriented neural networks [235]. They are robust and powerful classification techniques which disambiguate suspicious feature identification during a computer-aided diagnosis. They are derived from the newest biological discoveries aiming to imitate decision-making and sensory processing in biological systems. They have an advantage over other artificial intelligence methods since they directly incorporate brain-based mechanisms, and therefore make CAD systems more powerful.

There are also some drawbacks with the described methods: The statistical method cannot include structural information, the syntactic method does not operate based on adaptation rules, and the neural network approach doesn't contain semantic information in its architecture.

7.9. PIXEL LABELING USING NEURAL NETWORKS

Pixel labeling is a data-driven pixel classification approach leading to image segmentation. The goal is to assign a label to every pixel in the image. For a supervised classifier, the labels reflect the tissue types of the pixels. For an MRI image, appropriate labels will be gray matter, white matter, cerebrospinal fluid, bone, and fat. For an unsupervised classifier, the labels are arbitrarily assigned to pixels. A subsequent process will determine the correspondence between labels and tissue types.

7.9.1 Noncontextual Pixel Labeling

For this type of labeling, no pixel neighborhood (context information) is considered. Instead data obtained from several sources, so-called multispectral data, are used. For MRI, multispectral data are T_1, T_2, and proton density weighted images. All or only a subset of the multispectral data can be taken to generate a feature vector for each pixel location. For example, if T_1 and T_2 weighted images are used, then the vector $\mathbf{x}$ generated for a pixel at position (m, n) is

$$\mathbf{x}_{m,n} = \left(T_{1_{mn}}, T_{2_{mn}}\right) \tag{7.72}$$

Figure 7.17 shows an example of a feature vector for a single pixel.

The resulting feature vectors are classified by assigning a single label to each pixel position. This leads to cluster of points described by pixel feature vectors from a single tissue type. It has been shown [30] that even using multispectral data from MRI images, the characteristics of different tissue types overlap. This represents a problem in MRI segmentation.

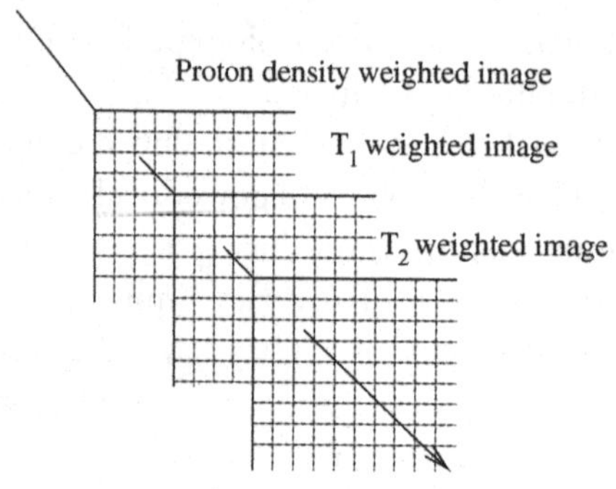

Figure 7.17 Multispectral data describing a feature vector for a pixel in an image.

7.9.2 Contextual Pixel Labeling

Noncontextual pixel labeling often leads to noisy results. An alternative technique involves the context information of pixels and uses the data in the spatial neighborhood of the pixel to be classified. Figure 7.18 visualizes this procedure. The feature vector may include both textural and contextual information. Usually, pixel intensity information across scales or measurements based on pixel intensity are used as features. In mammography, this pixel labeling is quite often used for tumor segmentation [77,410]. A multiscale approach for MRI segmentation was proposed in [125]. Each pixel in an image was assigned a feature vector containing a limited number of differential geometric invariant features evaluated across a range of different scales at the pixel location. The features measured at different scales are products of various first- and second-order image derivatives.

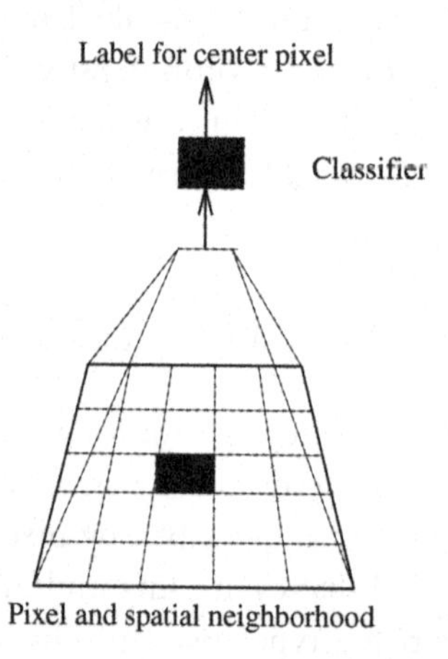

Figure 7.18 Contextual pixel labeling.

7.10. CLASSIFICATION STRATEGIES FOR MEDICAL IMAGES

There are two different strategies for classifying individual regions in medical images: region-based and pixel-based classification. For the region-based classification, the feature vector contains measurements of a set of features from a region in an image, while for the pixel-based classification, contextual or noncontextual information about every single pixel is obtained.

7.10.1 Pixel-Based Classification

Some features have been designed specifically for pixel-based classification in which a feature value is computed for each pixel. Each pixel in the image is then classified individually. Laws' texture energy measures are features which have been shown to discriminate well between various textured regions in an image when individual pixels are classified. These measures were successfully employed for spiculated lesion detection [393]. A pixel-level feature is considered a feature which is computed in the neighborhood of a pixel and is associated with that pixel. The neighborhood pixel will have a separate, and possibly different, feature value.

As stated before, classification at pixel level includes either contextual or noncontextual information of each image pixel. This technique is illustrated for MRI corresponding to acoustic neuromas [80]. A 7×7 pixel patch size, with a central square size of 5×5 pixels, was employed as the scheme to encode contextual information. The classification result achieved by an MLP for each pixel was a real value in the continuous range $[0, 1]$. The result values were used to construct result images whereby each pixel in the image being processed was replaced by its classification result value in the result image constructed. This is shown in Fig. 7.19.

The classification results achieved based on this method are shown in Fig. 7.20.

7.10.2 Region-Based Classification

In a region-based classification scheme, an object is first segmented from an image. The object is then classified based on features that are usually determined to describe the whole object.

Extracted features from regions are shape, texture, intensity, size, position within the image, contextual features, and fractal characteristics. In other words, an object-level feature is computed over an entire object.

Subsequent to image segmentation, the individual detected regions have to be classified in terms of anatomical structures. A set of features from each region is extracted, and then the regions are classified based on the generated feature vectors. This procedure is explained in Fig. 7.21.

A common procedure is to first extract a larger feature set and then refine this, such that all classes of anatomical structures are correctly represented. An application of this

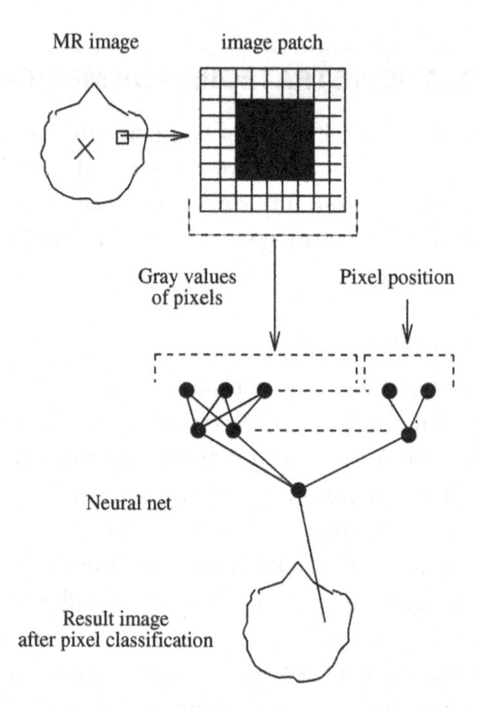

Figure 7.19 Pixel-based classification. An MLP is applied to the classification of all pixels in an image.

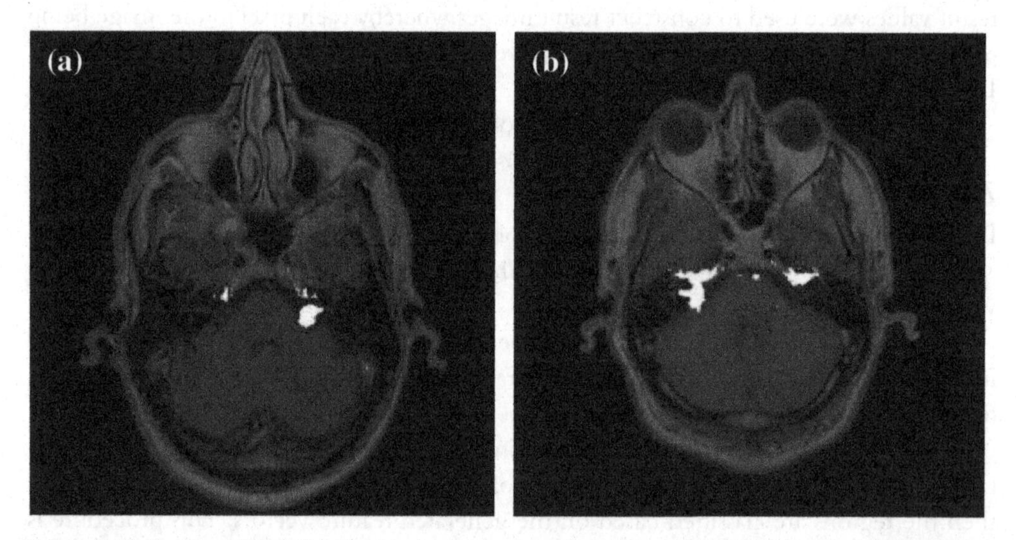

Figure 7.20 Pixel-level classification for detecting acoustic neuromas in MRI: (a) typical classification result; (b) fp-tumor adjacent to the acoustic neuroma [80]. *(Images courtesy of Prof. B. Thomas, University of Bristol.)*

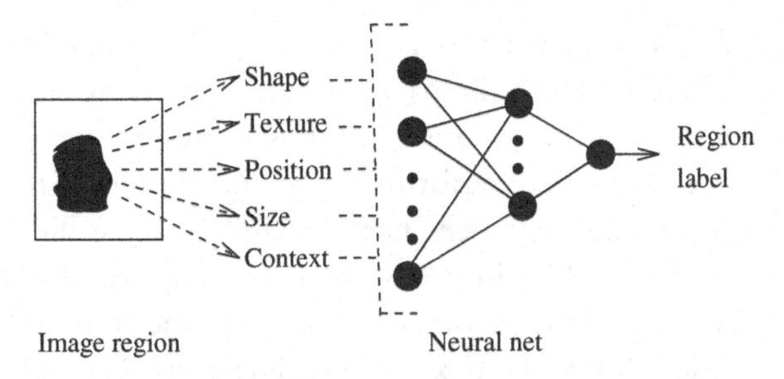

Figure 7.21 Region-based classification.

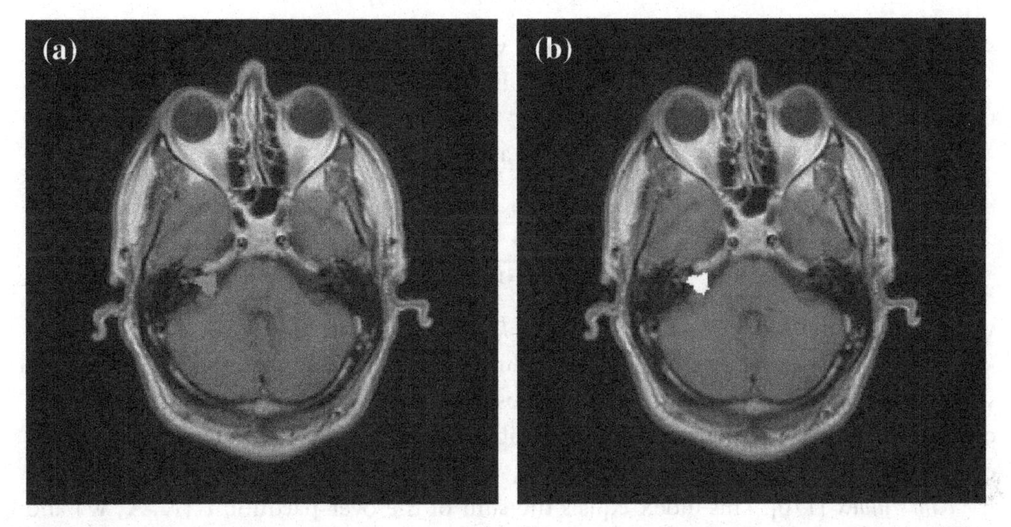

Figure 7.22 Region-level classification. A candidate tumor region cluster was erroneously classified as acoustic neuroma: (a) original image; and (b) classification result showing fp-tumor. *(Images courtesy of Prof. B. Thomas, University of Bristol.)*

method is visualized in Fig. 7.22. It shows a classification error whose structure is similar to that of an acoustic neuroma.

7.11. PERFORMANCE EVALUATION OF CLUSTERING TECHNIQUES

Determining the optimal number of clusters represents one of the most crucial classification problems. This task is known as cluster validity. The chosen validity function enables the validation of an accurate structural representation of the partition obtained by a clustering method. While a visual visualization of the validity is relatively simple for two-dimensional data, in case of multidimensional data sets this becomes very tedious. In this sense, the main objective of cluster validity is to determine the optimal number of clusters that provide the best characterization of a given multi-dimensional data set. An incorrect assignment of values to the parameter of a clustering algorithm results in a data partitioning scheme that is not optimal and thus leading to wrong decisions.

In this section, we evaluate the performance of the clustering techniques in conjunction with three cluster validity indices, namely Kim's index, Calinski Harabasz (CH) index, and the intraclass index. These indices were successfully applied before in biomedical time-series analysis [113]. In the following, we describe the above-mentioned indices.

Calinski Harabasz index [46]: This index is computed for m data points and K clusters as

$$\mathrm{CH} = \frac{[\mathrm{trace} B/(K-1)]}{[\mathrm{trace} W/(m-K)]} \tag{7.73}$$

where B and W represent the between and within cluster scatter matrices.

The maximum hierarchy level is used to indicate the correct number of partitions in the data.

Intraclass index [113]: This index is given as

$$I_W = \frac{1}{n}\sum_{k=1}^{K}\sum_{i=1}^{n_k} ||\mathbf{x}_i - \mathbf{w}_k||^2 \tag{7.74}$$

where n_k is the number of points in cluster k and $\mathbf{w}_k$ is a prototype associated with the kth cluster. I_W is computed for different cluster numbers. The maximum value of the second derivative of I_W as a function of cluster number is taken as an estimate for the optimal partition. This index provides a possible way of assessing the quality of a partition of K clusters.

Kim's index [176]: This index equals the sum of the over-partition $v_o(K, \mathbf{X}, \mathbf{W})$ and under-partition $v_u(K, \mathbf{X}, \mathbf{W})$ function measure

$$I_{Kim} = \frac{v_u(K) - v_{umin}}{v_{umax} - v_{umin}} + \frac{v_o(K) - v_{omin}}{v_{omax} - v_{omin}} \tag{7.75}$$

where $v_u(K)$ is the under-partitioned average over the cluster number of the mean intra-cluster distance and measures the structural compactness of each class, v_{umin} is its minimum while v_{umax} the maximum value. $v_u(K, \mathbf{X}, \mathbf{W})$ is given by the average of the mean intra-cluster distance over the cluster number K and measures the structural compactness of each and every class. $v_o(K, \mathbf{X}, \mathbf{W})$ is given by the ratio between the cluster number K and the minimum distance between cluster centers, describing intercluster separation. $\mathbf{X}$ is the matrix of the data points and $\mathbf{W}$ is the matrix of the prototype vectors. Similarly, $v_o(K)$ is the over-partitioned measure defined as the ratio between the cluster number and the minimum distance between cluster centers measuring the intercluster separation. v_{omin} is its minimum while v_{omax} the maximum value. The goal is to find the optimal cluster number with the smallest value of I_{Kim} for a cluster number $K = 2$ to K_{max}.

7.12. CLASSIFIER EVALUATION TECHNIQUES

There are several techniques for evaluating the classification performance in medical imaging problems. The most known are the confusion matrix, ranking order curves, and ROC curves.

7.12.1 Confusion Matrix

When conducting classification experiments, one possibility of evaluating the performance of a system is to determine the number of correctly and wrongly classified data. This gives a first impression of how correctly the classification was performed. In order to get a better understanding of the achieved classification, it is necessary to know which classes of data were most often misplaced. A convenient tool when analyzing results of classifier systems in general is the confusion matrix, which is a matrix containing information about the actual and predicted classes. The matrix is two-dimensional and has as many rows and columns as there are classes. The columns represent the true classifications and the rows represent the system classifications. If the system performs perfectly, there will be scores only in the diagonal positions. If the system has any misclassifications, these are placed in the off-diagonal cells. Table 7.3 shows a sample confusion matrix. Based on the confusion matrix, we can easily determine which classes are being confused with each other.

7.12.2 Ranking Order Curves

Ranking order curves represent a useful technique for estimating the appropriate feature dimension. They combine in a single plot several feature selection results for different classifiers where each of them visualized the function of error rate over the number of features.

Usually, we first see an improvement in classification performance when considering additional features. But after some time a saturation degree is reached, and the selection of new features leads to a deterioration of the classification performance because of overtraining. Based on ranking order curves, it is possible to determine the feature dependence. Some features alone do not necessarily deteriorate the classification performance. However, in combination with others they have a negative effect on the performance.

Table 7.3 Confusion matrix for a classification of three classes A, B, C.

	Output		
Input	***A%***	***B%***	***C%***
A	90	0	10
B	0	99	1
C	0	90	10

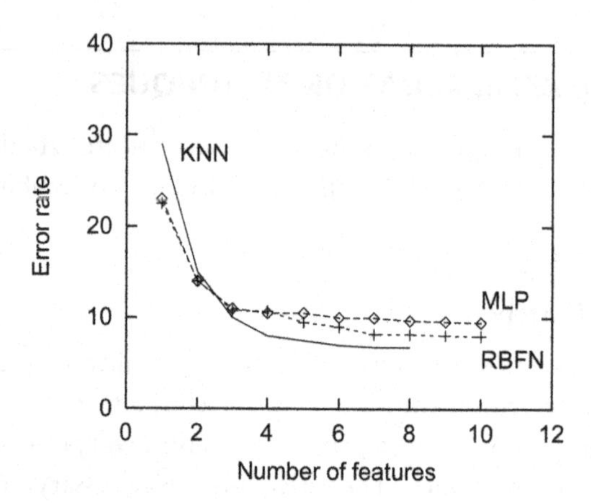

Figure 7.23 Example of ranking order curves showing feature selection results using three different classifiers (KNN, MLP, RBFN).

The classifier type plays an important role as well. In other words, an optimal performance can be achieved only as a combination of an appropriate classifier architecture and the best selected feature subset.

Ranking order curves are an important tool when it comes to determining potentially useful features, monitoring overtraining, and excluding redundant and computationally expensive features. This requires a vast number of simulations such that the ranking order of the available features becomes evident. Based on the determined histograms, it becomes possible to detect the redundant features so that in subsequent simulations they can be excluded. Figure 7.23 visualizes three feature ranking order curves for different classifiers.

7.12.3 ROC Curves

Receiver operating characteristic (ROC) curves have their origin in signal detection theory. Since the outcome of a particular condition in a yes-no signal detection experiment can be represented as an ordered pair of values (the hit and false-alarm rates), it is useful to have a way to graphically present and interpret them. This is usually done by plotting hit rate against false-alarm rate. Such a plot is called a receiver operating characteristic or ROC.

In pattern recognition ROC curves play an important role in providing information about the overlap between two classes. In many cases, the mean values of distinct classes may differ significantly, yet their variances are large enough to impede a correct class distinction.

Figure 7.24a illustrates the overlapping of two pdfs describing the distribution of a feature in two classes. The threshold in the figure is given by a perpendicular line. For a better understanding, one pdf is inverted as suggested in [364]. We will assume that all values on the left of the threshold belong to class ω_1, while those on the right belong

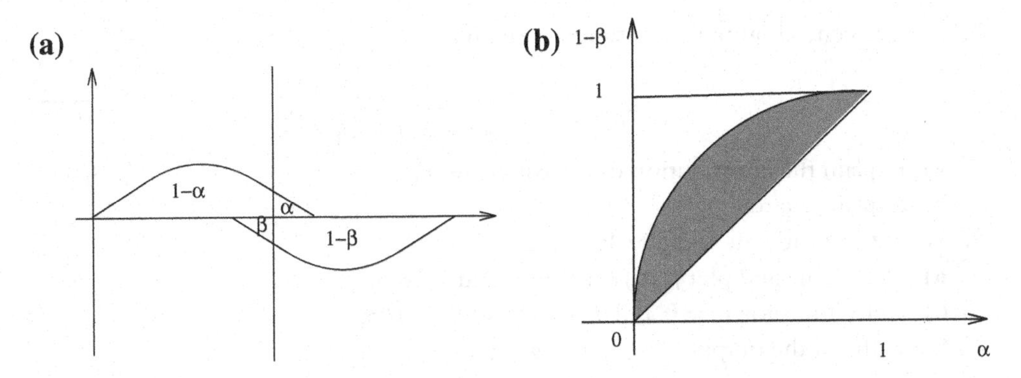

Figure 7.24 Example of (a) overlapping pdfs of the same feature in two classes and (b) the resulting ROC curve.

to class ω_2. However, this decision has an error probability of α while the probability of a correct decision is given by $1 - \alpha$. Let β be the probability of a wrong decision concerning class ω_2 while $1-\beta$ is the probability of a correct decision. The threshold can be moved in both directions, and this will lead for every position to different values for α and β. In case of perfect overlap of the two curves, for every single threshold position $\alpha = 1 - \beta$ holds. This corresponds to the bisecting line in Fig. 7.24b where the two axes are α and $1 - \beta$. If the overlap gets smaller, the corresponding curve in Fig. 7.24b moves apart from the bisecting line. In the ideal case, there is no overlap and we obtain $1-\beta = 1$. In other words, the less the class overlap, the larger the area between the curve and the bisecting line.

We immediately see that the area varies between zero for complete overlap and 1/2 (area of the upper triangle) for complete separation. This means we have been provided with a measure of the class separability of a distinct feature.

In practical problems, the ROC curve can be easily determined by moving the threshold and computing the correct and false classification rate over the available training vectors.

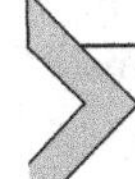

7.13. EXERCISES

1. Consider a biased input of the form

$$\tau_t(i) = \sum_k a_t(i) w_{ik} + b \tag{7.76}$$

and a sigmoidal activation function. What bias b is necessary for $f(0) = 0$? Is this acceptable? Are there any other possibilities?

2. A more general output activation is given by

$$o_j = f(\tau_j) = \frac{1}{1 + \exp -\{\frac{\tau_j - \theta_j}{\theta_0}\}} \tag{7.77}$$

   a) Explain the contribution of the parameter θ_j.
   b) Explain the role of θ_0.
3. For $f(\tau_j)$ given as in Exercise 4:
   a) Determine and plot $f'(\tau_j)$ for $\tau_j = 0$ and $\theta_0 = 5$.
   b) Repeat this for $\tau_j = 0$ and $\theta_0 = 50$ and $\theta_0 = 0.5$.
4. Show that if the output activation is given by

$$o_j = f(\tau_j) = \frac{1}{1 + \exp \tau_j} \tag{7.78}$$

   then we obtain for its derivative

$$\frac{\partial f(\tau_j)}{\partial \tau_j} = f'(\tau_j) = o_j(1 - o_j) \tag{7.79}$$

   Is it possible to have a τ_j such that we obtain $f(\tau_j) = 0$?
5. Design a feedforward network such that for given binary inputs A and B it implements
   a) The logical NAND function.
   b) The logical NOR function.
6. The necessary number of neurons in a hidden layer is always problem-oriented. Develop and discuss the advantages of removing or adding iteratively neurons from the hidden layer.
7. A method to increase the rate of learning yet to avoid the instability is to modify the weight updating rule

$$w_{ij}(n) = w_{ij}(n-1) + \eta \delta_{h_j} p_i^t \tag{7.80}$$

   by including a momentum term as described in [71]

$$\Delta w_{ij}(n) = \alpha \Delta w_{ij}(n-1) + \eta \delta_{h_j} p_i^t \tag{7.81}$$

   where α is a positive constant called the momentum constant. Describe why this is the case.
8. The momentum constant is in most cases a small number with $0 \leq \alpha < 1$. Discuss the effect of choosing a small negative constant with $-1 < \alpha \leq 0$ for the modified weight updating rule given in Exercise 7.
9. Set up two data sets, one set for training an MLP, and the other one for testing the MLP. Use a single-layer MLP and train it with the given data set. Use two possible nonlinearities: $f(x) = \sin x$ and $f(x) = e^{-x}$. Determine for each of the given nonlinearities:

a) The computational accuracy of the network by using the test data.
b) The effect on network performance of varying the size of the hidden layer.

10. Consider the updating rule for the Kohonen map given by eq. (7.16). Assume that the initial weight vectors $\mathbf{w}_j$ are not different. Comment on the following two distinct cases:
a) They are nonrandom but distinct from each other.
b) They are random but some weights can be the same.
11. In [183] it is stated that if the input patterns $\mathbf{x}$ have as a pdf a given $p(\mathbf{x})$, then the point density function of the resulting weight vectors $\mathbf{w}_j$ approximates $p(\mathbf{x})$. Describe based on eq. (7.16) why this is the case.
12. Comment on the differences and similarities between the Kohonen map and the "neural gas" network.
13. When is the Kohonen map preserving the topology of the input space? What condition must the neural lattice fulfill?
14. Consider a Kohonen map performing a mapping from a 2-D input onto a 1-D neural lattice of 60 neurons. The input data are random points uniformly distributed inside a circle of radius 1 centered at the origin. Compute the map produced by the neural network after 0, 25, 100, 1000, and 10,000 iterations.
15. Consider a Kohonen map performing a mapping from a 3-D input onto a 2-D neural lattice of 1,000 neurons. The input data are random points uniformly distributed inside a cube defined by $\{(0 < x_1 < 1), (0 < x_2 < 1), (0 < x_3 < 1)\}$. Compute the map produced by the neural network after 0, 100, 1,000, and 10,000 iterations.
16. Radial basis neural networks are often referred to as "hybrid" neural networks. Comment on this property by taking into account the architectural and algorithmic similarities among radial basis neural networks, Kohonen maps, and MLPs.
17. Find a solution for the XOR problem using an RBF network with two hidden units where the two radial basis function centers are given by $\mathbf{m}_1 = [1, 1]^T$ and $\mathbf{m}_2 = [0, 0]^T$. Determine the output weight matrix $\mathbf{W}$.
18. Find a solution for the XOR problem using an RBF network with four hidden units where four radial basis function centers are given by $\mathbf{m}_1 = [1, 1]^T$, $\mathbf{m}_2 = [1, 0]^T$, $\mathbf{m}_3 = [0, 1]^T$, and $\mathbf{m}_4 = [0, 0]^T$. Determine the output weight matrix $\mathbf{W}$.

CHAPTER EIGHT

Transformation and Signal-Separation Neural Networks

Contents

Pattern Recognition and Signal Analysis in Medical Imaging
http://dx.doi.org/10.1016/B978-0-12-409545-8.00008-X

8.1. INTRODUCTION

Neural networks are excellent candidates for feature extraction and selection, and for signal separation. The underlying architectures are mostly employing unsupervised learning algorithms and are viewed as nonlinear dynamical systems.

The material in this chapter is organized in three parts: the first part describes the neurodynamical aspects of neural networks, the second part deals with principal component analysis (PCA) and with related neural networks, and the third part deals with independent component analysis (ICA) and neural architectures performing signal separation.

Several neural network models such as the generalized Hebbian algorithm, adaptive principal component extraction, and the linear and nonlinear Oja's algorithms are reviewed. As an application for principal component analysis, the emerging area of medical image coding is chosen.

ICA is gaining importance in artifact separation in medical imaging. This chapter reviews the most important algorithms for ICA, such as the Infomax, FastICA, and topographic ICA. Imaging brain dynamics is becoming the key to the understanding of cognitive processes associated with the human brain. This chapter describes artifact separation based on ICA for two modalities of imaging brain dynamics, magnetoencephalographic (MEG) recordings and the functional magnetic resonance imaging (fMRI).

8.2. NEURODYNAMICAL ASPECTS OF NEURAL NETWORKS

In neural networks we deal with fields of neurons. Neurons within a field are topologically ordered, mostly based on proximity. The fields are related only by synaptic connections between them [76,183,390]. In mammalian brains, we find the topological ordering in volume proximity packing. An input field of neurons is defined with $\mathbf{F_X}$. We consider here only two-field neural networks and define with $\mathbf{F_Y}$ the output field. Let us assume that field $\mathbf{F_X}$ has n neurons and field $\mathbf{F_Y}$ has p neurons. There is a mapping defined from the input to the output field and described as $\mathbf{F_X} \rightarrow \mathbf{F_Y}$. Based on synaptic connections, we assume that there are m pairs of connected neurons $(\mathbf{x_i}, \mathbf{y_i})$. In this way, the function $f : \mathbf{R^n} \rightarrow \mathbf{R^p}$ generates the following associated pairs: $(\mathbf{x_1}, \mathbf{y_1}), \ldots, (\mathbf{x_m}, \mathbf{y_m})$. The overall system behaves as an adaptive filter enabling a data flow from the input to the output layer and vice versa. The feature data changes the network parameters. Let m_{ij} describe the feedforward connection between the ith neuron from field $\mathbf{F_X}$ and the jth neuron from field $\mathbf{F_Y}$. m_{ij} can be positive (excitatory), negative (inhibitory), or zero. The synaptic connections between the two fields can described by an $n \times p$ synaptic matrix $\mathbf{M}$. A similar matrix denoted by $\mathbf{N}$ and having $p \times n$ elements describes the feedback connections between the two layers. Figure 8.1 shows the structure of an interconnected two-layer field.

In field terminology, a neural network can be very conveniently described by the quadruple $(\mathbf{F_X}, \mathbf{F_Y}, \mathbf{M}, \mathbf{N})$. $\mathbf{F_X}$ and $\mathbf{F_Y}$ represent not only the collection of topological neurons, but also their activation and signal computational characteristics. A two-layer neural network is called heteroassociative, while one-layer neural networks are called autoassociative [183]. In general $\mathbf{M}$ and $\mathbf{N}$ are of different structures. In biological networks, $\mathbf{M}$ outnumbers $\mathbf{N}$, making such networks more feedforward networks. Artificial neural networks adopted the same concept, as can be seen from backpropagation-type neural networks and radial basis neural networks. Besides the bidirectional topologies, there also are unidirectional topologies where a neuron field synaptically intraconnects to itself as shown in Fig. 8.2.

The matrices $\mathbf{P}$ and $\mathbf{Q}$ intraconnect $\mathbf{F_X}$ and $\mathbf{F_Y}$. $\mathbf{P}$ is an $n \times n$ matrix and $\mathbf{Q}$ is a $p \times p$ matrix. In biological networks, $\mathbf{P}$ and $\mathbf{Q}$ are often symmetric and this symmetry reflects a lateral inhibition or competitive connection topology. $\mathbf{P}$ and $\mathbf{Q}$ are in most cases diagonal matrices with positive diagonal elements and negative or zero-off nondiagonal elements. An intra- and interconnected structure of neural fields is described mathematically as $(\mathbf{F_X}, \mathbf{F_Y}, \mathbf{M}, \mathbf{N}, \mathbf{P}, \mathbf{Q})$ and shown in Fig. 8.3.

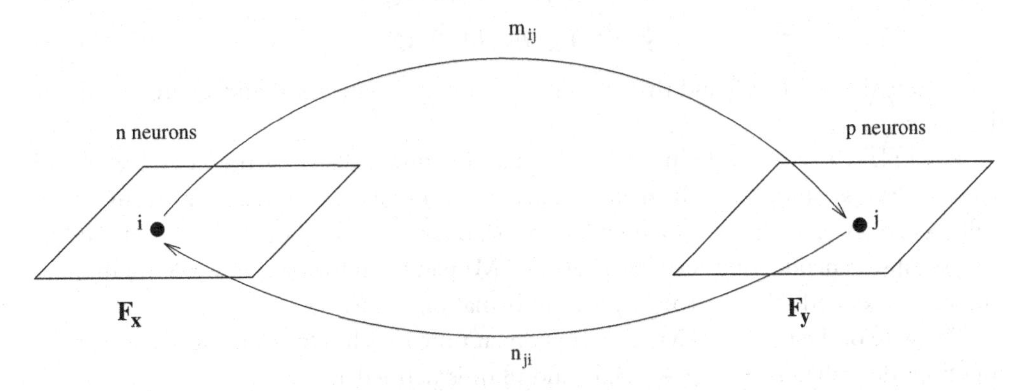

Figure 8.1 Neuronal structure between two neural fields.

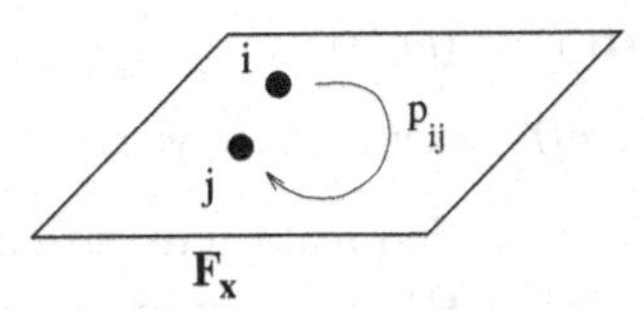

Figure 8.2 Unidirectional neural network.

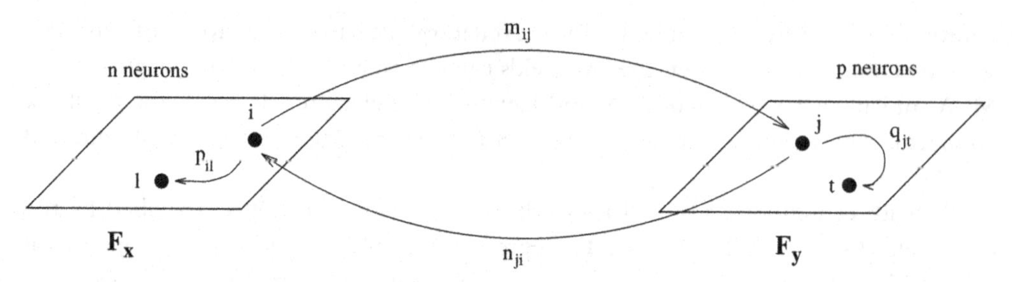

Figure 8.3 Intra- and interconnected neural fields.

The neural activity and the synaptic connections change over time, and this implies the existence of a neuronal dynamical system. Such a system is described by a set of first-order differential equations:

$$\dot{x}_i = g_i(\mathbf{F_X}, \mathbf{F_Y}, \mathbf{M}, \mathbf{P}, \mathbf{Q}) \tag{8.1}$$

$$\dot{m}_{ij} = l_i(\mathbf{F_X}, \mathbf{F_Y}, \mathbf{M}, \mathbf{P}, \mathbf{Q}) \tag{8.2}$$

$$\dot{y}_i = h_i(\mathbf{F_X}, \mathbf{F_Y}, \mathbf{M}, \mathbf{P}, \mathbf{Q}) \tag{8.3}$$

or, in vector notation,

$$\dot{\mathbf{x}} = \mathbf{g}(\mathbf{F_X}, \mathbf{F_Y}, \mathbf{M}, \mathbf{P}, \mathbf{Q}) \tag{8.4}$$

$$\dot{\mathbf{M}} = \mathbf{l}(\mathbf{F_X}, \mathbf{F_Y}, \mathbf{M}, \mathbf{P}, \mathbf{Q}) \tag{8.5}$$

$$\dot{\mathbf{y}} = \mathbf{h}(\mathbf{F_X}, \mathbf{F_Y}, \mathbf{M}, \mathbf{P}, \mathbf{Q}) \tag{8.6}$$

It is assumed that $\mathbf{N} = \mathbf{0}$ and that the intraconnection matrices $\mathbf{P}$ and $\mathbf{Q}$ are not time-dependent.

Time plays a critical role in neuronal dynamics: time is "fast" at the neural level and "slow" at the synaptic level. In mammalian brains, membrane fluctuations occur at the millisecond level, while synaptic fluctuations occur at the second or minute level. Therefore, synapses encode long-term memory (LTM) pattern information, while membrane fluctuations encode short-term memory information (STM).

The state of the neuronal dynamical system at time t with activation and synaptic time functions described by eqs. (8.4), (8.5), and (8.6) is defined as

$$\mathbf{x}(t) = (x_1(t), \ldots, x_n(t)) \tag{8.7}$$

$$\mathbf{m}(t) = (m_{11}(t), \ldots, m_{np}(t)) \tag{8.8}$$

$$\mathbf{y}(t) = (y_1(t), \ldots, y_p(t)) \tag{8.9}$$

The state space of field $\mathbf{F_X}$ is the extended real vector space $\mathbf{R^n}$, that of $\mathbf{F_Y}$ is $\mathbf{R^p}$, and of the two-layer neural network is $\mathbf{R^n} \times \mathbf{R^p}$. A point in the state space specifies a snapshot of all neural behavior. A trajectory defines the time evolution of the network activity.

Based on the fact that there is either the LTM or STM or even both that change over time, we consider different types of dynamical systems: neuronal dynamics (only the activation fluctuates over time) defined by eq. (8.8), synaptic dynamics (only synaptic changes are considered) defined by eq. (8.7), and neuro-synaptic dynamics (both activation and synapses change) defined by eqs. (8.7) and (8.8).

While most dynamical analysis has been focused on either neuronal dynamics [61,65,139] or synaptic dynamics [130,188,301], little is known about the dynamical properties of neuro-synaptic dynamics [10,247,249]. The dynamics of coupled systems with different timescales as found in neuro-synaptic dynamical systems is one of the most challenging research topics in the dynamics of neural systems.

To analyze the dynamical properties of the neural system, we adopt the same definitions as in [189]. Equilibrium is steady state (for fixed-point attractors). Convergence means synaptic equilibrium:

$$\dot{\mathbf{M}} = \mathbf{0} \tag{8.10}$$

Stability means neuronal equilibrium:

$$\dot{\mathbf{x}} = \mathbf{0} \tag{8.11}$$

And total stability is joint neuronal-synaptic steady state:

$$\dot{\mathbf{x}} = \dot{\mathbf{M}} = \mathbf{0} \tag{8.12}$$

In biological systems both neurons and synapses change as the feedback system samples fresh environmental stimuli. The neuronal and synaptic dynamical systems ceaselessly approach equilibrium and may never achieve it.

Neurons fluctuate faster than synapses fluctuate. In feedback systems this dynamical asymmetry creates the famous *stability convergence dilemma.*

8.2.1 Neuronal Dynamics

STM neural systems are systems that have a fluctuating neural activity over time but a time-constant synaptic fluctuation. Here, we consider a symmetric autoassociative neural network with $\mathbf{F_X} = \mathbf{F_Y}$ and a time-constant $\mathbf{M} = \mathbf{M}^\mathbf{T}$. This network can be described based on the Cohen-Grossberg [65] activity dynamics:

$$\dot{x}_i = -a_i(x_i)\left[b_i(x_i) - \sum_{j=1}^{n} f_j(x_j)m_{ij}\right] \tag{8.13}$$

where $a_j > 0$ describes an amplification function. b_i are essentially arbitrary, and the matrix m_{ij} is symmetric. The function f is nonlinear and increasing. We can choose a sigmoid function for f, $f_j(x_j) = \tan h\, x_j$.

From eq. (8.13) we can derive two subsystems, an additive and a multiplicative system. These systems are explained in the following.

8.2.1.1 *Additive Activation Dynamics*

The additive associative neural network is derived from eq. (8.13) by assuming $a_i(x_i)$ is a constant a_i and $b_i(x_i)$ is proportional to x_i,

$$\dot{x}_i = -a_i x_i + \sum_{j=1}^{n} f_j(x_j) m_{ij} + I_i \tag{8.14}$$

with

$$a_i(x_i) = 1 \tag{8.15}$$

$$b_i(x_i) = a_i x_i - I_i \tag{8.16}$$

I_i is an input term. Neurobiologically a_i measures the inverse cell membrane's resistance, and I_i the current flowing through the resistive membrane. Since the synaptic changes for the additive model are assumed nonexistent, the only way to achieve an excitatory and inhibitory effect is through the weighted contribution of the neighboring neuron outputs. The most famous representatives of this group are the Hopfield neural network [138] and the cellular neural network [61].

8.2.1.2 *Multiplicative Activation Dynamics*

The multiplicative or shunting neural network is derived from eq. (8.13) by assuming $a_i(x_i)$ is linear a_i and $b_i(x_i)$ is nonlinear:

$$\dot{x}_i = -a_i x_i - x_i \left(\sum_{j=1}^{n} f_j(x_j) m_{ij} \right) + I_i \tag{8.17}$$

The product term $x_i \left(\sum_{j=1}^{n} f_j(x_j) m_{ij} \right)$ gave the network its name [258].

8.2.2 Synaptic Dynamics

A neural network learns a pattern if the system encodes the pattern in its structure. Also, the network connections change as it learns the information.

For example, the neural network has learned the stimulus-response pair (x_i, y_i) if it responds with y_i when x_i is the stimulus (input). The stimulus-response pair (x_i, y_i) is a sample from the mapping function $f : \mathbf{R^n} \rightarrow \mathbf{R^p}$. The system has learned the function f, if it responds to every single stimulus x_i with its correct y_i. However, in most practical cases, only partial or approximative learning is possible. Learning involves a change in synapses and quantization. Only a subset of all patterns in the sampled pattern environment is learned. Since memory capacity is limited, an adaptive system such as a neural network has to learn efficiently by replacing old stored patterns with new patterns.

Learning can be either supervised or unsupervised. Supervised learning uses class-membership information while unsupervised learning does not. Biological synapses learn locally and without supervision on a single pass of noisy data. Local information is information available physically and briefly to the synapse. It involves synaptic properties or neuronal signal properties. Locally unsupervised synapses associate signals with signals. This leads to conjunctive, or correlation, learning laws constrained by locality.

Here, we will examine two unsupervised learning laws: signal Hebbian learning law, and competitive learning law or Grossberg law [116].

8.2.2.1 Signal Hebbian Learning Law

The deterministic Hebbian learning law correlates local neuronal signals:

$$\dot{m}_{ij} = -m_{ij} + f_i^{\mathbf{X}}(x_i) f_j^{\mathbf{Y}}(y_j) \tag{8.18}$$

The field notation **X** and **Y** can be omitted and we obtain

$$\dot{m}_{ij} = -m_{ij} + f_i(x_i) f_j(y_j) \tag{8.19}$$

m_{ij} is the synaptic efficacy along the axon connecting the ith neuron in field $\mathbf{F_X}$ with the jth neuron in field $\mathbf{F_Y}$. If $m_{ij} \geq 0$ then the synaptic injunction is excitatory, and it is inhibitory if $m_{ij} \leq 0$.

8.2.2.2 Competitive Learning Law

This law modulates the output signal $f_j(y_j)$ with the signal–synaptic difference $f_i(x_i) - m_{ij}$

$$\dot{m}_{ij} = f_j(y_j)[f_i(x_i) - m_{ij}] \tag{8.20}$$

The jth neuron in $\mathbf{F_Y}$ wins the competition at time t if $f_j(y_j(t)) = 1$, and loses it if $f_j(y_j(t)) = 0$. Competitive learning means that synapses learn only if their postsynaptic neurons win. In other words, postsynaptic neurons code for presynaptic signal patterns [189]. The neurons in $\mathbf{F_Y}$ compete for the activation induced by signal patterns from $\mathbf{F_X}$. They excite themselves and inhibit one another.

8.2.3 Neuro-Synaptic Dynamics

Biologically, neural networks model both the dynamics of neural activity levels, the short-term memory (STM), and the dynamics of synaptic modifications, the long-term memory (LTM). The actual network models under consideration may be considered extensions of Grossberg's shunting network [117] or Amari's model for primitive neuronal competition [9]. These earlier networks are considered pools of mutually inhibitory neurons with fixed synaptic connections. Here, we are looking at systems where the synapses can be modified by external stimuli.

The dynamics of competitive systems may be extremely complex, exhibiting convergence to point attractors and periodic attractors. For networks that model only the dynamics of the neural activity levels, Cohen and Grossberg [65] found a Lyapunov function as a necessary condition for the convergence behavior to point attractors.

Networks where both LTM and STM states are dynamic variables cannot be placed in this form since the Cohen-Grossberg equation (8.13) does not model synaptic dynamics. However, a large class of competitive systems have been identified as being "generally" convergent to point attractors even though no Lyapunov functions have been found for their flows. The emergent global properties of a network, rather than the behavior of the individual units and the local computation performed by them, describe the network's behavior. Global stability analysis techniques, such as Lyapunov energy functions, show the conditions under which a system approaches an equilibrium point in response to an input pattern. The equilibrium point is then the stored representation of the input. This property is termed the content addressable memory (CAM) property. Local stability, by contrast, involves the analysis of network behavior around individual equilibrium points.

Such a neuro-synaptic system is a laterally inhibited network with a deterministic signal Hebbian learning law [130] that is similar to the spatio-temporal system of Amari [10].

The general neural network equations describing the temporal evolution of the STM and LTM states for the jth neuron of an N-neuron network are

$$\dot{x}_j = -a_j x_j + \sum_{i=1}^{N} D_{ij} f(x_i) + B_j \sum_{i=1}^{p} m_{ij} y_i \tag{8.21}$$

$$\dot{m}_{ij} = -m_{ij} + y_i f(x_j) \tag{8.22}$$

where x_j is the current activity level, a_j is the time constant of the neuron, B_j is the contribution of the external stimulus term, $f(x_i)$ is the neuron's output, y_i is the external stimulus, and m_{ij} is the synaptic efficiency. The neural network is modeled by a system of deterministic equations with a time-dependent input vector rather than a source emitting input signals with a prescribed probability distribution. Our interest is to store patterns as equilibrium points in N-dimensional space. In fact, the formation of stable one-dimensional cortical maps under the aspect of topological correspondence and under the restriction of a constant probability of the input signal is demonstrated in [9].

In [249] it was shown that competitive neural networks with a combined activity and weight dynamics can be interpreted as nonlinear singularly perturbed systems [175,319]. A quadratic-type Lyapunov function was found for the coupled system, and the global stability of an equilibrium point representing a stored pattern was proven.

8.3. PCA-TYPE NEURAL NETWORKS

As previously stated, principal component analysis (PCA) represents an important method for redundant feature extraction and for data compression.

PCA has several applications in nuclear medical imaging [14,288], X-ray computed tomography [159] and other X-ray fields, and magnetic resonance imaging (MRI) [17,20,81,239,240].

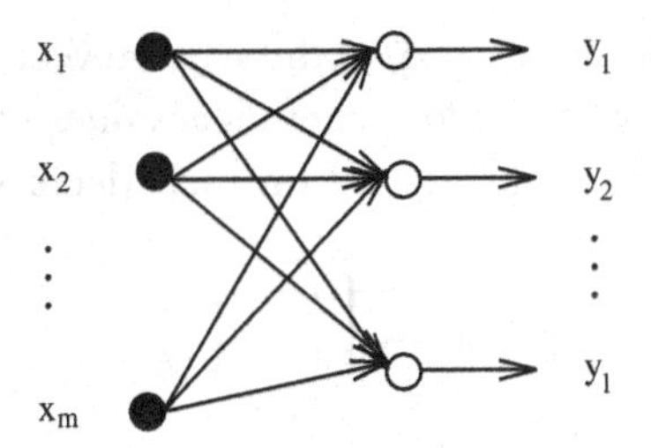

Figure 8.4 Feedforward neural network with a single layer of neurons.

In the following we will present neural network architectures which extract the principal components in a self-organized manner. In other words, unsupervised learning applied to a one-layer neural network performs PCA.

8.3.1 Generalized Hebbian Algorithm (GHA)

A simple feedforward neural network with a single layer of linear neurons as shown in Fig. 8.4 can compute the principal components of a data vector $\mathbf{x}$.

Two basic assumptions regarding the network architecture are necessary:

- The neurons of the output layer are linear.
- The network has m inputs and l outputs with $l < m$.

Only the feedforward connections $\{w_{ij}\}$ with $i = 1, 2, \ldots, m$ and $j = 1, 2, \ldots, l$ are subject to training for determining the principal components based on the generalized Hebbian algorithm (GHA).

The jth neuron output $y_i(n)$ at time t is given by

$$y_j(n) = \sum_{i=1}^{m} w_{ji}(n)x_i(n) \quad j = 1, 2, \ldots, l \tag{8.23}$$

The learning is performed based on a generalized Hebbian learning as shown in [327]:

$$\Delta w_{ji}(n) = \eta \left[y_j(n)x_i(n) - y_j(n) \sum_{k=1}^{j} w_{ki}(n)y_k(n) \right] \tag{8.24}$$

$$i = 1, 2, \ldots, m \quad j = 1, 2, \ldots, l \tag{8.25}$$

$\Delta w_{ji}(n)$ is the change applied to the connection weight $w_{ji}(n)$ at time n while η represents the learning rate.

Let us assume as shown in [56,327] that in the limit we have

$$\Delta \mathbf{w}_j(n) \to \mathbf{0} \quad \text{and} \quad \mathbf{w}_j(n) \to \mathbf{q}_j \quad \text{as} \quad n \to \infty \quad \text{for all} \quad j = 1, 2 \ldots, l \tag{8.26}$$

and

$$||\mathbf{w}_j(n)|| = 1 \quad \text{for all j} \tag{8.27}$$

Then the limiting values $\mathbf{q}_1, \mathbf{q}_2, \dots, \mathbf{q}_l$ of the synapse vectors of the neurons represent the normalized eigenvectors of the l dominant eigenvalues of the correlation matrix $\mathbf{R}$. The dominant eigenvalues are ranked based on their decreasing values. For the steady state, we obtain

$$\mathbf{q}_j^T \mathbf{R} \mathbf{q}_k = \begin{cases} \lambda_j, & k = j \\ 0, & k \neq j \end{cases} \tag{8.28}$$

with $\lambda_1 > \lambda_2 > \dots > \lambda_l$.

In the limit, for the jth neuron output we write

$$\lim_{n \to \infty} y_j(n) = \mathbf{x}^T(n)\mathbf{q}_j = \mathbf{q}_j^T \mathbf{x}(n) \tag{8.29}$$

This shows that in the limit the GHA acts as an eigenanalyzer of the input data.

We can also give a matrix representation of the GHA. Let $\mathbf{W}$ denote the $l \times p$ connection matrix of the feedforward network shown in Fig. 8.4. Additionally, we will assume that the learning-rate parameter of the GHA from eq. (8.24) is time-varying, such that in the limit it holds that

$$\lim_{n \to \infty} \eta(n) = 0 \quad \text{and} \quad \sum_{n=0}^{\infty} \eta(n) = \infty \tag{8.30}$$

Then the GHA can be rewritten in matrix form:

$$\Delta \mathbf{W}(n) = \eta(n)\{\mathbf{y}(n)\mathbf{x}^T(n) - \mathrm{LT}[\mathbf{y}(n)\mathbf{y}^T(n)]\mathbf{W}(n)\} \tag{8.31}$$

where LT[.] is a lower triangular matrix operator. In [327] it was shown that, by randomly initializing the weight matrix $\mathbf{W}(n)$, the GHA will converge, such that $\mathbf{W}^T(n)$ will be approximately equal to a matrix whose columns are the first l eigenvectors of the $m \times m$ correlation matrix $\mathbf{R}$ of the m-dimensional input vector. The l eigenvalues are ordered by decreasing values.

The practical value of the GHA is that it helps in finding the first l eigenvalues of correlation matrix $\mathbf{R}$, assuming that they are distinct. This is accomplished without even computing $\mathbf{R}$, while the l eigenvalues are computed by the algorithm directly from the input vector $\mathbf{x}$.

A description of the generalized Hebbian algorithm is given below:

1. **Initialization:** Initialize the synaptic weights of the network, w_{ij}, with small random numbers. Set η equal to a small positive value.
2. **Adaptation:** For $n = 1$ and $j = 1, 2, \dots, l$, and $i = 1, 2, \dots, m$, compute

$$y_j(n) = \sum_{i=1}^{m} w_{ji}(n)x_i(n)$$

$$\Delta w_{ji}(n) = \eta \left[y_j(n)x_i(n) - y_j(n) \sum_{k=1}^{j} w_{ki}(n)y_k(n) \right]$$

where $x_i(n)$ is the ith component of the m-dimensional input vector $\mathbf{x}(n)$ and l the prespecified number of principal components.

3. **Continuation:** Increment n by 1, and go to step 2 until the weights w_{ij} reach their steady-state values. Finally, the weight vectors will converge to the normalized eigenvectors of the correlation matrix $\mathbf{R}$ of the input vector $\mathbf{x}$.

8.3.2 Adaptive Principal Component Extraction (APEX)

The GHA is based only on variation of the feedforward synapses to determine the principal components. The adaptive principal component extraction (APEX) algorithm [79] described in this section employs both feedforward and feedback synapses for PCA. APEX is an iterative algorithm that determines the jth principal component from the $(j-1)$ previous ones. Figure 8.5 shows the neural architecture employed for derivation of the APEX algorithm. The input vector $\mathbf{x}$ is m-dimensional, and the neuron output functions are considered linear.

There are two types of synapses involved in the training:

- **Feedforward weights:** connect the input nodes with each of the neurons $1, 2, \ldots, j$, with $j < m$. A feedforward weight vector for the jth neuron is given by

$$\mathbf{w}_j = \left[w_{j1}(n), w_{j2}(n), \ldots, w_{jm}(n)\right]^T$$

  The feedforward weights are trained based on a Hebbian learning algorithm as presented in eq. (8.19), are excitatory, and provide self-amplification.

- **Lateral weights:** connect the jth neuron and the other $1, 2, \ldots, j-1$ neurons. A lateral weight vector for the jth neuron is given by

$$\mathbf{a}_j(n) = \left[a_{j1}(n), a_{j2}(n), \ldots, a_{j,j-1}(n)\right]^T$$

  The lateral weights are trained based on an anti-Hebbian learning algorithm and provide an inhibitory contribution.

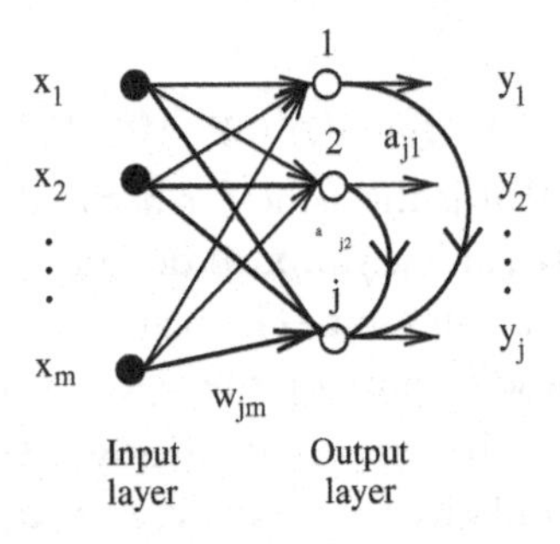

Figure 8.5 Adaptive principal component extraction (APEX) neural network architecture.

The jth neuron output $y_j(n)$ is given by

$$y_j(n) = \mathbf{w}_j^T(n)\mathbf{x}(n) + \mathbf{a}_j^T(n)\mathbf{y}_{j-1}(n) \tag{8.32}$$

where $\mathbf{w}_j^T(n)\mathbf{x}(n)$ describes the contribution of the feedforward synapses, and $\mathbf{a}_j^T(n)\mathbf{y}_{j-1}(n)$ that of the lateral synapses.

The feedback vector $\mathbf{y}_{j-1}(n)$ is defined by the outputs of neurons $1, 2, \ldots, j-1$:

$$\mathbf{y}_{j-1}(n) = [y_1(n), y_2(n), \ldots, y_{j-1}(n)]^T \tag{8.33}$$

Assume that the eigenvalues of the correlation matrix $\mathbf{R}$ are distinct and arranged in decreasing order:

$$\lambda_1 > \lambda_2 > \cdots > \lambda_m \tag{8.34}$$

Additionally, we assume that the neurons have already reached their stable steady states

$$\mathbf{w}_k(0) = \mathbf{q}_k, \quad k = 1, 2, \ldots, j-1 \tag{8.35}$$

and

$$\mathbf{a}_k(0) = \mathbf{0}, \quad k = 1, 2, \ldots, j-1 \tag{8.36}$$

where $\mathbf{q}_k$ corresponds to the kth eigenvalue of matrix $\mathbf{R}$. $n = 0$ denotes the iteration start. Based on eqs. (8.32), (8.33), (8.35), and (8.36), we obtain the following result:

$$\mathbf{y}_{j-1}(n) = \left[\mathbf{q}_1^T\mathbf{x}(n), \mathbf{q}_2^T\mathbf{x}(n), \ldots, \mathbf{q}_{j-1}^T\mathbf{x}(n)\right] = \mathbf{Q}\mathbf{x}(n) \tag{8.37}$$

$\mathbf{Q}$ is the $(j-1) \times m$ eigenvector matrix. This matrix has $j-1$ eigenvectors corresponding to the dominant eigenvalues $\lambda_1, \lambda_2, \ldots, \lambda_{j-1}$ of the correlation matrix $\mathbf{R}$:

$$\mathbf{Q} = [\mathbf{q}_1, \mathbf{q}_2, \ldots, \mathbf{q}_{j-1}]^T \tag{8.38}$$

It is required to use the jth neuron to determine the next largest eigenvalue λ_j of the correlation matrix $\mathbf{R}$.

The adaptation synaptic equations for the feedforward synaptic vector and lateral inhibition synaptic vector are defined as

$$\mathbf{w}_j(n+1) = \mathbf{w}_j(n) + \eta[y_j(n)\mathbf{x}(n) - y_j^2(n)\mathbf{w}_j(n)] \tag{8.39}$$

and

$$\mathbf{a}_j(n+1) = \mathbf{a}_j(n) - \eta[y_j(n)\mathbf{y}_{j-1}(n) + y_j^2(n)\mathbf{a}_j(n)] \tag{8.40}$$

η represents the learning rate that is the same for both adaptation equations since they implement the same type of dynamics. $y_j(n)\mathbf{x}(n)$ denotes Hebbian learning as shown in eq. (8.19), while $y_j(n)\mathbf{y}_{j-1}(n)$ denotes anti-Hebbian learning. The two residual terms, $y_j^2(n)\mathbf{w}_j(n)$ and $y_j^2(n)\mathbf{a}_j(n)$, impose stability on the algorithm.

In [79] it was shown that by choosing a sufficiently small eigenvalue that the weight vectors do not adjust too fast, and given the eigenvectors $\mathbf{q}_1, \mathbf{q}_2, \ldots, \mathbf{q}_{j-1}$, then neuron j in Fig. 8.5 computes the next largest eigenvalue λ_j and the corresponding eigenvector $\mathbf{q}_j$.

In practical applications, the learning rate η_j is set equal to

$$\eta_j = \frac{1}{\lambda_{j-1}} \tag{8.41}$$

λ_{j-1} is available, since it was computed by neuron $j-1$.

An algorithmic description of the APEX algorithm is given below:

1. **Initialization:** Initialize the synaptic weight vectors of the network, $\mathbf{w}_j$ and $\mathbf{a}_j$, with small random numbers. Set η equal to a small positive value.
2. **Adaptation, part I:** Consider the first neuron by setting $j = 1$, and compute for $n = 1, 2, \ldots$

$$\begin{aligned} y_1(n) &= \mathbf{w}_1^T(n)\mathbf{x}(n) \\ \mathbf{w}_1(n+1) &= \mathbf{w}_1(n) + \eta[y_1(n)\mathbf{x}(n) - y_1^2(n)\mathbf{w}_1(n)] \end{aligned}$$

   $\mathbf{x}(n)$ is the input vector. For large n, we have $\mathbf{w}_1(n) \rightarrow \mathbf{q}_1$, where $\mathbf{q}_1$ represents the eigenvector associated with the largest eigenvalue λ_1 of the correlation matrix $\mathbf{R}$.
3. **Adaptation, part II:** Consider now the second neuron by setting $j = 2$, and compute for $n = 1, 2, \ldots$

$$\begin{aligned} \mathbf{y}_{j-1}(n) &= \left[y_1(n), y_2(n), \ldots, y_{j-1}(n)\right]^T \\ y_i(n) &= \mathbf{w}_j^T(n)\mathbf{x}(n) + \mathbf{a}_j^T(n)\mathbf{y}_{j-1}(n) \\ \mathbf{w}_j(n+1) &= \mathbf{w}_j(n) + \eta[y_j(n)\mathbf{x}(n) - y_j^2(n)\mathbf{w}_j(n)] \\ \mathbf{a}_j(n+1) &= \mathbf{a}_j(n) - \eta[y_j(n)\mathbf{y}_{j-1}(n) + y_j^2(n)\mathbf{a}_j(n)] \end{aligned}$$

4. **Continuation:** Consider the next neuron, increment j by 1, and go to step 3 until $j = m$, where m is the desired number of principal components. For large n, we obtain $\mathbf{w}_j(n) \rightarrow \mathbf{q}_j$ and $\mathbf{a}_j(n) \rightarrow \mathbf{0}$, where $\mathbf{q}_j$ is the jth eigenvector of the correlation matrix $\mathbf{R}$.

8.3.3 Linear and Nonlinear Oja's Algorithm

There is an amazing correspondence between the self-organization principle in neuro-computation and principal component analysis in statistics. Oja showed in [271] that it is possible with a single linear neuron to find the first principal component of a stationary input. The weights of this neuron must be trained based on a Hebbian-type learning rule.

Let us assume that the input vector $\mathbf{x}$ and the weight vector $\mathbf{w}$ are m-dimensional, and the only neuron is linear and its output y is given as

$$y = \sum_{i=1}^{m} w_i x_i \tag{8.42}$$

For the synaptic adaptation we consider a Hebbian-type learning

$$w_i(n+1) = w_i(n) + \eta y(n) x_i(n) \tag{8.43}$$

where n is the discrete time and η the learning rate. Described by the above equation, the learning rule has the tendency to grow too much. Therefore, it is necessary to find a form of saturation for the learning rule, and thus ensure convergence. In [271], it was proposed to use the following normalization term:

$$w_i(n+1) = \frac{w_i(n) + \eta y(n) x_i(n)}{\left(\sum_{i=1}^{m} [w_i(n) + \eta y(n) x_i(n)]^2\right)^{\frac{1}{2}}} \tag{8.44}$$

If the learning rate η is small, then we can expand eq. (8.44) as a power series in η:

$$w_i(n+1) = w_i(n) + \eta y(n)[x_i(n) - y(n) w_i(n)] + O(\eta^2) \tag{8.45}$$

where $O(\eta^2)$ corresponds to the higher-order effects in η. For small η they can be ignored, and we get a new representation of the learning rule:

$$w_i(n+1) = w_i(n) + \eta y(n)[x_i(n) - y(n) w_i(n)] \tag{8.46}$$

$y(n)x_i(n)$ is the known Hebbian modification while $y(n)w_i(n)$ is for stabilization purposes.

Equation (8.46) representing the learning algorithm can also be expressed in matrix form. Let

$$\mathbf{x} = [x_1(n), x_2(n), \ldots, x_m(n)]^T$$

be an input vector and

$$\mathbf{w} = [w_1(n), w_2(n), \ldots, w_m(n)]^T$$

the synaptic weight vector. The output y can be represented as following

$$y(n) = \mathbf{x}^T(n)\mathbf{w}(n) = \mathbf{w}^T(n)\mathbf{x}(n) \tag{8.47}$$

Then eq. (8.46) can be rewritten as

$$\mathbf{w}(n+1) = \mathbf{w}(n) + \eta y(n)[\mathbf{x}(n) - y(n)\mathbf{w}(n)] \tag{8.48}$$

By substituting eq. (8.47) into eq. (8.46), we obtain a new representation of the learning rule:

$$\mathbf{w}(n+1) = \mathbf{w}(n) + \eta[\mathbf{x}(n)\mathbf{x}^T(n)\mathbf{w}(n) - \mathbf{w}^T(n)\mathbf{x}(n)\mathbf{x}^T(n)\mathbf{w}(n)\mathbf{w}(n)] \tag{8.49}$$

Let us assume that the correlation matrix $\mathbf{R}$ is positive definite with the largest eigenvalue λ_1 having multiplicity 1. In [271] it was shown that a Hebbian-based linear neuron has the following two important properties:

1. The variance of the model output approaches the largest eigenvalue of the correlation matrix $\mathbf{R}$:

$$\lim_{n\to\infty} \sigma^2(n) = \lambda_1 \tag{8.50}$$

2. The synaptic weight vector of the model approaches the associated eigenvector

$$\lim_{n\to\infty} \mathbf{w}(n) = \mathbf{q}_1 \tag{8.51}$$

with

$$\lim_{n\to\infty} ||\mathbf{w}(n)|| = 1 \tag{8.52}$$

Oja's single neuron learning can be generalized to networks having several neurons in the output layer. The proposed learning rule consists of Hebbian learning of the connection weights and a nonlinear feedback term that is used to stabilize the weights. Specifically, Oja's learning equation is of the form

$$\mathbf{W}(n+1) = \mathbf{W}(n) + \eta[\mathbf{x}(n)\mathbf{x}^{\mathbf{T}}\mathbf{W}(n) - \mathbf{W}(n)(\mathbf{W}^T(n)\mathbf{x}(n)\mathbf{x}^T(n)\mathbf{W}(n))] \tag{8.53}$$

When the input data are assumed to arrive as samples from some stationary pattern class distribution with autocorrelation matrix $\mathbf{R}$, eq. (8.53) can be averaged to the more convenient equation

$$\mathbf{W}(n+1) = \mathbf{W}(n) + \eta[\mathbf{I} - \mathbf{W}(n)\mathbf{W}^T(n)]\mathbf{x}(n)\mathbf{x}^T(n)\mathbf{W}(n) \tag{8.54}$$

In the matrix case, Oja has conjectured similar convergence properties as in the one-unit case. Those properties have been proved in [399] based on a mathematical analysis of the learning rule.

Let us assume that $\mathbf{R}$ has eigenvalues arranged in decreasing order $\lambda_1 > \lambda_2 > \cdots > \lambda_m$ and let $\mathbf{q}_i$ be the corresponding eigenvalue to the ith eigenvector. Let $\mathbf{Q}$ be an $m \times m$ eigenvector matrix

$$\mathbf{Q} = [\mathbf{q}_1, \ldots, \mathbf{q}_m] \tag{8.55}$$

Let $\mathbf{W}$ denote the $l \times m$ weight matrix of the feedforward network

$$\mathbf{W} = [\mathbf{w}_1, \ldots, \mathbf{w}_l]^T \tag{8.56}$$

Then Oja's algorithm shown in eq. (8.54) converges to a fixed point, which is characterized as follows:

1. The product of $\mathbf{W}^T(n)\mathbf{W}(n)$ tends in the limit to the identity matrix:

$$\lim_{n\to\infty} \mathbf{W}^T(n)\mathbf{W}(n) = \mathbf{I} \tag{8.57}$$

2. The columns of $\mathbf{W}(n)$ converge to a dominant eigenspace of the associated autocorrelation matrix:

$$\lim_{n\to\infty} \mathbf{w_i}(n) = \mathbf{q}_i \tag{8.58}$$

An algorithmic description of Oja's learning rule is given below:

1. **Initialization:** Initialize the synaptic weights of the network, w_{ij}, with small random numbers. Set η equal to a small positive value.
2. **Adaptation:** For $n = 1$ compute the weight matrix $\mathbf{W}$ based on

$$\mathbf{W}(n+1) = \mathbf{W}(n) + \eta[\mathbf{I} - \mathbf{W}(n)\mathbf{W}^T(n)]\mathbf{x}(n)\mathbf{x}^T(n)\mathbf{W}(n) \tag{8.59}$$

where $\mathbf{x}$ is an m-dimensional input vector.

3. **Continuation:** Increment n by 1, and go to step 2 until the weight matrix $\mathbf{W}$ reaches the steady-state value. Finally, the weight vectors will converge to the eigenvectors of the correlation matrix $\mathbf{R}$ of the input vector $\mathbf{x}$.

PCA has some limitations that make it less attractive from a neural network point of view. These limitations are: (1) PCA networks implement only linear input-output mappings, (2) slow convergence makes it not useful in large problems, (3) second-order statistics describe only Gaussian data completely, and (4) outputs of PCA networks are mutually uncorrelated but not independent.

For this reason, it is necessary to study nonlinear generalizations of PCA learning algorithms and networks. In [274], some nonlinear generalizations of Oja's learning rule in eq. (8.54) have been proposed:

$$\mathbf{W}(n+1) = \mathbf{W}(n) + \eta[\mathbf{I} - \mathbf{W}(n)\mathbf{W}^T(n)]\mathbf{x}(n)g(\mathbf{x}^T(n)\mathbf{W}(n)) \tag{8.60}$$

$$\begin{aligned}\mathbf{W}(n+1) = \mathbf{W}(n) + \eta[&\mathbf{x}(n)g(\mathbf{x}^T(n)\mathbf{W}(n)) \\ &-\mathbf{W}(n)g(\mathbf{W}^T(n)\mathbf{x}(n))g(\mathbf{x}^T(n)\mathbf{W}(n))]\end{aligned} \tag{8.61}$$

$$\mathbf{W}(n+1) = \mathbf{W}(n) + \eta[\mathbf{x}(n)\mathbf{x}^T(n)\mathbf{W}(n) - \mathbf{W}(n)\mathbf{W}^T(n)\mathbf{x}(n)g(\mathbf{x}^T(n)\mathbf{W}(n))] \tag{8.62}$$

g is usually a monotone odd nonlinear function. Here the function $g(t)$ is applied separately to each component of its argument vector. Generally, we choose for $g(t)$ the function $g(y(t)) = \tan h(\alpha y(t))$, where α is a scalar parameter. The advantage of using nonlinearities is that the neurons become more selective during the learning phase. They also introduce higher-order statistics into the computations in an implicit way and increase the independence of the outputs, so that the original signals can sometimes be roughly separated from their mixtures.

The nonlinear Hebbian learning rules proposed by Oja are not only related to PCA. Provided that all directions of the input space have equal variance, and the nonlinearity is $\tan h(y(t))$, then the learning rule minimizes the kurtosis in several directions, and it tends to find those directions along which the input data projections are statistically independent. Therefore these algorithms can be used for independent component analysis (ICA). Minimizing the kurtosis means, as we will see in Section 8.4.2, to find the directions of the weight vectors in the input space, on which the projection of the input data is heavily clustered and deviates strongly from the Gaussian distribution.

8.3.4 Medical Image Coding

In this section we complete the discussion of PCA-type neural networks by examining their use for solving medical image coding.

A set of four mammograms were selected from the publicly available MIAS database at http://skye.icr.ac.uk/misdb/miasdb.html. All 1024×1024 images are monochrome with 256 gray levels. The images were coded using a linear feedforward neural network with L input neurons and M linear neurons, where $M << L$ is performing the PCA based on three different algorithms. The output of the network $\mathbf{y}$ is defined by $\mathbf{y} = \mathbf{W}^T\mathbf{x}$. $\mathbf{x}$ is the input vector and $\mathbf{W}$ is the weight matrix of the network.

Depending on the type of the activation function (linear or nonlinear) three different PCA algorithms, generalized Hebbian algorithm (GHA) [327], Oja's symmetric algorithm [272], and nonlinear PCA (NLPCA) [165], were implemented.

Oja's symmetric algorithm will provide the convergence of the weight matrix to a subspace spanned by the most significant eigenvectors of the input data covariance matrix. The outputs of the network will then be decorrelated and their variance maximized, so the information content of the L-dimensional vectors will be maximally transferred by the M-dimensional output vectors in the mean square sense.

The weight adaptation equations for the feedforward weight matrix $\mathbf{W}$ for Oja's algorithm are given by eq. (8.54) and for the NLPCA algorithm are given by eq. (8.61).

The synaptic weights w_{ij} connecting the neuron i of the input layer (consisting of L neurons) with the neuron j of the output layer (consisting of M neurons) of the network are updated by the GHA algorithm as described in eq. (8.24).

To train the network, different block sizes, from 2×2, 4×4, and 8×8 pixels, were used.

In the following the convergence properties of the three PCAs are shown, along with the compression results achieved based on the PCAs without and with overlapping. To reduce the blocking effects associated with any transform-based algorithm, overlapping of neighboring image blocks by one pixel was allowed.

The convergence speed of the three algorithms is shown in Fig. 8.6. They were trained with 320,000 8×8 sample blocks. The graph 8.6 shows the number of iterations versus the

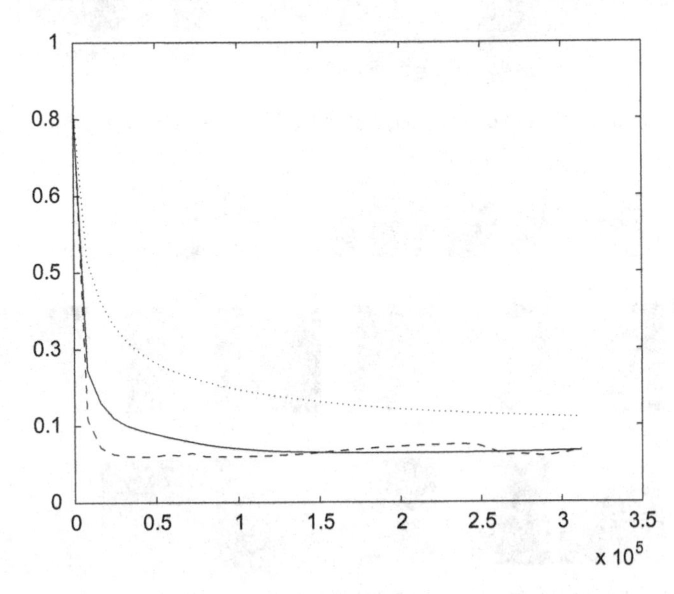

Figure 8.6 The three different implementations of the PCA algorithm: solid, GHA; dotted, Oja's symmetric algorithm; dashed, nonlinear PCA type learning. The x-axis shows the number of the training step; the corresponding value on the y-axis shows the normalized error.

output error. The results in the same figure show that the nonlinear PCA-type learning algorithm converges very fast (dashed curve), the GHA algorithm converges more slowly (solid curve) but more steadily, and Oja's symmetric algorithm takes longer than the other two but decreases monotonously.

Figures 8.7 and 8.8 show the $2 \times 2, 4 \times 4$, and 8×8 masks representing the synaptic weights learned by the network. Figure 8.7 shows the 4×4 weight matrices for two and four largest principal components for Oja's algorithm, which is the fastest of the three algorithms.

In Fig. 8.8 a comparison between the 8×8 weight matrices for Oja, GHA, and nonlinear PCA for the eight largest principal components is illustrated.

A mammogram from the MIAS database, which will serve as an original image for compression purposes, is shown in Fig. 8.9.

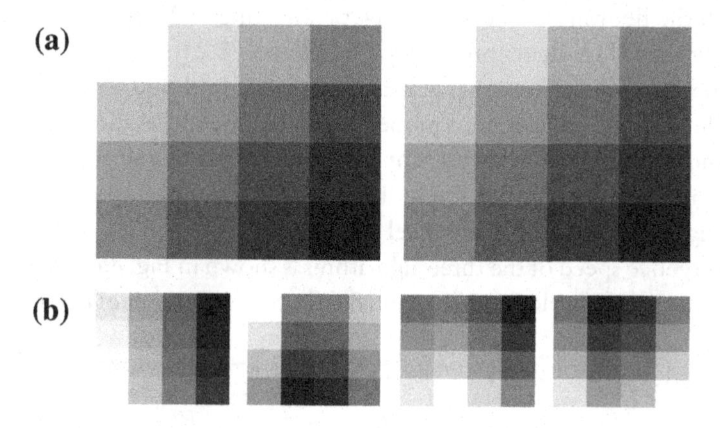

Figure 8.7 Weight matrices based on Oja's algorithm for (a) two and (b) four principal components.

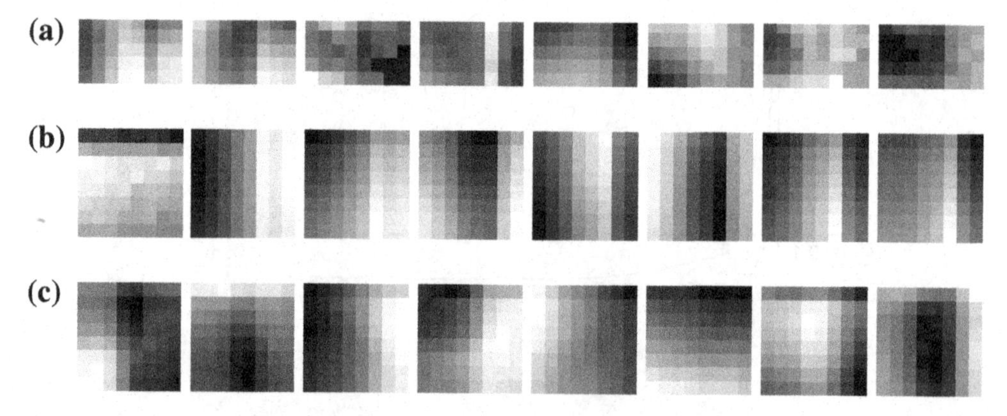

Figure 8.8 8×8 weight matrices based on (a) Oja, (b) GHA, and (c) nonlinear PCA, and eight largest principal components.

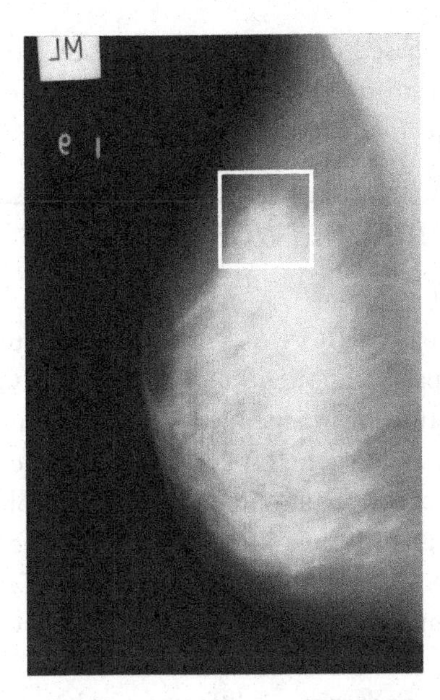

Figure 8.9 Original digitized mammogram from MIAS database. The white square indicates the area, which is subject to the compression operation.

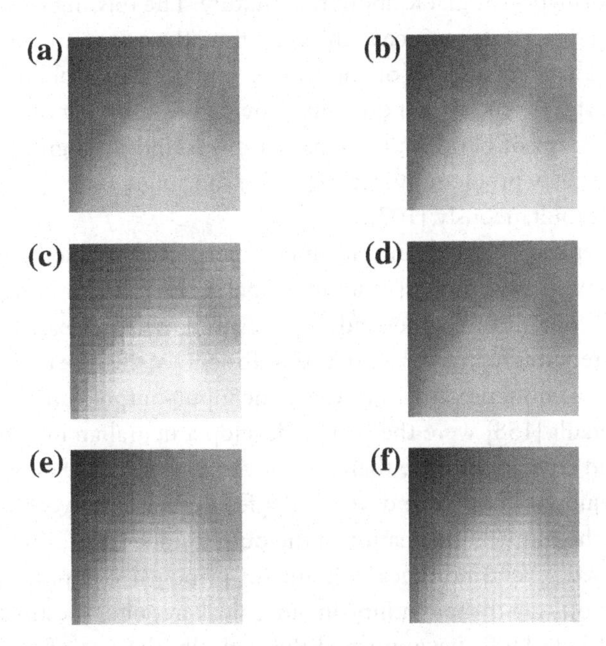

Figure 8.10 Cut-outs of the compressed test image. Compression results with and without overlapping for the three PCA algorithms. (a) Oja without overlapping, (b) Oja with one overpixel, (c) GHA without overlapping, (d) GHA with one overlapping pixel, (e) nonlinear without overlapping, and (f) nonlinear with one overlapping.

The compression results based on the three PCAs are shown in Fig. 8.10a, c, and e. The nonlinear PCA shows a strong blocking effect as a result of the compression process. The best image quality subjectively rated is achieved by Oja's algorithm.

8.4. ICA-TYPE NEURAL NETWORKS

8.4.1 Introduction

Independent component analysis (ICA) is viewed as an extension of the widely used statistical technique, principal component analysis (PCA). ICA has been developed in the context of blind separation of independent sources from their linear mixtures, and it has been applied to many areas such as speech, radar, medical instrumentation, mobile telephone communication, and hearing-aid devices. The problem is defined as the estimation of original source signals from a known and measurable sensor output when the sensor receives an unknown mixture of the source signals. ICA is closely related to blind source separation in communication, which consists of recovering unobserved signals or "sources" from several observed mixtures. The observed signals typically represent the sensors' outputs where each sensor receives a different combination (mixture) of the source signals. The term "blind" refers to the fact that (1) the source signals are not observed, and (2) there is no information available about the mixture. The missing mixture information is not considered a disadvantage since at the same time there is an underlying assumption about the statistical independence of the source signals. Thus "blindness" represents a strength, since it captures the spatial diversity provided by an array of sensors.

An interesting form of blind source separation is found in cognitive processing, the so-called "cocktail party problem," where a person can single out a specific speaker from a group speaking simultaneously [107].

Blind source separation represents an underdetermined problem for which a close solution does not exist, since the single source signal statistics, the mixing, and the transfer channels are all unknown. It goes beyond decorrelation, which is based on second-order statistics, and thus requires higher-order statistics. To achieve this, the learning process must employ some suitable nonlinearities. However, the input-output function remains linear.

Jutten and Herault [158] were the first to develop a neural architecture and learning algorithm for blind separation; since then several variants of it have appeared in the literature. Bell and Sejnowski [23] have developed a feedforward network and learning rule which minimizes the mutual information at the output nodes. Karhunen and Joutsensalo [165] have proposed several nonlinear variants of principal component analysis learning and have demonstrated their utility in sinusoidal frequency estimation. Oja [166] and Girolami and Fyfe [108] have justified theoretically the use of the nonlinear PCA algorithm in blind separation of source signals. In general, source separation is obtained by minimizing an appropriate contrast function, i.e., a function of the distribution of the estimated sources. Usually, the contrast functions are based on maximum likelihood [15,49,293], the infomax principle [24,212,315], and mutual information [11,66,144].

Although each model stems from different theoretical aspects, it is possible to generalize them under the maximum likelihood principle, leading to simple and efficient algorithms [49].

Summarizing, ICA can be viewed as an improvement of PCA with a high practical value. The coefficients of a linear expansion of the data vectors are no longer required to be uncorrelated for ICA, but mutually independent. However, although ICA provides in many cases a more meaningful representation of the data than PCA, it still sometimes employs PCA as a preprocessing and simplifying step.

8.4.1.1 The Data Model

Suppose that the sensor signals $x_i(t)(i = 1, \ldots, N)$ are generated by linearly weighting a set of N statistically independent random sources $s_j(t)$ with time-independent coefficients a_{ij}:

$$x_i(t) = \sum_{j=1}^{N} a_{ij} s_j(t) \tag{8.63}$$

This equation can be equivalently expressed in vector notation as

$$\mathbf{x}(t) = \mathbf{A}\mathbf{s}(t) \tag{8.64}$$

where $\mathbf{s}(t) = [s_1(t), \ldots, s_N(t)]^T, \mathbf{x}(t) = [x_1(t), \ldots, x_N(t)]$, and $\mathbf{A} = [a_{ij}]$.

The objective is to recover the source signals $s_j(t)(j = 1, \ldots, N)$ from the sensor signals $x_i(t)(i = 1, \ldots, N)$ in the absence of any special information about the properties of $\mathbf{A}$ and $\mathbf{x}(t)$. This definition of signal separation can be somewhat confusing: If $s_j(t)$ $(j = 1, \ldots, N)$ are source signals, then $d_1 s_{p1}(t), \ldots, d_N s_{pN}(t)$ may as well be considered source signals, where $\{p_1, \ldots, p_N\}$ is an arbitrary permutation of $\{1, \ldots, N\}$ and $d_1, \ldots, d_N$ are any constants unequal to zero. This is because $d_1 s_{p1}(t), \ldots, d_N s_{pN}(t)$ are statistically independent and $x_i(t)$ can be expressed by their linear combination. Since the sensor signals might be possibly recalled in a different order and re-scaled from the resulting vector $\bar{\mathbf{s}}(t)$, we define blind source separation as any procedure providing the following type of signal:

$$\bar{\mathbf{s}}(t) = \mathbf{D}\mathbf{P}\mathbf{s}(t) \tag{8.65}$$

where $\mathbf{P}$ is a permutation matrix and $\mathbf{D}$ is a diagonal matrix with nonzero elements.

The problem of obtaining the data model of ICA is to estimate the mixing matrix $\mathbf{A}$ using only the information contained in the sensor vector $\mathbf{x}$. In this perspective, the statistical model has two components: the mixing $\mathbf{A}$ and the probability distribution of the source vector $\mathbf{s}$.

The following assumptions [166] are necessary for the blind separation process:

1. Matrix $\mathbf{A}$ is constant and nonsingular. The mixing matrix is the parameter of interest, and its columns are assumed to be linearly independent such that it is invertible.
2. The source signals $s_j(t)(j = 1, \ldots, N)$ are mutually statistically independent signals. Although nothing is specified with regard to the distribution of each source,

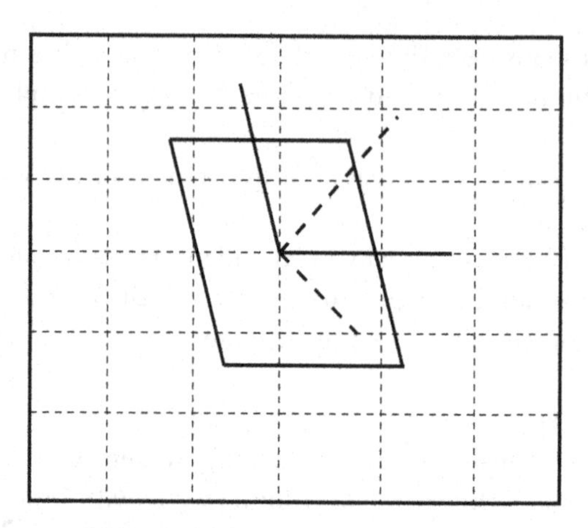

Figure 8.11 The basis vectors of ICA (solid lines) and PCA (dashed lines). The data vectors are distributed uniformly inside the parallelogram.

the assumption of mutual source independence states a great deal about their joint distribution.

3. Each source signal $s_j(t)$ is a stationary zero-mean stochastic process. Only one source signal can be Gaussian. Since the linear combination of Gaussian source signals is Gaussian, it becomes impossible to separate the initial sources from each other.

Under assumptions 1–3 it can be proved that the source signals can be uniquely determined from the sensor signals except the arbitrariness of the permutation matrix $\mathbf{P}$ and the diagonal matrix $\mathbf{D}$.

Here it is useful to compare ICA to standard PCA. Although in PCA, the data model has the same form as in eq. (8.63), PCA decorrelates only the signals by constraining the basis vectors $\mathbf{a}(t)$ to be mutually orthonormal. On the other hand, ICA separates the signals. Usually, the basis vectors $\mathbf{a}(t)$ are not mutually orthogonal. Figure 8.11 illustrates the differences between ICA and PCA [166].

In the figure, it is assumed that the data vectors are uniformly distributed within the sketched parallelogram. The solid lines correspond to the two basis vectors of ICA, and the dashed lines to the orthogonal basis vectors of PCA. It is obvious that the basis vectors of ICA capture the data much better.

The problem of estimating the data model of ICA is to estimate the mixing matrix $\mathbf{A}$ using only the information contained in the sensor vector $\mathbf{x}$. In the following sections we take a closer look at these aspects.

8.4.1.2 Definition of Independence

As we have seen so far, the basic assumption of mutual source independence represents the building block for blind source separation.

For a better understanding of the theory behind ICA estimation, we give in the following the mathematical definitions of independence and uncorrelatedness.

Given are two random variables y_1 and y_2. They are said to be independent if the occurrence or nonoccurrence of y_1 does not influence the occurrence or nonoccurrence of y_2, and vice versa.

Independence can be defined in terms of probability densities. Let $p(y_1, y_2)$ denote the joint probability density function (pdf) of y_1 and y_2. Also, let $p_1(y_1)$ denote the marginal pdf of y_1:

$$p_1(y_1) = \int p(y_1, y_2)dy_2 \tag{8.66}$$

and similarly for y_2. The two random variables y_1 and y_2 are statistically independent if and only if the joint pdf is factorizable:

$$p(y_1, y_2) = p_1(y_1)p_2(y_2) \tag{8.67}$$

This definition allows us to derive an important result pertaining to the product of two functions of these two independent random variables y_1 and y_2

$$E\{h_1(y_1)h_2(y_2)\} = E\{h_1(y_1)\}E\{h_2(y_2)\} \tag{8.68}$$

An interesting implication emerges from the definition of independence: Statistically independent variables are also uncorrelated, but the reverse is not necessarily true. It is said that two random variables y_1 and y_2 are uncorrelated if their covariance equals zero:

$$E\{y_1y_2\} = E\{y_1\}E\{y_2\} = 0 \tag{8.69}$$

In summary, if the variables are independent, then they are uncorrelated. This follows directly from eq. (8.68) by just taking $h_1(y_1) = y_1$ and $h_2(y_2) = y_2$. It is very important to mention that uncorrelatedness does not imply independence.

Because uncorrelatedness results from independence, many ICA methods constrain the estimation procedure such that it always first determines the uncorrelated estimates of the independent components. This simplifying preprocessing step reduces the number of free parameters. In other words, blind separation can be based on independence; but independence cannot be reduced to simple decorrelation. This is evident from the fact that there are only $N(N-1)/2$ such conditions while N^2 parameters are unknown.

The fundamental restriction for ICA is that only non-Gaussian independent components can be estimated. At most one independent component is allowed to be Gaussian. To understand this restriction, we have to take a close look at the mixing of Gaussian source signals. If the mixing matrix is orthogonal, it leads to sensor signals that are also Gaussian, uncorrelated, and of unit variance. Their joint pdf is completely symmetric, and so it does not contain any information on the directions of the columns of the mixing matrix. In other words, the mixing matrix $\mathbf{A}$ cannot be estimated. In conclusion, the mixing remains unidentifiable for Gaussian independent components.

8.4.1.3 Preprocessing for ICA

In order to make the ICA estimation simpler and more feasible for practical applications, it is useful to preprocess the signals. Second-order statistics (decorrelation) is employed to remove redundancy, and thus make ICA estimation simpler. The most used techniques are centering, and the whitening or sphering of a random signal.

By centering a signal $\mathbf{x}$, the mean vector $\mathbf{m} = E\{\mathbf{x}\}$ is subtracted from $\mathbf{x}$, in order to make both $\mathbf{x}$ and $\mathbf{s}$ zero-mean variables.

The other preprocessing technique is whitening, and it is applied after the centering and before the ICA estimation. By whitening we transform the observed vector $\mathbf{x}$ linearly into a new vector $\mathbf{z}$, which has as a covariance matrix the identity matrix:

$$E\{\mathbf{z}\mathbf{z}^T\} = \mathbf{I} \tag{8.70}$$

In other words, the components of $\mathbf{z}$ are uncorrelated and their variances equal unity. The whitening transformation can be performed based on the eigenvalue decomposition [84] of the covariance matrix $E\{\mathbf{x}\mathbf{x}^\mathbf{T}\} = \mathbf{Q}\mathbf{\Lambda}\mathbf{Q}^T$, where $\mathbf{Q}$ is the orthonormal matrix of the eigenvectors of $E\{\mathbf{x}\mathbf{x}^\mathbf{T}\}$ and $\mathbf{\Lambda}$ is the diagonal matrix of its eigenvalues, $\mathbf{\Lambda} = \text{diag}\{\lambda_i\}$. The transformation for whitening is thus given by

$$\mathbf{z} = \mathbf{Q}\mathbf{\Lambda}^{-\frac{1}{2}}\mathbf{Q}^T\mathbf{x} \tag{8.71}$$

Whitening leads to a new orthogonal mixing matrix $\tilde{\mathbf{A}}$,

$$\mathbf{z} = \mathbf{Q}\mathbf{\Lambda}^{-\frac{1}{2}}\mathbf{Q}^T\mathbf{A}\mathbf{s} = \tilde{\mathbf{A}}\mathbf{s} \tag{8.72}$$

with $E\{\tilde{\mathbf{x}}\tilde{\mathbf{x}}^T\} = \tilde{\mathbf{A}}\tilde{\mathbf{A}}^T = \mathbf{I}$. But $\tilde{\mathbf{A}}$ is at the same time a rotation matrix since it connects to spatially white vectors $\mathbf{z}$ and $\mathbf{s}$. In other words, "whitening" or "sphering" the data reduces the mixture to a rotation matrix.

The whitening also reduces the number of parameters to be estimated. The new orthogonal matrix $\tilde{\mathbf{A}}$ has about half the number of parameters of an arbitrary matrix. Additionally, whitening provides a dimension reduction. By employing PCA, the small eigenvalues can be discarded. The benefit lies not only in dimension reduction, it also prevents overlearning of the ICA algorithm and reducing of noise.

8.4.2 Techniques for ICA Estimation

As stated in the previous section, ICA estimation requires non-Gaussianity. This is different from classical statistical theory where random variables are assumed to have Gaussian distributions. Furthermore, the central limit theorem [285] states that the sum of independent random variables tends to a normal distribution. This represents a very important consequence for ICA estimation, as will be seen in the following.

We will now state the ICA estimation problem. Given N independent realizations of the sensor signals $\mathbf{x}$, determine an estimate of the inverse of the mixing matrix $\mathbf{A}$. In other words, find a demixing matrix $\mathbf{W}$, whose individual rows are a rescaling and permutation of those of the mixing matrix $\mathbf{A}$. Thus, the solution to the ICA estimation

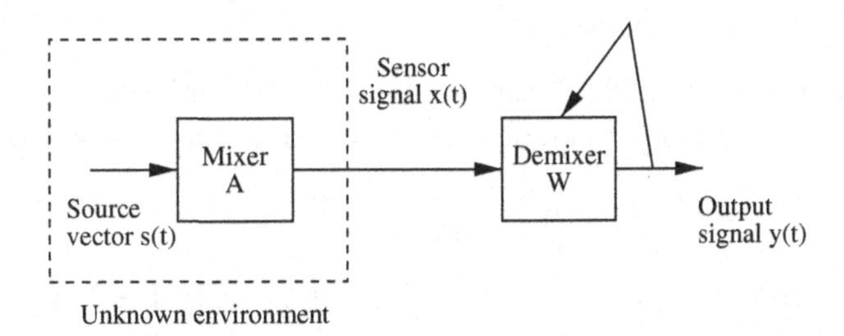

Figure 8.12 Block diagram for the blind source separation problem.

problem may be expressed in the form

$$\mathbf{y} = \mathbf{W}\mathbf{x} = \mathbf{W}\mathbf{A}\mathbf{s} \rightarrow \mathbf{D}\mathbf{P}\mathbf{s} \tag{8.73}$$

where $\mathbf{D}$ is a nonsingular diagonal matrix and $\mathbf{P}$ a permutation matrix. The block diagram of the ICA estimation problem is illustrated in Fig. 8.12.

We immediately see that y_i is a linear combination of the s_i. Based on the central limit theorem, y_i is more Gaussian than any of the s_i, and becomes least Gaussian when it in fact equals one of the s_i.

For the ICA estimation problem, we could take as $\mathbf{W}$ a matrix that maximizes the non-Gaussianity of $\mathbf{W}\mathbf{x}$.

We see that the following aspects are important:

- ICA must go beyond second-order statistics. A cost function with a penalty term involving only pairwise decorrelation would not lead to signal separation.
- Source separation can be achieved by optimizing a "contrast function" [49]. This contrast function represents a scalar measure describing the distributional property of the output $\mathbf{y}$ such as entropy, mutual independence, high-order decorrelations, and divergence between the joint distribution of $\mathbf{y}$ and a given model.
- Fast adaptation is feasible, even based on simple algorithms, and requires a small number of samples.

We see that "contrast functions" are objective functions for source separation. They are based either on high-order correlations or on information theory. They are a real function of the probability distribution of the output $\mathbf{y} = \mathbf{W}\mathbf{x}$, and they serve as objectives: They have to be designed such that the source separation is achieved when they reach their minimum. Contrast functions for source separation are usually denoted as $\phi[\mathbf{y}]$.

In [49] several such "contrast functions" are proposed. The most important groups are:

1. High-order approximations: kurtosis and negentropy.
2. Information-theoretic contrasts: matching distributions such as likelihood and infomax, and matching the structure such as mutual information.

For simplification reasons, we need to assume that y is a random variable of zero-mean (centered) and of unit variance (sphered):

$$E\{y\} = 0 \quad \text{and} \quad E\{y^2\} = 1 \tag{8.74}$$

8.4.2.1 Kurtosis

As a measure for non-Gaussianity, we can use simple higher-order statistics called kurtosis, which represents a fourth-order cumulant with zero time lags. The kurtosis of y is classically defined by

$$\text{kurt}\,(y) = E\{y^4\} - 3(E\{y^2\})^2 \tag{8.75}$$

If eq. (8.74) holds, then we get for the kurtosis $\text{kurt}(y) = E\{y^4\} - 3$, which represents a shifted version of the fourth moment $E\{y^4\}$. For a normal random variable the kurtosis is equal to zero, whereas for most non-Gaussian random variables it is nonzero, being either positive or negative. Negative kurtosis is typical for sub-Gaussian variables having a "flat" pdf that exhibits rather constant values around zero and otherwise is very small. An example of such a sub-Gaussian random variable is the uniform distribution

$$p(y) = \begin{cases} \frac{1}{2\sqrt{5}}, & \text{if} \quad |y| \leq \sqrt{5} \\ 0, & \text{otherwise} \end{cases} \tag{8.76}$$

Positive kurtosis is typical for super-Gaussian variables having a "spiky" pdf with long tails such as the Laplace distribution:

$$p(y) = \exp\,(-|y|) \tag{8.77}$$

Kurtosis is a linear operation satisfying the superposition principle. Thus, for any two random variables y_1 and y_2 and a scalar α it holds that

$$\text{kurt}(y_1 + y_2) = \text{kurt}(y_1) + \text{kurt}(y_2) \tag{8.78}$$

$$\text{kurt}(\alpha y_1) = \alpha^4 \text{kurt}(y_1) \tag{8.79}$$

To use kurtosis in practical problems means to start from some weight matrix $\mathbf{W}$ and compute the direction the kurtosis is growing most strongly (positive kurtosis) or decreasing most strongly (negative kurtosis). However, it was shown in [143] that kurtosis is sensitive to outliers and therefore is not considered a robust measure of non-Gaussianity.

An interesting interpretation to kurtosis is given in terms of contrast functions. High-order statistics can be employed to define contrast functions which are simple approximations to those derived from the maximum likelihood approach, as we will see in the next sections.

8.4.2.2 Negentropy

Negentropy is usually taken as another non-Gaussian measure and describes the information-theoretic quantity of (differential) entropy. The entropy of a random variable is viewed as a measure of the uncertainty of an event. The maximal entropy is achieved by a distribution that exhibits the highest probable randomness. The differential entropy

H of a random vector $\mathbf{y}$ with a given pdf $f(\mathbf{y})$ is defined as [285]

$$H(\mathbf{y}) = -\int f(\mathbf{y}) \log f(\mathbf{y}) d\mathbf{y} \tag{8.80}$$

In [285] it was proved that the Gaussian variable has the largest entropy among all random variables of equal variance. The implication of this fundamental result is that the entropy can serve as a measure of non-Gaussianity. The entropy is small for "spiky" pdfs or those pdfs being centered around certain values.

The goal is to obtain a measure of non-Gaussianity that is zero for a Gaussian variable and in general is nonnegative. Such a measure is the so-called negentropy J, defined as follows:

$$J(\mathbf{y}) = H(\mathbf{y}_{gauss}) - H(\mathbf{y}) \tag{8.81}$$

where $\mathbf{y}_{gauss}$ has a normal pdf and has the same mean and variance as $\mathbf{y}$. The negentropy has the following important properties: (1) It is always nonnegative, and it is zero if and only if $\mathbf{y}$ has a normal pdf, and (2) it is invariant to invertible linear transformations [66]. Its main disadvantage is that it is computationally expensive. This makes it necessary to approximate the negentropy, and the most useful approximation is given in [147]. By using a nonquadratic function G, we get the following approximation for J:

$$J(y) \propto [E\{G(y)\} - E\{G(\nu)\}]^2 \tag{8.82}$$

where ν is a Gaussian variable of zero mean and unit variance. y is a random variable of zero mean and unit variance. This holds for practically any nonquadratic function G. The choice of G is important: Only a slowly growing G ensures a robust estimation. The following choices of G are common [147]:

$$G_1(u) = \frac{1}{a_1} \log \cos h\, a_1 u, \quad G_2(u) = -\exp(-u^2/2) \tag{8.83}$$

where $1 \leq a_1 \leq 2$ is some suitable constant.

The proposed approximation of negentropy is thus simple, easy to compute, and robust.

8.4.2.3 Minimization of Mutual Information

The maximum likelihood principle leads to several contrast functions which can be derived from the Kullback divergence. The Kullback divergence is a measure of the distance between two probability density functions $p(\mathbf{x})$ and $\widehat{p}(\mathbf{x})$ on R^N and is given by

$$K(p|\widehat{p}) = \int_{\mathbf{x}} p(\mathbf{x}) \log \left\{ \frac{p(\mathbf{x})}{\widehat{p}(\mathbf{x})} \right\} d\mathbf{x} \tag{8.84}$$

assuming the existence of the integral. The divergence between two random vectors $\mathbf{w}$ and $\mathbf{z}$ is represented by $K[\mathbf{w}|\mathbf{z}] \geq 0$. The equality holds if and only if $\mathbf{w}$ and $\mathbf{z}$ have the same distribution.

Although K is not symmetric and cannot be considered as a distance, it represents a statistical measure for quantifying the closeness of two distributions.

The minimization of mutual information is another approach for ICA estimation and leads to finding most non-Gaussian directions [147]. The mutual information I between m (scalar) random variables, $y_i, i = 1, \dots, m$, is given by

$$I(y_1, y_2, \dots, y_m) = \sum_{i=1}^{m} H(y_i) - H(\mathbf{y}) \tag{8.85}$$

Mutual information determines the independence between random variables and has the following properties: (1) It is always nonnegative, and (2) it is zero if and only if the variables are statistically independent. Thus, mutual information gives the whole statistical picture.

A very important property of mutual information [285] is given for an invertible linear transformation $\mathbf{y} = \mathbf{Wx}$:

$$I(y_1, y_2, \dots, y_n) = \sum_{i} H(y_i) - H(\mathbf{x}) - \log|\det(\mathbf{W})| \tag{8.86}$$

In general, we write $H(\mathbf{y}) = H(\mathbf{Wx}) = H(\mathbf{x}) + \log|\det(\mathbf{W})|$ where $\det(\mathbf{W})$ is the determinant of $\mathbf{W}$.

We can easily deduce an important relation between negentropy and mutual information:

$$I(y_1, y_2, \dots, y_n) = C - \sum_{i} J(y_i) \tag{8.87}$$

The constant C is not a function of $\mathbf{W}$ if the y_i are uncorrelated.

This leads to an alternative ICA estimation as an invertible transformation based on the fact that we have to determine $\mathbf{W}$ from eq. (8.73) by minimizing the mutual information (statistical dependence) of the transformed components s_i. Or, from eq. (8.87), we see that ICA estimation by minimization of mutual information corresponds to maximizing the sum of non-Gaussianities of the estimates, when the estimates are uncorrelated.

An interesting interpretation of mutual information is given in terms of contrast functions. The idea is to look for a contrast function that also matches the unknown system structure. In other words, the observed data should be modeled by adjusting both the unknown system and the source distribution. In other words, the divergence $K[\mathbf{y}|\mathbf{s}]$ should be minimized with respect to $\mathbf{A}$ based on the distribution of $\mathbf{y} = \mathbf{Wx}$ and with respect to the model distribution of $\mathbf{s}$. A basic property of the Kullback divergence is

$$K[\mathbf{y}|\mathbf{s}] = K[\mathbf{y}|\tilde{\mathbf{y}}] + K[\tilde{\mathbf{y}}|\mathbf{s}] \tag{8.88}$$

where $\tilde{\mathbf{y}}$ is a random vector with independent entries and with the same distribution as $\mathbf{y}$. $\mathbf{s}$ is a vector with independent entries. Minimization of the above relationship is only given by the minimization of the first term $K[\mathbf{y}|\tilde{\mathbf{y}}]$ since the second term vanishes if

$\mathbf{s} = \tilde{\mathbf{y}}$. In other words, the contrast function

$$\phi_{MI}[\mathbf{y}] = K[\mathbf{y}|\tilde{\mathbf{y}}] \tag{8.89}$$

has to be minimized. The Kullback divergence $K[\mathbf{y}|\tilde{\mathbf{y}}]$ between a distribution and the closest distribution with independent entries is defined as the mutual information between the entries of $\mathbf{y}$. We always have $\phi_{MI}[\mathbf{y}] \geq 0$. The equality holds, if and only if $\mathbf{y} = \mathbf{W}\mathbf{x}$ is distributed as $\tilde{\mathbf{y}}$.

8.4.2.4 Maximum Likelihood Estimation

ICA estimation is also possible based on another information-theoretic principle, the maximum likelihood. From the derivation in Chapter 6, we recall that the likelihood function represents the probability density function of a data set in a given model, but is considered as a function of the unknown parameters of the model. In [293] it was shown that the log-likelihood for a noiseless ICA model is given by

$$L = \sum_{t=1}^{T} \sum_{i=1}^{n} \log p_i \left(\mathbf{w}_i^T \mathbf{x}(t)\right) + T \log |\det(\mathbf{W})| \tag{8.90}$$

where f_i are the density functions of the s_i, and $\mathbf{x}(t), t = 1, \ldots, T$, are the realizations of $\mathbf{x}$. We recall, that the pdf of $\mathbf{y} = \mathbf{W}\mathbf{x}$ is given by $p_{\mathbf{y}} = p_{\mathbf{x}}(\mathbf{W}^{-1}\mathbf{y})|\det(\mathbf{W}^{-1})|$, and based on this the above log-likelihood equation can be explained.

In [49] it was shown that the maximum likelihood principle is associated with a contrast function,

$$\phi_{ML}[\mathbf{y}] = K[\mathbf{y}|\mathbf{s}] \tag{8.91}$$

The maximum likelihood principle when applied to blind source separation states: Find a matrix $\mathbf{A}$ such that the distribution of $\mathbf{W}\mathbf{x}$ is as close as possible in terms of Kullback divergence to the hypothesized distribution of sources. From eqs. (8.88), (8.89), and (8.91), we obtain the following fundamental relationship showing the interconnection between maximum likelihood and mutual information [49]:

$$\phi_{ML}[\mathbf{y}] = \phi_{MI}[\mathbf{y}] + \sum_{i=1}^{n} K[y_i|s_i] \tag{8.92}$$

This equation states that in order to maximize the likelihood two components need to be minimized: The first component describes the deviation from independence, while the second component represents the mismatch between the source signals s_i and their estimation y_i.

8.4.2.5 The Information Maximization Estimation

In [285] it was shown that the mutual information represents a measure of the uncertainty about the output of a system that is resolved by observing the system input. The

Infomax principle is based on the maximization of the mutual information and produces a reduction in redundancy in the output compared to that in the input.

This idea was captured in a nonlinear neural network proposed in [24], and it was shown that the network is able to separate independent signals.

By assuming $\mathbf{x}$ is the input to the neural network whose output vector $\mathbf{y} = \mathbf{g}(\mathbf{W}^T\mathbf{x})$ is nonlinear, and $\mathbf{W}$ is the weight matrix, ICA estimation is based on the maximization of the output vector entropy

$$H[\mathbf{g}(\mathbf{y})] = H\left(g_1\left(\mathbf{w}_1^T\mathbf{x}\right), \ldots, g_n\left(\mathbf{w}_n^T\mathbf{x}\right)\right) \tag{8.93}$$

where the nonlinearities g_i are chosen as the cumulative distribution functions corresponding to the densities p_i:

$$g_i'(\cdot) = p_i(\cdot) \tag{8.94}$$

This ICA estimation based on the principle of network entropy maximization, or "infomax," corresponds to maximum likelihood estimation, as we will prove by introducing a contrast function.

In [49] was determined an associated contrast function for the infomax:

$$\phi_{IM}[\mathbf{y}] = -H[\mathbf{g}(\mathbf{y})] \tag{8.95}$$

The infomax idea yields the same contrast function as the likelihood:

$$\phi_{IM}[\mathbf{y}] = \phi_{ML}[\mathbf{y}] \tag{8.96}$$

This proves the connection between maximum likelihood and infomax [49].

8.4.3 FastICA

The previous sections dealt with the theoretical aspects of ICA estimation. In this section, we will focus on ICA implementation issues and present a practical algorithm for maximizing the contrast function. The data must be preprocessed by centering and whitening as described in the preceding section.

8.4.3.1 FastICA for One Unit or One Independent Component

We first consider the one-unit version. The unit can be an artificial neuron having a weight vector $\mathbf{w}$ that is updated based on a learning algorithm, such that the projection $\mathbf{w}^T\mathbf{x}$ maximizes non-Gaussianity. As a measure for non-Gaussianity the approximation of the negentropy as given by eq. (8.82) can be chosen [147].

The FastICA algorithm is a fixed-point iteration scheme for finding a maximum of the non-Gaussianity of $\mathbf{w}^T\mathbf{x}$. Let g be the derivative of the non-quadratic function used in eq. (8.82). Choices of such nonquadratic functions were given in eq. (8.83).

An algorithmic description of the FastICA algorithm for estimating one independent component is given below:

1. **Preprocessing:** Whiten the data to give $\mathbf{x}$.

2. **Initialization:** Initialize the weight vector $\mathbf{w}$ of unit norm with random numbers.
3. **Adaptation:** Compute the change of the weight vector $\mathbf{w}$ according to

$$\mathbf{w} \leftarrow E\{\mathbf{x}g(\mathbf{w}^T\mathbf{x})\} - E\{g'(\mathbf{w}^T\mathbf{x})\}\mathbf{w} \tag{8.97}$$

where g is defined as in eq. (8.83).
4. **Normalization:** Normalize the weight vector $\mathbf{w}$

$$\mathbf{w} \leftarrow \frac{\mathbf{w}}{||\mathbf{w}||} \tag{8.98}$$

5. **Continuation:** If the weight vector $\mathbf{w}$ is not converged, go back to step 3.

The algorithm converged only if the old and new values of $\mathbf{w}$ point in the same direction, i.e., their dot product is close to 1.

The derivation of this algorithm is simple and was given in [147]. The maxima of the approximation of the negentropy of $\mathbf{w}^T\mathbf{x}$ are obtained at certain optima of $E\{G(\mathbf{w}^T\mathbf{x})\}$. The Kuhn-Tucker conditions determine the optima of $E\{G(\mathbf{w}^T\mathbf{x})\}$, under the constraint $E\{(\mathbf{w}^T\mathbf{x})^2\} = ||\mathbf{w}||^2 = 1$, as points where

$$E\{\mathbf{x}g(\mathbf{w}^T\mathbf{x})\} - \beta\mathbf{w} = 0 = F(\mathbf{w}) \tag{8.99}$$

This equation can be solved by Newton's method, and we need the Jacobian matrix $JF(\mathbf{w})$:

$$JF(\mathbf{w}) = E\{\mathbf{x}\mathbf{x}^T g'(\mathbf{w}^T\mathbf{x})\} - \beta\mathbf{I} \tag{8.100}$$

An approximation of the above is given by

$$E\{\mathbf{x}\mathbf{x}^T g'(\mathbf{w}^T\mathbf{x})\} \approx E\{\mathbf{x}\mathbf{x}\}E\{g'(\mathbf{w}^T\mathbf{x})\} = E\{g'(\mathbf{w}^T\mathbf{x})\}\mathbf{I} \tag{8.101}$$

It was assumed that the data are whitened.

The Jacobian matrix becomes diagonal and can be easily inverted. The resulting Newton iteration is

$$\mathbf{w} \leftarrow -\frac{E\{\mathbf{x}g(\mathbf{w}^T\mathbf{x})\} - \beta\mathbf{w}}{E\{g'(\mathbf{w}^T\mathbf{x})\} - \beta} \tag{8.102}$$

This completes the proof of the FastICA algorithm.

8.4.3.2 FastICA for Several Units or Several Independent Components

To estimate several independent components, the one-unit FastICA is extended for several units (neurons) with weight vectors $\mathbf{w}, \ldots, \mathbf{w}_n$. One important fact needs to be mentioned: some of the weight vectors might converge to the same maxima. To avoid this, the outputs $\mathbf{w}_1^T\mathbf{x}, \ldots, \mathbf{w}_n^T\mathbf{x}$ have to be decorrelated after every iteration. For a whitened $\mathbf{x}$ this is equivalent to orthogonalization. There are several known methods to achieve this [147]. Here, only the symmetric decorrelation is considered. It has several benefits over other techniques: (1) the weight vectors $\mathbf{w}_i$ are estimated in parallel and not sequentially and (2) it does not perpetuate the errors from one weight vector to the next.

The symmetric orthogonalization of $\mathbf{W}$ can be accomplished by involving matrix square roots

$$\mathbf{W} = (\mathbf{W}\mathbf{W}^T)^{\frac{-1}{2}}\mathbf{W} \tag{8.103}$$

Numerical simplifications of this equation are given in [147].

An algorithmic description of the FastICA algorithm for estimating several independent components is given below:

1. **Preprocessing:** Whiten the data to give $\mathbf{x}$.
2. **Number of independent components:** Choose m as number of independent components to estimate.
3. **Initialization:** Initialize each the weight vector $\mathbf{w}_i, i = 1, \ldots, m$, of unit norm with random numbers.
4. **Adaptation:** Compute the change of every weight vector $\mathbf{w}_i, i = 1, \ldots, m$, according to

$$\mathbf{w}_i \leftarrow E\left\{\mathbf{x}g\left(\mathbf{w}_i^T\mathbf{x}\right)\right\} - E\left\{g'\left(\mathbf{w}_i^T\mathbf{x}\right)\right\}\mathbf{w}_i \tag{8.104}$$

   where g is defined as in eq. (8.83).
5. **Symmetric orthogonalization:** Do a symmetric orthogonalization of the matrix $\mathbf{W} = (\mathbf{w}_1, \ldots, \mathbf{w}_m)^T$ by

$$\mathbf{W} \leftarrow (\mathbf{W}\mathbf{W}^T)^{\frac{-1}{2}}\mathbf{W} \tag{8.105}$$

6. **Continuation:** If the weight matrix $\mathbf{W}$ is not converged, go back to step 4.

8.4.3.3 FastICA and Maximum Likelihood

It is interesting and useful to reveal the connection between FastICA and the well-known infomax and maximum likelihood algorithms presented in [23].

By writing $\mathbf{y} = \mathbf{W}\mathbf{x}$, $\beta_i = E\{y_i g(y_i)\}$, and $\Gamma = \text{diag}\{\beta_i - E\{g'(y_i)\}\}^{-1}$, the FastICA equation (8.102) can be equivalently written as

$$\mathbf{W} \leftarrow \Gamma[\text{diag}\{-\beta_i\} + E\{g(\mathbf{y})\mathbf{y}^T\}]\mathbf{W} \tag{8.106}$$

The matrix $\mathbf{W}$ needs to be orthogonalized after every step.

We can easily see the analogy between eq. (8.106) and the stochastic gradient method for maximizing likelihood [147]:

$$\mathbf{W} \leftarrow \mu[\mathbf{I} + g(\mathbf{y})\mathbf{y}^T]\mathbf{W} \tag{8.107}$$

where μ is the learning rate. FastICA can be considered as a fixed-point algorithm for maximum likelihood estimation of the ICA data model. The convergence of eq. (8.106) can be controlled by the adequate choice of the matrices Γ and $\text{diag}\{-\beta_i\}$. Another advantage over the ordinary maximum likelihood is the estimation of both super- and sub-Gaussian independent components.

8.4.3.4 Properties of the FastICA Algorithm

The presented FastICA and the chosen contrast functions enjoy several practical properties [147] compared with other ICA estimation methods:

- **Fast convergence:** Convergence is cubic compared to ordinary ICA methods where it is linear.
- **Simplicity:** No learning parameter is needed.
- **Generalization:** Finds, based on nonlinearity of choice, g directly independent components of any non-Gaussian distribution.
- **Optimized performance:** Based on the choice of a suitable nonlinearity g.
- **Parallel processing:** Learning algorithm is parallel, distributed, computationally simple, and requires little memory space.

8.4.4 The Infomax Neural Network

In [24] a self-organizing learning algorithm is described that maximizes the information transferred in a network of nonlinear units. It was shown that the neural network is able to perform ICA estimation and that higher-order statistics are introduced by the nonlinearities in the transfer function.

The network architecture is shown in Fig. 8.13.

The network has N input and output neurons, and an $N \times N$ weight matrix $\mathbf{W}$ connecting the input layer neurons with the output layer neurons. Assuming sigmoidal units, the neuron outputs are given by

$$\mathbf{y} = g(\mathbf{u}), \quad \text{with} \quad \mathbf{u} = \mathbf{W}\mathbf{x} \tag{8.108}$$

where g is a logistic function $g(u_i) = \frac{1}{1+\exp(-u_i)}$.

The idea of this algorithm is to find an optimal weight matrix $\mathbf{W}$ such that the output entropy $H(\mathbf{y})$ is maximized. The algorithm initializes $\mathbf{W}$ to the identity matrix $\mathbf{I}$. The elements of $\mathbf{W}$ are updated based on the following rule

$$\mathbf{W} \leftarrow -\eta \left(\frac{\partial H(\mathbf{y})}{\partial \mathbf{W}} \right) \mathbf{W}^T\mathbf{W} = -\eta(\mathbf{I} + f(\mathbf{u})\mathbf{u}^T)\mathbf{W} \tag{8.109}$$

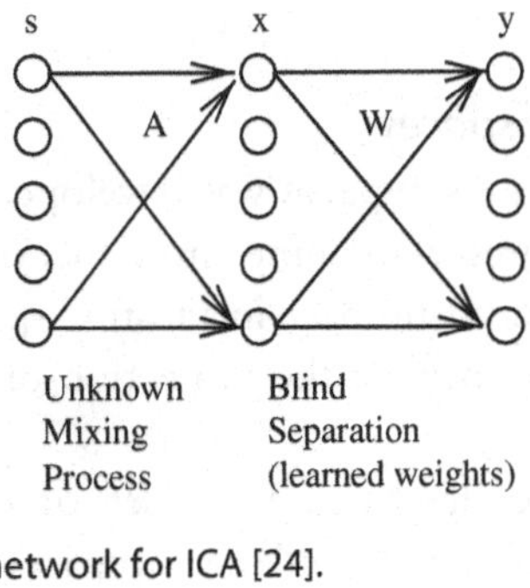

Figure 8.13 Feedforward neural network for ICA [24].

where η is the learning rate. The term $\mathbf{W}^T\mathbf{W}$ in eq. (8.109) was first proposed in [11] to avoid matrix inversions and to improve convergence. The vector-function f has the elements

$$f_i(u_i) = \frac{\partial}{\partial u_i} \ln g_i'(u_i) = (1 - 2y_i) \tag{8.110}$$

During learning, the learning rate is reduced gradually until the weight matrix $\mathbf{W}$ stops changing considerably. Equation (8.109) represents the so-called "infomax" algorithm.

An algorithmic description of the infomax ICA algorithm is given below:

1. **Preprocessing:** Whiten the data to give $\mathbf{x}$.
2. **Number of independent components:** Choose m as the number of independent components to estimate.
3. **Initialization:** Initialize each the weight matrix $\mathbf{W}$ to the identity matrix $\mathbf{I}$.
4. **Adaptation:** Compute the change of the weight matrix $\mathbf{W}$ according to

$$\mathbf{W} \leftarrow -\eta \left(\frac{\partial H(\mathbf{y})}{\partial \mathbf{W}} \right) \mathbf{W}^T\mathbf{W} = -\eta(\mathbf{I} + f(\mathbf{u})\mathbf{u}^T)\mathbf{W} \tag{8.111}$$

   where f is defined as in eq. (8.110). Reduce the learning rate gradually.
5. **Continuation:** If the weight matrix $\mathbf{W}$ is not converged, go back to step 4.

8.4.5 Topographic Independent Component Analysis

The topographic independent component analysis [146] represents a unifying model which combines topographic mapping with ICA.

Achieved by a slight modification of the ICA model, it can at the same time be used to define a topographic order between the components, and thus has the usual computational advantages associated with topographic maps.

The paradigm of topographic ICA has its roots in [145] where a combination of invariant-feature subspaces [185] and independent subspaces [50] is proposed. In the following, we will describe these two parts, which substantially reflect the concept of topographic ICA.

8.4.5.1 Invariant-Feature Subspaces

The principle of invariant-feature subspaces was developed by Kohonen [185] with the intention of representing features with some invariances. This principle states that an invariant feature is given by a linear subspace in a feature space. The value of the invariant feature is given by the squared norm of the projection of the given data point on that subspace.

A feature subspace can be described by a set of orthogonal basis vectors $\mathbf{w_j}$, $j = 1, \ldots, n$, where n is the dimension of the subspace. Then the value $G(\mathbf{x})$ of the feature G with the input vector $\mathbf{x}$ is given by

$$G(\mathbf{x}) = \sum_{j=1}^{n} \langle \mathbf{w_j}, \mathbf{x} \rangle^2 \tag{8.112}$$

In other words, this describes the distance between the input vector $\mathbf{x}$ and a general linear combination of the basis vectors $\mathbf{w_j}$ of the feature subspace [185].

8.4.5.2 Independent Subspaces

Traditional ICA works under the assumption that the observed signals $x_i(t)$, $(i = 1, \ldots, n)$ are generated by a linear weighting of a set of n statistically independent random sources $s_j(t)$ with time-independent coefficients a_{ij}. In matrix form, this can be expressed as

$$\mathbf{x}(t) = \mathbf{A}\mathbf{s}(t) \tag{8.113}$$

where $\mathbf{x}(t) = [x_1(t), \ldots, x_n(t)]^T$, $\mathbf{s}(t) = [s_1(t), \ldots, s_n(t)]$, and $\mathbf{A} = [a_{ij}]$.

In multidimensional ICA [50], the sources s_i are not assumed to be all mutually independent. Instead, it is assumed that they can be grouped in n-tuples, such that within these tuples they are dependent on each other, but are independent outside. This newly introduced assumption was observed in several image processing applications. Each n-tuple of sources s_i corresponds to n basis vectors given by the rows of matrix $\mathbf{A}$. A subspace spanned by a set of n such basis vectors is defined as an independent subspace. In [50] two simplifying assumptions are made: (1) Although s_i are not at all independent, they are chosen to be uncorrelated and of unit variance, and (2) the data are preprocessed by whitening (sphering) them. This means the $\mathbf{w_j}$ are orthonormal.

Let J be the number of independent feature subspaces and S_j, $j = 1, \ldots, J$ the set of indices that belong to the subspace of index j. Assume that we have T given observations $\mathbf{x}(t)$, $t = 1, \ldots, T$. Then the likelihood L of the data based on the model is given by

$$L(\mathbf{w_i}, i = 1, \ldots, n) = \prod_{t=1}^{T} \left[|\det \mathbf{W}| \prod_{j=1}^{J} p_j(\langle \mathbf{w_i}, \mathbf{x}(t) \rangle, i \in S_j) \right] \tag{8.114}$$

with $p_j(\cdot)$ being the probability density inside the jth n-tuple of s_i. The expression $|\det \mathbf{W}|$ is due to the linear transformation of the pdf. As always with ICA, $p_j(\cdot)$ need not be known in advance.

8.4.5.3 Fusion of Invariant Feature and Independent Subspaces

In [145] it is shown that a fusion between the concepts of invariant and independent subspaces can be achieved by considering probability distributions for the n-tuples of s_i being spherically symmetric, that is, depending on the norm. In other words, the pdf $p_j(\cdot)$ has to be expressed as a function of the sum of the squares of the s_i, $i \in S_j$ only. Additionally, it is assumed that the pdfs are equal for all subspaces.

The log-likelihood of this new data model is given by

$$\log L(\mathbf{w_i}, i = 1, \ldots, n) = \sum_{t=1}^{T} \sum_{j=1}^{J} \log p \left(\sum_{i \in S_j} \langle \mathbf{w_i}, \mathbf{x}(t) \rangle^2 \right) + T \log |\det \mathbf{W}| \quad (8.115)$$

$p\left(\sum_{i \in S_j} s_i^2\right) = p_j(s_i, i \in S_j)$ gives the pdf inside the jth n-tuple of s_i. Based on the prewhitening, we have $\log |\det \mathbf{W}| = 0$.

For computational simplification, set

$$G\left(\sum_{i \in S_j} s_i^2\right) = \log p \left(\sum_{i \in S_j} \langle \mathbf{w_i}, \mathbf{x}(t) \rangle^2 \right) \quad (8.116)$$

Since it is known that the projection of visual data on any subspace has a super-Gaussian distribution, the pdf has to be chosen to be sparse. Thus, we will choose $G(u) = \alpha\sqrt{u} + \beta$ yielding a multidimensional version of an exponential distribution. α and β are constants and enforce that s_i is of unit variance.

8.4.5.4 The Topographic ICA Architecture

Based on the concepts introduced in the preliminary subsections, this section describes the topographic ICA.

To introduce a topographic representation in the ICA model, it is necessary to relax the assumption of independence among neighboring components s_i. This makes it necessary to adopt an idea from self-organized neural networks, that of a lattice. It was shown in [146] that a representation which models topographic correlation of energies is an adequate approach for introducing dependencies between neighboring components.

In other words, the variances corresponding to neighboring components are positively correlated while the other variances are in a broad sense independent. The architecture of this new approach is shown in Fig. 8.14.

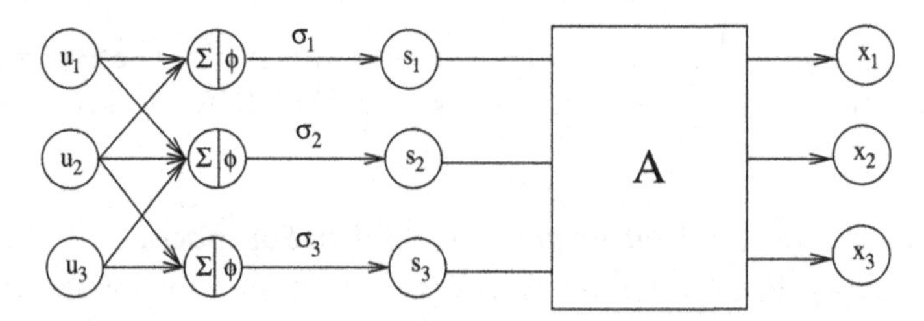

Figure 8.14 Topographic ICA model [146]. The variance generated variables u_i are randomly generated and mixed linearly inside their topographic neighborhoods. This forms the input to nonlinearity ϕ, thus giving the local variance σ_i. Components s_i are generated with variances σ_i. The observed variables x_i are obtained as with standard ICA from the linear mixture of the components s_i.

This idea leads to the following representation of the source signals:

$$s_i = \sigma_i z_i \tag{8.117}$$

where z_i is a random variable having the same distribution as s_i, and the variance σ_i is fixed to unity.

The variance σ_i is further modeled by a nonlinearity:

$$\sigma_i = \phi\left(\sum_{k=1}^{n} h(i,k)u_k\right) \tag{8.118}$$

where u_i are the higher-order independent components used to generate the variances, while ϕ describes some nonlinearity. The neighborhood function $h(i,j)$ can either be a two-dimensional grid or have a ring-like structure. Further u_i and z_i are all mutually independent.

The learning rule is based on the maximization of the likelihood. First, it is assumed that the data are preprocessed by whitening and that the estimates of the components are uncorrelated. The log-likelihood is given by:

$$\log L(\mathbf{w_i}, i = 1, \dots, n) = \sum_{t=1}^{T}\sum_{j=1}^{n} G\left(\sum_{i=1}^{n} \mathbf{w_i}^T\mathbf{x}(t)^2\right) + T\log|\det\mathbf{W}| \tag{8.119}$$

The update rule for the weight vector $\mathbf{w}_i$ is derived from a gradient algorithm based on the log-likelihood assuming $\log|\det\mathbf{W}| = 0$:

$$\Delta\mathbf{w}_i \propto E\{\mathbf{x}\left(\mathbf{w}_i^T\mathbf{x}\right) r_i\} \tag{8.120}$$

where

$$r_i = \sum_{k=1}^{n} h(i,k)g\left(\sum_{j=1}^{n} h(k,j)(\mathbf{w}_j^T\mathbf{x})^2\right) \tag{8.121}$$

The function g is the derivative of $G = -\alpha_1\sqrt{u} + \beta_1$. After every iteration, the vectors $\mathbf{w}_i$ in eq. (8.120) are normalized to unit variance and orthogonalized. This equation represents a modulated learning rule, where the learning term is modulated by the term r_i.

The classic ICA results from the topographic ICA by setting $h(i,j) = \delta_{ij}$.

8.4.6 Imaging Brain Dynamics

The analysis of magnetoencephalographic (MEG) recordings is important both for basic brain research and for medical diagnosis and treatment. ICA is an effective method for removing artifacts and separating sources of the brain signals from these recordings [380]. A similar approach is proving useful for analyzing functional magnetic resonance brain imaging (fMRI) data [239,240].

8.4.6.1 Separation of Artifacts in MEG Data

MEG is a technique to measure the activity of the cortical neurons. However, it is difficult to extract the essential features of the neuromagnetic signals from that of artifacts generally resembling pathological signals. In [380], a new method for separating brain activity from artifacts based on ICA is presented. The approach is based on the assumption that the brain activity and the artifacts, e.g., eye movements or blinks, are anatomically and physiologically separate processes and statistically independent signals.

Figure 8.15 shows a subset of 12 spontaneous MEG signals $x_i(t)$ from the frontal, temporal, and occipital areas. The figure also shows the helmet positions of the sensors. The considered artifacts were ocular (blinking and horizontal saccades) and muscular (biting) in nature.

The signal vector $\mathbf{x}$ in the ICA model (8.63) consists of the amplitude $x_i(t)$ of the recorded signals at a certain time point. The $\mathbf{x}(t)$ vectors were whitened using PCA and

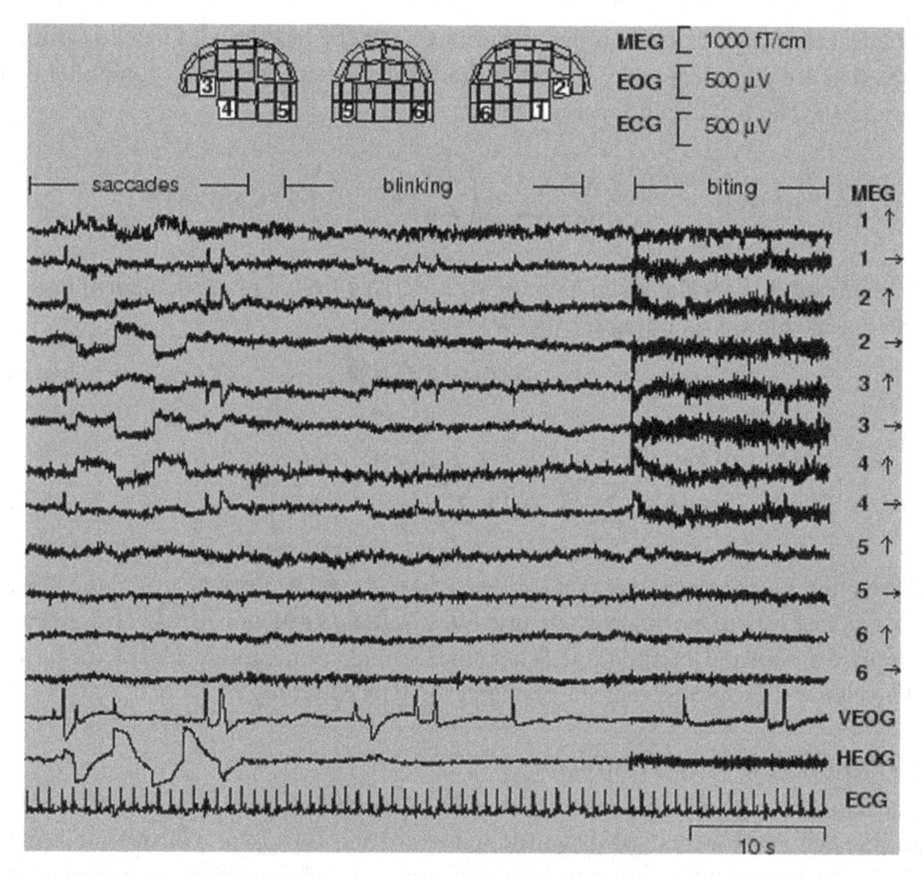

Figure 8.15 Samples of MEG signals, showing artifacts produced by blinking, saccades, biting, and cardiac cycle. For each of the six positions shown, the two orthogonal directions of the sensors are plotted. *(Image from [380] reprinted with permission from The MIT Press.)*

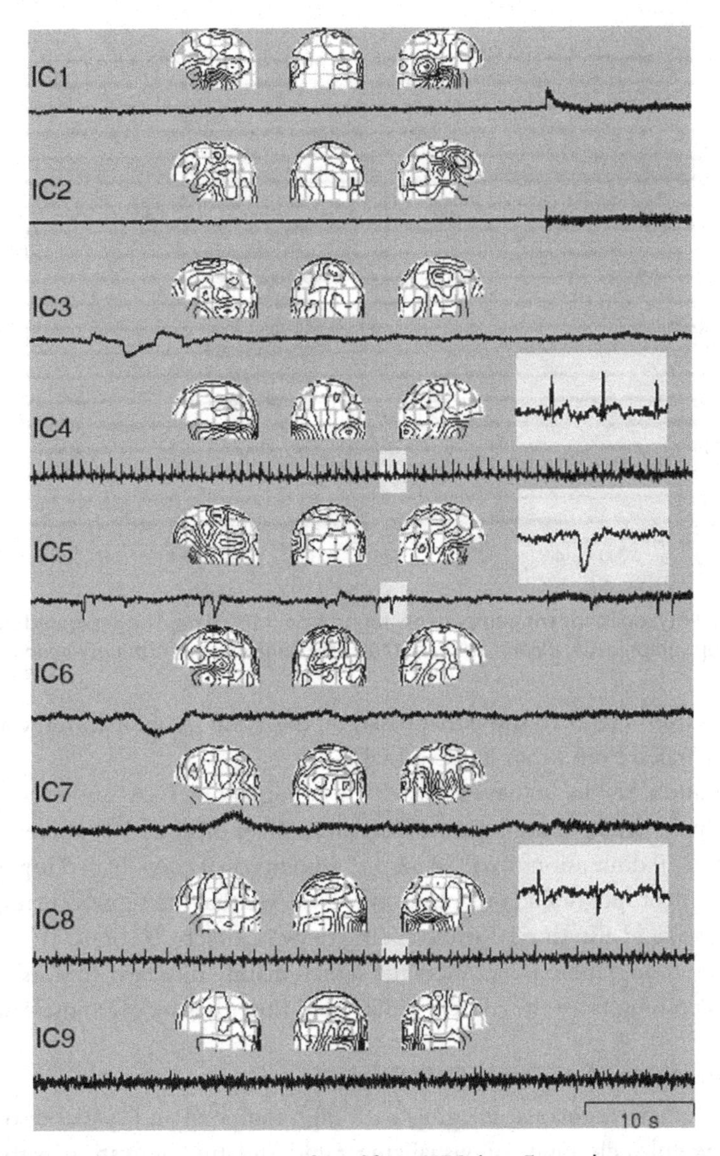

Figure 8.16 Nine independent components found from MEG data. For each component the left, back, and right views of the field patterns generated by these components are shown: the solid line stands for magnetic flux coming out of the head, and the dotted line for the flux inward. *(Image from [380] reprinted with permission from The MIT Press.)*

the dimensionality was reduced. Using FastICA, it became possible to determine the independent components. Figure 8.16 shows sections of nine independent components (IC) $s_i(t), i = 1, \dots, 9$ found from the recorded data together with the corresponding field pattern. The following ICs were determined: muscular activity (IC1,IC2), ocular activity (IC3,IC5), cardiac (IC4), and others (IC8,IC9).

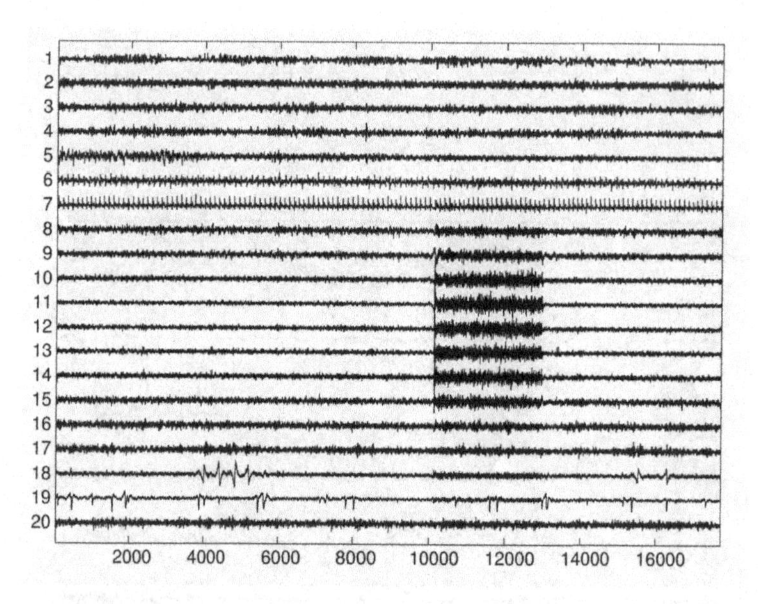

Figure 8.17 Twenty independent components found from MEG data. The separated source signals are found by topographic ICA. *(Reprinted from [146] with permission from Springer-Verlag.)*

The results from Fig. 8.16 show that based on the FastICA algorithm the artifacts can be reliably separated from other MEG signals.

The separation results achieved based on topographic ICA are very interesting. Figure 8.17 shows the resulting separated signals from an original 122-dimensional input data reduced to 20 dimensions by PCA. A one-dimensional ring-shaped topography was employed. For the topographic organization, we can see that the signals corresponding to bites (IC9–IC15) and also those corresponding to eye artifacts (IC18, IC19) are adjacent. In other words, topographic ICA orders the signals mainly into two clusters, one created by the signals coming from the muscle artifact and the other by eye muscle activity.

8.4.6.2 Separation of Artifacts in fMRI Data

Functional magnetic resonance imaging with high temporal and spatial resolution represents a powerful technique for visualizing rapid and fine activation patterns of the human brain [18,96,193,194,267]. As is known from both theoretical estimations and experimental results [33,194,266], an activated signal variation appears very low on a clinical scanner. This motivates the application of analysis methods to determine the response waveforms and associated activated regions. Generally, these techniques can be divided into two groups: Model-based techniques require prior knowledge about activation patterns, whereas model-free techniques do not. However, model-based analysis methods impose some limitations on data analysis under complicated experimental conditions. Therefore, analysis methods that do not rely on any assumed model of functional response are considered more powerful and relevant. We distinguish two groups of model-free methods: transformation-based and clustering-based. The first method, principal component analysis (PCA) [17,358] or independent component analysis (ICA) [239,240], transforms raw data into high-dimensional vector space to separate functional response and artifacts from each other. The major problem of PCA is that it only decorrelates the components. ICA can separate the components, but is still limited by stationary distribution and the linear mixture assumption [239].

The second method, fuzzy clustering analysis (FCA) [62,328] or selforganizing map (SOM) [94,261], attempts to group time signals of the brain into several clusters based on the temporal similarity among them.

Functional organization of the brain is based on two complementary principles, localization and connectionism. Localization means that each visual function is performed mainly by a small set of the cortex. Connectionism, on the other hand, means that the brain regions involved in a certain visual-cortex function are widely distributed, and thus the brain activity necessary to perform a given task may be the functional integration of activity in distinct brain systems. It is important to stress that in neurobiology the term "connectionism" is used in a different sense than in neural network terminology.

According to the principle of functional organization of the brain, it was suggested for the first time in [239] that the multifocal brain areas activated by performance of a visual task should be unrelated to the brain areas whose signals are affected by artifacts of physiological nature, head movements, or scanner noise related to fMRI experiments. Every single process mentioned earlier can be described by one or more spatially independent components, each associated with a single time course of a voxel and a component map. It is assumed that the component maps, each described by a spatial distribution of fixed values, represent overlapping, multifocal brain areas of statistically dependent

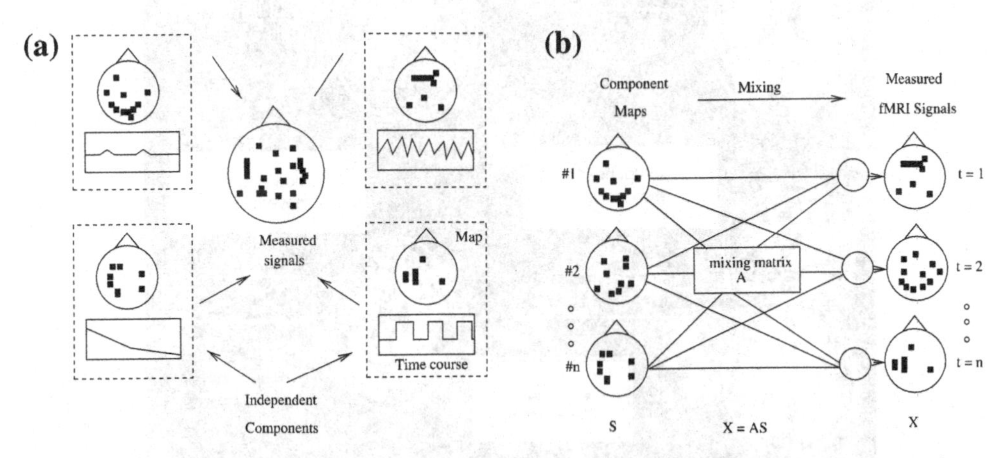

Figure 8.18 Visualization of ICA applied to fMRI data. (a) Scheme of fMRI data decomposed into independent components and (b) fMRI data as a mixture of independent components where the mixing matrix **M** specifies the relative contribution of each component at each time point [239].

fMRI signals. This aspect is visualized in Fig. 8.18. In addition, it is considered that the distributions of the component maps are spatially independent, and in this sense uniquely specified.

It was shown in [239] that these maps are independent if the active voxels in the maps are sparse and mostly nonoverlapping. Additionally it is assumed that the observed fMRI signals are the superposition of the individual component processes at each voxel. Based on these assumptions, ICA can be applied to fMRI time series to spatially localize and temporally characterize the sources of BOLD activation.

In [239] a new method for analyzing fMRI data based on the infomax ICA algorithm [24] is described. The fMRI data were acquired while a subject performed a Stroop color-naming task, and the data were then decomposed into spatially independent components.

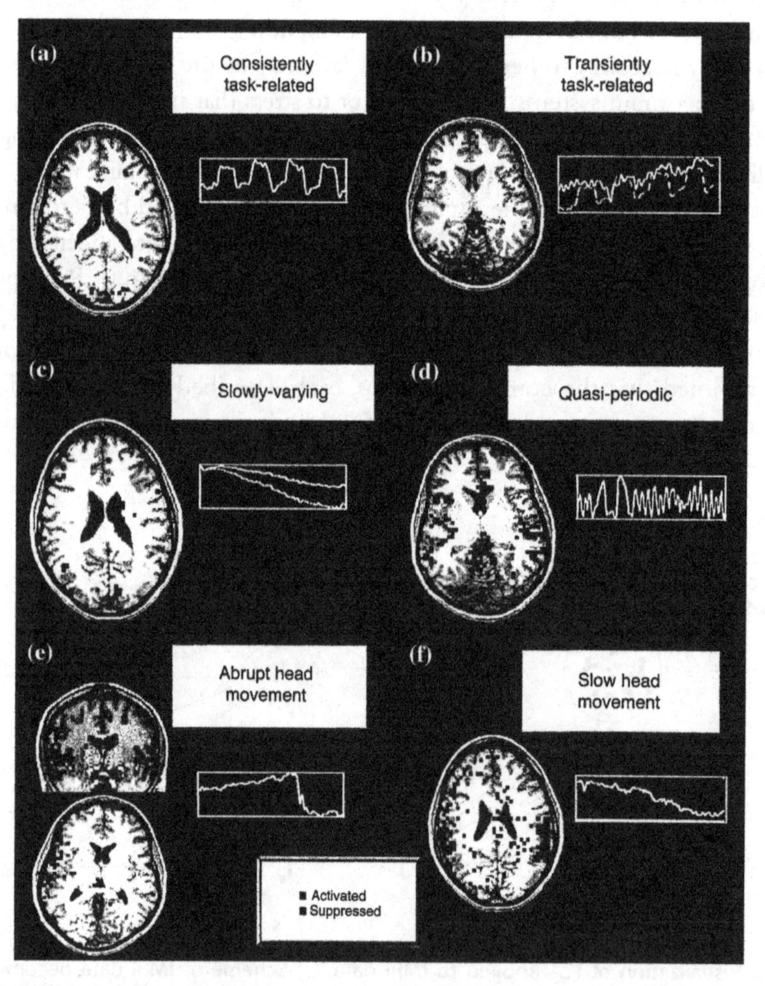

Figure 8.19 Different classes of components detected by ICA decomposition of Stroop task fMRI data. *(Reprinted from [240] with permission from National Academy of Sciences.)*

Each component consisted of voxel values at fixed three-dimensional locations (component "map"), and a unique associated time course of activation.

ICA derived one and only one component with a time course closely matching the time course of alternations between experimental and control tasks.

Time courses of other IC were transiently task-related, quasiperiodic, or slowly varying. Figure 8.19 illustrates the different classes of IC components.

The results in Fig. 8.19 show that ICA can be used to reliably separate fMRI data sets into meaningful constituent components.

In [277] a comparison is shown between different ICA techniques and PCA when applied to fMRI. For a visual task, unique task-related activation maps and associated time courses were obtained by different ICA techniques and PCA.

Figure 8.20 shows, for eight ICs, the component time course most closely associated with the visual task for three ICA techniques (infomax, FastICA, topographic ICA) and PCA. The best results are achieved by FastICA and topographic ICA, yielding almost identical reference functions, and a correlation coefficient of $r = 0.92$ between those two. PCA has a much lower correlation as does infomax.

However, the activation maps do not reveal much of a difference between the ICA techniques. Figure 8.21 shows the activation maps as a comparison of results obtained by the four techniques.

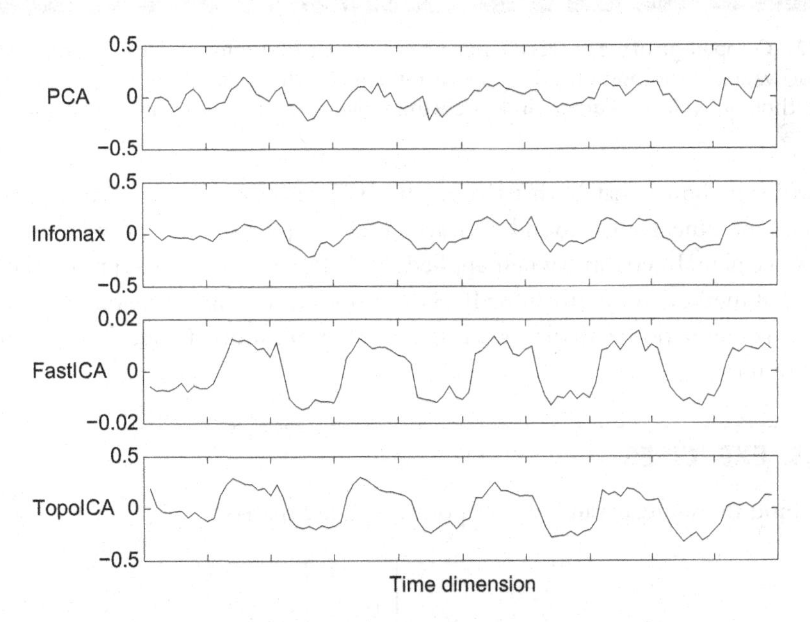

Figure 8.20 Computed reference functions for the four techniques, PCA, infoma x, FastICA, and topographic ICA, for 16 ICs. The correlation coefficients found are $r = 0.69$ for PCA, $r = 0.85$ for infomax, $r = 0.94$ for FastICA, and $r = 0.85$ for topographic ICA.

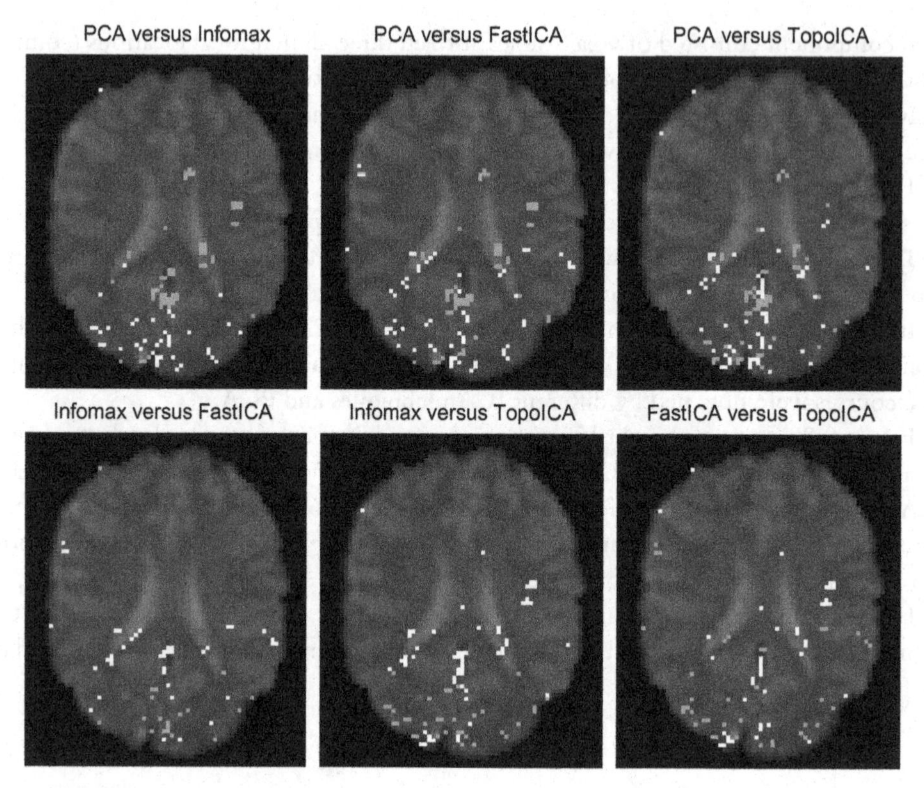

Figure 8.21 Comparison of the results for the four techniques, PCA, infomax, FastICA, and topographic ICA. The activation maps show only the pairwise differences. The darker pixels show active voxels for the first technique: while the lighter pixels show the active voxels for the second technique.

It has been shown that both topographic ICA and FastICA outperform, in terms of component time course found and activation maps, principal component analysis and infomax neural networks when applied to fMRI studies. This supports the theory that different methods for performing ICA decompositions employing different objective functions in conjunction with different criteria of optimization of these functions produce different results.

8.5. EXERCISES

1. Find the singular value decomposition of the matrix

$$X = \begin{bmatrix} 4 & 4 \\ 0 & 2 \\ 3 & 0 \\ 0 & 5 \end{bmatrix} \tag{8.122}$$

2. Determine the eigenvalues of the matrix

$$X = \begin{bmatrix} 1 & 3 \\ 4 & 2 \\ 1 & 3 \end{bmatrix} \tag{8.123}$$

Give the singular value decomposition of X.

3. Comment on the capability of PCA as an image compression mechanism. What are the strengths and weaknesses of each of the following robust PCA algorithms: Oja, NLPCA, and GHA?
4. Discuss the differences between GHA and the APEX algorithm. Why is the asymptotic convergence of APEX in theory exactly the same as that of GHA? Explain this intuitively!
5. Show that the following equality $\Phi_{MI}[\mathbf{y}] > 0$ holds, if and only if $\mathbf{y} = \mathbf{Wx}$ is statistically distributed as $\tilde{\mathbf{y}}$. $\Phi_{MI}[\mathbf{y}] > 0$ is given by eq. (8.68).
6. Show that the maximum likelihood principle is associated with the contrast function $\Phi_{[ML]} = K[\mathbf{y}|\mathbf{s}]$; see eq. (8.74).
7. Show that the infomax and the maximum likelihood contrast function have the same contrast function $\Phi_{IM}[\mathbf{y}] = \Phi_{ML}[\mathbf{y}]$, as shown in eq. (8.96).
8. Discuss the algorithmic differences between infomax and FastICA.
9. Compare topographical ICA with other topographic mappings.

CHAPTER NINE

Neuro-Fuzzy Classification

Contents

9.1. INTRODUCTION

In a medical imaging system, uncertainties can be present at any point resulting from incomplete or imprecise input information, ambiguity in medical images, ill-defined, or overlapping boundaries among the disease classes or regions, and indefiniteness in extracting features and relations among them. Any decision taken at a particular point will heavily influence the following stages. Therefore, an image analysis system must have sufficient possibilities to capture the uncertainties involved at every stage, such that the system's output results can be described with minimal uncertainty. In pattern recognition parlance, a pattern can be assigned to more than one class. Or in medical diagnosis, a patient can have a set of symptoms, which can be attributed to several diseases.

Pattern Recognition and Signal Analysis in Medical Imaging
http://dx.doi.org/10.1016/B978-0-12-409545-8.00009-1

The symptoms need not be strictly numerical. Thus, fuzzy variables can be both linguistic and/or set variables. An example of a fuzzy variable is a temperature ranging from 97°F to 103°F, which can be described as very high, or more or less high. This gives an idea of the concept of fuzziness.

The main difference between fuzzy and neural paradigms is that fuzzy set theory tries to mimic the human reasoning and thought process, whereas neural networks attempt to emulate the architecture and information representation scheme of the human brain. It is therefore meaningful to integrate these two distinct paradigms by enhancing their individual capabilities in order to build a more intelligent processing system. This new processing paradigm is known under the name of neuro-fuzzy computing. Compared to standard neural networks or simple fuzzy classifiers, it is a powerful computational paradigm: A neuro-fuzzy system can process linguistic variables compared to a classical neural network, or it learns overlapping classes better than a fuzzy classifier.

This chapter reviews the basic concepts of fuzzy sets and the definitions needed for fuzzy clustering, and it presents several of the best-known fuzzy clustering algorithms and fuzzy learning vector quantization. Applications of neuro-fuzzy classification to medical image compression and exploratory data analysis are also shown.

9.2. FUZZY SETS

Fuzzy sets represent a suitable mathematical technique for the description of imprecision and vagueness. Usually, vagueness describes the difficulty of obtaining concise affirmations regarding a given domain. On the other hand, in fuzzy set theory, the crisp alternative yes-no can be endlessly expanded. Thus, fuzzy set theory not only deals with ambiguity and vagueness, but represents a theory of nuance reasoning [85].

This chapter will review some of the basic notions and results in fuzzy set theory.

The two basic components of fuzzy systems are fuzzy sets and operations on fuzzy sets. Fuzzy logic defines rules, based on combinations of fuzzy sets by these operations. The notion of fuzzy sets has been introduced by Zadeh [404].

9.2.0.1 Crisp Sets

Definition 9.2.1 Let X be a nonempty set considered to be the universe of discourse. A crisp set A is defined by enumerating all elements $x \in X$

$$A = \{x_1, x_2, \ldots, x_n\} \tag{9.1}$$

that belong to A.

The membership function can be expressed by a function u_A, mapping X on a binary value described by the set $I = \{0, 1\}$:

$$u_A : X \to I, \quad u_A(x) = \begin{cases} 1 & \text{if } x \in A \\ 0 & \text{if } x \notin A \end{cases} \tag{9.2}$$

$u_A(x)$ represents the membership degree of x to A. Thus, an arbitrary x either belongs to A, or it does not; partial membership is not allowed.

For two sets A and B, combinations can be defined by the following operations:

$$\{A \cup B = \{x|x \in A \quad \text{or} \quad x \in B\} \tag{9.3}$$

$$A \cap B = \{x|x \in A \quad \text{and} \quad x \in B\} \tag{9.4}$$

$$\bar{A} = \{x|x \notin A, x \in X\} \tag{9.5}$$

Additionally, the following rules have to be satisfied:

$$A \cup \bar{A} = \varnothing, \quad \text{and} \quad A \cap \bar{A} = X \tag{9.6}$$

9.2.0.2 Fuzzy Sets

Definition 9.2.2 Let X be a nonempty set considered to be the universe of discourse. A fuzzy set is a pair (X, A), where $u_A : X \to I$ and $I = [0, 1]$.

Figure 9.1 illustrates an example of a possible membership function.

The family of all fuzzy sets on the universe x will be denoted by $L(X)$. Thus

$$L(X) = \{u_A | u_A : X \to I\} \tag{9.7}$$

$u_A(x)$ is the membership degree of x to A. It can be also interpreted as the plausibility degree of the affirmation "x belongs to A." If $u_A(x) = 0$, x is definitely not in A, and if $u_A(x) = 1$, x is definitely in A. The intermediate cases are fuzzy.

Definition 9.2.3 The fuzzy set A is called nonambiguous or crisp, if $u_A(x) \in \{0, 1\}$.

Definition 9.2.4 If A is from $L(X)$, the complement of A is the fuzzy set $\bar{A}$ defined as

$$u_{\bar{A}}(x) = 1 - u_A(x), \quad \forall x \in X \tag{9.8}$$

In order to manipulate fuzzy sets, we need to have operations that enable us to combine them. As fuzzy sets are defined by membership functions, the classical set-theoretic operations have to be replaced by function-theoretic operations.

For two fuzzy sets A and B on X, the following operations can be defined.

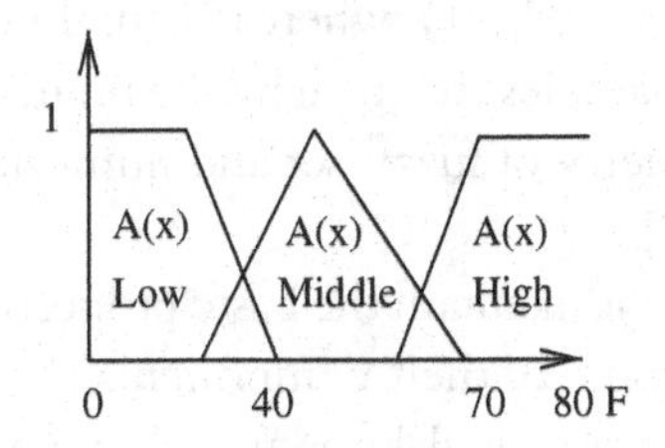

Figure 9.1 Example of a membership function of temperature.

Definition 9.2.5

$$\begin{aligned}
&\text{Equality} \quad A = B \Longleftrightarrow u_A(x) = u_B(x), \quad \forall x \in X \\
&\text{Inclusion} \quad A \subseteq B \Longleftrightarrow u_A(x) \leq u_B(x), \quad \forall x \in X \\
&\text{Product} \quad u_{(AB)}(x) = u_A(x) \cdot u_B(x), \quad \forall x \in X \\
&\text{Difference} \quad u_{(A-B)}(x) = \max(u_A(x) - u_B(x), 0), \quad \forall x \in X \\
&\text{Intersection} \quad u_{(A\cap B)}(x) = \min(u_A(x), u_B(x)), \quad \forall x \in X \\
&\text{Union} \quad u_{(A\cup B)}(x) = \max(u_A(x), u_B(x)), \quad \forall x \in X
\end{aligned} \tag{9.9}$$

Definition 9.2.6 The family $A_1, \ldots, A_n$, $n \geq 2$, of fuzzy sets is a fuzzy partition of the universe X if and only if the condition

$$\sum_{i=1}^{n} u_{A_i}(x) = 1 \tag{9.10}$$

holds for every x from X.

The above condition can be generalized for a fuzzy partition of a fuzzy set. By C we define a fuzzy set on X. We may require that the family $A_1, \ldots, A_n$ of fuzzy sets be a fuzzy partition of C if and only if the condition

$$\sum_{i=1}^{n} u_{A_i}(x) = u_C(x) \tag{9.11}$$

is satisfied for every x from X.

9.3. NEURO-FUZZY INTEGRATION

Neural networks and fuzzy systems are both learning systems aiming to estimate input-output relationships. They are not based on a mathematical model, and learning incorporates experience with sample data. The fuzzy system infers adaptively and modifies its fuzzy associations from representative numerical samples. Neural networks generate blindly and change fuzzy rules based on training data [189]. Fuzzy systems estimate functions with fuzzy-set samples (A_i, B_i), whereas neural networks use numerical-point samples (x_i, y_i). Both kinds of samples are given by the input-output product space $X \times Y$. Figure 9.2 illustrates the geometry of fuzzy-set and numerical-point samples taken from the function $f : X \rightarrow Y$ [189].

Thus, neuro-fuzzy algorithms maintain the basic properties and architectures of neural networks and simply fuzzify some of their components.

As shown in [282], there are several known techniques of combining neuro-fuzzy concepts. Neuro-fuzzy hybridization is possible in many ways as shown in Figs. 9.3–9.7:

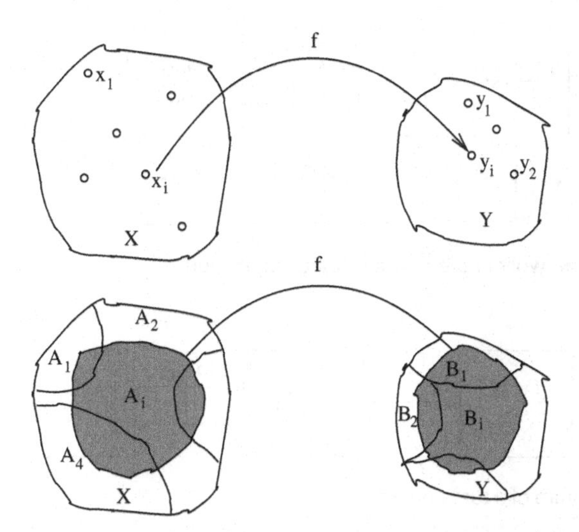

Figure 9.2 Function f maps X to Y. In the first illustration several numerical-point samples (x_i, y_i) are used for the function estimation $f : X \rightarrow Y$. In the second illustration only a few fuzzy subsets A_i of X and B_i of Y are used. The fuzzy association (A_i, B_i) describes the system structure as an adaptive clustering might infer. In practice, there are usually fewer different output associants or "rule" consequents B_i than input associants or antecedents A_i [189].

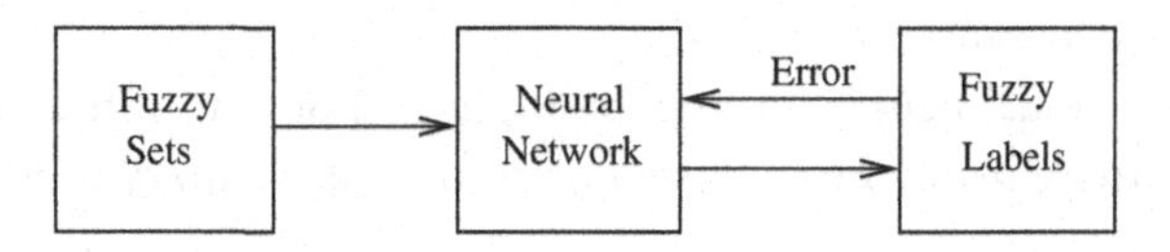

Figure 9.3 Neuronal network implementing fuzzy classifier.

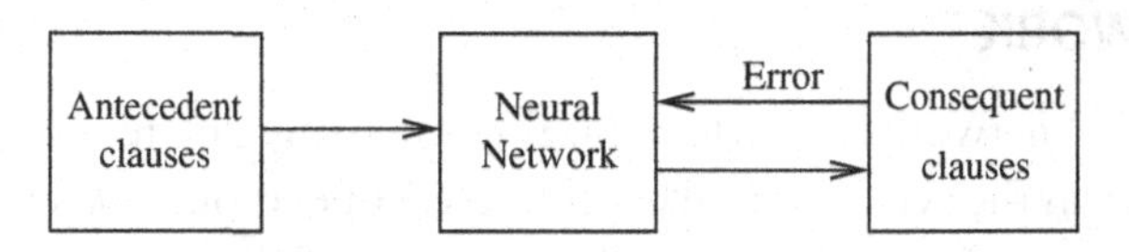

Figure 9.4 Neuronal network implementing fuzzy logic.

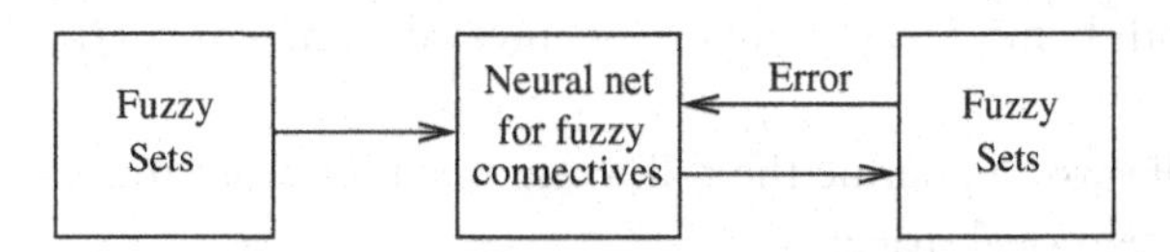

Figure 9.5 Neuronal network implementing fuzzy connectives.

- Incorporating fuzziness into the neural net framework: for example, building fuzzy neural network classifiers.
- Designing neural networks guided by fuzzy logic formalism: for example, a computational task within the framework of a preexisting fuzzy model.

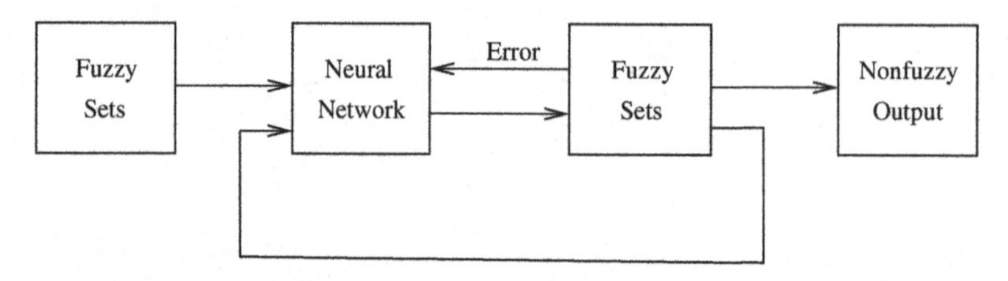

Figure 9.6 Neuronal network implementing self-organization.

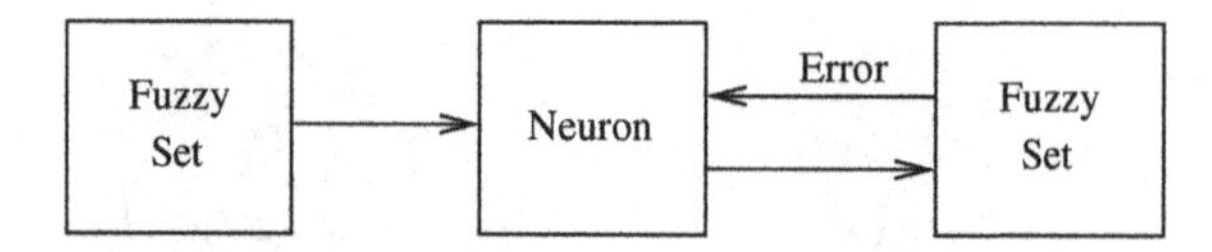

Figure 9.7 Block diagram of a fuzzy neuron.

- Changing the basic characteristics of the neurons: for example, when each node performs a fuzzy aggregation operation (fuzzy union, intersection).
- Using measures of fuzziness as the error or instability of a network: for example, incorporating various fuzziness measures in a multilayer network, thus performing unsupervised self-organization.
- Making the individual neurons fuzzy: the input and output of the neurons are fuzzy sets, and the activity involving the fuzzy neurons is also a fuzzy process.

9.4. MATHEMATICAL FORMULATION OF A FUZZY NEURAL NETWORK

A fuzzy neural network can include fuzziness in many different places such as at the input-output level, in the synapses, in the confluence operation, or even in the activation function. To give a precise mathematical formulation of the fuzzy neural network, we will adopt here the terminology used in [282]. $\mathbf{x}$ is the fuzzy input and $\mathbf{y}$ is the fuzzy output vector, both being fuzzy numbers or intervals. The connection weight vector is denoted by $\mathbf{W}$.

Mathematically, we can define the following mapping from the n-dimensional input space to the l-dimensional space:

$$\mathbf{x}(t) \in \mathbf{R^n} \rightarrow \mathbf{y}(t) \in \mathbf{R^l} \tag{9.12}$$

A confluence operation $\otimes$ determines the similarity between the fuzzy input vector $\mathbf{x}(t)$ and the connection weight vector $\mathbf{W}(t)$. For neural networks, the confluence operation represents a summation or product operation, whereas for a fuzzy neural network it describes an arithmetic operation such as fuzzy addition and fuzzy multiplication.

The output neurons implement the following nonlinear operation:

$$\mathbf{y}(t) = \psi[\mathbf{W}(t) \otimes \mathbf{x}(t)] \tag{9.13}$$

Based on the given training data $\{(\mathbf{x}(t), \mathbf{d}(t)), \mathbf{x}(t) \in \mathbf{R}^{\mathbf{n}}, \mathbf{d}(t) \in \mathbf{R}^{\mathbf{l}}, t = 1, \ldots, N\}$ the cost function can be optimized:

$$E_N = \sum_{t=1}^{N} d(\mathbf{y}(t), \mathbf{d}(t)) \tag{9.14}$$

where $d(\cdot)$ defines a distance in $\mathbf{R}^{\mathbf{l}}$.

The learning algorithm of the fuzzy neural network is given by

$$\mathbf{W}(t+1) = \mathbf{W}(t) + \epsilon \Delta \mathbf{W}(t) \tag{9.15}$$

and thus adjusts $N^{\mathbf{W}}$ connection weights of the fuzzy neural network.

9.5. FUZZY CLUSTERING

Traditional statistical classifiers assume that the pdf for each class is known or must somehow be estimated. Another problem is posed by the fact that sometimes clusters are not compact but shell-shaped. A solution to this problem is given by fuzzy clustering algorithms, a new classification paradigm intensively studied during the past three decades. The main difference between traditional statistical classification techniques and fuzzy clustering techniques is that in the fuzzy approaches an input vector belongs *simultaneously* to more than one cluster, while in statistical approaches it belongs *exclusively* to only one cluster.

Usually, clustering techniques are based on the optimization of a cost or objective function J. This predefined measure J is a function of the input data and of an unknown parameter vector set $\mathbf{L}$. Throughout this chapter, we will assume that the number of clusters n is predefined and fixed.

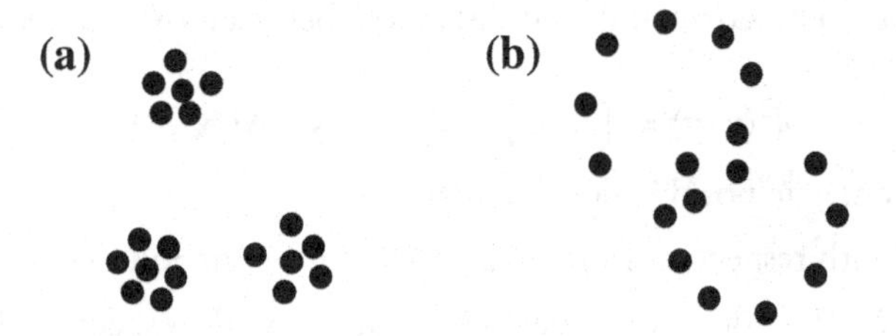

Figure 9.8 Different cluster shapes: (a) compact and (b) spherical.

A successful classification is based on estimating the parameter $\mathbf{L}$ such that the cluster structure of the input data is as good as possible determined. It is evident that this parameter depends on the cluster's geometry. We distinguish two different types of geometry: compact and spherical, as shown in Fig. 9.8.

Compact clusters are pretty well described by a set of n points $\mathbf{L}_i \in \mathbf{L}$ where each point describes such a cluster. Spherical clusters have two distinct parameters describing the center and the radius of the cluster. Thus the parameter vector set $\mathbf{L}$ is replaced by two new parameter vector sets, $\mathbf{V}$ describing the centers of the clusters, and $\mathbf{R}$ describing the radii of the clusters.

In the following, we will review the most important fuzzy clustering techniques, and show their relationship to nonfuzzy approaches.

9.5.1 Metric Concepts for Fuzzy Classes

Let $\mathbf{X} = \{\mathbf{x}_1, \mathbf{x}_2, \ldots, \mathbf{x}_p\}, \mathbf{x}_j \in \mathbf{R}^s$, be a data set. Suppose the optimal number of clusters in $\mathbf{X}$ is given and that the cluster structure of $\mathbf{X}$ may be described by disjunct fuzzy sets which combined yield $\mathbf{X}$.

Also, let C be a fuzzy set associated with a class of objects from $\mathbf{X}$ and $F_n(C)$ be the family of all n-member fuzzy partitions of C. Let n be the given number of subclusters in C. The cluster family of C can be appropriately described by a fuzzy partition P from $F_n(C), P = \{A_1, \ldots, A_n\}$.

Every class A_i is described by a prototype or a cluster center $\mathbf{L}_i$ which represents a point in an s-dimensional Euclidean space $\mathbf{R}^s$. The clusters' form can be either spherical or ellipsoidal. $\mathbf{L}_i$ represents the center of gravity of the fuzzy class A_i.

The fuzzy partition is typically described by an $n \times p$ membership matrix $\mathbf{U} = [u_{ij}]_{n \times p}$. Thus, the membership u_{ij} represents the degree of belonging of the pattern $\mathbf{x}_j$ to the ith class. For a crisp partition, the membership matrix $\mathbf{U}$ is binary-valued, whereas for a fuzzy partition its elements have values between 0 and 1. Here, we see the main difference between crisp and fuzzy partition: Given a fuzzy partition, a given data point $\mathbf{x}_j$ can belong to several classes with the degree of belonging specified by membership grades between 0 and 1. Given a crisp partition, this data point belongs to exactly one class. We will use the following notation hereafter: $u_{ij} = u_i(\mathbf{x}_j)$.

In the following, we will give the definition of a weighted Euclidean distance.

Definition 9.5.1 The norm-induced distance d between two data $\mathbf{x}$ and $\mathbf{y}$ from $\mathbf{R}^s$ is given by

$$d^2(\mathbf{x}, \mathbf{y}) = ||\mathbf{x} - \mathbf{y}|| = (\mathbf{x} - \mathbf{y})^T \mathbf{M} (\mathbf{x} - \mathbf{y}) \tag{9.16}$$

where $\mathbf{M}$ is a symmetric positive definite matrix.

The distance with respect to a fuzzy class is given by the following definition.

Definition 9.5.2 The distance d_i between $\mathbf{x}$ and $\mathbf{y}$ with respect to the fuzzy class A_i is given by

$$d_i(\mathbf{x}, \mathbf{y}) = \min(u_{A_i}(\mathbf{x}), u_{A_i}(\mathbf{y})) d(\mathbf{x}, \mathbf{y}), \quad \forall \mathbf{x}, \mathbf{y} \in \mathbf{X} \tag{9.17}$$

The distance d_i can be easily extended to the entire metric space Y. A_i^* denotes the extension of A_i to Y, where

$$u_{A_i^*}(\mathbf{x}) = \begin{cases} u_{A_i}(\mathbf{x}), & \text{if } \mathbf{x} \in \mathbf{X} \\ 1, & \text{else} \end{cases} \tag{9.18}$$

Definition 9.5.3 The extended distance is the distance $d_i^* : Y \times Y \rightarrow \mathbf{R}$, defined as

$$d_i^*(\mathbf{x}, \mathbf{y}) = \begin{cases} \min(u_{A_i}(\mathbf{x}), u_{A_i}(\mathbf{y}))d(\mathbf{x}, \mathbf{y}), & \text{if } \mathbf{x}, \mathbf{y} \in \mathbf{X} \\ u_{A_i}(\mathbf{x})d(\mathbf{x}, \mathbf{y}), & \text{if } \mathbf{x} \in \mathbf{X}, \mathbf{y} \notin \mathbf{X} \\ d(\mathbf{x}, \mathbf{y}), \quad \text{if } \mathbf{x}, \mathbf{y} \notin \mathbf{X} \end{cases} \tag{9.19}$$

9.5.2 Mean Value of a Fuzzy Class

The mean vector of all vectors from $\mathbf{X}$ is defined as the only point $\mathbf{R}^s$ for which the sum of the squared distances from the points in $\mathbf{X}$ is minimal.

Let us assume that $\mathbf{v}$ is a point in $\mathbf{R}^s$ and $\mathbf{v}$ is not from $\mathbf{X}$. The sum of the squared distances in A_i between $\mathbf{v}$ and the points in $\mathbf{X}$ is given by

$$F_i(\mathbf{v}) = \sum_{j=1}^{p} d_i^{*2}(\mathbf{x}_j, \mathbf{v}) = \sum_{j=1}^{p} u_{ij}^2 d^2(\mathbf{x}_j, \mathbf{v}) \tag{9.20}$$

The minimum point of the function F_i can be interpreted as the mean point of the fuzzy class A_i.

Theorem 9.5.1 *The function* $F_i : (\mathbf{R}^s - \mathbf{X}) \rightarrow \mathbf{R}$

$$F_i(\mathbf{v}) = \sum_{j=1}^{p} u_{ij}^2 ||\mathbf{v} - \mathbf{x}_j||^2 \tag{9.21}$$

has a unique minimum $\mathbf{m}_i$ *given by*

$$\mathbf{m}_i = \frac{1}{\sum_{j=1}^{p} u_{ij}^2} \sum_{j=1}^{p} u_{ij}^2 \mathbf{x}_j \tag{9.22}$$

9.5.3 Alternating Optimization Technique

Fuzzy clustering algorithms are based on finding an adequate prototype for each fuzzy cluster and suitable membership degrees for the data to each cluster. In general, the cluster algorithm attempts to minimize an objective function which is based on either an intraclass similarity measure or a dissimilarity measure.

Let the cluster substructure of the fuzzy class C be described by the fuzzy partition $P = \{A_1, \ldots, A_n\}$ of C. In Section 9.2 it was shown that this is equivalent to

$$\sum_{j=1}^{p} u_{ij} = u_C(\mathbf{x}_j), \quad j = 1, \ldots, p \tag{9.23}$$

Further, let $\mathbf{L}_i \in \mathbf{R}^s$ be the prototype of the fuzzy class A_i and a point from the data set $\mathbf{X}$. We then obtain

$$u_{A_i}(\mathbf{L}_i) = \max_j u_{ij} \tag{9.24}$$

The dissimilarity between a data point and a prototype $\mathbf{L}_i$ is given by

$$D_i(\mathbf{x}_j, \mathbf{L}_i) = u_{ij}^2 d^2(\mathbf{x}_j, \mathbf{L}_i) \tag{9.25}$$

The inadequacy $I(A_i, \mathbf{L}_i)$ between the fuzzy class A_i and its prototype is defined as

$$I(A_i, \mathbf{L}_i) = \sum_{j=1}^{p} D_i(\mathbf{x}_j, \mathbf{L}_i) \tag{9.26}$$

Assume $\mathbf{L} = (\mathbf{L}_1, \ldots, \mathbf{L}_n)$ is the set of cluster centers and describes a representation of the fuzzy partition P.

The inadequacy $J(P, \mathbf{L})$ between the partition P and its representation $\mathbf{L}$ is defined as

$$J(P, \mathbf{L}) = \sum_{i=1}^{n} I(A_i, \mathbf{L}_i) \tag{9.27}$$

Thus the following objective function $J : F_n(C) \times \mathbf{R}^{sn} \to R$ is obtained:

$$J(P, \mathbf{L}) = \sum_{i=1}^{n} \sum_{j=1}^{p} u_{ij}^2 d^2(\mathbf{x}_j, \mathbf{L}_i) = \sum_{i=1}^{n} \sum_{j=1}^{p} u_{ij}^2 ||\mathbf{x}_j - \mathbf{L}_i||^2 \tag{9.28}$$

The objective function is of the least squared error type, and the optimal fuzzy partition and its representation can be found as the local solution of the minimization problem

$$\begin{cases} \text{minimize} \quad J(P, \mathbf{L}) \\ \quad P \in F_n(C) \\ \quad \mathbf{L} \in \mathbf{R}^{sn} \end{cases} \tag{9.29}$$

Since an exact solution to the above problem is difficult to find, an adequate approximation based on an iterative method, the so-called alternating optimization technique [29], provides the solution by minimizing the functions $J(P, \cdot)$ and $J(\cdot, \mathbf{L})$.

In other words, the minimization problem from eq. (9.29) is replaced by two separate problems:

$$\begin{cases} \text{minimize} \quad J(P, \mathbf{L}) \rightarrow \min \\ \quad P \in F_n(C) \\ \quad \mathbf{L} \quad \text{is fixed} \end{cases} \tag{9.30}$$

and

$$\begin{cases} \text{minimize} \quad J(P, \mathbf{L}) \rightarrow \min \\ \quad \mathbf{L} \in \mathbf{R}^{sn} \\ \quad P \quad \text{is fixed} \end{cases} \tag{9.31}$$

For the solution to the first optimization problem we need to introduce the following notation:

$$I_j = \{i | 1 \leq i \leq n, \quad d(\mathbf{x}_j, \mathbf{L}_i) = 0\} \tag{9.32}$$

and

$$\bar{I}_j = \{1, 2, \ldots, n\} - I_j \tag{9.33}$$

We give without proof the following two theorems regarding the minimization of the functions $J(P, \cdot)$ or $J(\cdot, \mathbf{L})$ in eqs. (9.30) and (9.31).

Theorem 9.5.2 *$P \in F_n(C)$ represents a minimum of the function $J(\cdot, \mathbf{L})$ only if*

$$I_j = \varnothing \Rightarrow u_{ij} = \frac{u_C(\mathbf{x}_j)}{\sum_{k=1}^{n} \frac{d^2(\mathbf{x}_j, \mathbf{L}_i)}{d^2(\mathbf{x}_j, \mathbf{L}_k)}}, \quad \forall 1 \leq i \leq n; \quad 1 \leq j \leq p \tag{9.34}$$

and

$$I_j \neq \varnothing \Rightarrow u_{ij} = 0, \quad \forall i \in \bar{I}_j \tag{9.35}$$

and arbitrarily $\sum_{i \in I_j} u_{ij} = u_C(\mathbf{x}_j)$.

Theorem 9.5.3 *If $\mathbf{L} \in \mathbf{R}^{sn}$ is a local minimum of the function $J(P, \cdot)$, then $\mathbf{L}_i$ is the cluster center (mean vector) of the fuzzy class A_i, for every $i = 1, \ldots, n$*

$$\mathbf{L}_i = \frac{1}{\sum_{j=1}^{p} u_{ij}^2} \sum_{j=1}^{p} u_{ij}^2 \mathbf{x}_j \tag{9.36}$$

The alternating optimization (AO) technique is based on the Picard iteration of eqs. (9.34)–(9.36). This technique can also be used to obtain a local solution to a problem that will be addressed in eq. (9.29). The iterative technique is also known as the generalized fuzzy n-means (GFNM) algorithm.

9.5.4 Family of Fuzzy *n*-Means Objective Functions

Let us assume that $P = \{A_1, \ldots, A_n\}$ is a fuzzy partition of the data set $\mathbf{X}$ and $\mathbf{L}_i \in \mathbf{R}^s$ is the prototype of the fuzzy class A_i. In [29] a more general objective function is considered:

$$J_m(P, \mathbf{L}) = \sum_{i=1}^{n} \sum_{j=1}^{p} u_{ij}^m d^2(\mathbf{x}_j, \mathbf{L}_i) \tag{9.37}$$

with $m > 1$ being a weighting exponent, sometimes known as a fuzzifier, and d the norm-induced distance.

The prototypes are obtained as a solution to the optimization problem

$$J_m(P, \cdot) \rightarrow \min \tag{9.38}$$

and are given by

$$\mathbf{L}_i = \frac{1}{\sum_{j=1}^{p} u_{ij}^m} \sum_{j=1}^{p} u_{ij}^m \mathbf{x}_j \tag{9.39}$$

while the fuzzy partition is the solution of another optimization problem

$$J_m(\cdot, \mathbf{L}) \rightarrow \min \tag{9.40}$$

and is given by

$$u_{ij} = \frac{1}{\sum_{k=1}^{n} \left[\frac{d(\mathbf{x}_j, \mathbf{L}_i)}{d(\mathbf{x}_j, \mathbf{L}_k)} \right]^{\frac{2}{m-1}}}, \quad \forall 1 \le i \le n; \quad 1 \le j \le p \tag{9.41}$$

Since the parameter m can take infinite values, an infinite family of fuzzy clustering algorithms is obtained.

In case $m \rightarrow 1$, the fuzzy n-means algorithm converges to a hard n-means solution. The membership degrees take either 0 or 1, thus describing a crisp representation. In this case, each input vector or data point $\mathbf{x}_j$ belongs exclusively to a single cluster. Thus, the fuzzy n-means algorithm is an extension of the hard n-means clustering algorithm, which is based on a crisp clustering criterion.

The integer m works to eliminate noises, and as m becomes larger, more data with small degrees of membership are neglected.

If $m \rightarrow \infty$, we get for the membership degrees the maximum fuzziness degrees

$$u_{ij} \rightarrow \frac{1}{n}, i = 1, \ldots, n; \quad j = 1, \ldots, p \tag{9.42}$$

and all prototypes converge to the mean vector of $\mathbf{X}$,

$$\mathbf{L}_i \rightarrow m(\mathbf{X}) = \frac{\sum_{j=1}^{p} \mathbf{x}_j}{p}, \quad i = 1, \ldots, n \tag{9.43}$$

These ideas can be visualized by the following example, which illustrates the differences between crisp and fuzzy n-means clustering.

Example 9.5.1 Let $X = \{\mathbf{x}_1, \mathbf{x}_2, \mathbf{x}_3, \mathbf{x}_4\}$, where $\mathbf{x}_1 = [0,0]^T, \mathbf{x}_2 = [4,0]^T, \mathbf{x}_3 = [0,6]^T, \mathbf{x}_4 = [4,6]^T$. Let $\mathbf{L}_1 = [2,0]^T$ and $\mathbf{L}_2 = [2,6]^T$ be the cluster representatives. Assume that the distortion is based on the Euclidean distance between an input vector and a cluster representative. The membership matrix $\mathbf{U}$ for crisp n-means clustering is given by

$$\mathbf{U}_{hard} = \begin{bmatrix} 1 & 0 \\ 1 & 0 \\ 0 & 1 \\ 0 & 1 \end{bmatrix} \tag{9.44}$$

Based on eq. (9.37), we obtain $J_{hard}(P, L) = 8$. Assume now that $m = 1$ and that $u_{ij} \in [0, 1]$. Then we obtain for the objective function

$$J_1(P, L) = \sqrt{40}(u_{12} + u_{21} + u_{31} + u_{41}) + 2(u_{11} + u_{22} + u_{32} + u_{44}) > 8 \tag{9.45}$$

This can be easily obtained from $u_{i1} + u_{i2} = 1$. The conclusion we can draw is that hard clustering always yields better results than fuzzy clustering for $m = 1$. However, for $m \geq 2$, the resulting fuzzy objective function yields better results (<8) than that for hard clustering.

A reformulation of the clustering objective functions is given in [127] and directly incorporates the necessary optimum conditions satisfied by P.

The following objective function is considered

$$J_m(P, \mathbf{L}) = \sum_{i=1}^{n} \sum_{j=1}^{p} u_{ij}^m d^2(\mathbf{x}_j, \mathbf{L}_i) \tag{9.46}$$

The minimization of the function $J_m(\cdot, \mathbf{L}) \rightarrow \min$ leads to the membership values

$$u_{ij} = \frac{1}{\sum_{k=1}^{n} \left[\frac{d^2(\mathbf{x}_j, \mathbf{L}_i)}{d^2(\mathbf{x}_j, \mathbf{L}_k)} \right]^{\frac{1}{m-1}}}, \quad \forall 1 \leq i \leq n; \quad 1 \leq j \leq p \tag{9.47}$$

For u_{ij}^{m-1} we obtain

$$u_{ij}^{m-1} = \frac{1}{d^2(\mathbf{x}_j, \mathbf{L}_i)} \left[\sum_{k=1}^{n} \left(\frac{1}{d^2(\mathbf{x}_j, \mathbf{L}_k)} \right)^{\frac{1}{m-1}} \right]^{-(m-1)} \tag{9.48}$$

By using the above expression, we obtain a new expression for $J_m(P, \mathbf{L})$

$$J_m(P,\mathbf{L}) = \sum_{i=1}^{n}\sum_{j=1}^{p} u_{ij} u_{ij}^{m-1} d^2(\mathbf{x}_j, \mathbf{L}_i)$$
$$= \sum_{i=1}^{n}\sum_{j=1}^{p} u_{ij} \frac{1}{[\sum_{k=1}^{n} (\frac{1}{d^2(\mathbf{x}_j,\mathbf{L}_k)})^{\frac{1}{m-1}}]^{(m-1)}}$$

Taking into account that $\sum_{j=1}^{p} u_{ij} = 1, \forall j = 1, \ldots, p$, we thus obtain a new expression for $J_m(P, \mathbf{L})$

$$J_m(P,\mathbf{L}) = \sum_{j=1}^{p} \frac{1}{[\sum_{k=1}^{n} (\frac{1}{d^2(\mathbf{x}_j,\mathbf{L}_k)})^{\frac{1}{m-1}}]^{(m-1)}} \tag{9.49}$$

This adapted expression depends only on the $\mathbf{L}$, and it is described by a new function $R_m(\mathbf{L})$. We are now in position to give the reformulated version of the fuzzy n-means criterion:

$$R_m(\mathbf{L}) = \sum_{j=1}^{p}\left[\sum_{k=1}^{n}\left(\frac{1}{d^2(\mathbf{x}_j, \mathbf{L}_k)}\right)^{\frac{1}{m-1}}\right]^{1-m} \tag{9.50}$$

In [127] it was shown that the original and reformulated versions of fuzzy n-means objective functions are completely equivalent.

9.5.5 Generalized Fuzzy *n*-Means Algorithm

In this section we describe the GFNM algorithm based on the theoretical results derived in the previous section. We consider here compact clusters completely described by a point representative $\mathbf{L}_i$.

Let $\mathbf{X} = \{\mathbf{x}_1, \ldots, \mathbf{x}_p\}$ define the data set and C be a fuzzy set on $\mathbf{X}$. The following assumptions are made:

- C represents a cluster of points from $\mathbf{X}$.
- C has a cluster substructure described by the fuzzy partition $P = \{A_1, \ldots, A_n\}$.
- n is the number of known subclusters in C.

The basic GFNM requires a random initialization of the fuzzy partition. The prototypes $\mathbf{L}_1, \ldots, \mathbf{L}_n$ are computed based on eq. (9.36), while the new fuzzy partition of C is computed using eqs. (9.34) and (9.35). These equations are continuously updated until two successive fuzzy partitions do not change substantially.

In order to monitor the convergence of the algorithm, the $n \times p$ partition matrix $\mathbf{Q}^i$ is introduced to describe each fuzzy partition P^i at the ith iteration and is used to determine the distance between two fuzzy partitions. The matrix $\mathbf{Q}^i$ is defined as

$$\mathbf{Q}^i = \mathbf{U} \quad \text{at iteration} \quad i \tag{9.51}$$

The termination criterion for iteration m is given by

$$d(P^m, P^{m-1}) = ||\mathbf{Q}^m - \mathbf{Q}^{m-1}|| < \epsilon \tag{9.52}$$

where ϵ defines the admissible error and $||\cdot||$ is any vector norm.

An algorithmic description of the GFNM is given below:

1. **Initialization:** Choose the number n of subclusters in C and the termination criterion ϵ. P^1 is selected as a random fuzzy partition of C having n atoms. Determine $\mathbf{Q}^1$, the matrix representation of P^1.
2. **Adaptation, part I:** Determine the cluster prototypes $\mathbf{L}_i, i = 1, \ldots, n$ using

$$\mathbf{L}_i = \frac{1}{\sum_{j=1}^{p} u_{ij}^2} \sum_{j=1}^{p} u_{ij}^2 \mathbf{x}_j \tag{9.53}$$

3. **Adaptation, part II:** Determine a new fuzzy partition P^2 of C using the following rules:

$$I_j = \varnothing \Rightarrow u_{ij} = \frac{u_C(\mathbf{x}_j)}{\sum_{k=1}^{n} \frac{d^2(\mathbf{x}_j, \mathbf{L}_i)}{d^2(\mathbf{x}_j, \mathbf{L}_k)}}, \quad \forall 1 \le i \le n; \quad 1 \le j \le p \tag{9.54}$$

and

$$I_j \neq \varnothing \Rightarrow u_{ij} = 0, \quad \forall i \in I_j \tag{9.55}$$

and arbitrarily $\sum_{i \in I_j} u_{ij} = u_C(\mathbf{x}_j)$.

Determine $\mathbf{Q}^2$, the matrix representation of the fuzzy partition P^2.

4. **Continuation:** If the difference between two successive partitions is smaller than a predefined threshold $||\mathbf{Q}^1 - \mathbf{Q}^2|| < \epsilon$, then stop. Else set $P^1 = P^2, \mathbf{Q}^1 = \mathbf{Q}^2$ and go to step 2.

If we obtain $C = \mathbf{X}$, the GFNM algorithm converts to the fuzzy n-means (FNM) algorithm. Then the membership degrees in eqs. (9.54) and (9.55) are determined by the following new rules:

$$I_j = \varnothing \Rightarrow u_{ij} = \frac{1}{\sum_{k=1}^{n} \frac{d^2(\mathbf{x}_j, \mathbf{L}_i)}{d^2(\mathbf{x}_j, \mathbf{L}_k)}}, \quad \forall 1 \le i \le n; \quad 1 \le j \le p \tag{9.56}$$

and

$$I_j \neq \varnothing \Rightarrow u_{ij} = 0, \quad \forall i \in I_j \tag{9.57}$$

and arbitrarily assign $\sum_{i \in I_j} u_{ij} = 1$.

9.5.6 Generalized Adaptive Fuzzy *n*-Means Algorithm

In this section we describe an adaptive fuzzy technique based on using different metrics to allow the detection of cluster shapes ranging from spherical to ellipsoidal clusters.

To achieve this an adaptive metric is used. We define a new distance metric $d(\mathbf{x}_j, \mathbf{L}_i)$ from the data point $\mathbf{x}_j$ to the cluster prototype $\mathbf{L}_i$ as

$$d^2(\mathbf{x}_j, \mathbf{L}_i) = (\mathbf{x}_j - \mathbf{L}_i)^T \mathbf{M}_i (\mathbf{x}_j - \mathbf{L}_i) \tag{9.58}$$

where $\mathbf{M}_i$ is a symmetric and positive definite shape matrix and adapts to the clusters' shape variations. The growth of the shape matrix is monitored by a bound:

$$|\mathbf{M}_i| = \rho_i, \quad \rho_i > 0, \quad i = 1, \dots, n \tag{9.59}$$

Let $\mathbf{X} = \{\mathbf{x}_1, \dots, \mathbf{x}_p\}, \mathbf{x}_j \in \mathbf{R}^s$ be a data set. Let C be a fuzzy set on $\mathbf{X}$ describing a fuzzy cluster of points in $\mathbf{X}$, and having a cluster substructure which is described by a fuzzy partition $P = \{A_1, \dots, A_n\}$ of C. Each fuzzy class A_i is described by the point prototype $\mathbf{L}_i \in \mathbf{R}^s$. The local distance with respect to A_i is given by

$$d_i^2(\mathbf{x}_j, \mathbf{L}_i) = u_{ij}^2 (\mathbf{x}_j - \mathbf{L}_i)^T \mathbf{M}_i (\mathbf{x}_j - \mathbf{L}_i) \tag{9.60}$$

As an objective function we choose

$$J(P, \mathbf{L}, M) = \sum_{i=1}^{n} \sum_{j=1}^{p} d^2(\mathbf{x}_j, \mathbf{L}_i) = \sum_{i=1}^{n} \sum_{j=1}^{p} u_{ij}^2 (\mathbf{x}_j - \mathbf{L}_i)^T \mathbf{M}_i (\mathbf{x}_j - \mathbf{L}_i) \tag{9.61}$$

where $M = (\mathbf{M}_1, \dots, \mathbf{M}_n)$.

The objective function is again chosen to be of the least squared error type. We can find the optimal fuzzy partition and its representation as the local solution of the minimization problem:

$$\begin{cases} \text{minimize} \quad J(P, \mathbf{L}, M) \\ \sum_{i=1}^{n} u_{ij} = u_C(\mathbf{x}_j), \quad j = 1, \dots, p \\ |\mathbf{M}_i| = \rho_i, \quad \rho_i > 0, \quad i = 1, \dots, n \\ \mathbf{L} \in \mathbf{R}^{sn} \end{cases} \tag{9.62}$$

The following theorem regarding the minimization of the functions $J(P, \mathbf{L}, \cdot)$ is given without proof. It is known as the adaptive norm theorem.

Theorem 9.5.4 *Assuming the point prototype $\mathbf{L}_i$ of the fuzzy class A_i equals the cluster center of this class, $\mathbf{L}_i = \mathbf{m}_i$, and the determinant of the shape matrix $\mathbf{M}_i$ is bounded, $|\mathbf{M}_i| = \rho_i, \rho_i > 0, i = 1, \dots, n$, then $\mathbf{M}_i$ is a local minimum of the function $J(P, \mathbf{L}, \cdot)$ only if*

$$\mathbf{M}_i = [\rho_i |\mathbf{S}_i|]^{\frac{1}{s}} \mathbf{S}_i^{-1} \tag{9.63}$$

where $\mathbf{S}_i$ is the within-class scatter matrix of the fuzzy class A_i:

$$\mathbf{S}_i = \sum_{j=1}^{p} u_{ij}^2 (\mathbf{x}_j - \mathbf{m}_i)(\mathbf{x}_j - \mathbf{m}_i)^T \tag{9.64}$$

The above theorem can be employed as part of an alternating optimization technique. The resulting iterative procedure is known as the generalized adaptive fuzzy n-means (GAFNM) algorithm.

An algorithmic description of the GAFNM is given below:

1. **Initialization:** Choose the number n of subclusters in C and the termination criterion ϵ. P^1 is selected as a random fuzzy partition of C having n atoms. Set iteration counter $l = 1$.
2. **Adaptation, part I:** Determine the cluster prototypes $\mathbf{L}_i = \mathbf{m}_i, i = 1, \ldots, n$ using

$$\mathbf{L}_i = \frac{1}{\sum_{j=1}^{p} u_{ij}^2} \sum_{j=1}^{p} u_{ij}^2 \mathbf{x}_j \tag{9.65}$$

3. **Adaptation, part II:** Determine the within-class scatter matrix $\mathbf{S}_i$ using

$$\mathbf{S}_i = \sum_{j=1}^{p} u_{ij}^2 (\mathbf{x}_j - \mathbf{m}_i)(\mathbf{x}_j - \mathbf{m}_i)^T \tag{9.66}$$

Determine the shape matrix $\mathbf{M}_i$ using

$$\mathbf{M}_i = [\rho_i |\mathbf{S}_i|]^{\frac{1}{s}} \mathbf{S}_i^{-1} \tag{9.67}$$

and compute the distance $d^2(\mathbf{x}_j, \mathbf{m}_i)$ using

$$d^2(\mathbf{x}_j, \mathbf{m}_i) = (\mathbf{x}_j - \mathbf{m}_i)^T \mathbf{M}_i (\mathbf{x}_j - \mathbf{m}_i) \tag{9.68}$$

4. **Adaptation part III:** Compute a new fuzzy partition P^l of C using the following rules

$$I_j = \varnothing \Rightarrow u_{ij} = \frac{u_C(\mathbf{x}_j)}{\sum_{k=1}^{n} \frac{d^2(\mathbf{x}_j, \mathbf{m}_i)}{d^2(\mathbf{x}_j, \mathbf{m}_k)}}, \quad \forall 1 \le i \le n; \quad 1 \le j \le p \tag{9.69}$$

and

$$I_j \neq \varnothing \Rightarrow u_{ij} = 0, \quad \forall i \in I_j \tag{9.70}$$

and arbitrarily $\sum_{i \in I_j} u_{ij} = u_C(\mathbf{x}_j)$.

The standard notation is used

$$I_j = \{i | 1 \le i \le n, \quad d(\mathbf{x}_j, \mathbf{L}_i) = 0\} \tag{9.71}$$

and

$$\bar{I}_j = \{1, 2, \ldots, n\} - I_j \tag{9.72}$$

5. **Continuation:** If the difference between two successive partitions is smaller than a predefined threshold, $||P^l - P^{l-1}|| < \epsilon$, then stop. Else go to step 2.

An important issue for the GAFNM algorithm is the selection of the bounds of the shape matrix $\mathbf{M}_i$. They can be chosen as:

$$\rho_i = 1, \quad i = 1, \ldots, n \tag{9.73}$$

If we choose $C = \mathbf{X}$, we obtain $u_C(\mathbf{x}_j) = 1$ and thus we get the following membership degrees:

$$u_{ij} = \frac{1}{\sum_{k=1}^{n} \frac{d^2(\mathbf{x}_j, \mathbf{m}_i)}{d^2(\mathbf{x}_j, \mathbf{m}_k)}}, \quad \forall 1 \le i \le n; \quad 1 \le j \le p \tag{9.74}$$

The resulting iterative procedure is known as the adaptive fuzzy n-means (AFNM) algorithm.

9.5.7 Generalized Fuzzy *n*-Shells Algorithm

So far, we have considered clustering algorithms that use point prototypes as cluster prototypes. Therefore, the previous algorithms cannot detect clusters that can be described by shells, hyperspheres, or hyperellipsoids. The generalized fuzzy n-shells algorithm [72,73] is able to detect such clusters. The cluster prototypes used are s-dimensional hyperspherical shells, and the distances of data points are measured from the hyperspherical surfaces. Since the prototypes contain no interiors, they are referred to as shells. They are completely described by a set of centers $\mathbf{V}$ and a set of radii $\mathbf{R}$.

Let $\mathbf{X} = \{\mathbf{x}_1, \ldots, \mathbf{x}_p\}, \mathbf{x}_j \in \mathbf{R}^s$ be a data set. Let C be a fuzzy set on $\mathbf{X}$ describing a fuzzy cluster of points in $\mathbf{X}$. C has a cluster substructure that is described by a fuzzy partition $P = \{A_1, \ldots, A_n\}$ of C. Each fuzzy class A_i is described by the point prototype $\mathbf{L}_i \in \mathbf{R}^s$. It is assumed that the cluster resembles a hyperspherical shell, and thus the prototype $\mathbf{L}_i$ of A_i is determined by two parameters $(\mathbf{v}_i, \mathbf{r}_i)$, where $\mathbf{v}_i$ is the center of the hypersphere and r_i is the radius.

The hyperspherical shell prototype $\mathbf{L}_i$ with center $\mathbf{v}_i \in \mathbf{R}^s$ and radius $r_i \in R^+$ is defined by the set

$$\mathbf{L}_i(\mathbf{v}_i, r_i) = \{\mathbf{x} \in \mathbf{R}^s | (\mathbf{x} - \mathbf{v}_i)^T(\mathbf{x} - \mathbf{v}_i) = r_i^2\} \tag{9.75}$$

The distance d_{ij} between the point $\mathbf{x}_j$ and the prototype $\mathbf{L}_i(\mathbf{v}_i, r_i)$ is defined as

$$d_{ij}^2 = d^2(\mathbf{x}_j, \mathbf{L}_i) = [||\mathbf{x}_j - \mathbf{v}_i|| - r_i]^2 \tag{9.76}$$

The corresponding objective function $J(P, \mathbf{L})$ describing the deviation between the fuzzy partition and its representation $\mathbf{L}$ can be written as

$$J(P, \mathbf{L}) = \sum_{i=1}^{n} \sum_{j=1}^{p} D_i(\mathbf{x}_j, \mathbf{L}_i) = \sum_{i=1}^{n} \sum_{j=1}^{p} u_{ij}^2 d_{ij}^2 \tag{9.77}$$

L represents the set of centers

$$\mathbf{V} = \{\mathbf{v}_1, \ldots, \mathbf{v}_n | \mathbf{v}_i \in \mathbf{R}^s\} \tag{9.78}$$

while the set of radii is given by

$$R = \{r_1, \ldots, r_n | r_i \in R^+\} \tag{9.79}$$

This means that the objective function $J(P, \mathbf{L})$ is a function $J : F_n(\mathbf{X}) \times \mathbf{R}^{sn} \times \mathbf{R}^{+n} \rightarrow R$ and is defined as

$$J(P, \mathbf{L}) = J(P, V, R) = \sum_{i=1}^{n} \sum_{j=1}^{p} u_{ij}^2 [||\mathbf{x}_j - \mathbf{v}_i|| - r_i]^2 \tag{9.80}$$

The generalized shell clustering algorithm is based on an iterative procedure to solve the following minimization problem:

$$\begin{cases} \text{minimize} \quad J(P, V, R) \\ \sum_{i=1}^{n} u_{ij} = u_C(\mathbf{x}_j), \quad j = 1, \ldots, p \\ V \in \mathbf{R}^{sn}, R \in \mathbf{R}^{+n} \end{cases} \tag{9.81}$$

To solve this problem two theorems are needed. The first theorem is the theorem for optimal fuzzy partition.

Theorem 9.5.5 *Assume that* $\mathbf{X}$ *has at least* $n < p$ *distinct points, and define the following sets:*

$$I_j = \{i | 1 \leq i \leq n, \quad d(\mathbf{x}_j, \mathbf{L}_i) = 0\} \tag{9.82}$$

and

$$\bar{I}_j = \{1, 2, \ldots, n\} - I_j \tag{9.83}$$

A fuzzy partition P *represents the minimum of the objective function* $J(\cdot, \mathbf{V}, \mathbf{R})$ *only if*

$$I_j = \varnothing \Rightarrow u_{ij} = \frac{u_C(\mathbf{x}_j)}{\sum_{k=1}^{n} \frac{d_{ij}^2}{d_{kj}^2}} \tag{9.84}$$

and

$$I_j \neq \varnothing \Rightarrow u_{ij} = 0, \quad \forall i \in \bar{I}_j \tag{9.85}$$

and arbitrarily $\sum_{i \in I_j} u_{ij} = u_C(\mathbf{x}_j)$.

The next theorem gives the conditions for the optimal prototypes. The conditions describe a system of nonlinear equations to determine the radii and centers of the cluster prototypes.

Theorem 9.5.6 *Let P be a fixed fuzzy partition. Then* **V** *and* **R** *minimize the objective function* $J(P, \cdot, \cdot)$ *only if the centers and the radii equations*

$$\begin{cases} \sum_{j=1}^{p} u_{ij}^2 \frac{||\mathbf{x}_j - \mathbf{v}_i|| - r_i}{||\mathbf{x}_j - \mathbf{v}_i||}(\mathbf{x}_j - \mathbf{v}_i) = 0 \\ \sum_{j=1}^{p} u_{ij}^2 [||\mathbf{x}_j - \mathbf{v}_i|| - r_i] = 0 \end{cases} \tag{9.86}$$

hold, where $i = 1, \ldots, n$.

The preceding theorems can be used as the basis of an alternating optimization technique. The resulting iterative procedure is known as the generalized fuzzy n-shells (GFNS) algorithm.

An algorithmic description of the GFNS is given below:

1. **Initialization:** Choose the number n of subclusters in C and the termination criterion ϵ. P^1 is selected as a random fuzzy partition of C having n atoms. Set iteration counter $l = 1$.
2. **Adaptation, part I:** Determine the centers $\mathbf{v}_i$ and radii r_i by solving the system of equations

$$\sum_{j=1}^{p} u_{ij}^2 \frac{||\mathbf{x}_j - \mathbf{v}_i|| - r_i}{||\mathbf{x}_j - \mathbf{v}_i||}(\mathbf{x}_j - \mathbf{v}_i) = 0 \tag{9.87}$$

$$\sum_{j=1}^{p} u_{ij}^2 [||\mathbf{x}_j - \mathbf{v}_i|| - r_i] = 0 \tag{9.88}$$

   where $i = 1, \ldots, n$.
3. **Adaptation, part II:** Determine the distance d_{ij} of the point $\mathbf{x}_j$ from the prototype $\mathbf{L}_i(\mathbf{v}_i, r_i)$ defined as

$$d_{ij}^2 = d^2(\mathbf{x}_j, \mathbf{L}_i) = [||\mathbf{x}_j - \mathbf{v}_i|| - r_i]^2 \tag{9.89}$$

4. **Adaptation, part III:** Determine a new fuzzy partition P^l of C using the following rules

$$I_j = \varnothing \Rightarrow u_{ij} = \frac{u_C(\mathbf{x}_j)}{\sum_{k=1}^{n} \frac{d_{ij}^2}{d_{kj}^2}} \tag{9.90}$$

   and

$$I_j \neq \varnothing \Rightarrow u_{ij} = 0, \quad \forall i \in I_j \tag{9.91}$$

   and arbitrarily $\sum_{i \in I_j} u_{ij} = u_C(\mathbf{x}_j)$.

   Set $l = l + 1$.
5. **Continuation:** If the difference between two successive partitions is smaller than a predefined threshold, $||P^l - P^{l-1}|| < \epsilon$, then stop. Else go to step 2.

If we choose $C = \mathbf{X}$, we obtain $u_C(\mathbf{x}_j) = 1$ and thus we get the following fuzzy partition

$$I_j = \varnothing \Rightarrow u_{ij} = \frac{1}{\sum_{k=1}^{n} \frac{d_{ij}^2}{d_{kj}^2}} \tag{9.92}$$

and

$$I_j \neq \varnothing \Rightarrow u_{ij} = 0, \quad \forall i \in I_j \tag{9.93}$$

and arbitrarily $\sum_{i \in I_j} u_{ij} = 1$.

The resulting iterative procedure is known as the fuzzy n-shells (FNS) algorithm.

9.5.8 Generalized Adaptive Fuzzy *n*-Shells Algorithm

In this section we describe an adaptive procedure similar to the generalized adaptive fuzzy n-means algorithm, but for shell prototypes. As described in [73] hyperellipsoidal shell prototypes will be considered.

Let $\mathbf{X} = \{\mathbf{x}_1, \ldots, \mathbf{x}_p\}, \mathbf{x}_j \in \mathbf{R}^s$ be a data set. Let C be a fuzzy set on $\mathbf{X}$ describing a fuzzy cluster of points in $\mathbf{X}$. C has a cluster substructure, which is described by a fuzzy partition $P = \{A_1, \ldots, A_n\}$ of C. Each fuzzy class A_i is described by the point prototype $\mathbf{L}_i \in \mathbf{R}^s$. Let us assume that each cluster resembles a hyperellipsoidal shell.

The hyperellipsoidal shell prototype $\mathbf{L}_i(\mathbf{v}_i, r_i, \mathbf{M}_i)$ of the fuzzy class A_i is given by the set

$$\mathbf{L}_i(\mathbf{v}_i, r_i, \mathbf{M}_i) = \{\mathbf{x} \in \mathbf{R}^s | (\mathbf{x} - \mathbf{v}_i)^T \mathbf{M}_i (\mathbf{x} - \mathbf{v}_i) = r_i^2\} \tag{9.94}$$

with $\mathbf{M}_i$ representing a symmetric and positive definite matrix.

Distance d_{ij} between the point $\mathbf{x}_j$ and the cluster center $\mathbf{v}^i$ is defined as

$$d_{ij}^2 = d^2(\mathbf{x}_j, \mathbf{v}_i) = [(\mathbf{x} - \mathbf{v}_i)^T \mathbf{M}_i (\mathbf{x} - \mathbf{v}_i)]^{\frac{1}{2}} - r_i \tag{9.95}$$

Thus a slightly changed objective function is obtained:

$$J(P, V, R, M) = \sum_{i=1}^{n} \sum_{j=1}^{p} u_{ij}^2 d_{ij}^2 = \sum_{i=1}^{n} \sum_{j=1}^{p} u_{ij}^2 [[(\mathbf{x} - \mathbf{v}_i)^T \mathbf{M}_i (\mathbf{x} - \mathbf{v}_i)]^{\frac{1}{2}} - r_i]^2 \tag{9.96}$$

For optimization purposes, we need to determine the minimum of the functions $J(\cdot, V, R, M), J(P, \cdot, R, M)$, and $J(P, V, \cdot, M)$. It can be shown that they are given by the following propositions [85].

The following is the proposition for optimal partition.

Proposition 9.5.1 *The minimum of the function* $J(\cdot, \mathbf{V}, \mathbf{R}, \mathbf{M})$ *is given by the fuzzy partition* P *only if*

$$I_j = \varnothing \Rightarrow u_{ij} = \frac{u_C(\mathbf{x}_j)}{\sum_{k=1}^{n} \frac{d_{ij}^2}{d_{kj}^2}} \tag{9.97}$$

and

$$I_j \neq \varnothing \Rightarrow u_{ij} = 0, \quad \forall i \in I_j \tag{9.98}$$

and arbitrarily $\sum_{i \in I_j} u_{ij} = u_C(\mathbf{x}_j)$.

The next is the proposition for optimal prototype centers.

Proposition 9.5.2 *The optimal value of* **V** *with respect to the function* $J(P, \cdot, R, M)$ *is given by*

$$\sum_{j=1}^{p} u_{ij}^2 \frac{d_{ij}}{q_{ij}} (\mathbf{x}_j - \mathbf{v}_i) = 0, \quad i = 1, \ldots, n \tag{9.99}$$

where q_{ij} *is given by*

$$q_{ij} = (\mathbf{x}_j - \mathbf{v}_i)^T \mathbf{M}_i (\mathbf{x}_j - \mathbf{v}_i) \tag{9.100}$$

The following is the proposition for optimal prototype radii.

Proposition 9.5.3 *The optimal value of* R *with respect to the function* $J(P, V, \cdot, M)$ *is given by*

$$\sum_{j=1}^{p} u_{ij}^2 d_{ij} = 0, \quad i = 1, \ldots, n \tag{9.101}$$

Thus, we have given the necessary propositions to determine the minimum of the functions $J(\cdot, V, R, M)$, $J(P, \cdot, R, M)$, and $J(P, V, \cdot, M)$.

To ensure that the adaptive norm is bounded, we impose the constraint

$$|\mathbf{M}_i| = \rho_i, \quad \text{where} \quad \rho_i > 0, \quad i = 1, \ldots, n \tag{9.102}$$

The norm is given by the following theorem, the so-called adaptive norm theorem [85].

Theorem 9.5.7 *Let* $\mathbf{X} \subset \mathbf{R}^s$. *Suppose the objective function* J *already contains the optimal* P, V, *and* R. *If the determinant of the shape matrix* $\mathbf{M}_i$ *is bounded,* $|\mathbf{M}_i| = \rho_i, \rho_i > 0, i = 1, \ldots, n$, *then* $\mathbf{M}_i$ *is a local minimum of the function* $J(P, V, R, \cdot)$ *only if*

$$\mathbf{M}_i = [\rho_i |\mathbf{S}_{si}|]^{\frac{1}{s}} \mathbf{S}_{si}^{-1} \tag{9.103}$$

where $\mathbf{S}_{si}$ *represents the nonsingular shell scatter matrix of the fuzzy class* A_i

$$\mathbf{S}_{si} = \sum_{j=1}^{p} u_{ij}^2 \frac{d_{ij}}{q_{ij}} (\mathbf{x}_j - \mathbf{v}_i)(\mathbf{x}_j - \mathbf{v}_i)^T \tag{9.104}$$

In practice, the bound is chosen as $\rho_i = 1, i = 1, \ldots, n$.

The preceding theorems can be used as the basis of an alternating optimization technique. The resulting iterative procedure is known as the generalized adaptive fuzzy n-shells (GAFNS) algorithm.

An algorithmic description of the GAFNS is given below:

1. **Initialization:** Choose the number n of subclusters in C and the termination criterion ϵ. P^1 is selected as a random fuzzy partition of C having n atoms. Initialize $\mathbf{M}_i = \mathbf{I}, i = 1, \dots, n$ where $\mathbf{I}$ is an $s \times s$ unity matrix. Set iteration counter $l = 1$.
2. **Adaptation, part I:** Determine the centers $\mathbf{v}_i$ and radii r_i by solving the system of equations
$$\begin{cases} \sum_{j=1}^{p} u_{ij}^2 \frac{d_{ij}}{q_{ij}} (\mathbf{x}_j - \mathbf{v}_i) = 0 \\ \sum_{j=1}^{p} u_{ij}^2 d_{ij} = 0 \end{cases} \tag{9.105}$$
where $i = 1, \dots, n$ and $q_{ij} = (\mathbf{x}_j - \mathbf{v}_i)^T \mathbf{M}_i (\mathbf{x}_j - \mathbf{v}_i)$.
3. **Adaptation, part II:** Determine the shell scatter matrix $\mathbf{S}_{si}$ of the fuzzy class A_i
$$\mathbf{S}_{si} = \sum_{j=1}^{p} u_{ij}^2 \frac{d_{ij}}{q_{ij}} (\mathbf{x}_j - \mathbf{v}_i)(\mathbf{x}_j - \mathbf{v}_i)^T \tag{9.106}$$
where the distance d_{ij} is given by
$$d_{ij}^2 = [(\mathbf{x}_j - \mathbf{v}_i)^T \mathbf{M}_i (\mathbf{x}_j - \mathbf{v}_i)]^{1/2} - r_i \tag{9.107}$$
4. **Adaptation, part III:** Determine the approximate value of $\mathbf{M}_i$:
$$\mathbf{M}_i = [\rho_i |\mathbf{S}_{si}|]^{\frac{1}{s}} \mathbf{S}_{si}^{-1}, \quad i = 1, \dots, n \tag{9.108}$$
$\rho_i = 1$ or ρ_i is equal to the determinant of the previous $\mathbf{M}_i$.
5. **Adaptation, part IV:** Compute a new fuzzy partition P^l of C using the following rules:
$$I_j = \varnothing \Rightarrow u_{ij} = \frac{u_C(\mathbf{x}_j)}{\sum_{k=1}^{n} \frac{d_{ij}^2}{d_{kj}^2}} \tag{9.109}$$
and
$$I_j \neq \varnothing \Rightarrow u_{ij} = 0, \quad \forall i \in I_j \tag{9.110}$$
and arbitrarily $\sum_{i \in I_j} u_{ij} = u_C(\mathbf{x}_j)$.
Set $l = l + 1$.
6. **Continuation:** If the difference between two successive partitions is smaller than a predefined threshold, $||P^l - P^{l-1}|| < \epsilon$, then stop. Else go to step 2.

If we choose $u_C = \mathbf{X}$, we obtain $u_C(\mathbf{x}_j) = 1$ and thus we get the following fuzzy partition:
$$I_j = \varnothing \Rightarrow u_{ij} = \frac{1}{\sum_{k=1}^{n} \frac{d_{ij}^2}{d_{kj}^2}} \tag{9.111}$$

and

$$I_j \neq \varnothing \Rightarrow u_{ij} = 0, \quad \forall i \in I_j \tag{9.112}$$

and arbitrarily $\sum_{i \in I_j} u_{ij} = 1$.

The resulting iterative procedure is known as adaptive fuzzy n-shells (AFNS) algorithm. This technique enables one to identify the elliptical data substructure, and even to detect overlapping between clusters to some degree.

9.5.8.1 Simplified AFNS Algorithm

Newton's method can be employed to determine $\mathbf{v}_i$ and r_i from the system of eqs. (9.99) and (9.101). The drawback of the AFNS is its computational complexity. Therefore, some simplifications are necessary for a better practical handling. In [73] a simplification of the algorithm was proposed based on the fact that the radius r_i can be absorbed into the matrix $\mathbf{M}_i$ such that the hyperellipsoidal prototype $\mathbf{L}_i$ becomes the surface of the equation

$$(\mathbf{x} - \mathbf{v}_i)^T \mathbf{M}_i (\mathbf{x} - \mathbf{v}_i) = 1 \tag{9.113}$$

This means that eq. (9.101) is no longer necessary. The distance d_{ij} of the point $\mathbf{x}_j$ from the prototype $\mathbf{L}_i$ is now given by

$$d_{ij} = [(\mathbf{x} - \mathbf{v}_i)^T \mathbf{M}_i (\mathbf{x} - \mathbf{v}_i)]^{\frac{1}{2}} - 1 \tag{9.114}$$

This changes the objective function into a new one:

$$J(P, V, M) = \sum_{i=1}^{n} \sum_{j=1}^{p} u_{ij}^2 d_{ij}^2 = \sum_{i=1}^{n} \sum_{j=1}^{p} u_{ij}^2 [[(\mathbf{x} - \mathbf{v}_i)^T \mathbf{M}_i (\mathbf{x} - \mathbf{v}_i)]^{\frac{1}{2}} - 1]^2 \tag{9.115}$$

for $i = 1, \dots, n$. The centers are obtained as a solution of the system of equations

$$\sum_{j=1}^{p} u_{ij}^2 \frac{[(\mathbf{x} - \mathbf{v}_i)^T \mathbf{M}_i (\mathbf{x} - \mathbf{v}_i)]^{\frac{1}{2}} - 1}{d_{ij}} (\mathbf{x}_j - \mathbf{v}_i) = 0 \tag{9.116}$$

9.5.8.2 Infinite AFNS Family

By introducing a weighting exponent $m > 1$ as a fuzzifier, we obtain the infinite family J_m of objective functions

$$J(P, V, R, M) = \sum_{i=1}^{n} \sum_{j=1}^{p} u_{ij}^m d_{ij}^2 \tag{9.117}$$

Assume, that the prototype $\mathbf{L}_i$ is given by

$$(\mathbf{x} - \mathbf{v}_i)^T \mathbf{M}_i (\mathbf{x} - \mathbf{v}_i) = r_i \tag{9.118}$$

and the distance d_{ij} between the point $\mathbf{x}_j$ and the prototype $\mathbf{L}_i$ is now given by

$$d_{ij} = [(\mathbf{x} - \mathbf{v}_i)^T \mathbf{M}_i (\mathbf{x} - \mathbf{v}_i)]^{\frac{1}{2}} - r_i \tag{9.119}$$

Thus, the optimal fuzzy partition becomes

$$u_{ij} = \frac{1}{\sum_{k=1}^{n} \left[\frac{d_{ij}^2}{d_{kj}^2}\right]^{\frac{1}{m-1}}} \tag{9.120}$$

for $j = 1, \ldots, p$ and $i = 1, \ldots, n$.

The centers $\mathbf{v}_i$ and radii r_i are obtained as the solutions of the system of equations

$$\sum_{j=1}^{p} u_{ij}^m \frac{d_{ij}}{q_{ij}} (\mathbf{x}_j - \mathbf{v}_i) = 0 \tag{9.121}$$

$$\sum_{j=1}^{p} u_{ij}^m d_{ij} = 0 \tag{9.122}$$

where $i = 1, \ldots, n$ and $q_{ij} = (\mathbf{x}_j - \mathbf{v}_i)^T \mathbf{M}_i (\mathbf{x}_j - \mathbf{v}_i)$.

The shell scatter matrix $\mathbf{S}_{si}$ of the fuzzy class A_i becomes

$$\mathbf{S}_{si} = \sum_{j=1}^{p} u_{ij}^m \frac{d_{ij}}{q_{ij}} (\mathbf{x}_j - \mathbf{v}_i)(\mathbf{x}_j - \mathbf{v}_i)^T \tag{9.123}$$

and the matrix $\mathbf{M}_i, i = 1, \ldots, n$ is given by

$$\mathbf{M}_i = [\rho_i |\mathbf{S}_{si}|]^{\frac{1}{s}} \mathbf{S}_{si}^{-1}, \quad i = 1, \ldots, n \tag{9.124}$$

The constraint $|\mathbf{M}_i| = \rho_i, \rho_i > 0, i = 1, \ldots, n$, is imposed to achieve this.

The simplifications used for the AFNS can also be applied to the infinite AFNS family. This means that the prototypes can have equal radii $r_i = 1, i = 1, \ldots, n$, and the distance d_{ij} has the form

$$d_{ij} = [(\mathbf{x} - \mathbf{v}_i)^T \mathbf{M}_i (\mathbf{x} - \mathbf{v}_i)]^{\frac{1}{2}} - 1 \tag{9.125}$$

9.6. COMPARISON OF FUZZY CLUSTERING VERSUS PCA FOR fMRI

Recent work in fMRI [20] compared two paradigm-free (data-driven) methods: fuzzy clustering analysis (FCA) based on the fuzzy n-means algorithm versus principal component analysis (PCA) for two types of fMRI signal data. These are a water phantom with scanner noise contributions only and in vivo data acquired under null hypothesis conditions as shown in Fig. 9.9.

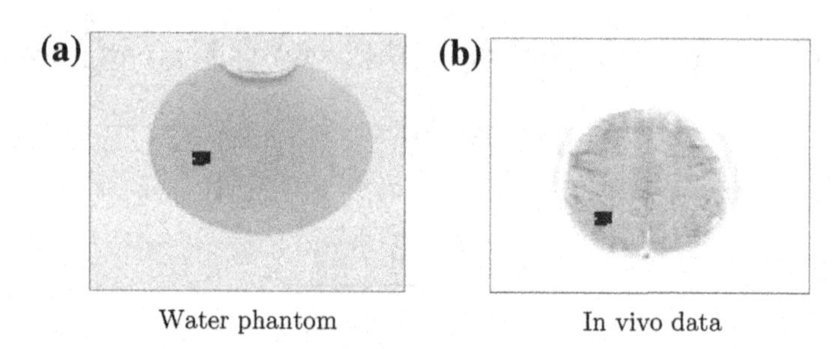

Figure 9.9 Test images with activation region overlaid: water phantom (a) and (b) in vivo data acquired under null hypothesis conditions. *(Images from [20] reprinted with permission from Elsevier Science.)*

Figure 9.10a shows the activation map determined by the FNM algorithm and added to the mean image of the time series acquired for the water phantom, while Fig. 9.10b shows that of the in vivo data. Figure 9.10c shows the activation map determined by PCA, added on the mean image for the same data set. The similarity between the activity maps calculated by the FNM algorithm and PCA is visible. Figure 9.10d, on the other hand, shows that PCA did not identify the activation focus, and produced well-scattered positives. In Fig. 9.10e the time courses obtained for water phantom data by both PCA and FNM algorithm are shown. Both exhibit a high correlation coefficient with respect to the simulated time course. For the in vivo data, only the FNM algorithm achieves a high correlation as shown in Fig. 9.10f, while the time course extracted by PCA is corrupted and fails to detect the simulated activation.

The experiments performed in [20] comparing the performance of FNM algorithm versus PCA showed the following:

- In the presence of scanner noise only, both methods show comparable performance.
- In the presence of other noise sources (artifacts), PCA fails to detect activation at lower contrast-to-noise ratios (CNR). This could be critical in fMRI. FNM outperforms PCA for all chosen CNR values.
- PCA globally fails to detect both positively and negatively correlated time courses, i.e., activation and deactivation. This is because of the arbitrary sign of the eigenvectors obtained from the correlation matrix decomposition. Only FNM preserves the original shapes of the time courses and yields immediately interpretable results.

9.7. FUZZY ALGORITHMS FOR LVQ

Let us consider the set $\mathbf{X}$ of samples from an n-dimensional Euclidean space and let $f(\mathbf{x})$ be the probability distribution function of $\mathbf{x} \in \mathbf{X} \in \mathbf{R}^n$. Learning vector

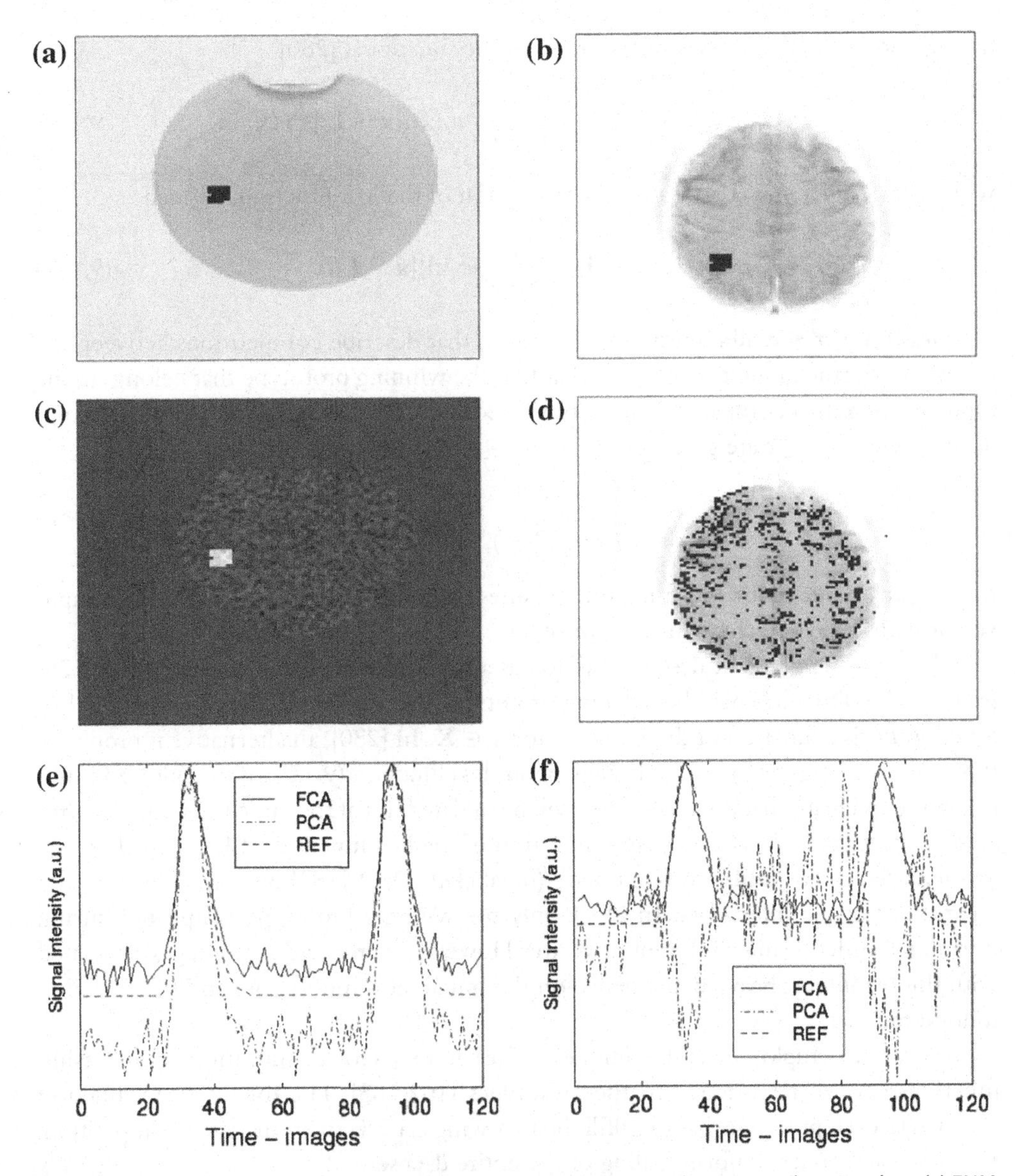

Figure 9.10 Results of applying FNM algorithm and PCA to water phantom and in vivo data. (a) FNM algorithm: Water phantom with the region extracted overlaid. (b) FNM: In vivo anatomy with the region extracted by FNM overlaid. (c) PCA: Activation map plus T_2^* anatomy. (d) PCA: Activation map plus T_2^* anatomy. (e) Time courses corresponding to FNM and PCA for water phantom: good agreement for both methods with the reference time course. (f) Time courses corresponding to FNM and PCA for in vivo data: only with FNM good agreement. *(Images from [20] reprinted with permission from Elsevier Science.)*

quantization is based on the minimization of the functional [280]

$$D(\mathbf{L}_1, \ldots, \mathbf{L}_c) = \int \cdots \int_{\mathbf{R}^n} \sum_{r=1}^{c} u_r(\mathbf{x}) ||\mathbf{x} - \mathbf{L}_r||^2 f(\mathbf{x}) d\mathbf{x} \quad (9.126)$$

with $D_{\mathbf{x}} = D_{\mathbf{x}}(\mathbf{L}_1, \ldots, \mathbf{L}_c)$ being the expectation of the loss function, defined as

$$D_{\mathbf{x}}(\mathbf{L}_1, \ldots, \mathbf{L}_c) = \sum_{r=1}^{c} u_r(\mathbf{x}) ||\mathbf{x} - \mathbf{L}_r||^2 \quad (9.127)$$

$u_r = u_r(\mathbf{x}), 1 \leq r \leq c$, all membership functions that describe competitions between the prototypes for the input $\mathbf{x}$. Supposing that $\mathbf{L}_i$ is the winning prototype that belongs to the input vector $\mathbf{x}$, that is, the closest prototype to $\mathbf{x}$ in the Euclidean sense the memberships $u_{ir} = u_r(\mathbf{x}), 1 \leq r \leq c$ are given by

$$u_{ir} = \begin{cases} 1, & \text{if} \quad r = i \\ u(\frac{||\mathbf{x}-\mathbf{L}_i||^2}{||\mathbf{x}-\mathbf{L}_r||^2}), & \text{if} \quad r \neq i \end{cases} \quad (9.128)$$

In the given context, the loss function measures the locally weighted error of each input vector with respect to the winning prototype.

Choosing the gradient descent method as a minimization technique for eq. (9.126) proves to be difficult, since the winning prototype involved in the definition of the loss function $D_{\mathbf{x}}$ is a function of the input vector $\mathbf{x} \in \mathbf{X}$. In [280], an alternative is proposed, that is, to use the gradient of the instantaneous loss function (9.127), if the pdf $f(\mathbf{x})$ is not known. This means that the prototypes will be updated not at the same time but one after another. If $u_{ir} = 0, \forall r \neq i$, then minimization of the loss function (9.127) based on the gradient descent is similar to Kohonen's (unlabeled) LVQ [183], which generates hard c-partitions of the input data. In LVQ, only the winning prototype is updated during learning in order to match the input vector. However, there are disadvantages associated with this technique: heavy dependence on the initial set of prototypes and susceptibility to local minima.

In order to employ fuzzy techniques for learning vector quantization, membership functions have to be assigned to the prototypes [162,163]. The membership function assigned to each prototype has to fulfill the following criteria to ensure a fair competition:

- Invariance under uniform scaling of the entire data set.
- Equal to 1 if the prototype is determined to be the winner.
- If the prototype is not the winner, it takes any value between 0 and 1.
- If it is not the winner, then it is close to 0, and its distance from the input vector approaches infinity.

Based on eq. (9.128) we can rewrite the loss function from eq. (9.127) in the form

$$D_{\mathbf{x}}(\mathbf{L}_1, \ldots, \mathbf{L}_c) = \sum_{r=1}^{c} u_r(\mathbf{x}) ||\mathbf{x} - \mathbf{L}_r||^2 = ||\mathbf{x} - \mathbf{L}_i||^2 + \sum_{r \neq i}^{c} u_{ir} ||\mathbf{x} - \mathbf{L}_r||^2 \quad (9.129)$$

The loss function is determined by the weighted contribution between membership functions and distance between input data and cluster prototypes. There is a relative contribution of the nonwinning prototype $\mathbf{L}_r$ with respect to the winning prototype $\mathbf{L}_i$ given by the ratio $u_{ir}\frac{||\mathbf{x}-\mathbf{L}_r||^2}{||\mathbf{x}-\mathbf{L}_i||^2}$.

To determine an adequate membership function, we have to require that the relative contribution of the nonwinning prototype $\mathbf{L}_r$ with respect to the winning prototype $\mathbf{L}_i$ be a function of the ratio $\frac{||\mathbf{x}-\mathbf{L}_i||^2}{||\mathbf{x}-\mathbf{L}_r||^2}$, that is,

$$u_{ir}\frac{||\mathbf{x}-\mathbf{L}_r||^2}{||\mathbf{x}-\mathbf{L}_i||^2} = p\left(\frac{||\mathbf{x}-\mathbf{L}_i||^2}{||\mathbf{x}-\mathbf{L}_r||^2}\right) \tag{9.130}$$

It can be seen immediately that the corresponding function $u(\cdot)$ is of the form $u(z) = zp(z)$. Thus, for eq. (9.127) we obtain the following result:

$$D_{\mathbf{x}} = ||\mathbf{x}-\mathbf{L}_i||^2\left[1+\sum_{r\neq i}^{c} p\left(\frac{||\mathbf{x}-\mathbf{L}_i||^2}{||\mathbf{x}-\mathbf{L}_r||^2}\right)\right] \tag{9.131}$$

The selection of membership functions of the form $u(z) = zp(z)$ implies that the winning prototype $\mathbf{L}_i$ is updated with respect to the input $\mathbf{x}$ by minimizing a weighted version of the squared Euclidean distance $||\mathbf{x}-\mathbf{L}_i||^2$. To satisfy the admissibility conditions for the membership function, the function $p(\cdot)$ has to satisfy the following properties [162,163]:

- $0 < p(z) < 1, \forall z \in (0,1)$.
- $p(z)$ approaches 1 as z approaches 0.
- $p(z)$ is a monotonically decreasing function in the interval $(0,1)$.
- $p(z)$ attains its minimum value at $z = 1$.

Several fuzzy learning vector quantization (FALVQ) algorithms can be determined based on minimizing the loss function (9.131).

The gradient of $D_{\mathbf{x}}$ with respect to the winning prototype $\mathbf{L}_i$ is

$$\begin{aligned}
\frac{\partial D_{\mathbf{x}}}{\partial \mathbf{L}_i} &= \frac{\partial}{\partial \mathbf{L}_i}\left[||\mathbf{x}-\mathbf{L}_i||^2 + \sum_{r\neq i}^{c}||\mathbf{x}-\mathbf{L}_i||^2 p\left(\frac{||\mathbf{x}-\mathbf{L}_i||^2}{||\mathbf{x}-\mathbf{L}_r||^2}\right)\right] \\
&= -2(\mathbf{x}-\mathbf{L}_i)\left\{1+\sum_{r\neq i}^{c} p\left(\frac{||\mathbf{x}-\mathbf{L}_i||^2}{||\mathbf{x}-\mathbf{L}_r||^2}\right)\right. \\
&\quad \left. +\frac{||\mathbf{x}-\mathbf{L}_i||^2}{||\mathbf{x}-\mathbf{L}_r||^2}p'\left(\frac{||\mathbf{x}-\mathbf{L}_i||^2}{||\mathbf{x}-\mathbf{L}_r||^2}\right)\right\}
\end{aligned} \tag{9.132}$$

The winning prototype $\mathbf{L}_i$ is adapted iteratively based on the following rule:

$$\Delta\mathbf{L}_i = -\eta'\frac{\partial D_\mathbf{x}}{\partial \mathbf{L}_i} = \eta(\mathbf{x}-\mathbf{L}_i)\left(1+\sum_{i\neq r}^{c} w_{ir}\right) \tag{9.133}$$

where $\eta = 2\eta'$ and

$$w_{ir} = p\left(\frac{||\mathbf{x}-\mathbf{L}_i||^2}{||\mathbf{x}-\mathbf{L}_r||^2}\right) + \frac{||\mathbf{x}-\mathbf{L}_i||^2}{||\mathbf{x}-\mathbf{L}_r||^2}p'\left(\frac{||\mathbf{x}-\mathbf{L}_i||^2}{||\mathbf{x}-\mathbf{L}_r||^2}\right) \tag{9.134}$$

Since $u(x) = xp(x)$, we obtain $u' = p(x) + xp'(x)$ and $w(x) = p(x) + xp'(x)$. Equation (9.134) can be similarly expressed as

$$w_{ir} = u'\left(\frac{||\mathbf{x}-\mathbf{L}_i||^2}{||\mathbf{x}-\mathbf{L}_r||^2}\right) = w\left(\frac{||\mathbf{x}-\mathbf{L}_i||^2}{||\mathbf{x}-\mathbf{L}_r||^2}\right) \tag{9.135}$$

For the nonwinning prototypes we obtain based on the minimization of $D_\mathbf{x}$

$$\begin{aligned}\frac{\partial D_\mathbf{x}}{\partial \mathbf{L}_j} &= \frac{\partial}{\partial \mathbf{L}_j}\left[||\mathbf{x}-\mathbf{L}_i||^2 + \sum_{r\neq i}^{c}||\mathbf{x}-\mathbf{L}_i||^2 p\left(\frac{||\mathbf{x}-\mathbf{L}_i||^2}{||\mathbf{x}-\mathbf{L}_r||^2}\right)\right]\\ &= -2(\mathbf{x}-\mathbf{L}_j)\left(\frac{||\mathbf{x}-\mathbf{L}_i||^2}{||\mathbf{x}-\mathbf{L}_j||^2}\right)^2 p'\left(\frac{||\mathbf{x}-\mathbf{L}_i||^2}{||\mathbf{x}-\mathbf{L}_j||^2}\right)\end{aligned} \tag{9.136}$$

The nonwinning prototype $\mathbf{L}_j \neq \mathbf{L}_i$ is adapted iteratively based on the following rule:

$$\Delta\mathbf{L}_j = -\eta'\frac{\partial D_\mathbf{x}}{\partial \mathbf{L}_j} = \eta(\mathbf{x}-\mathbf{L}_j)n_{ij} \tag{9.137}$$

where $\eta = 2\eta'$ and

$$n_{ij} = -\left(\frac{||\mathbf{x}-\mathbf{L}_i||^2}{||\mathbf{x}-\mathbf{L}_j||^2}\right)^2 p'\left(\frac{||\mathbf{x}-\mathbf{L}_i||^2}{||\mathbf{x}-\mathbf{L}_j||^2}\right) \tag{9.138}$$

Since $u'(x) = p(x) + xp'(x), x^2p'(x) = xu'(x) - xp(x) = xu'(x) - u(x)$. Thus we get for eq. (9.138) the following expression

$$\begin{aligned} n_{ij} &= u\left(\frac{||\mathbf{x}-\mathbf{L}_i||^2}{||\mathbf{x}-\mathbf{L}_j||^2}\right) - \frac{||\mathbf{x}-\mathbf{L}_i||^2}{||\mathbf{x}-\mathbf{L}_j||^2}u'\left(\frac{||\mathbf{x}-\mathbf{L}_i||^2}{||\mathbf{x}-\mathbf{L}_j||^2}\right)\\ &= n\left(\frac{||\mathbf{x}-\mathbf{L}_i||^2}{||\mathbf{x}-\mathbf{L}_j||^2}\right) = u_{ij} - \frac{||\mathbf{x}-\mathbf{L}_i||^2}{||\mathbf{x}-\mathbf{L}_j||^2}w_{ij}\end{aligned}$$

We also obtain $n(x) = -x^2p'(x) = u(x) - xu'(x)$.

The above strategy represents the framework for the derivation of fuzzy learning vector quantization algorithms as given in [163]. Table 9.1 shows the membership functions and interference functions $w(\cdot)$ and $n(\cdot)$ that generated three distinct fuzzy LVQ algorithms.

Table 9.1 Membership functions and interference functions for the FALVQ1, FALVQ2, and FALVQ3 families of algorithms.

Algorithm	$u(z)$	$w(z)$	$n(z)$
FALVQ1 $(0 < \alpha < \infty)$	$z(1+\alpha z)^{-1}$	$(1+\alpha z)^{-2}$	$\alpha z^2(1+\alpha z)^{-2}$
FALVQ2 $(0 < \beta < \infty)$	$z \exp(-\beta z)$	$(1-\beta z)\exp(-\beta z)$	$\beta z^2 \exp(-\beta z)$
FALVQ3 $(0 < \gamma < 1)$	$z(1-\gamma z)$	$1-2\gamma z$	γz^2

An algorithmic description of the FALVQ is given below:

1. **Initialization:** Choose the number c of prototypes, a fixed learning rate η_0, and the maximum number of iterations N. Set the iteration counter equal to zero, $\nu = 0$. Generate randomly an initial codebook $\mathbf{L} = \{\mathbf{L}_{1,0}, \ldots, \mathbf{L}_{c,0}\}$.
2. **Adaptation, part I:** Compute the updated learning rate $\eta = \eta_0\left(1 - \frac{\nu}{N}\right)$. Also set $\nu = \nu + 1$.
3. **Adaptation, part II:** For each input vector $\mathbf{x}$ find the winning prototype based on the equation

$$||\mathbf{x} - \mathbf{L}_{i,\nu-1}||^2 < ||\mathbf{x} - \mathbf{L}_{j,\nu-1}||^2, \quad \forall j \neq i \tag{9.139}$$

Determine the membership functions $u_{ir,\nu}$ using

$$u_{ir,\nu} = u\left(\frac{||\mathbf{x} - \mathbf{L}_{i,\nu-1}||^2}{||\mathbf{x} - \mathbf{L}_{r,\nu-1}||^2}\right), \quad \forall r \neq i \tag{9.140}$$

Determine $w_{ir,\nu}$ using

$$w_{ir,\nu} = u'\left(\frac{||\mathbf{x} - \mathbf{L}_{i,\nu-1}||^2}{||\mathbf{x} - \mathbf{L}_{r,\nu-1}||^2}\right), \quad \forall r \neq i \tag{9.141}$$

Determine $n_{ir,\nu}$ using

$$n_{ir,\nu} = u_{ir,\nu} - \left(\frac{||\mathbf{x} - \mathbf{L}_{i,\nu-1}||^2}{||\mathbf{x} - \mathbf{L}_{r,\nu-1}||^2}\right) w_{ir,\nu}, \quad \forall r \neq i \tag{9.142}$$

4. **Adaptation, part III:** Determine the update of the winning prototype $\mathbf{L}_i$ using

$$\mathbf{L}_{i,\nu} = \mathbf{L}_{i,\nu-1} + \eta(\mathbf{x} - \mathbf{L}_{i,\nu-1})\left(1 + \sum_{r \neq i}^{c} w_{ir,\nu}\right) \tag{9.143}$$

Determine the update of the nonwinning prototype $\mathbf{L}_j \neq \mathbf{L}_i$ using

$$\mathbf{L}_{j,\nu} = \mathbf{L}_{j,\nu-1} + \eta(\mathbf{x} - \mathbf{L}_{j,\nu-1})n_{ij,\nu} \tag{9.144}$$

5. **Continuation:** If $\nu = N$ stop, else go to step 2.

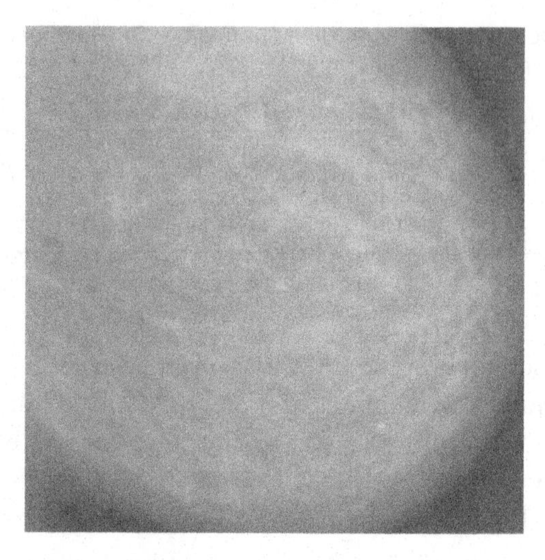

Figure 9.11 Original mammogram. *(Image from [281] reprinted with permission from Springer-Verlag.)*

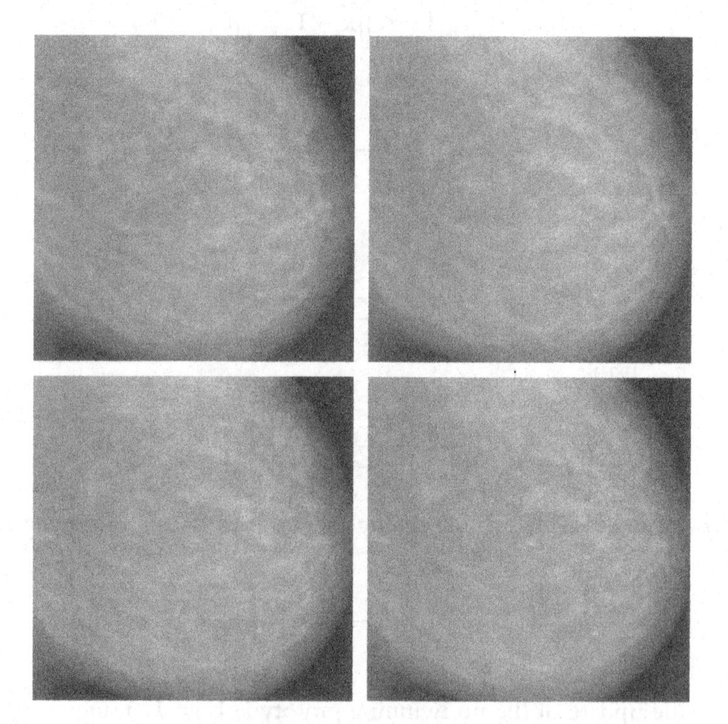

Figure 9.12 Reconstructed mammogram using FALVQ1, FALVQ2, FALVQ3, and LBG for codebook design. *(Images from [281] reprinted with permission from Springer-Verlag.)*

9.7.1 Medical Image Coding

In [281] is presented a novel image compression system designed for digital mammograms using wavelet-based image decomposition and vector quantization.

Codebook design was performed in the experiments by a family of FALVQ and the Linde-Buzo-Gray algorithm [218]. The FALVQ algorithms were tested against the LBG algorithm because of its implementation simplicity and satisfactory performance. The advantage of the FALVQ algorithms over the traditional LBG is that the initial codebook is generated randomly. However, the adaptation of both the winning and nonwinning prototypes is computationally extensive for the FALVQ algorithms. But this aspect can be controlled by the user, because the learning process terminates after a specified number of iterations. On average, the FALVQ and LBG algorithms are comparable in terms of their computational complexity and implementation simplicity. In [281] it was shown that the codebook design based on FALVQ algorithms outperformed the design based on the LBG algorithm as the length of the quantized vectors increased.

Figures 9.11 and 9.12 show an original mammogram and its reconstructed images based on a wavelet filter (Daubechies' 20 coefficient filter) and FALVQ1, FALVQ2, FALVQ3, and LBG used for codebook design. The original mammogram was compressed at 1.16 bpp. In the original mammogram, there is a very small black dot embedded inside a microcalcification called a sebaceous calcification. In each of these three images, the same area appeared to be identical.

9.8. EXERCISES

1. Prove eq. (9.22).
2. Prove the adaptive norm theorem for the generalized adaptive fuzzy n-means algorithm.
3. Prove for the generalized fuzzy n-shells algorithm both the theorem for optimal fuzzy partition and the theorem for optimal prototypes.
4. Prove the adaptive norm theorem for the generalized adaptive fuzzy n-shells algorithm.
5. Prove for the generalized adaptive fuzzy n-shells algorithm the propositions for optimal partition, for optimal prototype centers, and for optimal prototype radii.

Specialized Neural Networks Relevant to Bioimaging

Contents

10.1. INTRODUCTION

Neural networks have demonstrated a growing importance in the area of biomedical image processing and have been increasingly used for a variety of biomedical imaging tasks. The applications scan a wide spectrum: detection and characterization of disease patterns, analysis (quantification and segmentation), compression, modeling, motion estimation, and restoration of images from a variety of imaging modalities: magnetic resonance, positron emission tomography, ultrasound, radiography, mammography, and nuclear medicine. At the same time, traditional artificial neural networks and nonbiological

Pattern Recognition and Signal Analysis in Medical Imaging
http://dx.doi.org/10.1016/B978-0-12-409545-8.00010-8

image processing have demonstrated limited capabilities when applied to integrated medical imaging systems or to data analysis [78,94,161,171,190,263,328,394,397]. A better understanding of brain-based mechanisms of visual processing would potentially enable dramatic progress in clinical medical imaging. This fact has led to the development of new neural architectures derived from the basic architectures but with emphasis on incorporating visual processing mechanisms and cognitive decision-making [235].

In Chapter 7, we described the foundations of neural networks. This chapter presents some of the most important neural architectures relevant to bioimaging and their applications to several bioimaging tasks. In general, medical image patterns either possess a circular symmetric shape (e.g., nodules) or appear as small objects with a variety of geometric patterns (e.g., calcifications).

The applications in bioimaging considered most important and their underlying required neural architectures are as follows:

- Disease pattern classification: Based on convolutional neural networks [221,322], hierarchical pyramid neural networks [325], factorization MLPs, and radial basis neural networks [374].
- Medical image restoration: Based on a modified Hopfield neural network [170,279]
- Medical image segmentation: Based on RBF neural networks [410] and Hopfield neural networks with a priori image models [111].
- Tumor boundary detection: Based on a modified Hopfield neural network [408].
- Medical image coding and compression: Based on robust principal component analysis [160] and combined neural architectures.

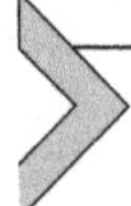

10.2. BASIC ASPECTS OF SPECIALIZED NEURAL NETWORK ARCHITECTURES

Specialized neural network architectures implement processing techniques relevant to most bioimaging applications, such as invariant and context-based classification, optimization, and elastic contour models.

10.2.1 Invariant Neural Networks

Since most medical image patterns either possess a circular symmetric shape or appear as small objects with a variety of geometric patterns, image pattern recognition does not call on top-down or left-right geometry as classification criteria. Therefore, one can take advantage of this characteristic as an invariance. In other words, we can rotate and/or shift the input vector two-dimensionally and maintain the same output assignments for the training. This method has two fundamental effects on the neural network:

- It trains the neural network that the rotation and shift of the input vector would receive the same classification result.
- It increases the training data and thus improves the performance of the neural network.

Neural networks in bioimaging operate directly on images without extracting features out of these images. It is known that the neighborhood correlation is usually higher than the long-distance correlation between two pixels on an image. It is conceivable that features associated with nearby pixels should be emphasized. In neural network terms, the local interactions should be included in the training algorithm rather than the nonlocal interactions. A good example of this type of approach is found in [221,322,406].

10.2.2 Context-Based Neural Networks

An important problem in image analysis is finding small objects in large images. This problem is challenging for bioimaging, because:

- Searching a large image is computationally intensive.
- Small objects, in most cases on the order of a few pixels in size, have relatively few distinctive features which enable them to be distinguished from nontargets.

By embedding multiresolution pyramid processing in a neural network architecture, we gain several advantages over traditional neural processing: (1) Both training and testing can be performed at a reduced computational cost based on a coarse-to-fine paradigm and (2) learning of low-resolution contextual information facilitates the detection of small objects. A good example of this type of approach is found in [325].

10.2.3 Optimization Neural Networks

Image segmentation can also be achieved using unsupervised clustering algorithms [122,30]. A typical algorithm, such as c-means [189], determines a class partitioning by minimizing the following objective function:

$$J = \sum_{i=1}^{d} ||\mathbf{x_i} - \mathbf{m}_{c(\mathbf{x_i})}||^2 \tag{10.1}$$

where $\mathbf{x_i}$ is the pattern vector, d the number of pattern vectors, $\mathbf{m}_{c(\mathbf{x_i})}$ the closest cluster centroid to $\mathbf{x_i}$, and C the number of classes. Using this type of algorithm, image segmentation can be formulated as an optimization problem. The Hopfield neural network model and its variants operate by minimizing an energy function. The topology of the Hopfield network employed in medical image segmentation can be visualized as a cube of nodes with width, depth, and height as shown in Fig. 10.1. The cube can be visualized as a three-dimensional grid of nodes where the number of nodes in the plane described by the width and depth corresponds to the pixels in the image, while the height of the cube, in terms of nodes, corresponds to the predefined number of clusters necessary for segmentation of this image. In order to minimize the energy function of the neural network, only one of the M neurons has to be active for each pixel. When convergence of the neural networks is achieved, the label represented by each pixel's active node was

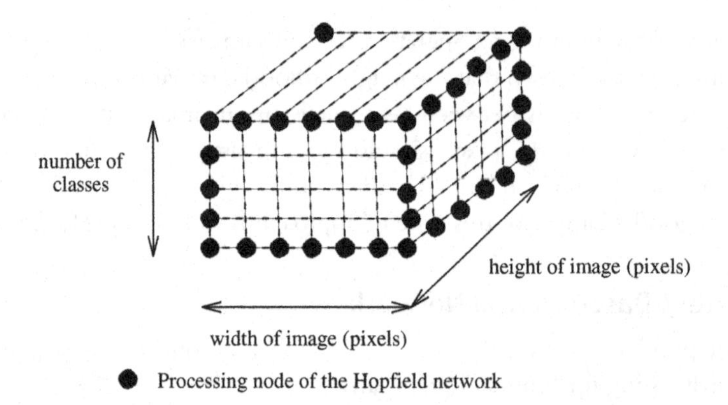

Figure 10.1 Image segmentation based on Hopfield network. The width and depth of the cube, in terms of nodes, corresponds to the number of pixels in a given image, while the height of the cube, in terms of nodes, corresponds to the prespecified number of clusters into which the image must be segmented.

assigned to the pixels to achieve a segmentation of the image. A good example of this type of approach is found in [12,111].

10.2.4 Active Contour Models

An important problem in bioimaging is to extract the boundary of an area of interest. The boundary detection problem can be formulated as an optimization problem based on an active contour model. The boundary is detected by seeking an active contour that minimizes an energy functional. The basic active contour model is an energy-minimizing spline [169] which is influenced by internal contour forces, image forces, and external forces. The internal forces serve to impose a piecewise smoothness constraint. The image forces push the contour toward salient image features such as lines, edges, and subjective contours such that it becomes a good fit of the image data features. The external forces are responsible for ensuring that the contour reaches the desired local minimum. The energy functional has to be carefully chosen so that it captures the relevant image properties. The goal is to achieve an optimized model such that a set of particular features in an image can be located. Two major factors dramatically influence the success of the elastic model: the initial position and the external energy. The assumed initial position is critical, since it has a dramatic impact on the convergence properties. Thus if it is not close to the desired features, the contour converges to the wrong solution. A robust solution can be achieved by employing a Hopfield neural network as shown in [408]. The most relevant applications for active contour models can be found in PET and MRI [161,173].

10.3. CONVOLUTION NEURAL NETWORKS (CNNs)

Convolution neural networks represent a well-established method in medical image processing [221,322]. Although it has a multilayer architecture like the MLP, when applied to image classification the CNN works directly with images and not with extracted features like the MLP. The CNN approach is based on the idea of neighborhood correlation: The correlation of adjacent pixels is usually higher than of nonadjacent. Therefore, it makes sense to emphasize features associated with adjacent pixels. In neural network terms, the local interactions should be included in the training algorithm rather than the nonlocal interactions. The basic structure of a CNN is shown in Fig. 10.2, which represents a four-layer CNN with two input images, three image groups in the first hidden layer, two groups in the second hidden layer, and a real-valued output [322]. The number of layers and the number of groups in each layer are implementation-oriented. The more complicated the disease patterns, the more layers are required to distinguish high-order information of image structures. The idea behind the convolution kernels is to emphasize important image characteristics rather than those less correlated values obtained from feature spaces for input. These characteristics are: (1) the horizontal versus vertical information; (2) local versus nonlocal information; and (3) the image processing (filtering) versus signal propagation.

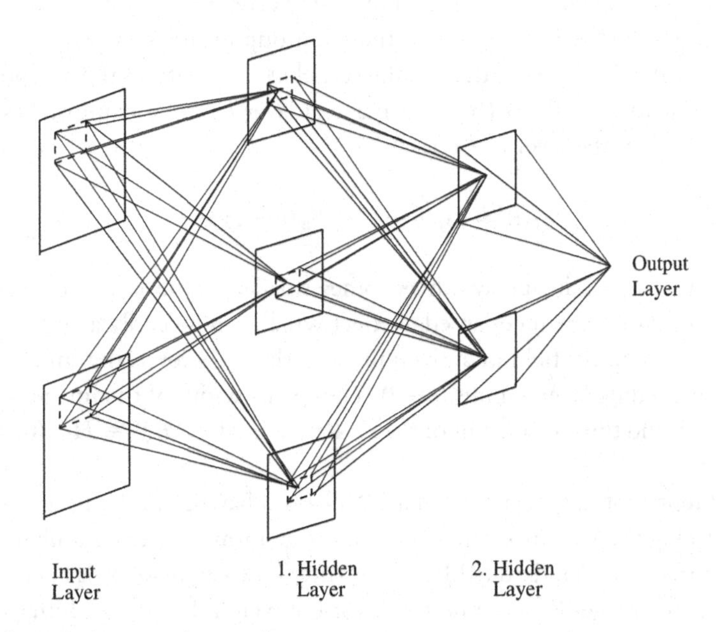

Figure 10.2 Convolution neural network.

The image propagates through the neural network from input layer to output layer by means of convolution with trainable weight kernels. In the following, we will describe the forward and backward propagation through the CNN.

10.3.1 Forward Propagation

Forward propagation pertains to the image propagation in the CNN from the input layer ($l = 1$) to the output layer ($l = L$) [322]. Let $H_{l,g}$ define the gth image group at layer l, and let $N(l)$ describe the number of such groups. The image $H_{l,g}(l \geq 2)$ is determined by applying a pointwise sigmoid nonlinearity to an intermediate image $I_{l,g}$, that is,

$$H_{l,g}(i,j) = \frac{1}{1 + \exp(-I_{l,g}(i,j))}, \quad g = 1, \ldots, N(l) \tag{10.2}$$

The intermediate image $I_{l,g}$ represents a superposition of the images obtained from the convolution of $H_{l-1,g'}$ at layer $l-1$ with trainable kernel of weights $w_{l-1,g,g'}$. Specifically, we obtain $I_{l,g}$ from

$$I_{l,g} = \sum_{g'=1}^{N(l-1)} H_{l-1,g'} {**} w_{l-1,g,g'} \tag{10.3}$$

where $**$ defines a 2-D convolution with the 2-D kernel $w_{l-1,g,g'}$ of weights connecting the g'th group in the $(l-1)$th layer with the gth group in the lth layer.

The spatial width $S_w(l-1)$ of the weight kernel $w_{l-1,g,g'}$ defines the receptive field for the layer l. The spatial width $S_H(l)$ of an image at layer l can be expressed as a function of the image width at the layer $l-1$ as

$$S_H(l) = S_H(l-1) - S_w(l-1) + 1 \tag{10.4}$$

It is evident that the image width becomes smaller as the layer number increases. However convolution introduces an edge effect which can be easily avoided by defining the width of the receptive field of a given node in the lth layer as the sum of the kernel widths of the preceding layers minus $(l-2)$. The spatial width of the image at the output layer ($l = L$) is 1, and thus the output of the CNN, defined as $O(g) \equiv H_{L,g}(0,0)$, becomes a real number.

The MLP represents a special case of a CNN, both having similar fundamental equations. By replacing the weight kernels and the image groups by real numbers, the CNN architecture turns into a standard MLP architecture. The weight kernels become ordinary weights, while the images become nodes. If for the weight kernels and image groups in a CNN we substitute real numbers, then we get ordinary MLP weights for the weight kernels and nodes for the images.

10.3.2 Backpropagation

For the training of the CNN, we employ the backpropagation algorithm. To each training image p (or set p of training images in case the input layer processes more than one image) there is a correspondent desired-output value $O_d^{(p)}(g)$, where $g = 1, \ldots, N(L)$ describes the output node number. As in case of the MLP, for each iteration t, training images are presented in a random fashion to the CNN and the resulting CNN outputs $O_a^{(p)}[t]$ are determined based on eqs. (10.2) and (10.3). The resulting CNN output error for a given training image p at iteration t is given by

$$E^p[t] = \frac{1}{2}\sum_{g=1}^{N(L)}(O_d^{(p)}(g) - O_a^{(p)}(g)[t])^2 \tag{10.5}$$

while the total CNN error at iteration t is given by

$$E[t] = \sum_{p=1}^{P} E^{(p)}[t] \tag{10.6}$$

where p represents the total number of training samples.

It is easy to see that for a CNN, the weight adjustment is based on a backpropagation process. [322] gives the derivation of the backpropagation algorithm for the CNN.

The kernel weights $w_{l,g,g'}$ are derived by the generalized delta rule

$$w_{l,g,g'}(i,j)[t+1] = w_{l,g,g'}(i,j)[t] - \eta\Delta w_{l,g,g'}(i,j)[t] \tag{10.7}$$

with

$$\Delta w_{l,g,g'}(i,j)[t] = \frac{\partial E[t]}{\partial w_{l,g,g'}(i,j)[t]} \tag{10.8}$$

The initialization process for the CNN is based on assigning the kernel weights and the other weighting factors a normalized random number. For the simulations, it is necessary to normalize the pixel values of the training and test images such that a 1.0 is assigned to the highest pixel value, and a 0 to the lowest pixel value.

The two-dimensional convolution operation represents an emulation of the radiologists' viewing of a suspected area, while the output side models their decision-making process. The neural network is trained based on a backpropagation algorithm such that it extracts from the center and the surroundings of an image block relevant information describing local features.

Example 10.3.1 An application of a CNN to mammograms is shown in [222]. The mammograms were digitized with a computer format of $2048 \times 2500 \times 12$ bits per image. The study pertains only to microcalcification detection and utilizes only the central region of 16×16 pixels as an input to evaluate the performance of four different neural networks:

Table 10.1 Performance of neural networks in the detection of clustered microcalcifications using group A_m as training set and group B_m as testing set [222].

Neural networks	**DYSTAL**	**BP/0H**	**BP/1H**	**CNN**
A_Z (area under the ROC curve)	0.78	0.75	0.86	0.97
Detection accuracy				
% true-positive detection / # false-positive per image	70 / 4.3	70 / 4.5	75 / 3.5	90 / 0.5

Table 10.2 Performance of neural networks in the detection of clustered microcalcifications using group B_m as training set and group A_m as testing set [222].

Neural networks	**DYSTAL**	**BP/0H**	**BP/1H**	**CNN**
A_Z (area under the ROC curve)	0.76	0.77	0.84	0.90
Detection accuracy				
% true-positive detection / # false-positive per image	70 / 4.3	70 / 4.2	75 / 3.7	90 / 0.5

dynamic stable associate learning (DYSTAL) [7], conventional backpropagation neural network with (BP/1H) and without (BP/0H) a hidden layer, and the CNN. The DYSTAL uses the winner-take-all approach of propagating maximum similarity and employs a threshold value for deciding if the maximum similarity of a new pattern vector is less than this value. If so, the new pattern becomes the representative of a false or true disease class.

Thirty-eight digital mammograms including 220 true and 1132 false subtle microcalcifications were considered for the study. The mammograms were divided into two sets: A_m, 19 images (containing 108 true and 583 false image blocks), and B_m, another set of 19 images (containing 112 true and 549 false image blocks). The central region of 16×16 pixels was considered as an input for the CNN, while the convolution kernel had the size of 5×5. This is perfectly sufficient since the microcalcifications are very small compared to other diseases. Tables 10.1 and 10.2 show the performance resulting from the three neural systems [222].

The best A_Z of 0.97 was achieved by the CNN, thus demonstrating the superiority of CNN over the other classifiers.

10.4. HIERARCHICAL PYRAMID NEURAL NETWORKS

One of the most challenging tasks in biomedical imaging is to find small disease patterns in large images. A special architecture, the so-called hierarchical pyramid neural network, has proved to be superior to both conventional neural network architectures and

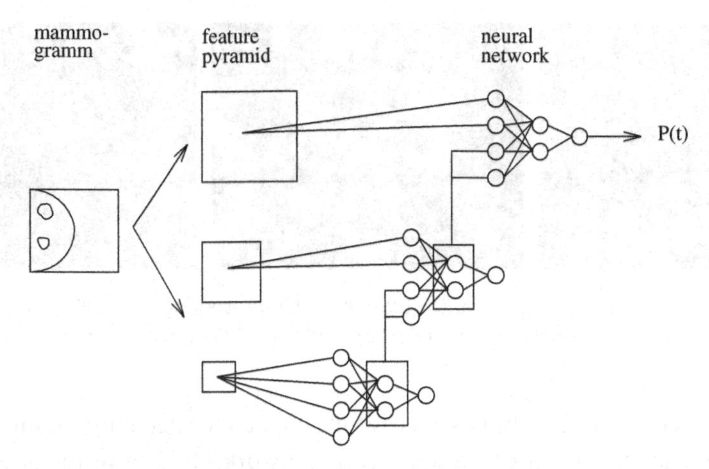

Figure 10.3 Hierarchical pyramid neural network architecture for learning context. The hidden units of low-resolution networks propagate hierarchically the context, such that the output of the highest resolution network is a probability estimator for the presence of a target [325].

standard statistical classification techniques [325]. By embedding multiresolution pyramid processing in a neural network architecture, several advantages over traditional neural processing can be gained: (1) both training and testing can be performed at a reduced computational cost based on a coarse-to-fine paradigm and (2) learning of low-resolution contextual information enables the detection of small objects.

The hierarchical pyramid neural network architecture is shown in Fig. 10.3.

A hierarchy of networks is trained in a coarse-to-fine paradigm on a feature set derived from a pyramid decomposition of the image. Low-resolution networks are first trained to detect small objects, which might be absent at this level of the pyramid. In order to detect small objects, low-resolution networks must learn the context in which these small objects exist. To include the context information with the feature information of the small objects, the outputs of the hidden units from the low-resolution networks are propagated hierarchically as inputs to the corresponding higher resolution neural networks. The neural nets in the HPNN are multilayer perceptrons, having one hidden layer with four hidden units. The inputs to the networks are features at different levels of an image pyramid with outputs $P(t)$, representing the probability that a target is present at a given location in the image.

The error function used for the training is chosen as a cross-entropy error:

$$E = -\sum_{i} d_i \log y_i - (1 - d_i) \log(1 - y_i) \tag{10.9}$$

where $d \in \{0, 1\}$ represents the desired output. To avoid overtraining, a regularization term of the form $r = \frac{\lambda}{2} \sum_i w_i^2$ is added to the total error on the training examples. λ is adjusted to minimize the cross-validation error.

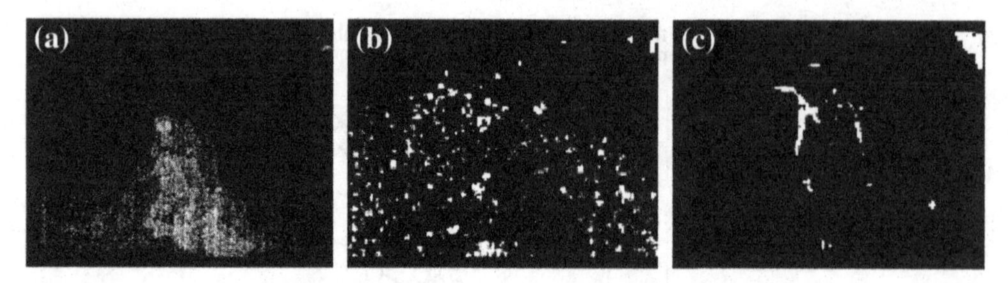

Figure 10.4 (a) Digitized mammogram, (b) hidden unit representing point-like structure, and (c) hidden unit representing elongated structure. *(Reprinted from [325] with permission from SPIE.)*

Example 10.4.1 The HPNN was successfully applied for detection of microcalcifications [325]. It outperforms the Chicago neural network [405] in terms of its ability to reduce the number of false-positive ROIs generated by the computer.

Figure 10.4 [325] shows how the system takes advantage of the context information by looking at the representations developed by various hidden units in the network. Two classes of hidden units were found. The first one shown in Fig. 10.4b describes point-like structure similar to that of individual microcalcifications. The second class of hidden units in Fig. 10.4c represents ductal location and appears to be tuned for longer, extended, and oriented structures. This example proves the HPNN's ability to automatically extract and process context information.

10.5. PROBLEM FACTORIZATION

The complexity of training can be dramatically reduced by using the problem factorization approach. This is a collection of small interconnected MLPs such that a factorization of the classification task is achieved by considering separate classification stages. The information is thus processed hierarchically. Also an overtraining can be avoided based on this method.

Example 10.5.1 The problem factorization approach can be used for the classification of candidate tumor regions. In [80] they are employed to classify clusters of regions that correspond to acoustic neuromas. Figure 10.5 shows a collection of three small interconnected MLPs that reflect a factorization of the classification task into three separate stages. The first two stages capture only the shape and the intercluster position (ICP), with each of them having four distinct features. The final stage combines the results from the previous two stages with two additional features, namely compactness and global angle.

By factorizing into the three stages, the classifier was reduced to three MLPs that were of lower complexity than a single MLP. Each of the three MLPs were trained separately. The data available constrain each of the three MLPs to a greater extent than single MLPs are constrained. Therefore, overfitting using a hierarchy of MLPs is highly

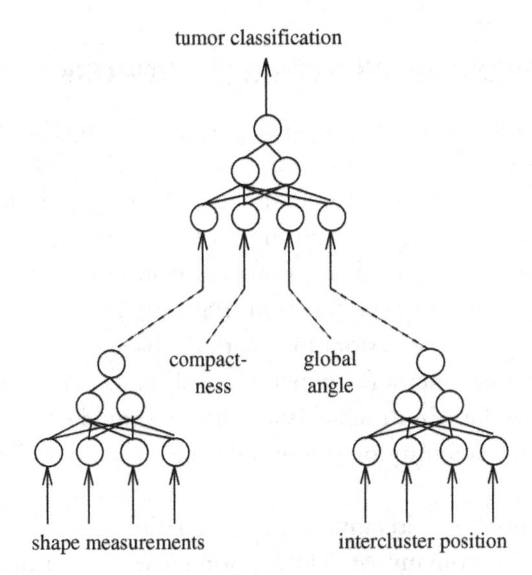

Figure 10.5 Classifier structure of the problem factorization approach to candidate tumor region combination classification.

Table 10.3 The best performance of the three types of classifiers against the testing data set. The training scheme employed was both early stopping as well as no early stopping.

Classifier structure	No early stopping	Early stopping
Single MLP	6.67% (5.71%)	0% (17.14%)
Committee of MLPs	13.33% (5.71%)	6.67% (14.29%)
Factorized MLPs:		
... shape	0% (20.00%)	6.67% (17.14%)
... Intercluster Position	0% (17.14%)	0% (22.86%)
... Final	0% (5.71%)	0% (5.71%)

Values not in parentheses are the error rates against tumor regions that correspond to acoustic neuromas, and values in parentheses are the error rates against tumor regions that did not correspond to acoustic neuromas [80].

reduced compared to use of a single larger MLP. For the simulations reported in [80], the data available were split into only two data sets, one for the training and the other for testing. The training data set consisted of 35 tumor regions and 70 nontumor. The testing data set contained 15 tumor regions and 35 nontumors. Table 10.3 shows a performance comparison among three different types of MLP architectures: single MLP versus committee of MLPs and versus factorized MLPs. The classifier that achieved the best results used the problem factorization approach.

10.6. MODIFIED HOPFIELD NEURAL NETWORK

The problem of restoring noisy-blurred images is important for many medical image applications, especially in confocal microscopy [92, 170, 357].

The goal of image restoration is to "undo" the blurring and noise artifacts that are imposed on the image during image acquisition. In other words, image-restoration algorithms try to restore the "original" image from the acquired image by deconvolving the blurring imposed by the point spread function (PSF) and by reducing the noise imposed by the image recording. Linear restoration methods have a major drawback: They are incapable of restoring frequencies for which the PSF has a zero response. Furthermore they cannot restrict the domain in which the solution should be found. This property is a major drawback as the intensity of an imaged object represents light energy, which is nonnegative.

For these reasons, nonlinear iterative image-restoration algorithms are the best candidates for restoring images contaminated by Poisson noise. Nonlinear iterative algorithms mostly follow two approaches: either the constrained Tikhonov restoration approach or the maximum likelihood restoration approach. They all tackle the above-mentioned problems in exchange for a considerable increase in computational complexity. To overcome both the increase in complexity and the linear methods' drawbacks, neural-network-based image-restoration algorithms were proposed and successfully applied to image-restoration problems in many application fields.

Often, the image-degradation can be adequately modeled by a linear blur and an additive white Gaussian process. Then the degradation model is given by

$$\mathbf{z} = \mathbf{D}\mathbf{x} + \eta \tag{10.10}$$

where $\mathbf{x}$, $\mathbf{z}$, and η represent the original and degraded images and the additive noise. The matrix $\mathbf{D}$ describes the linear spatially invariant or spatially varying distortion. Specifically in confocal microscopy, the blurring matrix represents the PSF of the microscope.

The purpose of digital image restoration is to invert the degradation process by turning the degraded image $\mathbf{z}$ into one that is as close to the original image $\mathbf{x}$ as possible, subject to a suitable optimality criterion. A common optimization problem is:

$$\text{minimize} \quad \mathbf{f}(\mathbf{x}) = \frac{1}{2}\mathbf{x}^{\mathrm{T}}\mathbf{T}\mathbf{x} - \mathbf{b}^{\mathrm{T}}\mathbf{x} \tag{10.11}$$

$$\text{subject to} \quad 0 \leq x_i \leq 255, \quad i = 1, \ldots, n \tag{10.12}$$

where x_i denotes the ith element of the vector $\mathbf{x}$, $\mathbf{b} = \mathbf{D}^{\mathrm{T}}\mathbf{z}$, and $\mathbf{T}$ is a symmetric, positive semidefinite matrix equal to

$$\mathbf{T} = \mathbf{D}^{\mathrm{T}}\mathbf{D} + \lambda\mathbf{C}^{\mathrm{T}}\mathbf{C} \tag{10.13}$$

In eq. (10.13), $\mathbf{C}$ represents a highpass filter and λ the regularization parameter, describing the interplay between deconvolution and noise smoothing.

Comparing eqs. (10.11) and (10.13), it becomes evident that looking for the minimum of the function $f(\mathbf{x})$ for the restoration problem corresponds to finding the minimum of the energy function E_{hs} of the Hopfield neural network, if we choose $\mathbf{W} = -\mathbf{T}$ and $\mathbf{x} = \mathbf{v}$.

The Hopfield neural network operates with binary patterns, and in order to employ it for image restoration, we have to transform the image gray values into binary state variables. A solution proposed in [170] is to represent them as a simple sum of the binary neuron state variables. The main disadvantage of this strategy is the resulting tremendous storage and number of interconnections. This problem can be overcome by defining a nonelementary neuron that takes discrete values between 0 and 255, instead of binary values [279]. In this case, the interconnections are based on pixel locations and not on gray values.

10.6.1 Updating Rule:

The modified Hopfield network for image restoration that was proposed in [279] is shown in Fig. 10.6 and is given by the following equations:

$$x_i(t+1) = g(x_i(t) + \Delta x_i), \quad i = 1, \ldots, n \tag{10.14}$$

where

$$g(v) = \begin{cases} 0, & v < 0 \\ v, & 0 \leq v \leq 255 \\ 255, & v > 255 \end{cases} \tag{10.15}$$

$$\Delta x_i = d_i(u_i) = \begin{cases} -1, & u_i < -\theta_i \\ 0, & -\theta_i \leq u_i \leq \theta_i \\ 1, & u_i > \theta_i \end{cases} \tag{10.16}$$

with

$$\theta_i = \frac{1}{2} t_{ii} > 0 \quad \text{and} \quad u_i = b_i - \sum_{j=1}^{n} t_{ij} x_j(t) \tag{10.17}$$

The degraded image $\mathbf{z}$ is used as the initial condition for $\mathbf{x}$. x_i are the states of neuron that take discrete values between 0 and 255, instead of binary values. The symbol s_i denotes a switch, which describes either a sequential or asynchronous updating mode.

In the following, an algorithm is presented that sequentially updates each pixel value according to the updating rule. Let $l(t)$ denote a partition of the set $\{1, \ldots, n\}$. The algorithm has the following form:

Algorithm 2

1. $\mathbf{x}(0) = \mathbf{D}^T\mathbf{z}$; $t := 0$ and $i := 1$.
2. Check termination.
3. Choose $l(t) = \{i\}$.

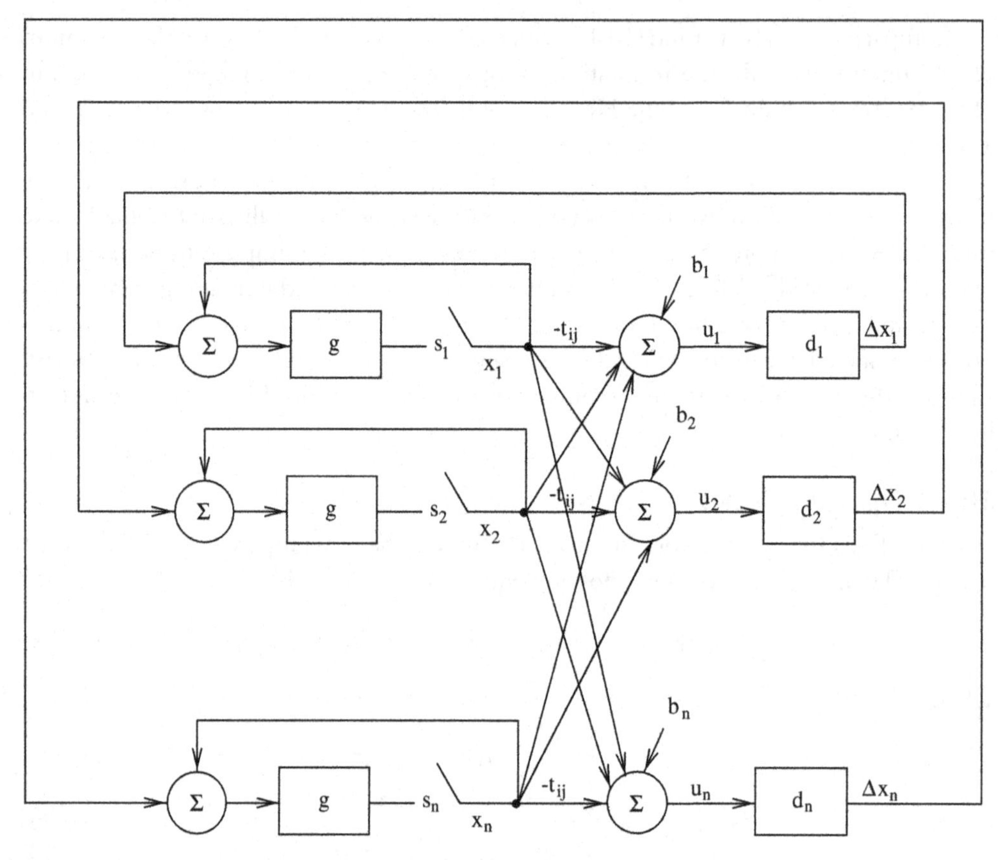

Figure 10.6 Block diagram of the modified Hopfield network model applied to image restoration.

4. $temp = g(x(t) + \Delta x_i e_i)$ where Δx_i is given by eq. (10.16).
5. If $temp \neq x(t)$ then $x(t+1) := temp$ and $t := t+1$.
6. $i; = i+1$ (if $i > n, i = i - n$) and go to step 1.

In step 3 of the preceding algorithm, the function $g(.)$ is applied separately to each component of its argument vector. In this case $g(\mathbf{x}) = [g(x_1), \ldots, g(x_n)]$, where $g(x_i)$ is defined by eq. (10.15.)

Example 10.6.1 Modified Hopfield neural networks are often applied in nuclear medicine image restoration [305]. Figure 10.7 shows an application of the algorithm to the image-restoration problem by using an artificial phantom source image study. An image of 50 × 50 pixels in size represents a cylindrical object with a diameter of 7 pixels as reported in [305]. This image simulates a cylindrical source of activity. The blurred image is restored by a modified Hopfield neural network.

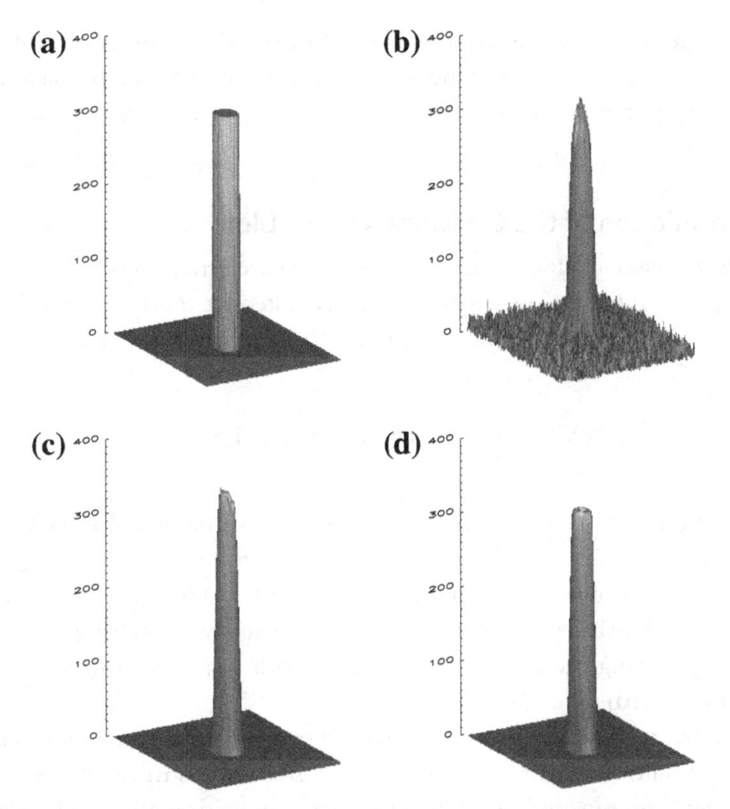

Figure 10.7 Image restoration based on a modified Hopfield neural network using artificial data. (a) Original image with a signal simulating a cylindrical source of activity, (b) blurred image by the PSF and additive noise, and (c) restoration result achieved by the modified Hopfield neural network. *(Reprinted from [305] with permission from Kluwer.)*

10.7. HOPFIELD NEURAL NETWORK USING A PRIORI IMAGE INFORMATION

The Hopfield neural network can also be applied to image segmentation, if the segmentation is posed as an optimization problem. In [111] a general framework for the design of the optimization criterion is proposed consisting of two main parts: one based on the observed image, and another based on an a priori image model. The approach exploits a well-known fact in statistical image segmentation technique, namely that neighboring pixels in an image are "similar" in some sense. Therefore, by not only using information about observed data but also including the local spatial interaction between pixel classes, a gain in segmentation results can be achieved.

The pixel classification problem can be viewed as an optimization problem: A feature vector, containing the gray values and additional data such as edge information and local correlation values, describes each discrete pixel location within an $N \times N$ image. Let $\mathbf{p}^i$ with $1 \leq i \leq N^2$ be such an M-dimensional feature vector for each image pixel. If we assume L a priori known underlying classes, then each pixel belongs to one of these L classes.

10.7.1 Formulation of the Classification Problem

Given a known number of classes L, assign each pixel of the observed image to one of these classes.

Since a priori information has to be incorporated in the image models, a special optimization function for the L-class segmentation problem is defined:

$$J(X) = \sum_{i=1}^{N^2} J(\mathbf{p}^i, \mathbf{m}_1, \ldots, \mathbf{m}_L, \mathbf{x}_i) - \lambda f(X) \tag{10.18}$$

with $X = \{\mathbf{x}_1, \ldots, \mathbf{x}_{N^2}\}$ being the set of binary L-dimensional class vectors $\mathbf{x}_i$. In addition, we impose that $x_{il} \in \{0, 1\}$ and $\sum_l x_{il} = 1$ for $1 \leq i \leq N^2, 1 \leq l \leq L$. $\mathbf{m}_l$ with $l = 1, \ldots, L$ represent the cluster centers. $f(X) = f(\{\mathbf{x}_1, \ldots, \mathbf{x}_{N^2}\})$ describe the prior available information about the underlying image, with λ being a regularization parameter emphasizing the a priori information with respect to the data-independent term $J(\mathbf{p}^i, \mathbf{m}_1, \ldots, \mathbf{m}_L, \mathbf{x}^i)$.

The minimum of the function $J(X)$ represents the optimal solution of the classification problem. As a result, we obtain the required L cluster centers $\mathbf{m}_l$ with $1 \leq l \leq L$.

The optimization function, in its general form, can be simplified and becomes at the same time more feasible for applications. A good choice is [111]

$$J(X) = \sum_{l=1}^{L} ||\mathbf{p}_i - \mathbf{m}_l||_2^2 x_{il} + \lambda \sum_{j \in \eta_i} \mathbf{x}_j^T \mathbf{x}_i \tag{10.19}$$

η_i is the set of all pixels which are neighbors of pixel i. A first-order neighborhood contains the four nearest neighbors of a pixel.

10.7.2 Algorithm for Iterative Clustering

The number of neurons in the Hopfield neural network corresponds to the number of pixels in the image. The energy of an $N \times N$-neuron Hopfield neural network is defined as

$$E = -\sum_{i=1}^{N^2} \sum_{j=1}^{N^2} w_{ij} x_i(t) x_j(t) - \sum_{i=1}^{N^2} b_i x_i(t) \tag{10.20}$$

w_{ij} are the weights, and x_i is the state of the ith neuron. b_i describes the bias input to the ith neuron.

The energy of a stable Hopfield neural network is decreasing over time. This basic fact can be used for solving the L-class pixel classification problem based on eq. (10.18). The weights and the bias inputs can be determined from eqs. (10.18), (10.19), and (10.20):

$$w_{ij} = 2\lambda, \quad \forall j \in \eta_i,\ i \neq j \tag{10.21}$$

and

$$b_l^i = \frac{1}{L}\sum_{l=1}^{L} \delta_l^i - \frac{8\lambda}{L} - \delta_l^i \tag{10.22}$$

with $\delta_l^i \equiv ||(\mathbf{p}_i - \mathbf{m}_l)||_2^2$.

The activation function is determined by

$$x_i = \begin{cases} 1, & \text{if} \quad \tau_i > 0 \\ 0, & \text{if} \quad \tau_i \leq 0 \end{cases} \tag{10.23}$$

where τ_i is given by eq. (7.66).

The *optimization algorithm of the Hopfield neural network using a priori image information* is iterative and described as follows [111]:

Algorithm 3

1. **Initialization:** Choose random values for the cluster centers $\mathbf{m}_l$ and the neuron outputs x_i.
2. **Forward computation part I:** At each iteration k and for each neuron i compute: (a) the input to the neuron using eqs. (10.21) and (10.22) and (b) the new state based on eq. (10.23).
3. **Forward computation part II:** If $x_i(k) \neq x_i(k-1) \forall i$ go to step (2), else go to step (4).
4. **Adaptation:** Compute new cluster centers $\{\mathbf{m}_l\}$ using $x_i(k)$, with $i = 1, \ldots, N^2$. Go to step (2).
5. **Continuation:** Repeat until the cluster centers do not change.

Example 10.7.1 The neural model was applied in [111] to segment masses in mammograms. Each pixel of the ROI image describing extracted masses belongs to either the mass or the background tissue and defines such a two-class classification problem. The gray levels of the pixels are used as the input feature. Figures 10.8 and 10.9 show the segmentation results obtained with a Hopfield network without ($\lambda = 0$) and with a priori information ($\lambda \neq 0$). As expected, including a priori information yields a smoother segmentation compared to $\lambda = 0$.

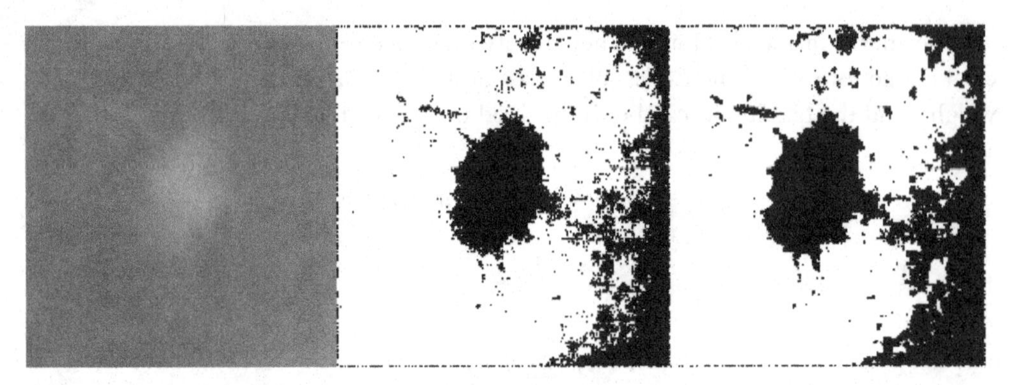

Figure 10.8 Segmentation results of a two-class classification problem: (left) original ROI, (center) segmentation result using a neural network with $\lambda = 0$, (right) segmentation result using a neural network with a priori information $\lambda = 1$. *(Reprinted from [111] with permission from IEEE.)*

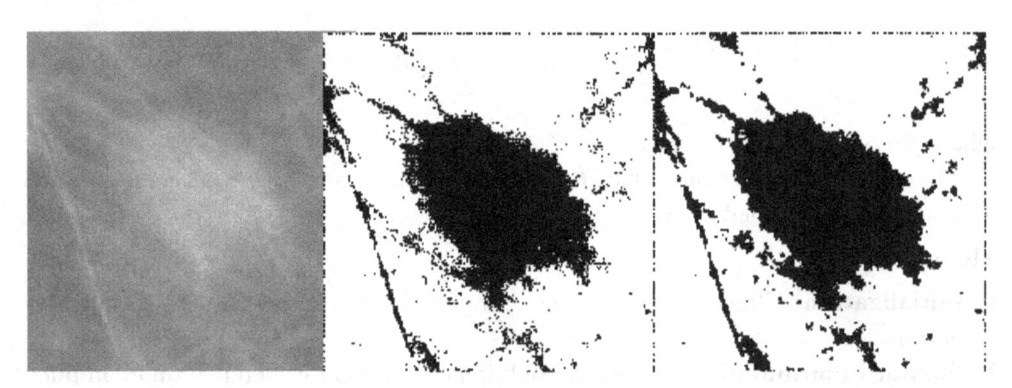

Figure 10.9 Segmentation results of a two-class classification problem: (left) original ROI, (center) segmentation result using a neural network with $\lambda = 0$, (right) segmentation result using a neural network with a priori information $\lambda = 2$. *(Reprinted from [111] with permission from IEEE.)*

10.8. HOPFIELD NEURAL NETWORK FOR TUMOR BOUNDARY DETECTION

The extraction of tumor boundaries in 3-D images poses a challenging problem for bioimaging. In MRI, they provide useful insights into the internal structures of the human body. Several attempts have been made to apply neural network architectures to brain image analysis [12,57,408]. One of the most efficient approaches is described in [408]. It yields results comparable to those of standard snakes, and at the same time, it is less computationally intensive. The flow diagram in Fig. 10.10 describes the implemented diagram from [408]. In the first step, the image is enhanced; and the second step involves

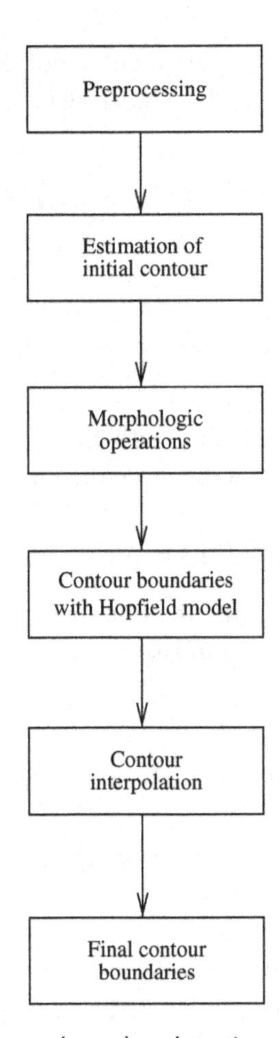

Figure 10.10 Block diagram of the tumor boundary detection method.

initial boundary detection and searching grids estimation; while the last step detects the tumor boundaries using the Hopfield neural network.

The boundary detection is also formulated as an optimization problem that seeks the boundary points to minimize an energy functional based on an active contour model. A modified Hopfield network is constructed to solve the optimization problem.

10.8.1 Mathematical Model of the Active Contour

The basic active contour model is an energy-minimizing spline. For any point $v(s) = (x(s), y(s))$ of the contour, an energy function can be defined as [169]

$$E = \int_0^1 \alpha||v'(s)||^2 + \beta||v''(s)||^2 - P(v(s))ds \tag{10.24}$$

The first two terms describe the internal energy of the contour, while the final term refers to the image forces. The first- and second-order derivatives of $v(s)$ are approximated as follows:

$$\left|\frac{dv_i}{ds}\right|^2 \approx |v_i - v_{i-1}|^2 = (x_i - x_{i-1})^2 + (y_i - y_{i-1})^2 \tag{10.25}$$

and

$$\left|\frac{d^2 v_i}{ds^2}\right|^2 \approx |v_{i-1} - 2v_i + v_{i+1}|^2 = (x_{i-1} - 2x_i + x_{i+1})^2 + (y_{i-1} - 2y_i + y_{i+1})^2 \tag{10.26}$$

x_i and y_i are the x- and y-coordinates of the ith boundary point v_i. In [408], the contour has to be attracted to edge points, and therefore the image forces depend on the gradient $g(v_i)$ of the image at point v_i

$$P(v_i) = -g(v_i) \tag{10.27}$$

The basic idea proposed in [408] is that the boundary is detected by iteratively deforming an initial approximation of the boundary through minimizing the following energy functional:

$$\begin{aligned} E_{Snake} = \sum_{i=1}^{N} \{ & \alpha[(x_i - x_{i-1})^2 + (y_i - y_{i-1})^2] \\ & + \beta[(x_{i-1} - 2x_i + x_{i+1})^2 + (y_{i-1} - 2y_i + y_{i+1})^2] \\ & - \gamma g_i \} \end{aligned} \tag{10.28}$$

Before employing the active contour model, an initial contour must be estimated. In [408], the contour for the first slice is estimated based on morphological operations.

10.8.2 Algorithm for Boundary Detection

Contour detection in an image is achieved by employing a binary Hopfield neural network that implements a two-dimensional $N \times M$ neuronal lattice. N is equal to the number of sampling points of the initial contour, while M equals the number of grid points along each grid line. Each neuron is described by a point (i, k) with $1 \leq i \leq N$ and $1 \leq k \leq M$ and is considered as a hypothetical boundary point. Each neuron output is either equal to 1 (shows presence of boundary elements) or equal to 0 (shows absence of boundary elements), see Fig. 10.11.

The energy function of this special Hopfield neural network is

$$E = -\frac{1}{2} \sum_{i}^{N} \sum_{k}^{M} \sum_{j}^{N} \sum_{l}^{M} w_{ikjl} x_{ik} x_{jl} - \sum_{i}^{N} \sum_{k}^{M} b_{ik} x_{ik} \tag{10.29}$$

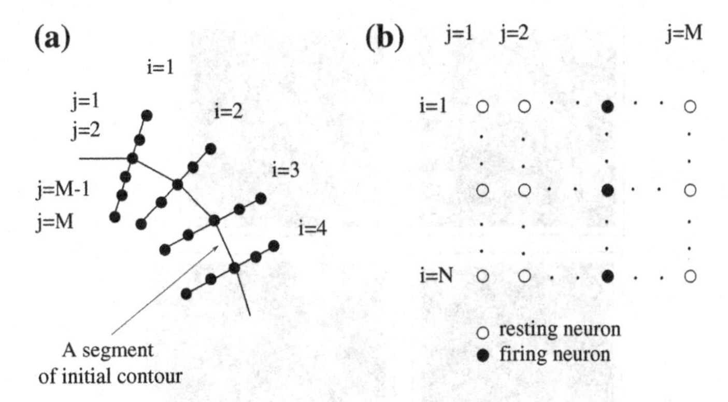

Figure 10.11 Correspondence between grid points and network architecture [408]. (a) Segment of the initial contour and search grid points and (b) network row-column assignment.

x_{ik} is the neuron state at point (i, k), and w_{ikjl} is the synaptic weight connecting a neuron at point (i, k) with one at point (j, l). The propagation rule for the neuron at (i, k) is

$$x_{ik}(t+1) = \sum_{j}^{N} \sum_{l}^{M} w_{ikjl} x_{jl}(t) + b_{ik} \tag{10.30}$$

and the corresponding activation function is

$$x_{ik} = \begin{cases} 1, & \text{if} \quad x_{ik} = \max(x_{ih};\, h = 1, \ldots, M) \\ 0, & \text{else} \end{cases} \tag{10.31}$$

Equation (10.31) constrains the M mutually exclusive neurons representing a boundary point such that only one neuron in each row is firing while the others are resting.

By comparing eqs. (10.28) and (10.29), the connection weights w_{ikjl} and neurons inputs b_{ik} are determined by

$$w_{ikjl} = \begin{cases} -[(4\alpha + 12\beta) - (2\alpha + 8\beta)\delta_{i+1j} \\ \quad - (2\alpha + 8\beta)\delta_{i-1j} + 2\beta\delta_{i+2j} \\ \quad + 2\beta\delta_{i-2j}] \cdot [x_{ik}x_{jl} + y_{ik}y_{jl}], & \text{if} \quad i = j \\ [(2\alpha + 8\beta)\delta_{i+1j} + (2\alpha + 8\beta)\delta_{i-1j} \\ \quad - 2\beta\delta_{i+2j} - 2\beta\delta_{i-2j}] \cdot [x_{ik}x_{jl} \\ \quad + y_{ik}y_{jl}], & i \neq j \end{cases} \tag{10.32}$$

and

$$b_{ik} = \gamma g_{ik} \tag{10.33}$$

Based on eqs. (10.30) and (10.31), the next state of each neuron can be determined. While the neuron states change, the optimal boundary points are detected. Convergence

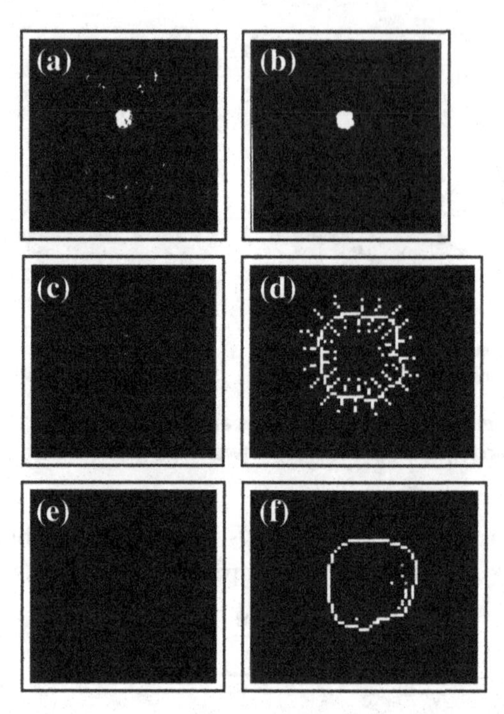

Figure 10.12 Example of a detected contour: (a) initial slice 39 (thresholded); (b) dilation of the original image; (c) initial contour; (d) search grid points and initial contour; (e) example of slice 49 containing a brain tumor; and (f) detected contour of slice 41 from (e). *(Reprinted from [408] with permission from IEEE.)*

is achieved if one and only one of the outputs of the neurons in each row is equal to 1. The location of the firing neurons corresponds to the detected boundary points.

Example 10.8.1 Figure 10.12 [408] shows the performance of the described network on a slice of T1-weighted MR axial brain image.

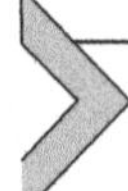

10.9. CASCADED SELF-ORGANIZED NEURAL NETWORK FOR IMAGE COMPRESSION

10.9.1 Image Compression

Digital image processing enables easy image retrieval, efficient storage, rapid image transmission for off-site diagnoses, and the maintenance of large image banks for purposes of teaching and research. To use digital signal techniques, analog signals such as X-rays must be converted into a digital format, or they must be already available in digital form. The sampling process produces a loss of information and can lead to a deterioration of the signal. On the other hand, because of technological improvements in analog-to-digital converters, the quantities of digital information produced can overwhelm available

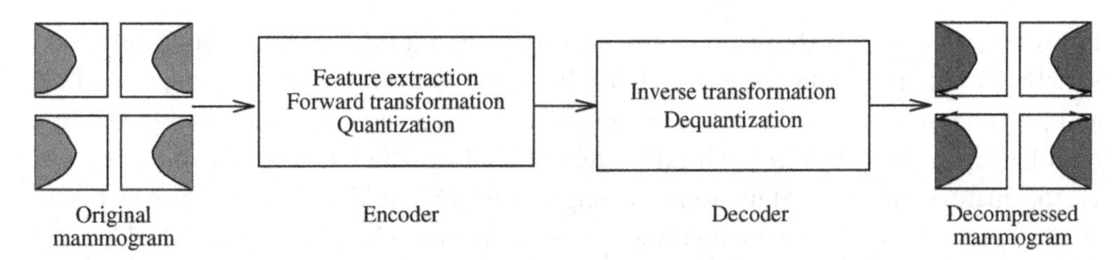

Figure 10.13 Scheme of the compression mechanism illustrated for digital mammography. Four images per patient are produced by a single screening.

resources. For example, a typical digitized mammogram with 4500 × 3200 pixels with 50-micron spot size and 12-bit-per-pixel depth is approximately 38 megabytes of data. Extensive medical studies using digitized data format can easily require unacceptably long transmission times and can lead to major data management problems in local disk storage. A solution solely based on advancing technologies for transmission and storage is not sufficient. Data compression is desirable and often essential for improving cost and time for storage and communication. The typical digital compression system has three main steps: (1) signal decomposition based on a Fourier and wavelet transform, (2) coefficient quantization, and (3) lossless or entropy coding such as Huffman or arithmetic coding. Decompression is defined as the inverse operation, and if quantization is employed, decompression is lossy in the sense that the original image cannot be perfectly reconstructed. Figure 10.13 shows a compression scheme applied in digital mammography, where the transmission time of four images per patient per screening is critical. Lossless or invertible compression allows perfect reconstruction of a digital image, but the obtained compression ratios range from 2:1 to 3:1 on still-frame gray scale medical images. For many medical applications this modest compression is not adequate. Lossy compression can provide excellent quality at a fraction of the bit rate. The bit rate of a compression system is the average number of bits produced by the encoder for each image pixel. If the original image has 12 bits per pixel (bpp) and the compression method yields a rate of R bpp, then the compression ratio is given by 12:R.

The first studies in medical lossy compression were based on applying variations on the standard discrete cosine transform (DCT) coding algorithm combined with scalar quantization and lossless coding. These are variations of the international standard Joint Photographic Experts Group (JPEG) compression algorithm [52]. The American College of Radiology-National Electrical Manufacturers Association (ACR-NEMA) standard has not yet specifically named a standard compression technique, but transform coding methods are suggested. New studies of efficient lossy image compression algorithms have been based on subband or wavelet coding in combination with scalar or vector quantization [323,392]. The main benefits of these techniques over traditional Fourier-type decompositions are better concentration of energy and decorrelation for a larger class of signals.

However, the main disadvantage represents the errors introduced by a resolution-based variable filter length. They are not well localized and represent a ringing artifact. In [28] and [291] a compression algorithm (Shapiro's embedded zero-tree algorithm [337]) was used based on the subband/pyramid/wavelet encoding. New forms of multiresolution vector quantization algorithms were investigated in [68] and [216]. The vector generalizations of the embedded zero-tree wavelet technique used in [68] introduced additional complexity and made this approach less attractive than the scalar wavelet coding selected. The nonwavelet multiresolution technique proposed in [216] yielded significant improvements at reasonable complexity for low-resolution images, e.g., for reduced-size medical images for progressive viewing during the displaying of the full image.

The main concern to be addressed in conjunction with lossy image compression is to specify the type and amount of distortion necessary for enough accuracy for diagnostic, educational, archival, or other purposes.

The tools for measuring image quality and their performance have been surveyed in two papers [3,88]. Besides these general surveys, there are several investigations that focus on evaluating the quality of compressed medical images. A detailed study [67] compares three approaches in this field and notes that (1) SNR is quick and cheap to obtain, but does not take into account the medical nature of the images; (2) receiver operating characteristic (ROC) analyses are serious attempts to capture the medical interest of the images through their diagnostic values; and (3) subjective rating shows a different trend from the actual diagnostic quality, which can reassure physicians that diagnostic utility is retained even when a compressed image is perceptually distinguishable from the original.

10.9.2 Neural Network Architecture

Transform-based algorithms for lossy compression transform an image to concentrate signal energy into a few coefficients, quantize the coefficients, and then entropy code them after assigning special symbols for runs of zeros. The use of principal component analysis (PCA) for image coding is known to be an optimal scheme for data compression based on the exploitation of correlation between neighboring pixels or groups of pixels [259]. Additional data compression can be achieved by using vector quantization subsequent to a robust PCA [55,259,355,375]. The basic idea of this improved compression scheme is visualized in Fig. 10.14.

An image is divided in blocks of the size $N \times N$. M relevant features, where $M < N$, are extracted from each block and then quantized based on a vector quantization (VQ) algorithm. VQ is defined as a mapping that assigns each feature vector to one of Q codewords from a finite subset, called a codebook. In the encoding process, a distance measure is evaluated to determine the closest codeword for each input vector. Then, the address corresponding to the codeword is assigned to every feature vector and transmitted. Compression is achieved by transmitting the index of the codeword instead of the vector itself.

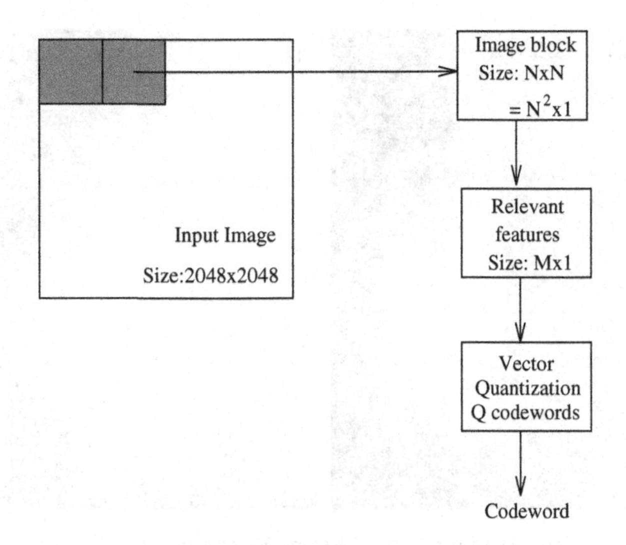

Figure 10.14 Compression scheme combining PCA with vector quantization.

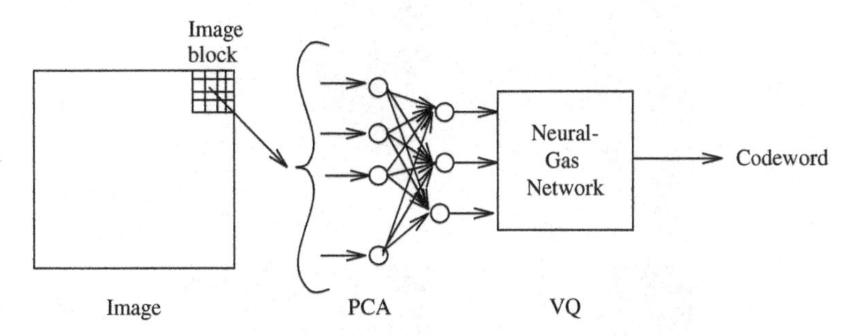

Figure 10.15 Architecture of the combined network (Oja's algorithm and "neural gas") for compression.

Figure 10.15 shows a cascaded architecture of self-organizing neural networks for lossy image compression combining a PCA-type neural network implementing Oja's algorithm with a vector quantizer realized by the "neural-gas" network. Oja's algorithm is applied to image blocks that are treated like data vectors and performs a projection of the data vector on a lower dimensional space. Additional data compression can be achieved by using the "neural-gas" network on the coder output to produce its quantized version. The combined network for compression is shown in Fig. 10.15.

As an indicator for comparison of information loss due to image compression, we choose the "classical" approach for image quality. Image quality is typically quantified objectively by average distortion or peak signal-to-noise ratio (PSNR). The PSNR

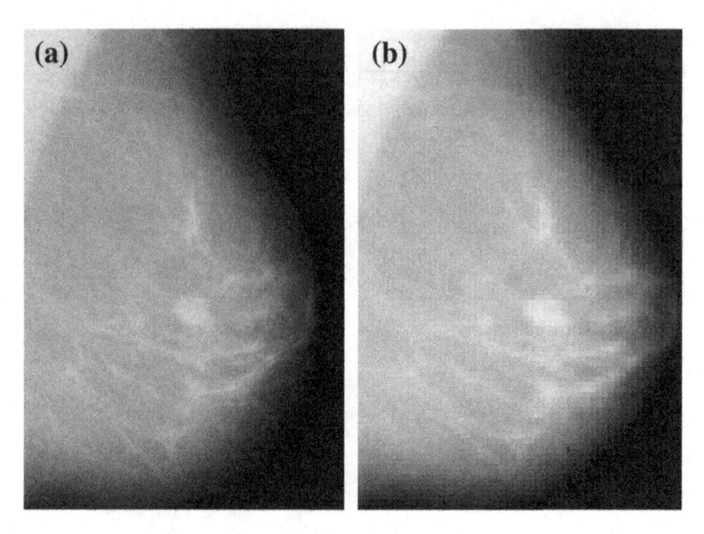

Figure 10.16 Original mammogram (a) and (b) same mammogram compressed by the proposed algorithm, compression rate = 64, PSNR = 23.5 dB.

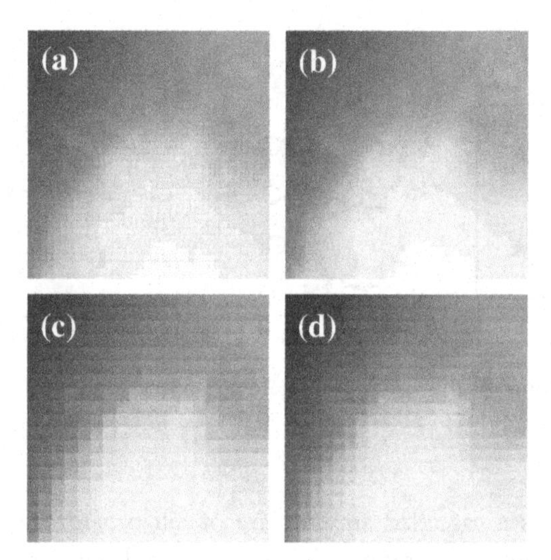

Figure 10.17 Cut-outs of the quantized test image. Compression results with and without overlapping depending on different block sizes. (a) 4×4, no overlapping, (b) 4×4 overlapping, (c) 8×8 no overlapping, and (d) 8×8 overlapping.

definition is $10 \log_{10} 256^2/\text{MSE}$, where MSE denotes the average squared error;

$$\text{MSE} = \frac{1}{1024 \times 1024} \sum_{i=1}^{1024} \sum_{j=1}^{1024} (x_{i,j} - y_{i,j})^2 \tag{10.34}$$

with $x_{i,j}$ the gray value of the original and $y_{i,j}$ that of the decompressed image for the pixel at the position i, j in a 1024×1024 image.

Example 10.9.1 This neural architecture can be applied to compress digital mammograms. To study the efficiency of our algorithm, mathematical phantom features are blended into clinically proved cancer-free mammograms. The influence of the neural compression method on the phantom features and the mammographic image is not perceptible to a mammogram reader up to a compression rate of 48:1. Figure 10.16 shows the achieved compression results for a block size of 8×8, 4 PCA components at a compression ratio of 64 and PSNR = 23.5. Oja's algorithm [272] was chosen for computing the principal components.

The quantization results for a 1024 codebook are shown in Fig. 10.17. Again the advantage of the overlapping over nonoverlapping is evident. With overlapping and a smaller block size annoying effects such as blocking are less accentuated.

CHAPTER ELEVEN

Spatio-Temporal Models in Functional and Perfusion Imaging

Contents

Medical imaging techniques often produce time series of three-dimensional images, where an object of interest is followed over time. The aim of such approaches is typically to locate and to classify functional or diseased tissue. To this end, often voxel-wise time series are analyzed.

An example for such imaging series is functional magnetic resonance imaging (fMRI), where scans of the brain of a subject are acquired during on- and off-phases of an external stimulus (sound, visual signals, pain). Series of images are also acquired to study perfusion in tissue, for example for Dynamic contrast-enhanced magnetic resonance imaging (DCE-MRI) or DCE computer tomography (DCE-CT) in oncology, dynamic susceptibility contrast imaging (DSC-MRI) in neuroimaging, myocardial perfusion MRI, and many techniques in nuclear medical imaging.

The analysis of imaging time series should, however, not only rely on the voxel-wise time curves. It can easily be seen that using the inherent spatial structure of images can be helpful to gain better results than an independent voxel-wise analysis. Using spatial structures can not only help to produce more robust results by "borrowing strength" from neighboring voxels, but also overcome the arising multiple test problem, as discussed below.

Pattern Recognition and Signal Analysis in Medical Imaging
http://dx.doi.org/10.1016/B978-0-12-409545-8.00011-X

Any analysis of imaging time series needs a careful registration of the single scans, to account for motion of the subject. After registration, in each scan the same voxel should actually present the same tissue [121].

In this chapter, we will explore spatio-temporal approaches for models applied on voxel basis. We use three different underlying frameworks for the voxel model:

(a) linear models, as used in fMRI,

(b) nonlinear models derived from a set of differential equations,

(c) nonparametric models, allowing a more flexible voxel model.

(b) and (c) are typically used in perfusion imaging MR, but also in Positron Emission Tomography (PET), and Single-Photon Emission CT (SPECT). In all these applications, a tracer in the tissue is followed over time. The contrast agent concentration over time can then be described by a set of differential equations.

In the following, spatial modeling is based on Gaussian Markov random fields GMRFs, see Section 2.5.3. GMRFs are widely used in imaging, for example in functional MRI [112,289,396], ultrasound perfusion imaging [389], DCE-MRI [333], and diffusion tensor imaging [131].

11.1. SPATIO-TEMPORAL LINEAR MODELS

In function MR imaging, typically linear models are used. They can also be found in PET, SPECT, and in electroencephalography (EEG) and magnetoencephalography (MEG). In neuroimaging, fMRI is a standard tool for classifying brain regions according to their function. [200,351]. A great advantage of fMRI is the noninvasive procedure in which the images can be retrieved by using MR scanners. The change of blood flow in the brain is measured while the subject receives some external stimulation. The neural system of the brain reacts by sending the information from neuron to neuron. Resulting from this are magnetic resonance images which visualize the change in blood oxygenation, the so-called *Blood Oxygenation Level Dependent* (BOLD) effect. To obtain fMRI data, a time series of MR scans of the subjects is acquired, during which the subject is exposed to one or more external stimuli.

One standard approach to analyze fMRI data are Statistical Parametric Maps (SPM), originally proposed by Friston et al. [97]. In the SPM approach, a linear model is formulated for the signal at voxel $i = 1, \dots, I$ at time $t = 1, \dots, T$

$$y_{it} = \mathbf{x}_t^\top \alpha_i + z_t \beta_i + \epsilon_{it}, \quad \epsilon_{it} \sim N(0, \sigma_i^2) \tag{11.1}$$

where $\epsilon_{it}, t = 1, \dots, T, i = 1, \dots, I$ are independent Gaussian errors with variance σ_i^2. The covariate $\mathbf{x}_t$ is used to cover arbitrary trends in the data. Typically the external stimulus occurs in on-off intervals, described by a variable s_t. In order to account for the time delay between external stimulus and the cerebral blood flow, a *hemodynamic response function* $h(s; \theta)$ is used instead of the original stimulus [112]. Therefore, the independent variable

z_{it} is obtained through a convolution of the function h and the on/off stimulus s_{t-d_i-s}:

$$z_{it} = \sum_{j=0}^{t-d_i} h(j;\,\theta_i) s_{t-d_i-j} \tag{11.2}$$

Using the linear model (11.1) the effect of the external stimulus on a voxel i can be tested by a hypothesis test $H_0 : \beta_i = 0$. However, testing β in many voxels leads to a multiple testing problem [104]. Friston et al. proposed to use a Gaussian random field to utilize spatial information and therefore to overcome the multiple test problem to some extent [97]. As an alternative, Posterior Probability Maps (PPM) based on Bayesian models were proposed, where no multiple test problem arises [112].

11.1.1 Posterior Probability Maps

For the PPM model, a *Hierarchical Bayesian model* (HBM) approach is used, see Section 6.6.1. That is, the model is formulated in three stages:

Stage 1: Data model.
Stage 2: Model of spatial correlation.
Stage 3: Prior specifications.

Stage 1 is the GLM (11.1). In stage 2, a GMRF is used as prior distribution in order to smooth the effects, and, on the other hand, to "borrow strength" from neighboring voxels, hence to allow a more robust parameter estimation. That is, the conditional distribution of β_i given its neighbors $\beta_{\partial(i)}$ is

$$\beta_i | \beta_{i\sim j}, \tau^2, \mathbf{w} \sim N\left(\sum_{i\sim j} \frac{w_{ij}\beta_j}{w_{i+}}, \frac{\tau^2}{w_{i+}}\right)$$

where $i \sim j$ means voxel i is neighbor of voxel j and $w_{i+} = \sum_{j\sim i} w_{ij}$ is the sum of locally adaptive smoothing weights between i and all nodes adjacent to i. Each weight w_{ij} determines the "smoothness" between i and j. High weights indicate that the activation β is similar in both voxels. In contrast small weights imply that the activation in both voxels can be seen as independent of each other.

From the full conditional the joint Gaussian distribution with precision matrix $\mathbf{K}$ can be derived:

$$p(\beta|\mathbf{w}, \tau^2) \propto \tau^{(I-1)/2} \left(\prod_{i=1}^{I-1} \lambda_i\right)^{1/2} \exp\left(-\frac{1}{2\tau^2}\beta^{\top}\mathbf{K}\beta\right) \tag{11.3}$$

with the precision matrix $\mathbf{K}$ defined as follows

$$K_{ij} = \begin{cases} w_{i+} & i = j \\ -w_{ij} & i \sim j \\ 0 & \text{otherwise} \end{cases} \tag{11.4}$$

Here, λ_i are the nonnegative eigenvalues of $\mathbf{K}$. That is, $\beta|\mathbf{w},\tau^2$ has an (improper) Gaussian distribution with expectation vector 0 and precision matrix $(1/\tau^2)\mathbf{K}$.

In stage 3, prior specification for the variance parameters σ_i^2, τ is assigned. These are typically conjugate inverse gamma prior distributions with parameters a, b and c, d, respectively. The weights w_{ij} can either be chosen/estimated upfront or, using some prior distribution, estimated along with the other parameters.

Using Bayes' theorem, the posterior pdf can be computed up to a constant

$$p(\beta,\tau^2,\sigma^2,\mathbf{w}|\mathbf{X}) \propto \mathbf{f}(\mathbf{X}|\beta,\sigma^2)\mathbf{p}(\beta|\tau^2,\mathbf{w})\mathbf{p}(\sigma^2)\mathbf{p}(\tau^2)\mathbf{p}(\mathbf{w}) \qquad (11.5)$$

Parameter estimates can be gained using an MCMC algorithm, see Section 2.6. From this, maps of the voxel-wise parameters can be drawn.

11.1.2 PPM with Global Smoothing Parameter

In the simplest approach, all the weights are set to one, $w_{ij} = 1$. In this case the β parameter maps are globally smoothed. The MCMC algorithm reduces to series of Gibbs steps:

- β is drawn voxel-wise or block-wise from a (multivariate) Gaussian distribution.
- τ^2 is drawn from an Inverse Gamma distribution.
- σ^2 is drawn from an Inverse Gamma distribution.

This model fits in the general class of structured additive regression (STAR) models [89]. Alternative inference methods for such models are available, for example Empirical Bayes or Integrated Nested Laplace Approximation (INLA) [35,317].

11.1.3 Adaptive Smoothing

However, global smoothing can oversmooth borders and other sharp features in the parametric map. Therefore adaptive GMRF approaches were proposed. Brezger et al. proposed to estimate the weights w_{ij} along with the other parameters in a fully Bayesian approach [34]. They use independent Gamma priors for the weights. The computation of the posterior pdf is straightforward. However, the update of the weight in the MCMC algorithms is computationally very expensive: a Metropolis-Hastings step has to be used, where in each iteration for each weight the eigenvalues of an $n \times n$-dimensional matrix (with n the number of voxels) have to be computed.

Apart from the computational burden, the model has some nice features. It not only allows to retain borders and other sharp features in the smoothness maps, it also allows to detect these borders via the estimated weights. By plotting the weights, one can detect differences from nonactivated and activated areas easily. Even locating the exact voxels accountable for large discrepancy is feasible. Conclusively, the estimation with an adaptive Gauss prior allows a more accurate and reliable estimation.

11.2. SPATIAL APPROACHES FOR NONLINEAR MODELS

In perfusion imaging, a tracer in the tissue is followed over time. For example, for DCE-MRI, a magnetic contrast agent is injected and the subject is scanned several times using T_1-weighted sequences. The reduction in T_1 relaxation time caused by a contrast agent is the dominant enhancement observed [287]. Typically, a low molecular weight contrast agent (less than 1000 Da in molecular weight, often a Gadolinium complex such as Gd-DTPA, Gd-DOTA, or Gd-HP-DO3A [114]) is used for such scans. The contrast agent concentration C_t can be computed from the T_1 signal by converting the signal into T1 relaxation time values using proton density weighted images and data from calibration phantoms with known T_1 relaxation times [42]. DCE-MRI has been proven to be useful in imaging a range of different tumor types. Liu et al. presented a Phase 1 study of a novel angiogenesis inhibitor with patients with different cancer types including breast, thyroid, renal cell, lung, squamous cell, ocular melanoma, Merkel cell, adenoid cystic prostate, and colorectal carcinoma [220]. They showed that kinetic parameters gained by DCE-MRI scans after two days were a useful indicator of drug pharmacology.

Analysis of perfusion imaging is typically done in five steps: (1) Computation of contrast concentration. To assess the kinetics of the contrast agent and therefore the blood, the contrast concentration-time curve has to be computed. (2) Determination of the Arterial Input Function (AIF). As the contrast agent is transported through the blood stream, the kinetic in the tissue is driven by the arrival of contrast agent in the tissue, described by an input function. Although the contrast agent is usually injected as a bolus, transport time through the blood stream leads to dispersion of the input function. Therefore, kinetic models consist of the convolution of the input function with the local kinetic model. Hence, an accurate determination of the input function is of importance when examining the kinetic in the tissue. (3) Determination of the local kinetic model. Kinetic models are a simplification of the physiological processes in the tissue, so not every aspect of the kinetic process can be considered. Model choice is restricted by temporal and spatial resolution, which directly depend on the contrast agent and the imaging modality. Typically compartment models are considered. (4) Model fitting for kinetic parameter estimation. This is usually done using least-squares methods or in a Bayesian framework. (5) Interpretation of kinetic parameters. Most kinetic parameters have a direct physiological interpretation. Often kinetic parameters are summarized for an image, losing the spatial distribution of the parameter.

11.2.1 Compartment Models

Most kinetic models are based on the idea of one or more compartments in the tissue, which exchange contrast agent via perfusion through the walls between the compartments.

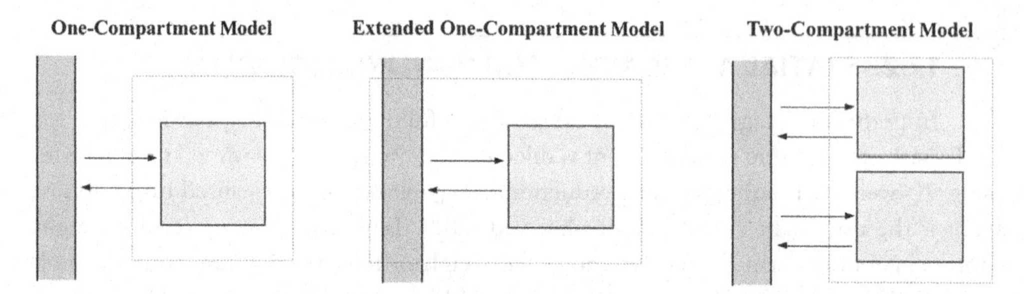

Figure 11.1 Different compartment models. The contrast agent is washed into the vascular space (left). The vascular compartment exchanges with the tissue compartments (right). The dashed line represents the field of view; only in the extended compartment model the vascular space is considered.

In the simplest compartment model the contrast agent is washed in by vascular compartment and exchange between the vascular compartment and the main compartment is driven by two parameters, in DCE-MRI referred to as K^{trans} and k_{ep}. The single compartment model was originally developed by Kety [174]. For PET Koeppe et al. describe a model based on the Kety [181], later Larsson et al. and Tofts et al. independently developed similar models for DCE-MRI [202,203,369] (see Fig. 11.1).

In DCE-MRI, the standard compartment model to describe the arterial influx of Gd-DTPA into the extra-vascular extra-cellular space (EES) and its venous efflux is derived from the differential equation

$$\frac{dC_t}{dt} = K^{trans} C_p - k_{ep} C_t \tag{11.6}$$

Here, $C_t(t)$ denotes the concentration of the contrast agent at time t, $C_p(t)$ denotes the arterial input function, and K^{trans} represents the volume transfer constant between blood plasma and EES, whereas k_{ep} represents the rate constant between EES and blood plasma.

With the initial conditions $C_p(0) = C_t(0) = 0$, the solution of (11.6) is

$$C_t(t) = K^{trans}[C_p(t) \star \exp(-k_{ep}t)] \tag{11.7}$$

$$= K^{trans} \int C_p(\tau) \exp[-k_{ep}(t - \tau)] d\tau \tag{11.8}$$

In DCE-MRI literature, the model is often referred to as the (simple) Tofts model [368]. It assumes that the volume to blood plasma is negligible. However, the model can easily be extended to allow for significant contributions of plasma space. The so-called "extended Tofts model" is

$$C_t(t) = v_p C_p(t) + K^{trans} \int C_p(\tau) \exp[-k_{ep}(t - \tau)] d\tau \tag{11.9}$$

The third parameter v_p represents the fraction of tissue occupied by blood. The volume of the EES per unit volume of tissue v_e can be computed by

$$v_e = K^{trans}/k_{ep} \tag{11.10}$$

and there remaining fraction $v_i = 1 - v_p - v_e$ is the fraction of intracellular space.

The arterial input function (AIF) describes the input of the contrast agent to the tissue. The AIF can be explicitly measured given a suitable image acquisition protocol. Where a measurement of the AIF is not available a standard AIF can be used. A popular standard AIF was proposed by Tofts et al. [369]:

$$C_p(t) = D\sum_{i=1}^{2} a_i \exp(-m_i t) \tag{11.11}$$

with previously determined values a_1, a_2, m_1 and m_2, and D the actual dose per body weight.

More complex compartment models have been described in the literature. For example, Port et al. describe a physiological model with additional compartments in the tumor [300]. They proposed a Kety model plus an additional exponential function and claimed that the additional compartment is necessary to describe the kinetic in tumor types, where well and poorly vascularized areas within a given tumor are present. More recently, a two-compartment exchange model (2CXM) was proposed by Sourbron and Buckley [345]. However, more-compartment models suffer from redundancy issues [342].

In PET studies typically one compartment is not appropriate as physiological model for the tissue. For the analysis of PET data two-compartment [51,142] as well as three-compartment models [32,148] have been proposed. Cunningham and Jones investigated a model where the number of compartment is unknown and has to be estimated from the data [70].

The Kety model assumes that the contrast agent is instantaneously and well mixed in each of the compartments. Distributed parameter models overcome these assumptions. The most prominent distributed parameter model is the tissue homogeneity model proposed by Johnson and Walker [155]. An adiabatic approximation to the tissue homogeneity (AATH) model was proposed by St. Lawrence and Lee [346]. In the AATH model, the observed contrast concentration can be written as convolution of the AIF with a response function $f(t)$,

$$C_t(t) = C_p(t) \star f(t) \tag{11.12}$$

and

$$f(t) = F_p \cdot \begin{cases} 0 & \text{for } t < 0 \\ 1 & \text{for } 0 \le t < T_c \\ E\exp[-(t - T_c)EF_p/v_e] & \text{for } t \ge T_c \end{cases} \tag{11.13}$$

where T_c is the transit time through the capillary, $v_e = K^{trans}/k_{ep}$ is the volume fraction of EES, E is the extraction fraction, and $F_p = K^{trans}/E$ is the mean plasma flow. Instead of a global mean transit time Koh et al. assumed a continuous distribution of transit times for the multiple pathways and proposed the distributed-capillary ATH model [182].

As the dynamic time series is always observed at discrete time points Murase proposed to discretize the extended Kety model [257]. In this case the first column of the design matrix is a vector of the integrated AIF, the second column is the integrated CTC, and the third column is the input function, each of that at time points $t_1, \ldots, t_n$, respectively. Thus, the kinetic parameters can be computed from the regression parameters. However, this method only works for data with high temporal resolution.

11.2.2 Estimating Kinetic Parameters from a Nonlinear Model

By carrying out the convolution in (11.9) the following nonlinear model can be derived from the extended Tofts model:

$$C_t(t) = v_p C_p(t) + DK^{trans} \sum_{i=1}^{2} \frac{a_i\{\exp(-m_i t) - \exp[-k_{ep} t]\}}{k_{ep} - m_i} \tag{11.14}$$

Estimation of the kinetic parameters K^{trans} and k_{ep}, and the volume v_p can be done using standard nonlinear least-squares methods, i.e., the minimization of the sum of squared errors (SSE) between model and observed data. Minimization of SSE has the intrinsic assumption that the distribution of the observation error is symmetric on the mean. In MRI the observation error of the T_1 signal is known to have a Rice distribution, which is approximately Gaussian for high signal-to-noise ratio [119]. However, the contrast agent concentration in DCE-MRI is a nonlinear transformation of the signal, hence, the actual structure of the observation error is unknown.

Optimization of nonlinear models is difficult and often sensitive to starting properties of the optimization algorithm. Levenberg-Marquardt, MINPACK-1, Simplex minimization, and quasi-Newton bounded minimization algorithms are used for optimization [2,41,141]. No algorithm is superior, and as Ahearn et al. point out, multiple search start point algorithms necessary to gain reliable estimates [2].

As an alternative, Bayesian inference can be used. To this end, prior information about the unknown parameters is needed. In a kinetic model, for example, one has some prior information about the kinetic parameters in the tissue under study. Schmid et al. used a log-normal prior distribution for K^{trans} and k_{ep} in a simple Tofts model [333]. With this prior information, the number of biologically impossible parameter estimates is significantly reduced. Orton et al. however used the fact that $v_e < 0$, i.e., $K^{trans} < k_{ep}$, hence assuming a uniform prior distribution for k_{ep} between 0 and a threshold k_U, and a uniform prior distribution for K^{trans} between 0 and k_{ep} [275]. Kärcher et al. use a similar approach in a two-compartment model, but use the fact that the volume of each compartment $v_{t,1/2} = K^{trans}_{1,2}/k_{ep,1/2}$ and, hence, use the formulation $v_{t,1/2} \sim U[0, 1]$ [164]. Additionally, they introduce a constraint $k_{ep,1} < k_{ep,2}$ by use of the prior $k_{ep,1} \sim U[0, k_{ep,2}]$.

The optimization procedures can be used either on contrast concentration averaged over a region of interest (ROI) or on a voxel level. ROIs can either cover the whole tumor or "hot spots" in the tumor [278,128]. ROI analyses have the advantage of increased contrast-to-noise ratio (CNR). Voxel-wise analysis can give more detailed information about the spatial distribution of the kinetic parameters. That is, maps of kinetic parameters can be drawn and the spatial distribution of the kinetic parameters can be interpreted [128].

11.2.3 Spatial Nonlinear Modeling

Voxel-wise analysis suffers however from a low CNR, often leading to unstable estimates or estimates with hight uncertainty, that is, large confidence or credible intervals. In order to strengthen voxel-wise analysis, spatial smoothness assumptions can be used. For example, a Gaussian Markov random field (GMRF) prior can be used, see Section 2.5.3. Here, the kinetic parameters can be seen as latent, i.e., unobserved variables, which follow a (more or less) smooth GMRF. Typically, such approaches are done using hierarchical Bayesian modeling.

As an example, we use a Tofts model on the data stage [333]. That is, the observation model for the observed contrast agent concentration Y_{it} in voxel i at time point t

$$Y_{it} = C_{t,i}(t) + \epsilon_{it}, \epsilon_{it} \sim N(0, \sigma^2) \text{ for each } i, t \tag{11.15}$$

with

$$C_{t,i}(t) = DK^{trans}_i \sum_{i=1}^{2} \frac{a_i\{\exp(-m_i t) - \exp[-k_{ep,i}t]\}}{k_{ep,i} - m_i} \tag{11.16}$$

similar to (11.14).

In the second stage of the HBM, priors on the kinetic parameters are constructed. To this end, the logarithms of the kinetic parameters are used, in order to assure positiveness. On the log-parameters $\gamma_i = \log(K^{trans}_i)$ and $\theta_i = \log(k_{ep,i})$ a GMRF prior is used. This implies the assumption of smoothness on the log scale, that is, areas with higher kinetic parameters (e.g., tumors) are less smooth than areas with lower kinetic parameters (e.g., normal tissue). The GMRF priors can be formulated as follows:

$$\gamma \sim N_J(0, (\tau^2\mathbf{Q})^{-1}) \tag{11.17}$$

$$\theta \sim N_J(0, (\nu^2\mathbf{Q})^{-1}) \tag{11.18}$$

Here, $\mathbf{Q}$ is the precision matrix describing the neighborhood structure of the voxels, see Section 2.5.3. The smoothing parameters τ^2 and ν^2 can be estimated along with the kinetic parameter maps from the data. For this, prior pdfs on this parameters have to be specified, typically using the conjugate Inverse Gamma prior.

The computation of the joint posterior pdf is straightforward up to a constant. The resulting Markov Chain Monte Carlo algorithm, however, is not trivial. It is composed of Metropolis-Hastings steps for the kinetic parameters and Gibbs steps for the noise variance and the smoothing parameters. A software package called `dcemriS4` for the statistical software `R` is publicly available, which allows to estimate the spatial model for any data set [363,386].

Similar approaches have been proposed for other models. Sommer et al. used a GMRF on a two-compartment model [344]. The spatial prior allowed to overcome redundancy issues they discovered in the two-compartment model. DePasquale et al. proposed a Bayesian framework for the analysis of the signal intensity in DCE-MRI [75]. They use two different approaches, the first one fitting a nonparametric curve with a smoothing prior for either temporal and spatial signal changes, estimating semiquantitative values. The second approach is an adhoc parametric function for the contrast agent time series, with a spatial smoothing prior on the parameters. The results of both approaches are then used for classification of malignant and benign tumor voxels. Using a penalized likelihood approach, Sommer et al. used spatial information on a multicompartment model similar to the PET model introduced by Cunningham et al. [343,70].

11.2.4 Adaptive Spatial Nonlinear Modeling

Schmid et al. applied an adaptive GMRF in the context of kinetic models for DCE-MRI [333]. The adaptive model allows locally homogeneous regions and quite sharp boundaries between drastically different tissue types, for example normal tissue versus tumor, but also sharp features in the tumor.

Here, local smoothing weights for both the K^{trans} and the k_{ep} maps are estimated. This increases the computational cost of the MCMC algorithm compared to the adaptive GMRF model for fMRI in Section 11.1.3. However, the model retains the advantages of the previous adaptive model: the weights can be drawn separately and show the border of the tumor and sharp features in the tumor.

11.3. NONPARAMETRIC SPATIAL MODELS

The nonlinearity of kinetic models introduces several issues while analyzing data from perfusion imaging. Even in a simple one-compartment ROI analysis using a least-squares approach, the results depend on the starting values and the algorithm, and the algorithms are not guaranteed to converge. To this end, nonparametric models using smooth functions were proposed.

Compartment models can typically be described by the convolution of a (known) input function with an unknown response function,

$$Y(t) = A(t) \star f(t) \tag{11.19}$$

For example, a more general expression of the simple Tofts model (11.8) can be written as

$$C_t(t) = C_p(t) \star f(t) \tag{11.20}$$

and

$$f(t) = K^{trans} \exp(-k_{ep}t) \tag{11.21}$$

This formulation can be used in many perfusion imaging applications. For example, analysis of first-pass perfusion cardiovascular MRI is typically performed via deconvolution of the myocardial signal with an input function measured in the left ventricular (LV) blood pool [153]. First-pass perfusion MRI provides insight into the perfusion of the coronary artery, in particular into the effect of microvascular diseases. Usually, patients are scanned twice, once under rest and once under drug-induced stress to identify tissue with restricted myocardial blood flow due to obstructive coronary lesions [283]. In myocardial perfusion applications, the deconvolution of signal and input functions is typically performed on segments of the myocardium, see [53] for definition of segments. Aggregating the data in the segments has the advantage of increased SNR, but loses spatial information. In the following we therefore look into methods of performing deconvolution on a voxel level.

11.3.1 Discrete Deconvolution using Regularized B-Splines

Assuming that $Y(t)$ and $f(t)$ are constant over small intervals Δt, a discretized form of (11.19) is given by

$$Y(\tau_i) = \sum_{j=1}^{n} A(\tau_i - t_j)f(t_j)\Delta t = \sum_{j=1}^{T} A_{ij} f(t_j) \tag{11.22}$$

where Y is measured on discrete time points $\tau_1, \ldots, \tau_T$, with T the number of scans in the time series and n the number of time points at which the input function is evaluated. The $T \times n$ matrix $\mathbf{A}$ may be interpreted as a convolution operator and is defined by

$$A_{ij} = \begin{cases} A(t_{n_i - j+1})\Delta t & \text{if } \tau_i \leq t_j \\ 0 & \text{otherwise} \end{cases} \tag{11.23}$$

where n_i is the maximum index j for which $\tau_i \leq t_j$ holds. In most applications $T = n$, but the input function might also be available in a better temporal resolution. For example, when an analytic form of the input function $f(t)$ is given and evaluated on a narrow grid.

By solving (11.22), the response function $f(t)$ can be deconvolved from $Y(t)$. However, this system may be numerically unstable, i.e., the deconvolved response function can be very susceptible to noise. To overcome this problem, one assumes that $f(t)$ is a smooth, k-times differentiable function. B-Splines for example, fulfill this property. The use of B-Splines was proposed by Jerosch-Herold et al. in myocardial first-pass perfusion MRI [154].

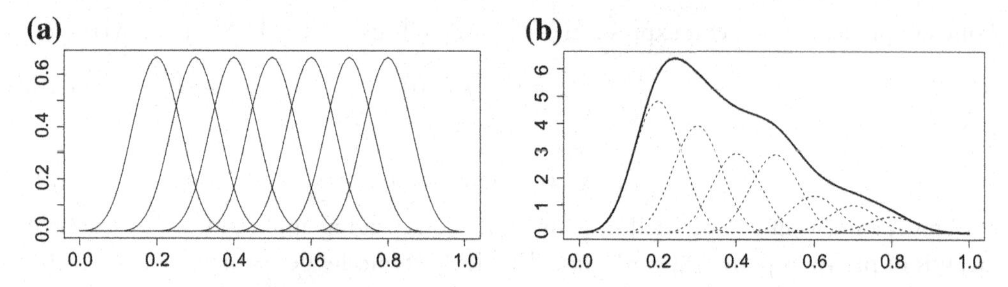

Figure 11.2 (a) B-Spline basis functions of 4th order on eleven equidistant knots on [0, 1]. (b) Weighted basis function (dashed lines) and the B-Spline, the sum of the weighted basis function (solid line).

For B-Splines basis functions are defined on a number of knots on the timescale, typically with equidistant distribution, see Fig. 11.2. Each basis function is a polynomial of kth order over $k+2$ knots. The B-Spline is then a weighted sum of the basis functions

$$f(t) = \sum_{j=1}^{p} \beta_j B_j(t) \tag{11.24}$$

where $B_j(t)$ is basis function j evaluated at time t, and β_j is an unknown weight, which has to be estimated from the data, the spline regression parameters.

In vector notation with $\mathbf{B} = (B_{ij})_{i=1,\ldots,n;j=1,\ldots,p}$, $B_{ij} = B_i(t_j)$ and $\mathbf{f} = (f(t_1), \ldots, f(t_n))^{\top}$, (11.24) can be written as

$$\mathbf{f} = \mathbf{B}\beta \tag{11.25}$$

Accordingly, (11.22) can be reformulated in matrix notation,

$$\mathbf{Y} = \mathbf{Af} = \mathbf{AB}\beta = \mathbf{D}\beta \tag{11.26}$$

where $\mathbf{D} = \mathbf{AB}$ is a $n \times p$ design matrix, representing the (discrete) convolution of the input function with the B-Spline polynomial basis functions.

The resulting B-Spline depends on the number and the distribution of knots. A large number of knots naturally leads to a good fit to the data, but to a rough function. In contrast, a low number of knots leads to a rather smooth function and a worse fit to the data. To this end, it was proposed to use a rather large number of knots and to penalize or regularize the Spline regression parameters β, the so-called penalized Spline or P-Spline approach [86]. Additionally, this approach can enhance the numerical stability of the system.

In myocardial perfusion MR, first and second differences were proposed as penalization [154,330]. That is, the penalization terms are

$$pen(\beta) = \lambda \sum_{j=2}^{p} (\beta_j - \beta_{j-1})^2 \tag{11.27}$$

or

$$pen(\beta) = \lambda \sum_{j=2}^{p} ((\beta_j - \beta_{j-1}) - (\beta_{j-1} - \beta_{j-2}))^2 \tag{11.28}$$

respectively. In both cases differences between temporal neighboring Spline regression parameters are penalized in order to gain a smooth $f(t)$. In matrix notation, the penalization can be written as

$$pen(\beta) = \lambda \beta^\top \mathbf{P} \beta \tag{11.29}$$

with $\mathbf{P}$ a penalization matrix. This approach can actually be seen as a one-dimensional special case of a Gaussian Markov random field energy function, see Section 2.5.3. The penalization of the roughness of the function $f(t)$ leads to a denoising of the curve. It also leads to a stabilization of the deconvolution procedure.

Estimation of β can be done using a penalized least-squares approach. That is, the penalised least square criterion

$$penLS(\beta) = (\mathbf{Y} - \mathbf{D}\beta)^\top (\mathbf{Y} - \mathbf{D}\beta) + pen(\beta) \tag{11.30}$$

the sum of the squared differences between observed concentration Y and estimated observation plus the penalization term is to be minimized. As an alternative, fully Bayesian methods can be used [332].

The penalty parameter λ acts as a smoothing parameter. A high λ, that is, a high penalization of the sum of differences leads to a smooth function. The choice of λ is therefore a crucial point and should be done data driven. Jerosch-Herold et al. suggest to use the L-curve method to determine the smoothing parameter in a myocardial perfusion application [154,156]. In a DCE-MRI application Schmid et al. estimate the smoothing parameter in a fully Bayesian model using MCMC [34]. In a more general framework, Fahrmeir et al. suggest to use an Empirical Bayes (EB) approach, estimating the smoothness parameter with Restricted Maximum Likelihood (REML) [90]. The smoothness parameter can also be chosen using Akaike Information Criterion (AIC) or Generalized Cross-Validation (GCV) [86,237].

The response function can show rather rapid changes, particularly if the first few seconds, in contrast to a rather constant behavior later in the series. Therefore, a more adaptive approach on smoothing may be necessary. Adaptive approaches typically involve a separate smoothing parameter for each difference, for example for first differences (11.27) it is rewritten as

$$pen(\beta) = \lambda_j \sum_{j=2}^{p} (\beta_j - \beta_{j-1})^2 \tag{11.31}$$

The smoothing parameters are then typically estimated using Bayes methods. For this, either independent priors on the smoothing parameters or a Spline approach on the smoothing parameters can be used [31,199].

11.3.2 Spatial Regularization for Spline-Based Deconvolution

Depending on the imaging application, voxel level deconvolution of the signal and input function might not be possible without further constraints. For example, in myocardial perfusion Goldstein et al. use assumptions on the shape of residue curves in order to perform a robust deconvolution [109]. A more elegant way is to use spatial information, that is, information from neighboring voxels to gain more robust estimates.

Let us assume the following hierarchical Bayes model: In each voxel i, the observed signal intensity Y_{it} at time t is the unknown true signal intensity $S_i(t)$ plus a Gaussian observation error

$$Y_{it} \sim N(S_i(t), \sigma_i^2) \text{ for all } i, t \tag{11.32}$$

Similar to Section 11.3.1 the true signal $S_i(t)$ is the convolution of $A(t)$ and $f(t)$

$$S_i(t) = A(t) \star f_i(t) \tag{11.33}$$

and $f(t)$ is modeled using B-Splines

$$f_i(t) = \sum_{p=1}^{P} \beta_{ip} B_p(t) \tag{11.34}$$

Using discretization, per voxel the model can be written as

$$\begin{aligned} Y_i &\sim N(\mathbf{Af}_i, \sigma^2 \mathbf{I}) && (11.35) \\ &\sim N(\mathbf{AB}\beta_i, \sigma^2 \mathbf{I}) && (11.36) \\ &\sim N(\mathbf{D}\beta_i, \sigma^2 \mathbf{I}) && (11.37) \end{aligned}$$

Using the differences penalty in a Bayesian approach, that is, defining the penalty (11.29) as negative log priori, we gain the Gaussian prior

$$\beta_i \sim N(0, \mathbf{P}^{-1}) \tag{11.38}$$

with $\mathbf{P}$ a penalization matrix, which can include adaptive smoothing parameters.

Assuming that adjacent voxel share tissue, and therefore their response functions should have similar shapes, a GMRF approach can be used for including the spatial correlation, see Section 2.5.3. Here, the GMRF can be used on the spline regression parameter at one knot based on the neighborhood structure,

$$\beta_p \sim N(0, \mathbf{Q}^{-1}) \tag{11.39}$$

where $\mathbf{Q}$ is a precision matrix defined by the neighborhood structure as defined in (2.104). Due to the relatively large gaps between slices in medical imaging neighborhoods should only be defined in two-dimensional slices. Similar to Section 11.1.3, the precision matrix $\mathbf{Q}$ can include adaptive smoothing weights. The combination of the temporal penalization and spatial correlation can be done using ideas from spatio-temporal interactions in regularized regression [26,180]. This can be done using the Kronecker matrix sum of the precision matrices, i.e., the precision matrix of the joint Gaussian prior of β is

$$\mathbf{K} = \mathbf{Q} \otimes \mathbf{I}_P + \mathbf{I}_N \otimes \mathbf{P} \tag{11.40}$$

where $\mathbf{I}_N$ is the identity matrix with dimension $N \times N$ and $\otimes$ is the Kronecker matrix product. This combines the spatial regularization of the Spline regression parameters with the temporal penalization [332].

As an alternative, the spatial MRF can be defined on the temporal differences. That is, we define

$$\delta_j = \beta_j - \beta_{j-1} \tag{11.41}$$

and use a GMRF on δ. This leads to the precision matrix

$$\tilde{\mathbf{K}} = \mathbf{Q} \otimes \tilde{\mathbf{P}} + \mathbf{I}_N \otimes \mathbf{P} \tag{11.42}$$

where $\tilde{\mathbf{P}}$ is the temporal penalization matrix, but with a fixed global smoothing parameter [331].

Parameter estimation can be done using a fully Bayesian approach. For this, priors on the error variance σ^2 and on the (adaptive) smoothing parameters τ^2 have to be chosen, typically inverse Gamma distribution. The computation of the joint posterior pdf is straightforward, given by Bayes' formula

$$p(\beta, \tau^2, \sigma^2|\mathbf{Y}) \propto f(\mathbf{Y}|\beta, \sigma^2)p(\beta|\tau^2)p(\tau^2)p(\sigma^2) \tag{11.43}$$

where $f(\mathbf{Y}|\beta, \sigma^2)$ is the Gaussian likelihood of the data given the model parameters. Inference is based on an MCMC algorithm, see Section 2.6:

- Draw β from a multivariate Gaussian distribution using efficient algorithms for sparse matrices (see Section 2.6 and [316]).
- Draw σ^2 from an inverse Gamma distributions.
- Draw all τ^2 using Metropolis-Hastings steps, cf. Section 11.1.3.

CHAPTER TWELVE

Analysis of Dynamic Susceptibility Contrast MRI Time-Series Based on Unsupervised Clustering Methods

Contents

12.1. INTRODUCTION

Cerebrovascular stroke is the third leading cause of mortality in industrial countries after cardiovascular disease and malignant tumors. Therefore, the analysis of cerebral circulation has become an issue of enormous clinical importance.

Novel magnetic resonance imaging (MRI) techniques have emerged during the past two decades that allow for rapid assessment of normal brain function as well as cerebral pathophysiology. Both diffusion-weighted imaging and perfusion-weighted imaging have already been used extensively for the evaluation of patients with cerebrovascular disease [74]. They represent promising research tools that provide data about infarct evolution as well as mechanisms of stroke recovery. Combining these two techniques with high-speed MR angiography leads to improvements in the clinical management of acute stroke subjects [276].

Measurement of tissue perfusion yields important information about organ viability and function. Dynamic susceptibility contrast MR imaging, also known as contrast agent bolus tracking, represents a noninvasive method for cerebrovascular perfusion analysis [381]. In contrast to other methods to determine cerebral circulation such as iodinated contrast media in combination with dynamic X-ray computed tomography (CT) [16] and the administration of radioactive tracers for positron emission tomography (PET)

Pattern Recognition and Signal Analysis in Medical Imaging
http://dx.doi.org/10.1016/B978-0-12-409545-8.00012-1

blood flow quantification studies [133], it allows high spatial and temporal resolution and avoids the disadvantage of patient exposure to ionizing radiation.

MR imaging allows assessment of regional cerebral blood flow (rCBF), regional cerebral blood volume (rCBV), and mean transit time (MTT), for definitions see e.g., [308].

In clinical praxis, the computation of rCBV, rCBF, and MTT values from the MRI signal dynamics has been demonstrated to be relevant, even if its underlying theoretical basis may be weak under pathological conditions [74]. The conceptual difficulties with regard to parameters MTT, rCBV, and rCBF arise from the four basic constraints: (i) homogeneous mixture of the contrast agent and blood pool, (ii) negligible contrast agent injection volume, (iii) hemodynamic indifference of the contrast agent, and (iv) strict intravascular presence of the indicator substance. Conditions (i)–(iii) are usually satisfied in dynamic susceptibility contrast MRI using intravenous bolus administration of gadolinium compounds. Condition (iv), however, requires an intact blood-brain barrier. This prerequisite is fulfilled in examinations of healthy subjects. These limitations for the application of the indicator dilution theory have been extensively discussed in the literature on MRI [290,308] and nuclear medicine [204]. If, nevertheless, absolute flow quantification by perfusion MRI should be performed, the additional measurement of the arterial input function is needed, which is difficult to obtain in clinical routine diagnosis.

However, clinicians agree that determining parameter images based on the MRI signal dynamics represents a key issue in clinical decision-making, bearing a huge potential for diagnosis and therapy.

The analysis of perfusion MRI data by unsupervised clustering methods provides the advantage that it does not imply speculative presumptive knowledge on contrast agent dilution models, but strictly focuses on the observed complete MRI signal time-series. In this chapter, the applicability of clustering techniques is demonstrated as tools for the analysis of dynamic susceptibility contrast MRI time-series and the performance of five different clustering methods is compared for this purpose.

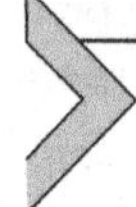

12.2. MATERIALS AND METHODS

12.2.1 Imaging Protocol

The study group consisted of four subjects: (i) two men aged 26 and 37 years without any neurological deficit, history of intracranial abnormality, or previous radiation therapy. They were referred to clinical radiology to rule out intracranial abnormality. (ii) two subjects (one man and one woman aged 61 and 76 years, respectively) with subacute stroke (symptoms 2–4 days, respectively) who underwent MRI examination as a routine clinical diagnostic procedure. All four subjects gave their written consent. Dynamic susceptibility contrast MRI was performed on a 1.5 T system (Magnetom Vision, Siemens, Erlangen, Germany) using a standard circularly polarized head coil for radio-frequency transmission and detection. First, fluid-attenuated inversion recovery, T2-weighted spin echo, and diffusion-weighted MRI sequences were obtained in transversal slice orientation enabling initial localization and evaluation of the cerebrovascular insult in the subjects with stroke. Then dynamic susceptibility contrast MRI was performed using a 2-D gradient-echo echo-planar imaging (EPI) sequence employing 10 transversal slices with a matrix size of 128×128 pixels, pixel size 1.88×1.88 mm, and a slice thickness of 3.0 mm (TR = 1.5 s, TE = 0.54 s, FA = 90°). The dynamic study consisted of 38 scans with an interval of 1.5 s, between each scan. The perfusion sequence and an antecubital vein bolus injection (injection flow 3 ml/s) of gadopentetate dimeglumine (0.15 mmol/kg body weight, Magnevist^{TM}, Schering, Berlin, Germany) were started simultaneously in order to obtain several (more than six) scans before cerebral first pass of the contrast agent. The registration of the images was performed based on the automatic image alignment (AIR) algorithm [395].

12.2.2 Data Analysis

In an initial step, a radiologist excluded by manual contour tracing the extracerebral parts of the given data sets. Manual presegmentation was used for simplicity, as this study is designed to examine only a few MRI data sets in order to demonstrate the applicability of the perfusion analysis method.

For each voxel, the raw gray-level time-series $S(\tau), \tau \in \{1, \ldots, 38\}$ was transformed into a pixel time course (PTC) of relative signal reduction $x(\tau)$ by

$$x(\tau) = \left(\frac{S(\tau)}{S_0}\right)^{\alpha} \tag{12.1}$$

where S_0 denotes the precontrast gray level and $\alpha > 0$ a distortion exponent. The effect of the native signal intensity was eliminated prior to contrast agent application. If time-concentration curves are not computed according to the above equation, i.e., avoid dividing the raw time-series data by the precontrast gray level before clustering, implicitly use is made of additional tissue-specific MR imaging properties that do not directly relate to perfusion characteristics alone.

In the study, S_0 was computed as the average gray level at scan times $\tau \in \{3, 4, 5\}$, excluding the first two scans. There exists an exponential relationship between the relative signal reduction $x(\tau)$ and the local contrast agent tissue concentration $c(\tau)$ [93,172, 256, 312]:

$$c(\tau) = -\ln x(\tau) = -\alpha \ln\left(\frac{S(\tau)}{S_0}\right) \tag{12.2}$$

where $\alpha > 0$ is an unknown proportionality constant. Based on eq. (12.2), the concentration-time curves (CTCs) are obtained from the signal PTCs.

Conventional data analysis was performed by computing MTT, rCBV, and rCBF parameter maps employing the relations (e.g., [16,350,409])

$$\text{MTT} = \frac{\int \tau \cdot c(\tau) d\tau}{\int c(\tau) d\tau}, \quad \text{rCBV} = \int c(\tau) d\tau, \quad \text{rCBF} = \frac{\text{rCBV}}{\text{MTT}} \tag{12.3}$$

Methods for analyzing perfusion MRI data require presumptive knowledge of contrast agent dynamics based on theoretical ideas of contrast agent distribution that cannot be confirmed by experiment: e.g., determination of relative CBF, relative CBV, or MTT computation from MRI signal dynamics. Although these quantities have been shown to be very useful for practical clinical purposes, their theoretical foundation is weak, as the essential input parameters of the model cannot be observed directly. On the other hand, methods for absolute quantification of perfusion MRI parameters do not suffer from these limitations [290]. However, they are conceptually sophisticated with regard to theoretical assumptions and they require additional measurement of arterial input characteristics which sometimes may be difficult to perform in clinical routine diagnosis. At the same time, these methods require computationally expensive data post-processing by deconvolution and filtering. For example, deconvolution in the frequency domain is very sensitive to noise. Therefore, additional filtering has to be performed, and heuristic constraints with regard to smoothness of the contrast agent residual function have to be introduced. Although other methods, such as singular value decomposition (SVD), could be applied, a so-called gamma variate fit [303,365] was used in this context.

The limitations with regard to perfusion parameter computation based eqs. (12.3) are addressed in the literature, e.g., [308,385].

12.2.3 Evaluation of the Clustering Methods

The following section is dedicated to presenting the algorithms and evaluating the discriminatory power of unsupervised clustering techniques. These are Kohonen's self-organizing map (SOM), fuzzy clustering based on deterministic annealing, "neural gas" network, and fuzzy c-means algorithm. These techniques are based on grouping image pixels together based on the similarity of their intensity profile in time (i.e., their time courses).

Let n denote the number of subsequent scans in a perfusion MRI study, and let K be the number of pixels in each scan. The dynamics of each pixel $\mu \in \{1, \ldots, K\}$, i.e., the sequence of signal values $\{\mathbf{x}^{\mu}(1), \ldots, \mathbf{x}^{\mu}(n)\}$, can be interpreted as a vector $\mathbf{x}^{\mu}(i) \in \mathbf{R}^{n}$ in the n-dimensional feature space of possible signal time-series at each pixel (PTC). For perfusion MRI, the feature vector represents the PTC.

The chosen parameters for each technique are given in the following. For SOM is chosen: (1) a one-dimensional lattice, and (2) the maximal number of iterations. For the fuzzy clustering based on deterministic annealing a batch Expectation Maximization (EM) version [248] of fuzzy clustering based on deterministic annealing is used in which the computation of CVs $\mathbf{w}_j$ (M-step) and assignment probabilities a_j (E-step) is decoupled and iterated until convergence at each annealing step characterized by a given "temperature" $T = 2\rho^2$. Clustering was performed employing 200 annealing steps corresponding to approximately 8×10^3 EM iterations within an exponential annealing schedule for ρ. The constant α in eq. (12.1) was chosen to be $\alpha = 3$. For "neural gas" network is chosen:

(1) the learning parameters $\epsilon_i = 0.5$ and $\epsilon_f = 0.005$, and (2) the lattice parameters λ_i equal to half the number of classes and $\lambda_f = 0.01$ and (3) the maximal number of iterations equal to 1000. For the fuzzy algorithms [29], the fuzzy factor = 1.05, and the maximal number of iterations equal to 120 is chosen.

The performance of the clustering techniques was evaluated by (i) qualitative visual inspection of cluster assignment maps, i.e., cluster membership maps according to a minimal distance criterion in the metric of the PTC feature space shown exemplarily only for the "neural gas" network, (ii) qualitative visual inspection of corresponding cluster-specific CTCs for the "neural gas" network, (iii) quantitative analysis of cluster-specific CTCs by computing cluster-specific relative perfusion parameters (rCBV, rCBF, MTT), (iv) comparison of the best-matching cluster representing the infarct region from the cluster assignment maps for all presented clustering techniques with conventional pixel-specific relative perfusion parameter maps, (v) quantitative assessment of asymmetry between the affected and a corresponding nonaffected contralateral brain region based on clustering results for a subject with stroke in the right basal ganglia, (vi) cluster validity indices, and (vii) receiver operating characteristic (ROC) analysis.

The implementation of a quantitative ROC analysis demonstrating the performance of the presented clustering paradigms is reported in the following: besides the four clustering techniques, "neural gas" network, Kohonen's self-organizing map (SOM), fuzzy clustering based on deterministic annealing, and fuzzy *c*-means vector quantization, for the last one two different implementations are employed: fuzzy *c*-means with unsupervised codebook initialization (FSM), and the fuzzy *c*-means algorithm (FVQ) with random codebook initialization. The two relevant parameters in an ROC study, sensitivity and specificity, are explained in the following for evaluating the dynamic perfusion MRI data. In the study, sensitivity is the proportion of the activation site identified correctly, and specificity is the proportion of the inactive region identified correctly. Both sensitivity and specificity are a function of the two threshold values Δ_1 and Δ_2, representing the thresholds for the reference and compared partition, respectively. Δ_2 is varied over its whole range while Δ_1 is kept constant. By plotting the trajectory of these two parameters (sensitivity and specificity), the ROC curve is obtained. In the ideal case, sensitivity and specificity are both one, and thus any curve corresponding to a certain method closest to the uppermost left corner of the ROC plot will be the method of choice. The results of quantitative ROC analysis presented in Fig. 12.14 show large values of the areas under the ROC curves as a quantitative criterion of diagnostic validity, i.e., agreement between clustering results and parametric maps.

The threshold value Δ_1 in Table 12.1 was carefully determined for both performance metrics, regional cerebral blood volume (rCBV) (left column) and mean transit time (MTT): the Δ_1 was chosen as the one that maximizes the AUC of the ROC curves of experimental series. The optimal threshold value Δ_1 is given individually for each data set, see Table 12.1, and corresponds to the maximum of the sum over all ROC areas for each possible threshold value.

Table 12.1 Optimal threshold value Δ_1 for the data sets #1 to #4 based on rCBV and MTT.

	rCBV	MTT
Data set #1	0.30	21.0
Data set #2	0.30	28.0
Data set #3	0.30	18.7
Data set #4	0.20	21.5

Reprinted from [246] with permission from IEEE.

The ground truth used for the ROC analysis is given by the segmentation obtained for the parameter values of the time-series of each individual pixel, i.e., the conventional analysis. The implemented procedure is: (a) Select a threshold Δ_1. (b) Then, determine the ground truth: for the time-series of each individual pixel, compare the MTT value to Δ_1. If the MTT value of this specific pixel is less than Δ_1, assign this pixel to the active ground truth region, otherwise, assign it to the inactive one. (c) Select a threshold Δ_2 independently of Δ_1. Determine all the clusters whose cluster-specific concentration-time curve reveals a MTT less than Δ_2. Assign all the pixels belonging to these clusters to the active region found by the method. Plot the (sensitivity, specificity) point for the chosen value of Δ_2 by comparing with the ground truth. (d) Repeat (c) for different values of Δ_2.

Thus, for each Δ_2, a single (sensitivity, specificity) point is obtained. For each Δ_1, however, a complete ROC curve is obtained by variation of Δ_2, where Δ_1 remains fixed. This means that, for different values of Δ_1, in general, different ROC curves are obtained. Δ_1 is chosen for each data set in such a way, that the area under the ROC curve (generated by variation of Δ_2) is maximal. The corresponding values for Δ_1 are given in Table 5.1.

12.3. RESULTS

In this section, the clustering results of the pixel time courses based on the presented methods are presented.

To elucidate the clustering process in general and thus to obtain a better understanding of the techniques, the cluster assignment maps and the corresponding cluster-specific concentration-time curves belonging to the clusters exemplarily only for the "neural gas" network are shown.

Clustering results for a 38 scan dynamic susceptibility MRI study in a subject with a subacute stroke affecting the right basal ganglia are presented in Figs. 12.1 and 12.2. After discarding the first two scans, a relative signal reduction time-series $x(\tau), \tau \in \{1, \ldots, n\}, n = 36$ can be computed for each voxel according to eq. (12.1). Similar PTCs form a cluster. Figure 12.1 shows the "cluster assignment maps" overlaid onto a EPI-scan of the perfusion sequence. In these maps, all the pixels are highlighted that belong to a specific cluster. The decision on assigning a pixel ν characterized by the

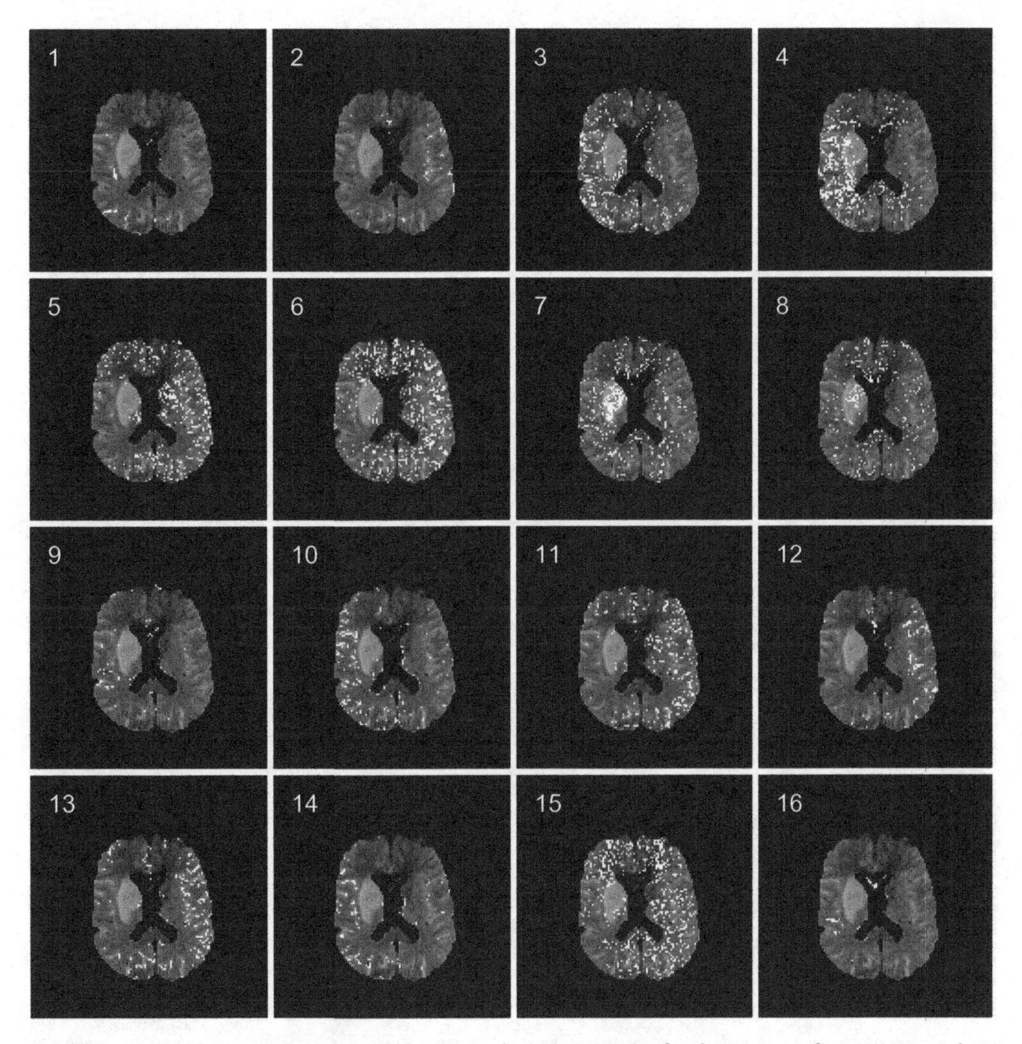

Figure 12.1 Cluster assignment maps for "neural gas" network of a dynamic perfusion MRI study in a subject with stroke in the right basal ganglia. Self-controlled hierarchical neural network clustering of PTCs $x(\tau)$ was performed by "neural gas" network employing 16 CVs, i.e., a maximal number of 16 separate clusters at the end of the hierarchical VQ procedure. For a better orientation, an anatomic EPI-scan of the analyzed slice is underlaid [201].

PTC $\mathbf{x}_\nu = (x_\nu(\tau)), \tau \in \{1, \ldots, n\}$ to a specific cluster j is based on a minimal distance criterion in the n-dimensional time-series feature space, i.e., ν is assigned to cluster j, if the distance $\|\mathbf{x}_\nu - \mathbf{w}_j\|$ is minimal, where $\mathbf{w}_j$ denotes the CV belonging to cluster j. Each CV represents the weighted mean value of all the PTCs belonging to this cluster.

Self-controlled hierarchical neural network clustering of PTCs $x(\tau)$ was performed by "neural gas" network employing 16 CVs, i.e., a maximal number of 16 separate clusters at the end of the hierarchical VQ procedure as shown in Fig. 12.1.

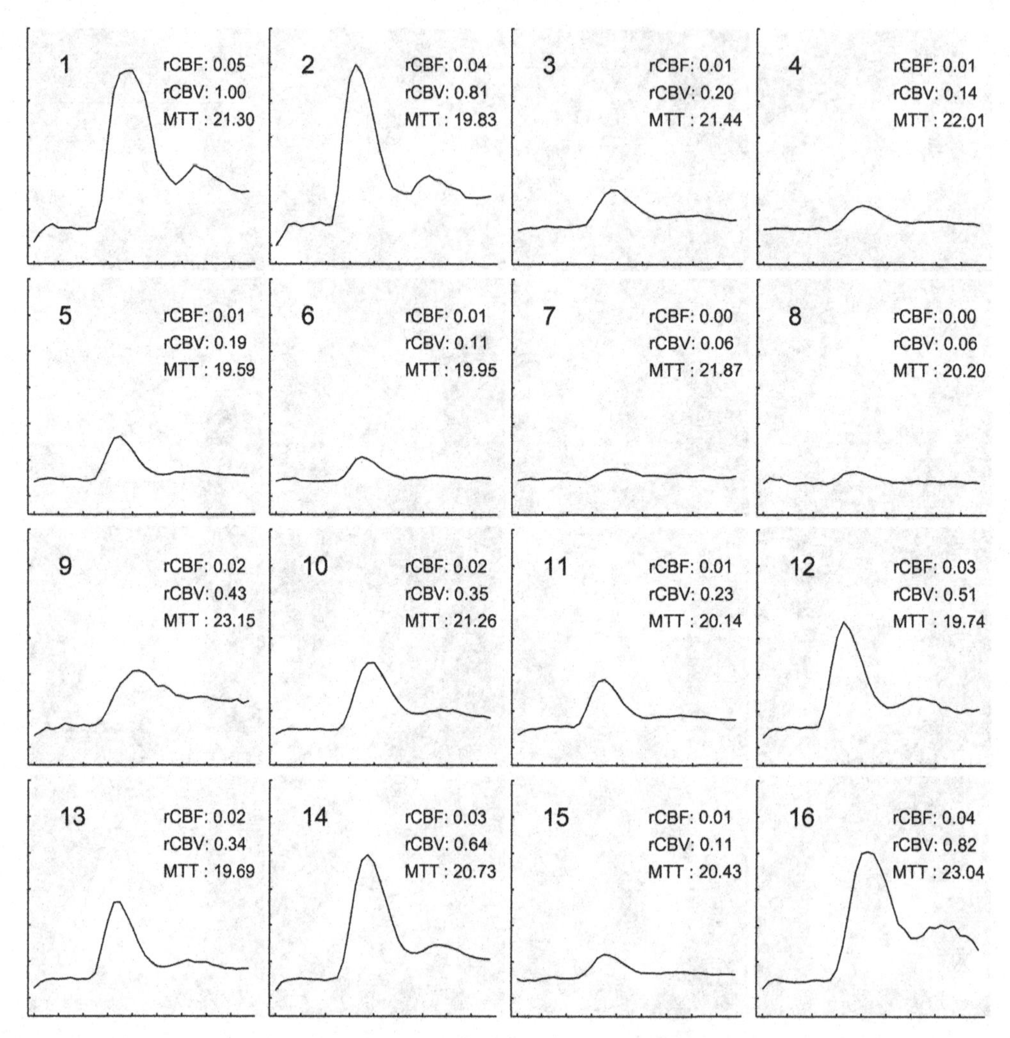

Figure 12.2 Cluster-specific concentration-time curves for "neural gas" network of a dynamic perfusion MRI study in a subject with stroke in the right basal ganglia. Cluster numbers correspond to Fig. 12.1. MTT values are indicated as multiples of the scan interval (1.5 s), rCBV values are normalized with respect to the maximal value (cluster #1). rCBF values are computed from MTT and rCBV by eq. (12.3). The x-axis represents the scan number while the y-axis is arbitrary [201].

Figure 12.2 shows the prototypical cluster-specific CTCs belonging to the pixel clusters of Fig. 12.1. These can be computed from eq. (12.2), where the pixel-specific PTC $x(\tau)$ is replaced by the cluster-specific CV.

The area of the cerebrovascular insult in the right basal ganglia for subject 1 is clearly represented mainly by cluster #7 and by cluster #8 containing other essential areas. The small CTC amplitude is evident, i.e., the small cluster-specific rCBV, rCBF as well as the large MTT. Clusters #3 and #4 contain peripheral and adjacent regions. Clusters #1, #2, #12, #14, and #16 can be attributed to larger vessels located in the sulci. Figure 12.2 shows the large amplitudes and apparent recirculation peaks in the corresponding cluster-specific CTCs.

Further, clusters #2, #12, and #11 represent large, intermediate, and small parenchymal vessels of the nonaffected left side showing subsequently increasing rCBV, and smaller recirculation peaks. The clustering technique unveils even subtle differences of contrast agent first-pass times: small time-to-peak differences of clusters #1, #2, #12, #14, and #16 enable discrimination between left and right side perfusion. Pixels corresponding to regions supplied by a different arterial input tend to be collected into separate clusters: For example, clusters #6 and #11 contain many pixels that can be attributed to the supply region of the left middle cerebral artery, whereas clusters #3 and #4 include regions supplied by the right middle cerebral artery. Contralateral clusters #6 and #11 vs. #3 and #4 show different cluster-specific MTT as an evidence for an apparent perfusion deficit at the expense of the right-hand side.

The diffusion-weighted image in Fig. 12.3a visualizes the structural lesion. Figures 12.3b–d represent the conventional pixel-based MTT, rCBF, and rCBV maps

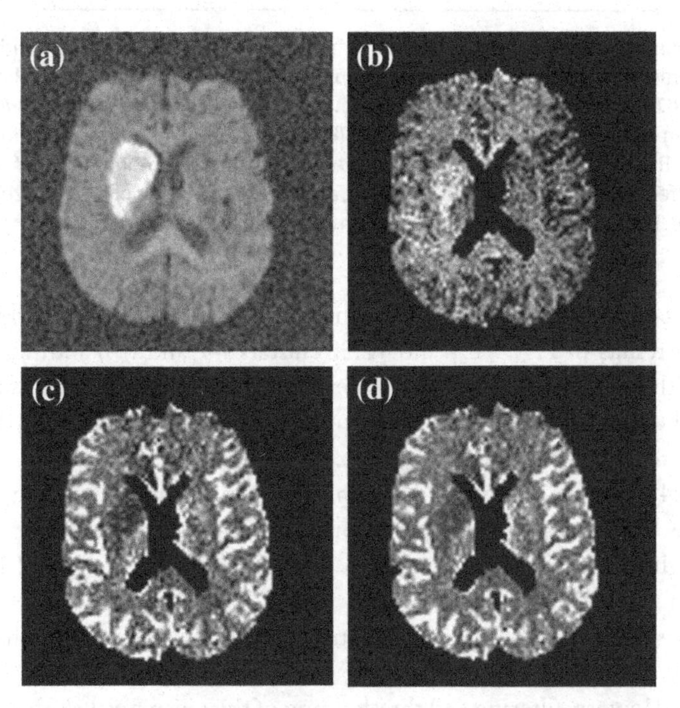

Figure 12.3 Diffusion-weighted MR image and conventional perfusion parameter maps of the same patient as in Figs. 12.1 and 12.2. (a) Diffusion-weighted MR image, (b) MTT map, (c) rCBV map, and (d) rCBF map [201].

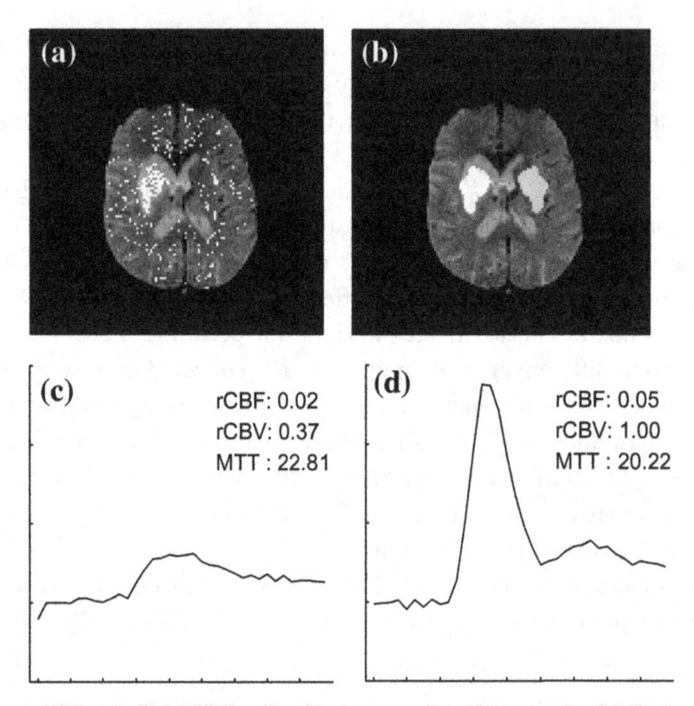

Figure 12.4 Quantitative analysis of the results for "neural gas" network in Fig. 12.1 with regard to side asymmetry of brain perfusion. (a) Best-matching cluster #7 of Fig. 12.1 representing the infarct region, (b) contiguous ROI constructed from (a) by spatial lowpass filtering and thresholding (white) and a symmetrical ROI at an equivalent contralateral position (light gray), (c) average concentration-time curve of the pixels in the ROI of the affected side, and (d) average concentration-time curve of the pixels in the ROI of the nonaffected side. For a better orientation, an anatomic EPI-scan of the analyzed slice is underlaid in (a) and (b). The x-axis represents the scan number while the y-axis is arbitrary for (c) and (d) [201].

at the same slice position in the region of the right basal ganglia. A visual inspection of the clustering results in Figs. 12.1 and 12.2 (clusters #7 and #8) shows a close correspondence with the findings of these parameter maps. In addition, the unsupervised and self-organized to clustering of pixels with similar signal dynamics allows a deeper insight into the spatiotemporal perfusion properties.

Figure 12.4 visualizes a method for comparative analysis of clustering results with regard to side differences of brain perfusion. The best-matching cluster #7 with the diffusion-weighted image corresponding to the infarct region in Fig. 12.1 is shown in Fig. 12.4a.

To better visualize the perfusion asymmetry between the affected and the nonaffected side, a spatially connected region of interest (ROI) can be obtained from the clustering results by spatial lowpass filtering and thresholding of the given pixel cluster. The resulting ROI is shown in Fig. 12.4b (white region). In addition, a symmetrical contralateral ROI can be determined (light gray region). Then, the mean CTCs value of all the pixels in the ROIs is determined and visualized in Fig. 12.4d together with the corresponding quantitative perfusion parameters: the difference between the affected (Fig. 12.4c) and the nonaffected (Fig. 12.4d) side with regard to CTC amplitude and dynamics is visualized, in agreement with highly differing corresponding quantitative perfusion parameters. Comparative quantitative analysis for fuzzy clustering based on deterministic annealing, self-organizing map, and the fuzzy *c*-means vector quantization is shown in Figs. 12.5–12.7.

The power of the clustering techniques is demonstrated also for a perfusion study in a control subject without evidence of cerebrovascular disease, as shown in Figs. 12.8 and 12.9. The conventional perfusion parameter maps together with a transversal T2-weighted scan at a corresponding slice position are presented in Fig. 12.10. Clusters #1, #3, #4, and #15 represent larger vessels primarily located in the cerebral sulci, while most of the other clusters seem to correspond to parenchymal vascularization. The important difference to the results of the stroke subject data in Figs. 12.1–12.3 and 12.5 is evident: the side asymmetry with regard to both the temporal pattern and the amplitude of brain perfusion is here nonexistent. This fact becomes obvious since each cluster in Fig. 12.1 contains pixels in roughly symmetrical regions of both hemispheres, different from the

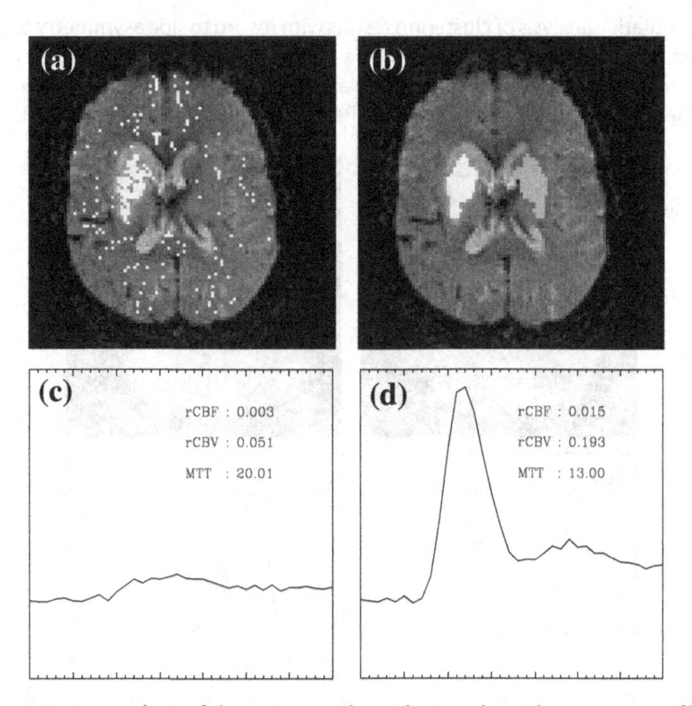

Figure 12.5 Quantitative analysis of clustering results with regard to side asymmetry of brain perfusion in analogy to Fig. 12.4 for vector quantization by fuzzy clustering based on deterministic annealing. For a better orientation, an anatomic EPI-scan of the analyzed slice is underlaid in (a) and (b). The x-axis represents the scan number while the y-axis is arbitrary for (c) and (d). Reprinted from [246] with permission from IEEE.

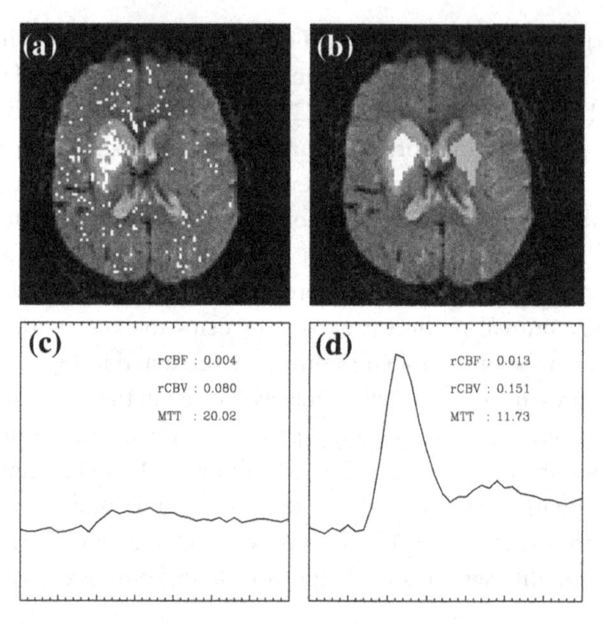

Figure 12.6 Quantitative analysis of clustering results with regard to side asymmetry of brain perfusion in analogy to Fig. 12.4 for vector quantization by a self-organizing map. For a better orientation, an anatomic EPI-scan of the analyzed slice is underlaid in (a) and (b). The x-axis represents the scan number while the y-axis is arbitrary for (c) and (d). Reprinted from [246] with permission from IEEE.

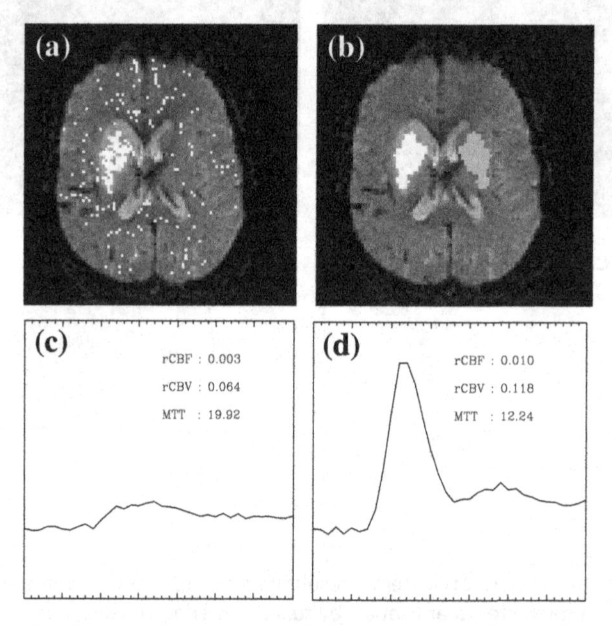

Figure 12.7 Quantitative analysis of clustering results with regard to side asymmetry of brain perfusion in analogy to Fig. 12.4 for fuzzy c-means vector quantization. For a better orientation, an anatomic EPI-scan of the analyzed slice is underlaid in (a) and (b). The x-axis represents the scan number while the y-axis is arbitrary for (c) and (d). Reprinted from [246] with permission from IEEE.

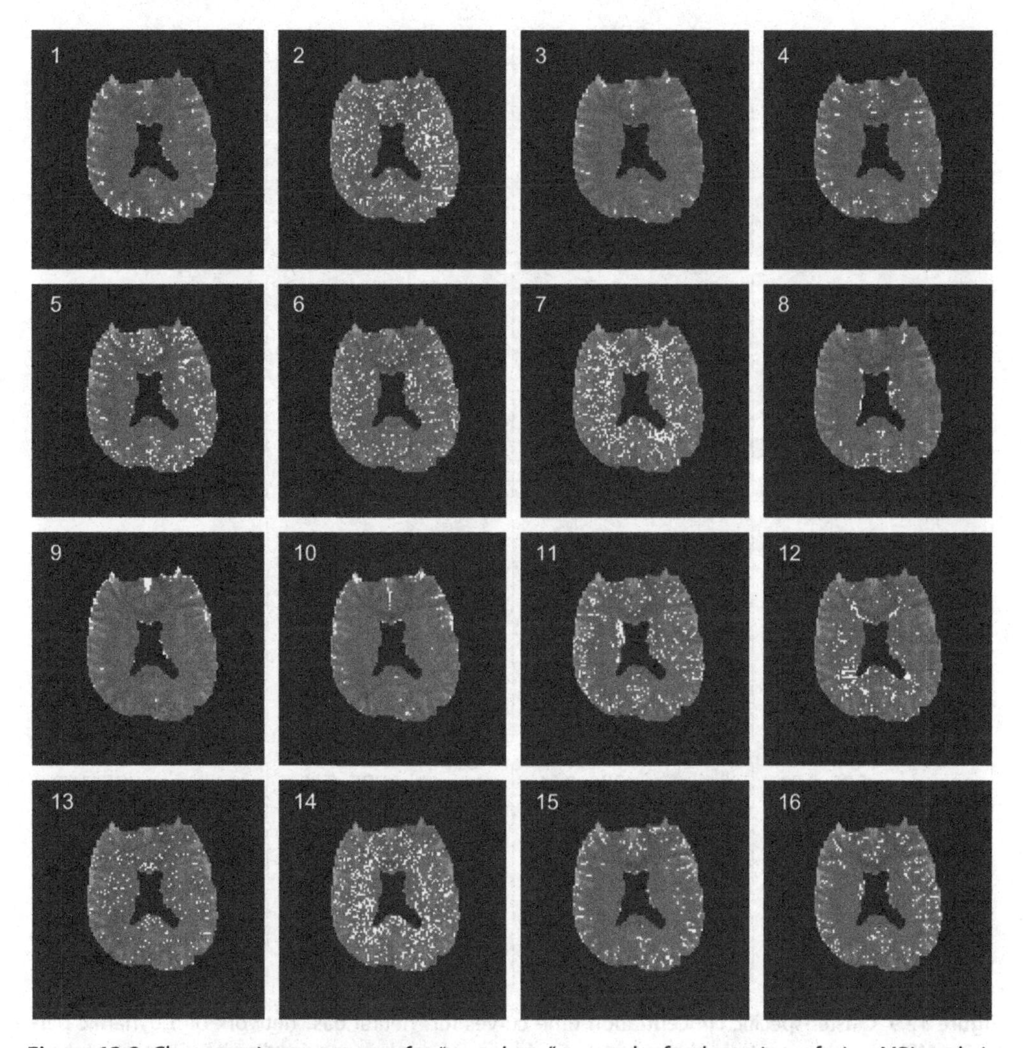

Figure 12.8 Cluster assignment maps for "neural gas" network of a dynamic perfusion MRI study in a control subject without evidence of cerebrovascular disease. For a better orientation, an anatomic EPI-scan of the analyzed slice is underlaid [201].

situation visualized in Fig. 12.1. In addition, no localized perfusion deficit results from the clustering. The clustering results of Figs. 12.8 and 12.9 match with the information derived from the conventional perfusion parameter maps in Figs. 12.10b–d.

The effectiveness of the different cluster validity indices and clustering methods in automatically evolving the appropriate number of clusters is demonstrated experimentally in the form of cluster assignment maps for the perfusion MRI data sets with the number of clusters varying from 2 to 36.

Table 12.2 shows the optimal cluster number K^* obtained for each perfusion MRI data set based on the different cluster validity indices.

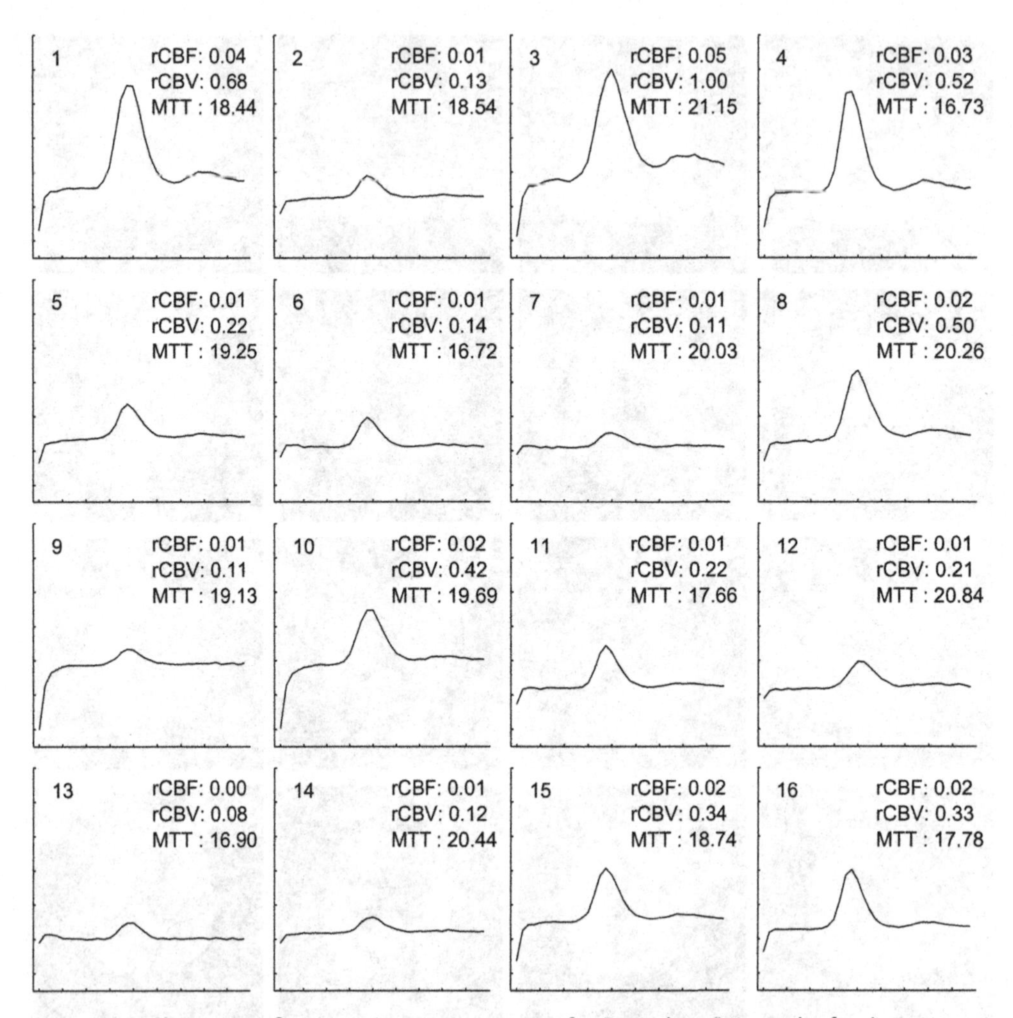

Figure 12.9 Cluster-specific concentration-time curves for "neural gas" network of a dynamic perfusion MRI study in a control subject without evidence of cerebrovascular disease. Cluster numbers correspond to Fig. 12.8. The x-axis represents the scan number while the y-axis is arbitrary [201].

As an example, Figs. 12.11 and 12.12 show results for cluster validity analysis for data set #1, representing the minimal rCBV obtained by the minimal free energy VQ, and the values of the three cluster validity indices depending on cluster number. The cluster-dependent curve for the rCBVs was determined based on the minimal obtained rCBV value as a result of the clustering technique for fixed cluster numbers. For each of the twenty runs of the partitioning algorithms, the minimal codebook-specific rCBV was computed separately. The cluster whose CTC shows the minimal rCBV is selected for the plot. The MTT of this CTC is indicated in the plot as well. The bottom part of the

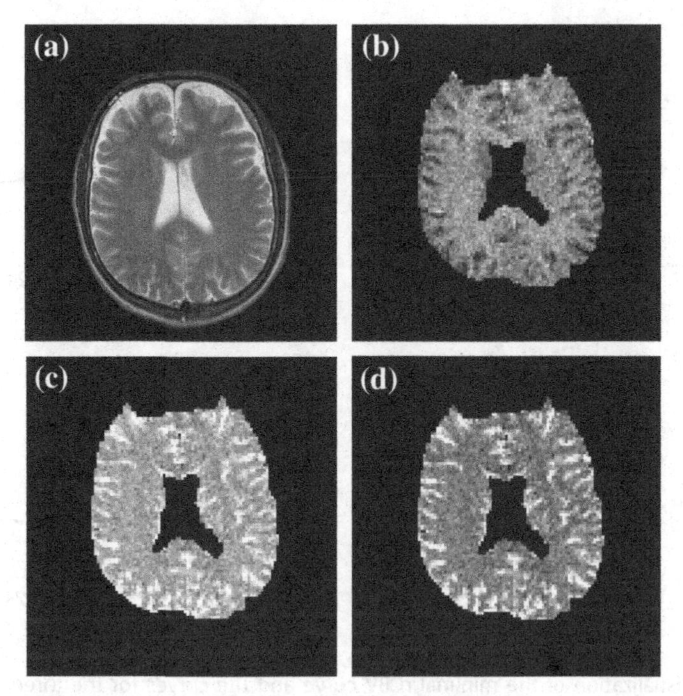

Figure 12.10 T2-weighted MR image and conventional perfusion parameter maps of the same subject as in Figs. 12.8 and 12.9. (a) T2-weighted MR image, (b) MTT map, (c) rCBV map, and (d) rCBF map. Reprinted from [246] with permission from IEEE.

Table 12.2 Obtained optimal cluster number K^* for the data sets #1 to #4 based on different cluster validity indices. The detailed curve for the cluster validity indices for data set #1 is shown exemplary in Figs. 12.11 and 12.12.

Index	Data set #1	Data set #2	Data set #3	Data set #4
K^*_{Kim}	18	6	10	12
K^*_{CH}	24	4	19	21
$K^*_{intraclass}$	3	3	3	3

Reprinted from [391] with permission from IEEE.

figure shows the cluster assignment maps for different cluster numbers corresponding to the optimal cluster number K^* and $K = K^* \pm 1$. The cluster assignment maps correspond to the cluster-specific concentration-time curves exhibiting the minimum rCBV.

The results show that based on the indices K_{Kim} and $K_{Intraclass}$ a larger number of clusters is needed to represent the data sets #1, #3, and #4.

In the following, the results of the quantitative ROC analysis are presented. An ROC curve for subject 1 in Fig. 12.13 using the "neural gas" network with $N = 16$ codebook vectors as the clustering algorithm is exemplarily shown.

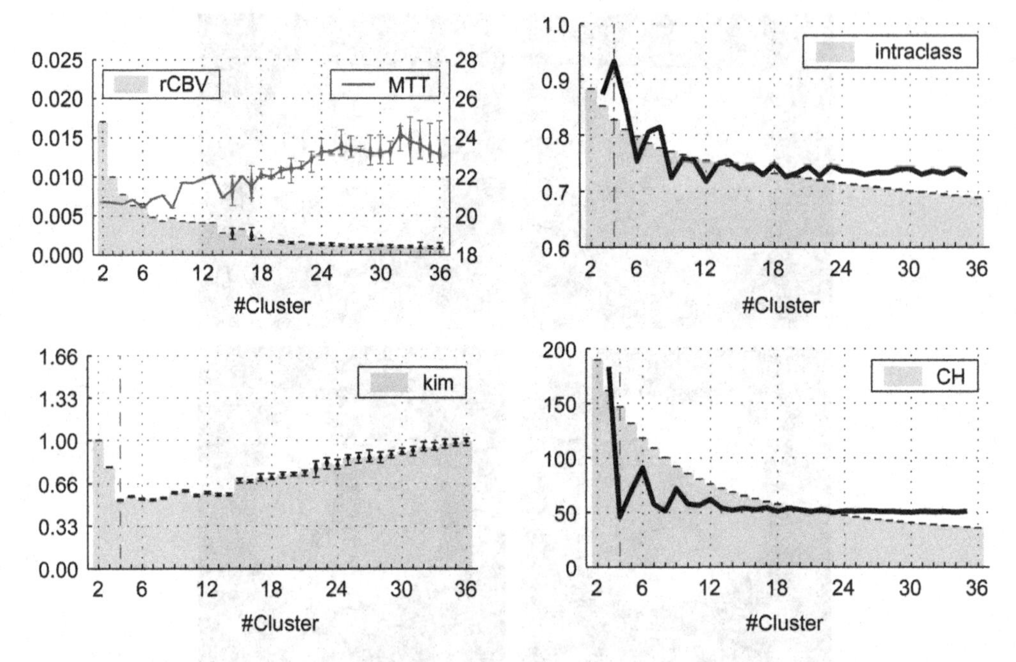

Figure 12.11 Visualization of the minimal rCBV curve and the curves for the three cluster validity indices, namely Kim's index, Calinski Harabasz (CH) index, and the intraclass index for data set #1 and as a result of classification based on the minimal free energy VQ. The cluster number is varied from 2 to 36. The average, minimal, and maximal values of 20 different runs using the same parameters but different-algorithms' initializations are plotted as vertical bars. For the intraclass and Calinski Harabasz validity indices, the second derivative of the curve is plotted as a solid line. Reprinted from [391] with permission from IEEE.

The clustering results are given for four subjects: subject 1 (stroke in the right basal ganglia), subject 2 (large stroke in the supply region of the middle cerebral artery (left hemisphere), and subjects 3 and 4 (both with no evidence of cerebrovascular disease). The codebook vectors from 3 to 36 for the proposed algorithms are varied and an ROC analysis using two different performance metrics is performed: the classification outcome regarding the discrimination of the concentration-time curves based on the rCBV-value and the discrimination capability of the codebook vectors based on their MTT-value. The ROC performances for the four subjects are shown in Fig. 12.14. The figure illustrates the average area under the curve and its deviations for 20 different ROC runs using the same parameters but different algorithms' initializations. The performed ROC analysis shows that rCBV outperforms MTT with regard to its diagnostic validity when compared to the conventional analysis serving as the gold standard in this study, as it can be seen from the larger area under the ROC curve for rCBV.

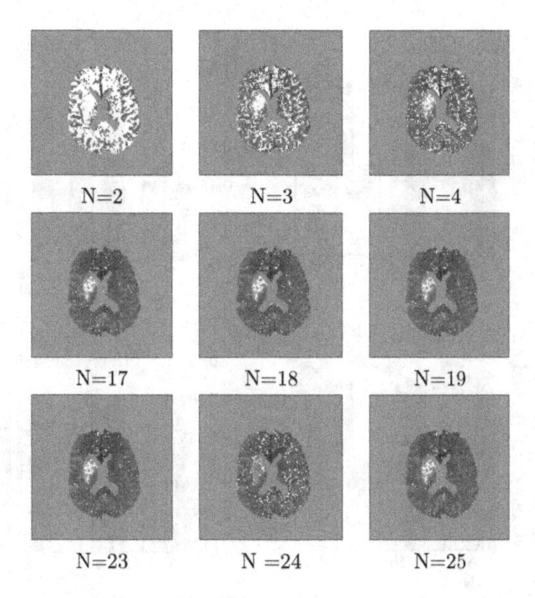

Figure 12.12 Cluster assignment maps for different cluster numbers corresponding to the optimal cluster numbers K^* and $K = K^* \pm 1$. The cluster assignment maps correspond to the cluster-specific concentration-time curves exhibiting the minimum rCBV. Reprinted from [391] with permission from IEEE.

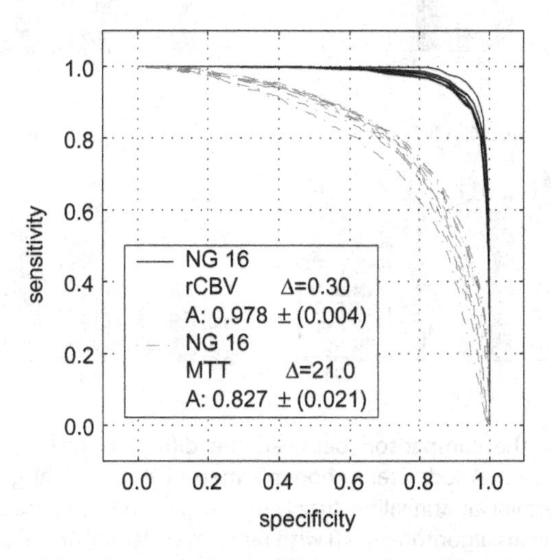

Figure 12.13 ROC curve of the cluster analysis of data set for subject 1 analyzed with the "neural gas" network for $N = 16$ codebook vectors. "A" represents the area under the ROC curve and Δ the threshold for rCBV/MTT. Reprinted from [246] with permission from IEEE.

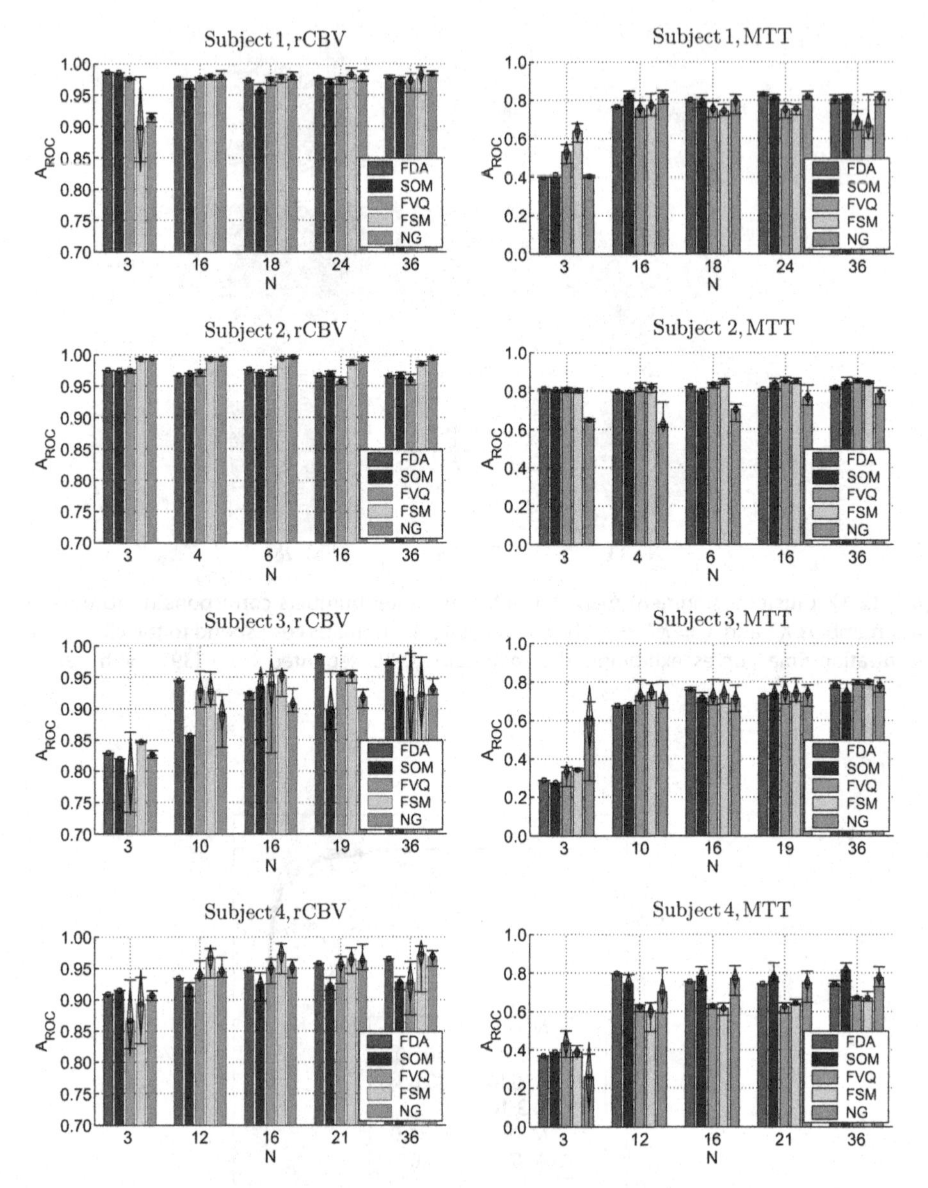

Figure 12.14 Results of the comparison between the different clustering analysis methods on perfusion MRI data. These methods are: Kohonen's map (SOM), "neural gas" network (NG), fuzzy clustering based on deterministic annealing, fuzzy *c*-means with unsupervised codebook initialization (FSM), and the fuzzy *c*-means algorithm (FVQ) with random codebook initialization. The average area under the curve and its deviations are illustrated for 20 different ROC runs using the same parameters but different algorithms' initializations. The number of chosen codebook vectors for all techniques is between 3 and 36 and results are plotted for four subjects. Subjects 1 and 2 had a subacute stroke, while subjects 3 and 4 gave no evidence of cerebrovascular disease. The ROC analysis is based on two performance metrics: regional cerebral blood volume (rCBV) (left column) and mean transit time (MTT) (right column) [201].

12.4. GENERAL ASPECTS OF TIME-SERIES ANALYSIS BASED ON UNSUPERVISED CLUSTERING IN DYNAMIC CEREBRAL CONTRAST-ENHANCED PERFUSION MRI

The benefits of unsupervised self-organized clustering over the conventional and single extraction of perfusion parameters are

(i) relevant information given by the signal dynamics of MRI time-series is not discarded,

(ii) a nonbiased interpretation as it results by the indicator-dilution theory of nondiffusible tracers only for an intact blood-brain barrier.

Nevertheless, clustering results support the findings from the indicator-dilution theory, since conventional perfusion parameters like MTT, rCBV, and rCBF values can be directly derived from the resulting prototypical cluster-specific CTCs.

The proposed clustering techniques were able to unveil regional differences of brain perfusion characterized by subtle differences of signal amplitude and dynamics. They could provide a rough segmentation with regard to vessel size, detect side asymmetries of contrast agent first pass, and identify regions of perfusion deficit in subjects with stroke.

In general, a minimal number of clusters is necessary to obtain a good partition quality of the underlying data set, which leads to a higher area under the ROC curve. This effect can clearly be seen for the subjects 3 and 4. For the data sets of subjects 1 and 2, the cluster number does not seem to play a key role. A possible explanation of this aspect is because of the large extent of the infarct area. Thus even with a smaller number of codebook vectors, it becomes possible to obtain a good separation of the stroke areas from the rest of the brain. Any further partitioning, obtained by increasing the number of codebook vectors, is not of crucial importance—the area under the curve does not change substantially. Also, for the patients without evidence of a cerebrovascular disease the area under the ROC curve is smaller than that for the subjects with stroke.

Three important aspects remain to be discussed: the interpretation of the codebook vector, the normalization of the signal-time curves, and the relatively high MTT values.

A codebook vector can be specified as a time-series representing the center (i.e., average) of all the time-series belonging to a cluster. Here, a cluster represents a set of pixels whose corresponding time-series are characterized by similar signal dynamics. Thus, "codebook vectors" as well as "clusters" are defined in an operational way that—at a first glance—does not refer to any physiological implications. However, it is common practice in the literature to conjecture [94] that similar signal characteristics may be induced by similar physiological processes or properties, although this cannot be proven definitely. It is very interesting to observe that the average values for the areas under the ROC curves seem to be higher for the patients with stroke in comparison to the patients without stroke. So far, no explanation can be given for this, however, this may be an important subject for further examination in future work. The different numbers of codebook vectors used for different subjects can be explained as follows: 16 and 36 codebook vectors were used for clustering in all data sets. In addition, the optimal number of clusters was determined by a detailed analysis using several so-called "cluster-validity criteria," namely Kim [176], Calinski-Harabasz (CH) [46], and intraclass [113].

In biomedical MRI time-series analysis considered here, a similar problem is faced: It is certainly not possible to interpret all details of the signal characteristics of the time-series belonging to each pixel of the data set as known physiological processes. Nevertheless, it may be a useful hypothesis to interpret the time-series of at least some clusters in the light of physiological meta-knowledge, although, a definite proof of such an interpretation will be missing. Hence, such an approach is certainly biased by subjective interpretation on the part of the human expert performing this interpretation of the resulting clusters, and may, thus, be subject to error. In summary, it is not claimed that a specific cluster is well correlated with physiological phenomena related to changes of brain perfusion, although one cannot exclude that a subjective interpretation of some of these clusters by human experts may be useful to generate hypotheses on underlying physiological processes in the sense of exploratory data analysis. These remarks are in full agreement with the whole body of literature dealing with unsupervised learning in MRI time-series analysis, such as [62,94].

The normalization of signal-time curves represents an important issue where the concrete choice depends on the observer's focus of interest. If cluster analysis is to be performed with respect to signal dynamics rather than amplitude, clustering should be preceded by time-series normalization. While normalization may lead to noise amplification in low-amplitude CTCs, in cluster analysis of signal time-series preceding normalization is not prohibitive. However, CTC amplitude unveils important clinical and physiological information, and therefore it forms the basis of the reasoning of not normalizing the signal-time curves before they undergo clustering.

In order to provide a possible explanation of the relatively high MTT values obtained in the results, the following should be mentioned: The rationale for using eq. (12.3) for computing MTT is that the arterial input function was not determined which is difficult to obtain in routine clinical diagnosis. The limitations of such an MTT computation have been addressed in detail in the theoretical literature on this topic, e.g., [409]. In particular, it has been pointed out that the signal intensity changes measured with dynamic MR imaging are related to the amount of contrast material remaining in tissue, not to the efflux concentration of contrast material. Therefore, if a deconvolution approach using the experimentally acquired arterial input function, e.g., according to [204,385], is not performed, eq. (12.3) can only be used as an approximation for MTT. However, this approximation has been widely used in the literature on both myocardial and cerebral MRI perfusion studies, e.g., [123,307,388].

In summary, the study shows that unsupervised clustering results are in good agreement with the information obtained from conventional perfusion parameter maps, but may sometimes unveil additional hidden information e.g., disentangle signals with regard to different vessel sizes. In this sense, clustering is not a competitive, but a complementary additional method that may extend the information extracted from conventional perfusion parameter maps by taking into account fine-grained differences of MRI signal dynamics in perfusion studies. Thus, the presented techniques can contribute to exploratory visual analysis of perfusion MRI data by human experts as a complementary approach to conventional perfusion parameter maps. It provides computer-aided support to appropriate data processing in order to assist the neuroradiologist, and not to replace his/her interpretation. In addition, following further piloting on larger samples, the nature of additional information can be better clarified, as the proposed techniques should be applicable in a larger group to assess validity and reliability. In conclusion, clustering is a useful extension to conventional perfusion parameter maps.

CHAPTER THIRTEEN

Computer-Aided Diagnosis for Diagnostically Challenging Breast Lesions in DCE-MRI

Contents

13.1. INTRODUCTION

Breast cancer is the most common cancer among women, but has an encouraging cure rate if diagnosed at an early stage. Thus, early detection of breast cancer continues to be the key for effective treatment. Magnetic resonance (MR) imaging is an emerging and promising new modality for detection and further evaluation of clinically, mammographically, and sonographically occult cancers [134,402]. Acquisition of temporal sequences of between three and six MR images depicting the kinetics of contrast agent molecules in the breast tissue allows for detecting and assessing suspicious tissue disorders with high sensitivity, even in the mammographically dense breasts of young women. Yet, the

Pattern Recognition and Signal Analysis in Medical Imaging
http://dx.doi.org/10.1016/B978-0-12-409545-8.00013-3

multitemporal nature of the three-dimensional image data poses new challenges to radiologists as the key-information, reflected by subtle temporal changes of the signal intensity, is only perceivable if all images of the temporal sequence are considered simultaneously.

In conventional X-ray mammography, computer-aided diagnosis (CAD) systems are being developed to expedite diagnostic and screening activities and are today moving from research to routine application in daily clinical practice. With breast cancer being an issue of enormous clinical importance with obvious implications to healthcare politics, much effort is spent today on research of similar techniques to aid or even automatize diagnosis in breast MRI.

The success of CAD in conventional X-ray mammography motivated the research of automated diagnosis techniques in breast MRI to expedite diagnostic and screening activities.

A standard multilayer perceptron (MLP) was applied to the classification of signal-time curves from dynamic breast MRI in [225]. Breast MR segmentation and lesion detection are accomplished based on cellular neural networks in [87] and a 100% detection sensitivity is reported. In [360], the performance of a backpropagation neural network based on kinetic, morphologic, and combined MR features, was shown to be comparable to that of an expert radiologist. The same type of neural network was used for breast MRI lesion classification in [242]. As inputs, a subset of 13 features out of a total of 42 features describing lesion shape, texture, and enhancement kinetics was selected. The main result was that the performance of the human readers significantly improved when aided by a CAD system. It could be shown that specificity at a sensitivity of 90% was 0.505 for lesion classification without CAD assistance and 0.807 for classification with CAD assistance.

Mean shift clustering in connection with automated selection of the most suspicious cluster resulted in accurate ROIs in breast MRI lesions, as shown in [354]. In [58], a fuzzy c-means clustering-based technique was tested for automatically identifying characteristic kinetics from breast lesions. By using four features extracted from these curves (maximum contrast enhancement, time to peak, uptake rate, and washout rate of the lesion kinetics), it was demonstrated that the prototype curves determined by the fuzzy classifier outperform those determined based on averaging over an ROI determined by an experienced radiologist. It was shown that the quantitative classifiers can support the radiologist in the diagnosis of breast lesions.

The computer-assisted interpretation of time-signal series as measured during a dynamic contrast-enhanced MR (DCE-MR) examination for each image voxel represents one of the major steps in designing CAD systems for breast MRI. Kuhl et al. have shown that the shape of the time-signal intensity curve represents an important criterion in differentiating benign and malignant enhancing lesions in DCE-MR imaging [191]. The results indicate that the enhancement kinetics, as represented by the time-signal intensity curves visualized in Fig. 13.1, differ significantly for benign and malignant enhancing lesions and thus represent a basis for differential diagnosis: plateau or

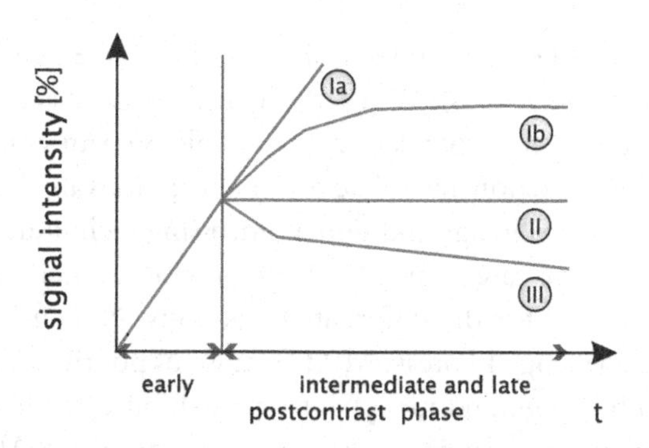

Figure 13.1 Schematic drawing of the time-signal intensity (SI) curve types [191]. Type I corresponds to a straight (Ia) or curved (Ib) line; enhancement continues over the entire dynamic study. Type II is a plateau curve with a sharp bend after the initial upstroke. Type III is a washout-time course. In breast cancer, plateau or washout-time courses (type II or III) prevail. Steadily progressive signal intensity time courses (type I) are exhibited by benign enhancing lesions.

washout-time courses (type II or III) prevail in cancerous tissue. Steadily progressive signal intensity time courses (type I) are exhibited by benign enhancing lesions, albeit these enhancement kinetics are shared not only by benign tumors but also by fibrocystic changes.

Even though the time-signal courses enable radiologists to distinguish different tissue states, assessing the signal characteristics is a time-consuming task which becomes further complicated due to the heterogeneity of lesion tissue causing the signal characteristics to vary spatially. Also this spatial variation of the signal characteristics reflects specific tissue properties which should be taken into account for assessing the state of lesions.

Morphologic criteria have also been identified as valuable diagnostic tools [334]. Visual assessment of morphological properties is a highly inter-observer variable [353], while automated computation of features leads to more reproducible indices and thus to a more standardized and objective diagnosis. Recently, combinations of different dynamic and morphologic characteristics have been reported [1,359] that can reach diagnostic sensitivities up to 97% and specificities up to 76.5%. Many of these studies were performed in the preoperative staging of patients with suspicious lesions (BI-RADS 4 and 5) including predominantly tumors with an extension greater than 2 cm. In such cases, magnetic resonance imaging (MRI) reaches a very high sensitivity in the detection of invasive breast cancer due to both, the typical appearance (ill-defined shape, stellate borders, and rim enhancement) of malignant tumors and characteristic SI time courses of contrast enhancement. Recent clinical research has shown that DCIS with small invasive carcinoma can be adequately visualized in MRI [378] and that MRI provides an accurate estimation of invasive breast cancer tumor size, especially in tumors of 2 cm or smaller [115].

Based on morphology and type of enhancement, lesions are classified according to the Breast Imaging Reporting and Data System (BIRADS) lexicon into: mass enhancement (three-dimensional tumor that has either a round, oval, lobular, or irregular shape), focus (tiny spot of enhancement less than 5 mm), and non-mass-like enhancement (enhancement of an area that is not a mass). While the diagnosis of masses is based on typical characteristic parameters such as spiculation (morphology), rim enhancement (texture), and washout kinetics, foci and non-mass-like enhancing lesions are diagnostically far more challenging and require novel image and signal processing techniques to be integrated in an automated system. In the case of mass-enhancing lesions, there are several BIRADS descriptors that can be used for the differential diagnosis. Non-mass-enhancing lesions represent a diagnostic challenge in breast MRI because of poorly defined boundaries and because lesions of both benign and malignant type exhibit considerable kinetic overlap compared to mass-enhancing lesions, see Fig. 13.2. Existing BIRADS descriptors have proven to be insufficient to aid in the automated differential diagnosis of these lesions.

However, more than 40% of the false-negative MR diagnosis are associated with non-mass-like enhancing lesions and thus indicating a lower sensitivity of MRI for these cases. It has been shown that double reading achieves a higher sensitivity but is time-consuming and as an alternative a computer-assisted system was suggested [265]. The success of CAD in conventional X-ray mammography [120,192,370,371,372] motivates furthermore the research of similar automated diagnosis techniques in breast MRI.

Non-mass-enhancing lesions exhibit a heterogeneous appearance in breast MRI with high variations in kinetic characteristics and typical morphological parameters [313,326,398], and have a lower reported specificity and sensitivity than mass-enhancing lesions. The diagnosis of non-mass-like enhancement lesions is thus far more challenging. Lesions of both benign and malignant type exhibit considerable kinetic overlap compared to mass-enhancing lesions, see Fig. 13.2. Malignant lesions such as ductal carcinoma in situ (DCIS) and invasive lobular cancer (ILC) exhibit a segmental or linear enhancement pattern and benign lesions such as fibrocystic changes present as well a non-mass-like enhancement [376]. However, a systematic classification of non-mass-like enhancing lesions is not in place. A classification of such lesions would be highly beneficial since they may reduce the biopsies' numbers. The morphological parameters with the highest predictive value in non-mass-enhancing lesions were reported in [326] as segmental distribution, clustered ring enhancement, and a clumped internal architecture. Another study has shown that including kinetic data on dynamic contrast-enhanced imaging and diffusion-weighted MRI imaging in addition to morphological characteristics showed a high diagnostic accuracy in characterization of these lesions [398].

Research initiatives have been focused on automated analysis of mass lesions [58,87,226,250,360] while very few studies investigated the characterization of the morphology and/or enhancement kinetic features of non-mass lesions [150,152,260,376]. The studies showed a much lower sensitivity and specificity for non-mass-like enhancement lesions compared with masses and suggested the need for more advanced algorithms for the diagnosis of non-mass-like enhancement.

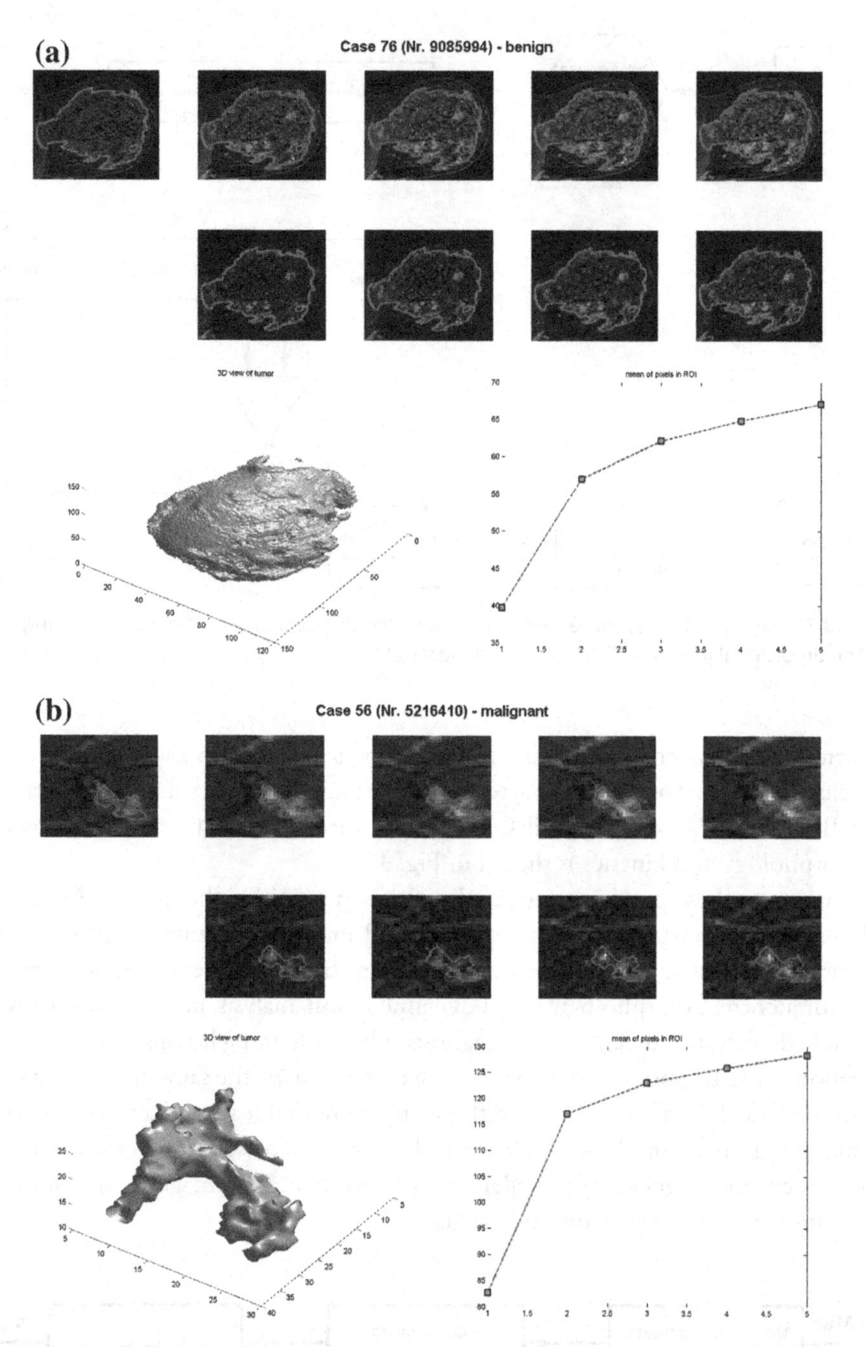

Figure 13.2 Morphological and dynamic representations of segmented benign (diffusely enhancing glandular tissue) and malignant (invasive ductal carcinoma) non-mass-like-enhancing lesions. The time scans in the first row are without motion compensation while those in the second row are motion-corrected. (a) Benign and (b) Malignant .

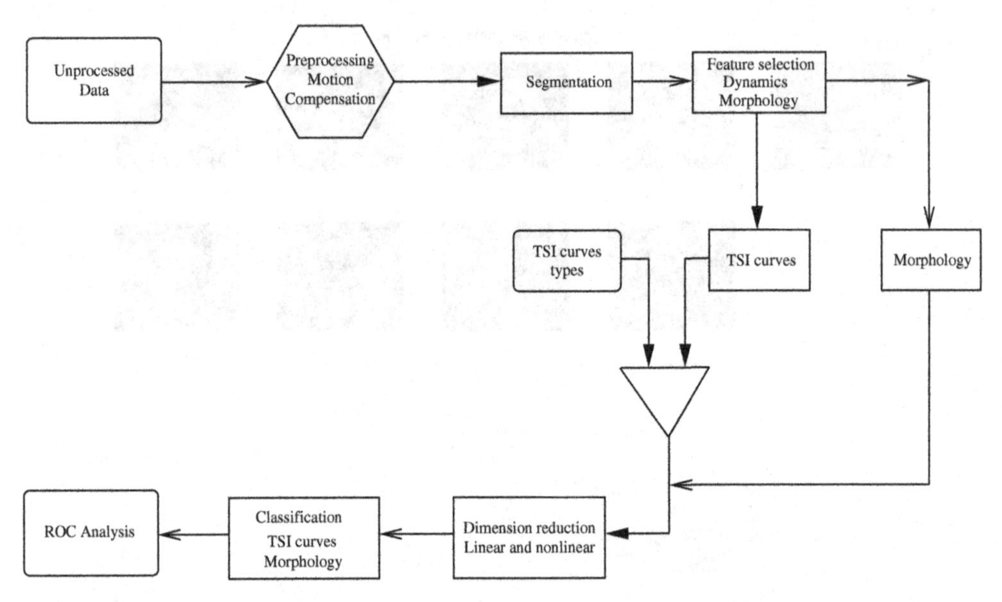

Figure 13.3 Diagram of a computer-assisted system for the evaluation of mass-enhancing lesions based on time-signal intensity (TSI) curves and morphology.

Current CAD systems are specialized for automated detection and diagnosis of mass-enhancing lesions that are well characterized by shape and kinetics descriptors according to the BIRADS lexicon. A typical CAD system for mass-enhancing lesions based on both morphology and kinetics is shown in Fig. 13.3.

To overcome these problems and revolutionize the state-of-the-art in CAD in breast MRI, we need to focus on correctly capturing and analyzing the unique spatio-temporal behavior of non-mass-enhancing lesions. Image registration and segmentation are fundamental for a correct morphological and dynamic lesion analysis and dramatically impact the correct detection and diagnosis of diagnostically challenging lesions. Therefore, spatial registration has to be performed before feature extraction. At the same time, accurate segmentation of the lesion is critical since the spatio-temporal features have to be extracted from tumor region. Figure 13.4 visualizes the flow diagram of a comprehensive computer-assisted system for diagnostically challenging lesions including image registration, lesion segmentation, feature, extraction, and evaluation.

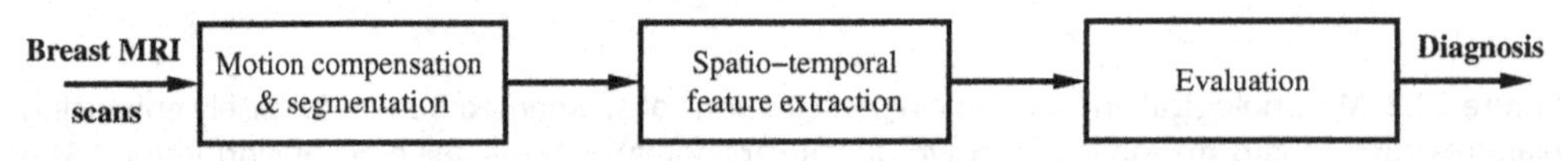

Figure 13.4 Diagram of a computer-assisted system for the evaluation of diagnostically challenging contrast enhancing lesions.

This chapter describes some important CAD systems for diagnostically challenging breast lesions in breast MRI. A CAD system for small lesion detection [348,349] using integrated morphologic and dynamic characteristics and one for non-mass-like-enhancing lesions based on spatio-temporal features [136] is described.

13.2. MOTION COMPENSATION

Automatic motion correction represents an important prerequisite to a correct automated challenging lesion evaluation [22, 135]. Motion artifacts are caused either by the relaxation of the pectoral muscle or involuntary patient motion and invalidate the assumption of same spatial location within the breast of the corresponding voxels in the acquired volumes for assessing lesion enhancement. Due to the elasticity and heterogeneity of breast tissue, only nonrigid image registration methods are suitable.

A common nonrigid motion compensation method [284] is based on the Horn and Schunck method and represents a variational method for computing the displacement field, the so-called optical flow u, in an image sequence with movement in between the image acquisitions.

The motion compensation algorithm represents a variational method for computing the displacement field, the so-called optical flow, in an image sequence. It is based on two typical assumptions for variational optical flow methods, the brightness constancy and smoothness assumption.

In this context, the MR image sequence f_0 is a differentiable function of brightness values on a four-dimensional spatio-temporal image domain Ω:

$$f_0 : \underbrace{\Omega}_{\subset R^3} \times R_+ \to R_+ \tag{13.1}$$

From this image sequence, we want to compute a dense vector field $\mathbf{u} = (u_1, u_2, u_3)^\top$: $\Omega \to R^{\{2,3\}}$ that describes the motion between the precontrast image at time point t and a postcontrast image at time point $t + k$, either for all three dimensions (R^3) or only in one transversal slice (R^2).

The initial image sequence f_0 is preprocessed and convolved with a Gaussian K_σ of a standard deviation σ:

$$f = K_\sigma * f_0 \tag{13.2}$$

The brightness constancy assumption dictates that under the motion $\mathbf{u}$, the image brightness values of the precontrast image at time t and the postcontrast image at time $t + k$ remain constant in every pixel:

$$f(\mathbf{x} + \mathbf{u}(\mathbf{x}), t + k) = f(\mathbf{x}, t) \quad \forall \mathbf{x} \in \Omega \tag{13.3}$$

Naturally, this condition by itself is not sufficient to describe the motion field $\mathbf{u}$ properly, since for a brightness value in an image voxel in the precontrast image, there are generally many voxels in the postcontrast image with the same brightness value, or, in the presence of noise, possibly even none at all. Therefore, we include the smoothness assumption, dictating that neighboring voxels should move in the same direction, which is expressed as the gradient magnitude of the flow field components is supposed to be 0.

$$|\nabla u_{\{1,2,3\}}(\mathbf{x})| = 0, \quad \forall \mathbf{x} \in \Omega \tag{13.4}$$

This constraint by itself would force the motion field to be a rigid translation, which is not the case in MR images. However, if we use both the brightness constancy assumption and the smoothness assumption as weak constraints in an energy formulation, the motion field **u** that minimizes this energy matches the postcontrast image to the precontrast image and is spatially smooth.

The variational method is based on the minimization of the continuous energy functional which penalizes all deviations from model assumptions

$$E(\mathbf{u}) = \int_{\Omega} \underbrace{(f(\mathbf{x} + \mathbf{u}(\mathbf{x}), t + k) - f(\mathbf{x}, t))^2}_{\textit{Data term}} \tag{13.5}$$

$$+ \alpha \underbrace{\left(|\nabla u_1(\mathbf{x})|^2 + |\nabla u_2(\mathbf{x})|^2 + |\nabla u_3(\mathbf{x})|^2\right)}_{\textit{Smoothness term}} \mathrm{d}\mathbf{x} \tag{13.6}$$

The weight term $\alpha > 0$ represents the regularization parameter where larger values correspond to smoother flow fields. This technique is a global method where the filling-in-effect yields dense flow fields and no subsequent interpolation is necessary as with the technique proposed in [132]. This method works within a single variational framework.

Given the computed motion from the precontrast image to a postcontrast image, the postcontrast image is being registered backwards before its difference image with the precontrast image is being computed for tumor classification:

$$f_{\text{post-registered}}(\mathbf{x}) = f_{\text{post}}(\mathbf{x} + \mathbf{u}(\mathbf{x})) \tag{13.7}$$

Breast MR images are mostly characterized by brightness, since bright regions are created by fatty tissue or contrast agent enhancing tumor tissue while dark regions describe glandular tissue and background. Tumors are mainly located in the glandular tissue and proliferate either into the fatty tissue (invasive) or along the boundary inside the glandular tissue (non-invasive).

Two major concerns have to be addressed when applying the optical flow approach to breast MRI: (1) The constancy assumption does not hold for objects appearing from one image to the next such as lesions for which the contrast agent enhancement is much stronger than in the surrounding tissue and (2) the lack of constant grid size in all directions since voxel size is smaller than the slice thickness. The first concern is alleviated by masking suspicious areas by a radiologist and by detecting the sharp gradients in the

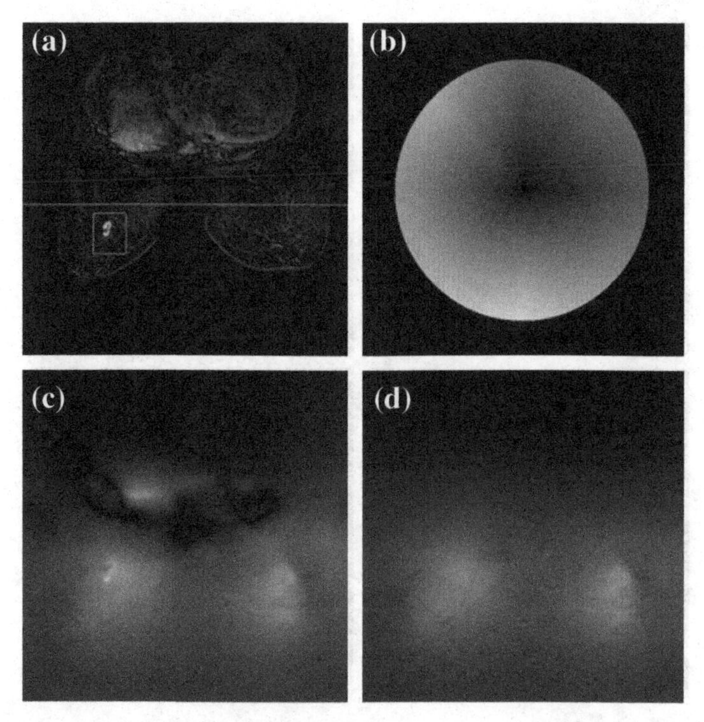

Figure 13.5 Motion detection on a transverse image. (a) Masking the data term: The green lines separate the boundary between masked and unmasked areas. (b) Color code describing motion from the interior of the image. (c) Motion in two directions determined without a mask, and (d) based on the mask from (a). The values for the standard deviation of the Gaussian presmoothing kernel and for the smoothness term are $\sigma = 3$ and $\alpha = 500$. (For interpretation of the references to color in this figure legend, the reader is referred to the web version of this book.)

motion field in the unmasked image. Figure 13.5 shows the masking of the entire upper image and visualizes the motion in the inner slice by a color code. This color code describes the motion direction based on the hue and the motion magnitude based on the brightness and thus identifies suspicious regions by detecting the sharp gradients.

The mask $m : \Omega \to \{0, 1\}$ can be easily incorporated in the energy formulation and it forces the data term to disappear in suspicious regions:

$$E(\mathbf{u}) = \int_{\Omega} m(\mathbf{x}) \left(f(\mathbf{x} + \mathbf{u}(\mathbf{x}), t + k) - f(\mathbf{x}, t) \right)^2 \tag{13.8}$$

$$+ \alpha \left(|\nabla u_1(\mathbf{x})|^2 + |\nabla u_2(\mathbf{x})|^2 + |\nabla u_3(\mathbf{x})|^2 \right) \mathrm{d}\mathbf{x} \tag{13.9}$$

In addition, there can be gaps between the slices where no nuclei are being excited in order to avoid overlapping of the slices.

Figure 13.6 shows an example of a tumor considerably shifting its position on adjacent transversal slices.

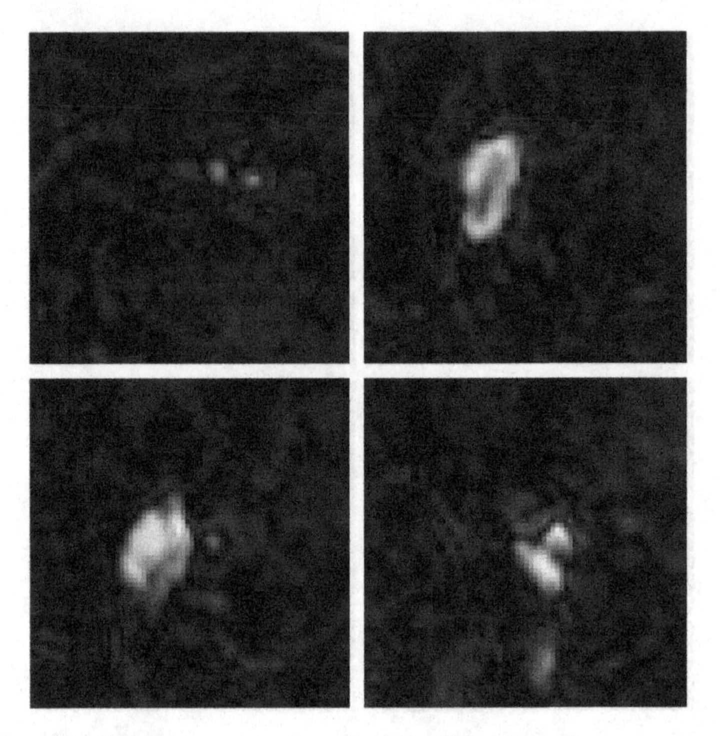

Figure 13.6 Tumor in adjacent transverse slices of a 512 × 512 × 32 image.

To overcome the second concern, it is important to decide whether the motion in transverse direction is having a significant impact. The present research has shown that there is no significant difference in visual quality if motion is computed in two or three directions. In the following notation, we will consider the motion in three directions, the one in two directions is analogous. This technique is a global method where the filling-in-effect yields dense flow fields and no subsequent interpolation is necessary as with the technique proposed in [132]. This method works within a single variational framework. This technique overcomes the aperture problem, provides subpixel accuracy, and can be easily enhanced and adapted.

The optimal motion correction results were achieved for motion compensation in two directions for mostly small standard deviations of the Gaussian kernel and smoothing parameter [136]. Alternatively, a motion compensation algorithm based on the technique described in [40] can be employed and separate robustification in the data term can be used.

We show in Fig. 13.7 an example of how the motion compensation algorithm works. The flow field between the pre- and postcontrast image as well as the motion compensated postcontrast image are visualized. Here the three-dimensional image is represented by the middle transverse slice as usual. Although the brightness and thus magnitude of the

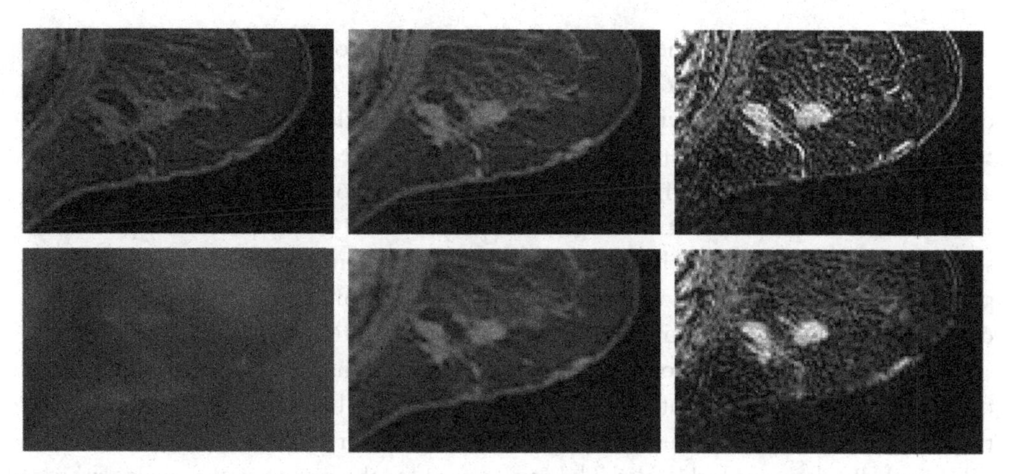

Figure 13.7 Example results of the motion compensation algorithm. Top left: Middle slice of pre-contrast image of a tumor. Top middle: Original second postcontrast image. Top right: Subtraction image of original images (linearly re-scaled). Bottom left: Color code of computed flow field. Bottom middle: Motion compensated postcontrast image. Bottom right: Subtraction image using the motion compensated image (linearly re-scaled).

depicted displacement field is relatively small, it is still very useful to remove small artifacts in the subtraction image. Due to the motion compensation the images are more aligned with each other and thus the boundary of the breast becomes also less visible since it is not enhanced. Similar improvements can also be observed in the tumor itself and the surrounding tissue. Note that the images as well as the flow consist of three dimensions. This means the algorithm tries to find the best possible displacement in all three directions. This is the reason why the structures in the original and the motion compensated image can look differently when only considering one slice, because parts from different slices, not visible in the example, can have an influence on the depicted slice and vice versa.

13.3. LESION SEGMENTATION

Tumor segmentation represents the correct identification of the spatial location of a tumor. Manual segmentation performed by a radiologist is considered the gold standard. However, expert segmentation is not highly precise, prone to interobserver and intra-observer variability and it might include also nonenhancing tissue. It is time-consuming by viewing both spatial and temporal profiles and thus examining many series of enhanced data and profiles of pixels while determining the lesion boundary.

Several approaches are known in the literature for breast lesion segmentation. In [105], the segmentation is based on a user-defined seed point and surrounding sphere containing weighted voxels of background and lesion followed by a threshold maximizing the interclass variance of these voxels. A fuzzy-based clustering segmentation method is developed in [58] while mean shift clustering is used for ROI selection in connection with a connected threshold analysis. In [241], a segmentation method for masses is described based on the following steps: interactive lesion selection and threshold estimation, connected component analysis, and hole-filling and leakage removal. However the algorithm is suitable only for mass-enhancing lesions and would require modifications for nonmass lesions. The segmentation algorithm proposed in [407] requires a rough manual segmentation and refines this based on a graph-cut based energy minimization. In [339] a level set segmentation method was applied in combination with fuzzy c-means (FCM) clustering for both initial segmentation and level set evolution.

In an interactive region growing algorithm, each MR image has to be segmented into two regions, the region of interest (ROI), i.e., the voxels belonging to the tumor, and the background. Thus, a binary mask is created: the tumor voxels are assigned the true value, and all other voxels the false. The image used for the region growing algorithm is the difference image of the second postcontrast image and the native precontrast image. The center of the lesion is interactively marked on one slice of the subtraction images and then a region growing algorithm includes all adjacent contrast-enhancing voxels as well as those from neighboring slices. Thus a 3-D form of the lesion is determined. An interactive ROI is necessary whenever the lesion was connected with diffuse contrast enhancement, as it is the case in mastopathic tissue.

Figure 13.8 shows a transverse image of a tumor in the right breast and its binary segmentation, created with region growing.

To overcome the problems of these previous methods, an automatic segmentation method can be employed alternatively based on an active contour segmentation without edges as proposed by Chan and Vese [54] and improve the algorithm by taking into account the three-dimensional image sequence and by adding a smoothness assumption to level set function:

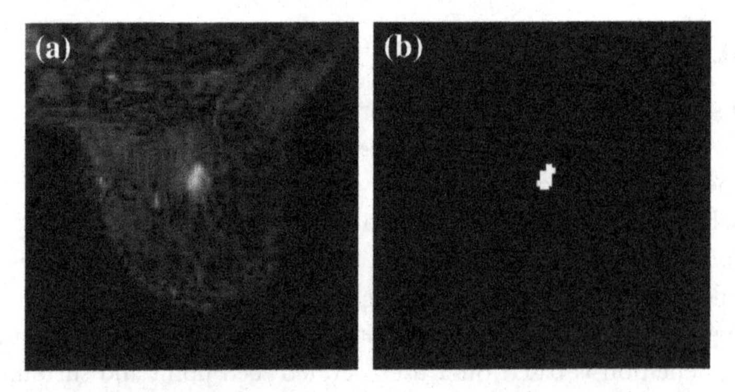

Figure 13.8 Example of an MR image segmentation showing the transverse image of a tumor in the right breast (a) and the binary segmentation of the tumor (b).

Using FDM (finite difference methods) and solving the following Hamilton-Jacobi equation (Euler-Lagrange equation corresponding to active contour without edges) give the boundary of target with given initial level set function $\phi_0(x, y, z)$:

$$\begin{aligned} \frac{\partial \phi}{\partial t} &= \delta_\epsilon \left[\mu \nabla \cdot \left(\frac{\nabla \phi}{|\nabla \phi|} \right) - \lambda_1 \frac{1}{n} \sum_{d=1}^{n} (f_d - c_d^{in})^2 + \frac{1}{n} \sum_{d=1}^{n} \lambda_2 (f_d - c_d^{out})^2 \right] \\ &\quad + \theta \nabla \cdot \left(\frac{\nabla \phi}{|\nabla \phi|} \right) = 0 && \text{in } (0, \infty) \times \Omega \\ \phi(0, x, y, z) &= \phi_0(x, y, z) && \text{in } \Omega \\ \frac{\delta_\epsilon(\phi)}{|\nabla \phi|} \frac{\partial \phi}{\partial \mathbf{n}} &= 0 && \text{on } \partial\Omega \end{aligned} \tag{13.10}$$

where $\{f_d\}_{d=1,\dots,n}$ is the image sequence for a given case consisting of n three-dimensional images. In our case $n = 5$. $c^{in} = (c_1^{in}, \dots, c_n^{in})^T$ are the average gray values for each image for the inside region, and similar $c^{out} = (c_1^{out}, \dots, c_n^{out})$ for the outside region. Since the contrast agent has a certain in-take-time, the consecutive images will be weighted more. The weighting factors $\gamma_1, \gamma_n \geq 0$ take into consideration accordingly this information content of an image. We also use $\delta_\epsilon(x) := \frac{1}{2}\left(1 + \frac{2}{\pi} \arctan\left(\frac{x}{\epsilon}\right)\right)$, $\mathbf{n} :=$ outer normal direction of $\partial\Omega$, and scalar parameters $\mu, \lambda_1, \lambda_2$.

Let $\phi^N(x, y, z)$ be the Nth step evolution of $\phi(t, x, y, z)$ in FDM. Then the curve γ for detected boundary is defined

$$\gamma := \{(x, y, z) \subset R^3 : \phi^N(x, y, z) \text{ for } \|\phi^N(x, y, z) - \phi^{N-1}(x, y, z)\| < \tau\} \tag{13.11}$$

where τ is the tolerance. To get a good approximation of a global optimizer of this nonconvex problem we make use of a coarse-to-fine multigrid implementation of the algorithm. The parameter $\eta \in [0.5, 1)$ thereby determines the downsampling factor.

For the numerical simulation we choose the following parameters: $\lambda_1 = 1, \lambda_2 = 1$, $(\gamma_1, \gamma_2, \gamma_3, \gamma_4)^T = (1, 3, 2, 1)^T, \eta = 0.8, \theta = 0.1$. An example of the proposed segmentation method can be seen in Fig. 13.9.

13.4. FEATURE EXTRACTION

13.4.1 Contour Features

To represent the shape of the tumor contour, the tumor voxels having non-tumor voxel as a neighbor can be extracted to represent the contour of the tumor. In this context, neighbor voxels include diagonally adjacent voxels, but not voxels from a different transverse slice. Due to the different grid sizes in the three directions of the MR images and possible gaps between transverse slices, the tumor contour in one transverse slice does not necessarily continue smoothly into the next transverse slice. Considering tumor contours between transverse slices therefore introduces contour voxels that are completely in the tumor

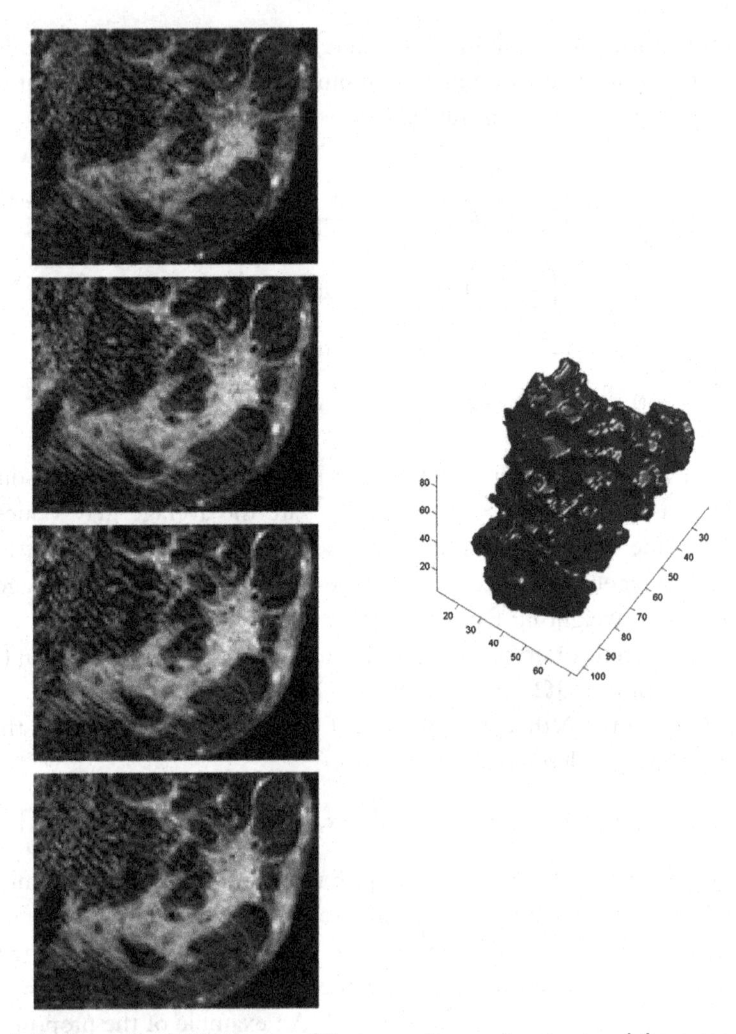

Figure 13.9 Left: Four subtraction images of the respective postcontrast and the precontrast image. The result of the segmentation is shown as a red line. ($\mu = 100.0, \alpha = 15.0$) Right: 3-D view of the segmented tumors. (For interpretation of the references to color in this figure legend, the reader is referred to the web version of this book.)

interior in one slice. This is illustrated in Fig. 13.10: the dark voxels are contour voxels and the arrows indicate the computed contour chain. If voxels in the tumor having at least one non-tumor voxel as a neighbor on an adjacent transverse slice were considered part of the contour, in this example, the crossed-out voxels would belong to the contour. Figure 13.11 shows an example for a tumor where the contour shifts considerably from one transverse slice to another.

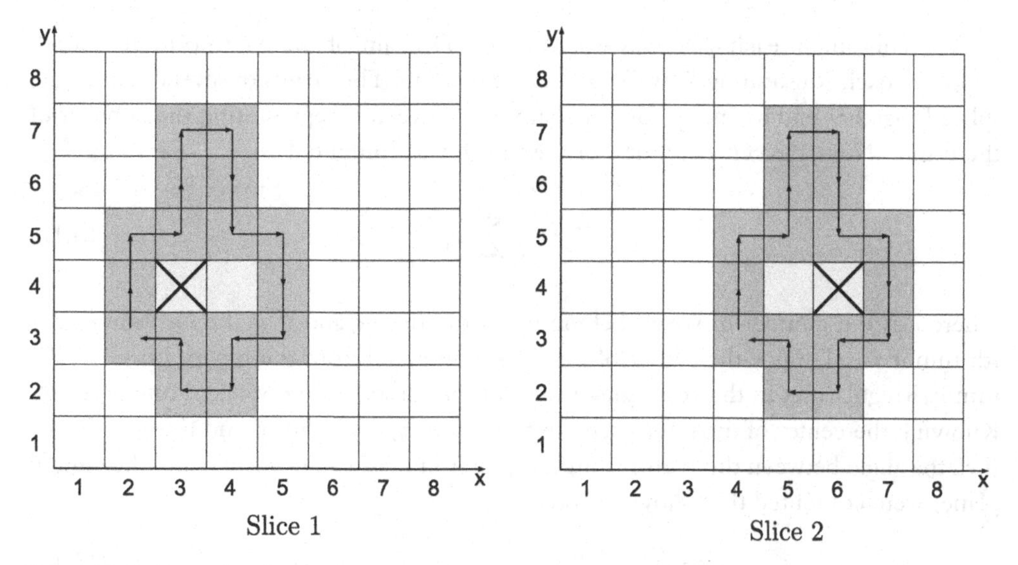

Figure 13.10 Example of contour computation.

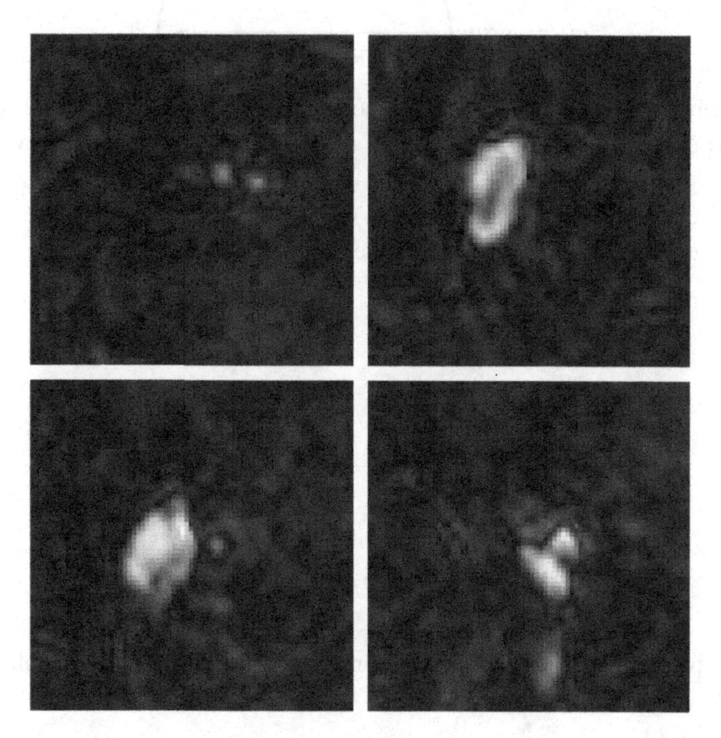

Figure 13.11 Left to Right: Tumor in adjacent transverse slices of a 512 × 512 × 32 voxel MR image.

The contour in each slice was stored as a 1-D chain of the 3-D position of each contour voxel, constituting a "walk" along the contour. The chains of several slices were spliced together end to end to form a chain of 3-D vectors representing the contour of the tumor. Next, the center of mass of the tumor was computed as

$$\overline{v} := \frac{1}{n} \sum_{i=1}^{n} v_i \tag{13.12}$$

where n is the number of voxels belonging to the tumor, and v_i is the location of the ith tumor voxel. Since the center of mass was computed from the binary image of the tumor, irregularities in the voxel gray values of the tumor were not taken into account. Knowing the center of mass, for each contour voxel c_i, the radius r_i and the azimuth ω_i (i.e., the angle between the vector from the center of mass to the voxel c_i and the sagittal plane) were computed the following way:

$$r_i := \|c_i - \overline{v}\|_2 \tag{13.13}$$

$$\omega_i := arcsin\left(\frac{c_{ix} - \overline{v}_x}{(c_{ix} - \overline{v}_x)^2 + \left(c_{iy} - \overline{v}_y\right)^2}\right) \tag{13.14}$$

where the subscripts x and y denote the position of the voxel in sagittal and coronal direction respectively. ω_i was also extended to the range from $-\pi$ to π by taking into account the sign of $\left(c_{iy} - \overline{v}_y\right)$.

From the chain of floating point values $r_1, \ldots, r_m$, the minimum value r_{min} and the maximum value r_{max} can be computed, as well as

$$\text{the mean value} \quad \overline{r} := \frac{1}{m} \sum_{i=1}^{m} r_i \tag{13.15}$$

$$\text{the standard deviation} \quad \sigma_r := \sqrt{\frac{1}{m} \sum_{i=1}^{m} (r_i - \overline{r})^2} \tag{13.16}$$

$$\text{and the entropy} \quad h_r := -\sum_{i=1}^{100} p_i \cdot \log_2 (p_i) \tag{13.17}$$

The entropy h_r is computed from the normalized distribution of the values into 100 "buckets", where p_i is defined as follows:

For $0 \leq i \leq 99$:

$$p_i := \frac{\left|\left\{ r_j \,\middle|\, i \leq \frac{(r_j - r_{min})}{r_{max} - r_{min}} \cdot 100 < i+1 \right\}\right|}{m} \tag{13.18}$$

From the radius, $r_{min}, r_{max}, \bar{r}, \sigma_r$, and h_r can be used as morphological features of the tumor. From the azimuth, only the entropy h_ω (computed for ω as in (13.17) and (13.18)) is used as a feature, since the values ω_{min} and ω_{max} are always around π and $-\pi$, respectively, and the value σ_ω is not invariant under rotation of the tumor image.

An additional measurement describing the compactness of the tumor represents the number of contour voxels, divided by the number of all voxels belonging to the tumor.

13.4.2 Enhancement Kinetic Features

While mass-enhancing lesions exhibit a typical kinetic behavior that is distinctive for malignant and benign lesions as shown in Fig. 13.1, non-mass-enhancing lesions have kinetic characteristics that are far less well characterized and of limited accuracy in discriminating between malignant and benign behavior [152]. As a dynamical feature, the slope of the relative signal intensity enhancement (RSIE) is used in most current CAD systems. Very few studies exist for analyzing the kinetics of non-masses.

In this section, we present techniques to extract enhancement kinetic features from diagnostically challenging lesions. They range from standard methods like the slope of the enhancement curve to regression techniques and spatio-temporal methods like the scaling index.

13.4.2.1 Slope of Mean Values

A simple but important feature is the slope of the enhancement. For a given image f the mean value inside the tumor region is considered:

$$\mu = \frac{1}{|T|} \sum_{\mathbf{p} \in \mathbf{T}} f_{\mathbf{p}}$$

This is performed for each of the I images belonging to a tumor and yields the mean values $\mu_1, \ldots, \mu_I$, from which the difference between subsequent values is considered

$$d_i = \frac{\mu_{i+1} - \mu_i}{\Delta t} \qquad i = 1, \ldots, I - 1$$

where Δt denotes the time between the images. These values are taken as additional features. This is the same as computing the average of all slopes between single pixels.

13.4.2.2 Regression Methods

Regression methods offer an elegant modality to determine descriptors for the parameters of a chosen model which approximate the given points best. Instead of just considering the single mean values, the mean values of the postcontrast images in relation to the precontrast image are considered. In this sense, we define the relative enhancement (RE) as

$$s_i = \frac{\mu_i - \mu_1}{\mu_1} \qquad i = 1, \ldots, I$$

Given the points $\{(t_i, s_i)^\top\}_{i=1,\dots,I}$ with $t_i = i\Delta t$ we try to apply regression with different functions described below.

Linear Function

This feature is also known as the elative Signal Intensity Enhancement Relative Signal Intensity Enhancement [210]. This paradigm is derived from observing the kinetic behavior of tumors: in the beginning both benign and malignant tumors are showing a rapid enhancement. However, the temporal behavior can be highly diagnostic for the type of the tumor. While benign mass tumors tend to have a further increase of the enhancement, malignant tumors are prone to have a rapid washout effect. Experimental and subsequent evaluation results have to prove if this idea can be transferred to the non-mass-like tumors. The idea is to approximate the last three values of the relative enhancement by a linear function $g(t) = at + b$ in order to describe the enhancement behavior toward the end. I is set equal to the number of slices. The parameters a and b have to be optimized in order to fit the given values. The optimal value can be computed as the solution of the following *least-squares regression*:

$$\underset{a,b}{\operatorname{argmin}} \left\| \underbrace{\begin{pmatrix} t_3 & 1 \\ \vdots & \vdots \\ t_I & 1 \end{pmatrix}}_{=:A} \begin{pmatrix} a \\ b \end{pmatrix} - \underbrace{\begin{pmatrix} s_3 \\ \vdots \\ s_I \end{pmatrix}}_{=:\mathbf{y}} \right\|$$

which yields as solution:

$$\begin{pmatrix} a \\ b \end{pmatrix} = (A^\top A)^{-1} A^\top y$$

After some computations we get for a the following result which is used as a feature:

$$a = \frac{\sum_{i=3}^{I} t_i \sum_{i=3}^{I} s_i - I \sum_{i=3}^{I} (t_i s_i)}{\left(\sum_{i=3}^{I} t_i\right)^2 - I \sum_{i=3}^{I} (t_i)^2}$$

Exponential Function

Exponential functions represent a good fit for the points of the relative enhancement. In [151] the following exponential approximation is used:

$$g(t) = A \cdot (1 - e^{-\alpha(t-1)}) \cdot e^{-\beta(t-1)}$$

with the parameters A, α, and β. An example of the fitted function can be seen in Fig. 13.12. It is now possible to fit the values instead of approximating them as it is the case with linear functions. The parameters of the nonlinear fitting function can be

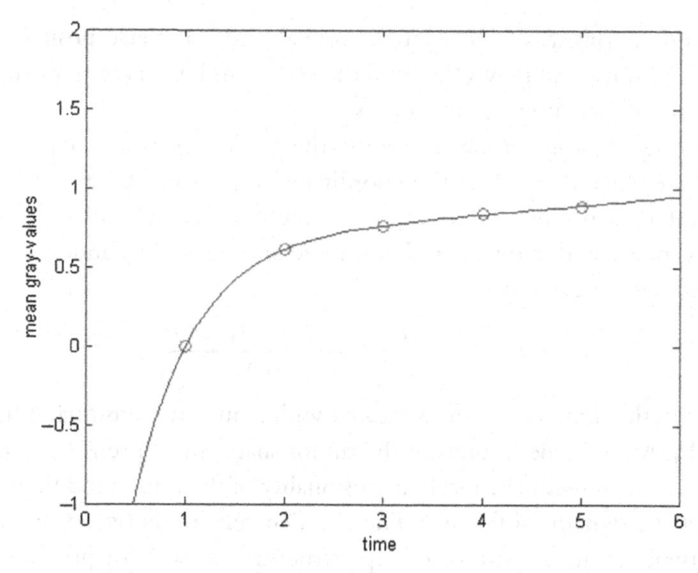

Figure 13.12 Example for fitting of the RE values (red circles) by the function proposed by Jansen et al. ($A = 0.69, \alpha = 1.74$, and $\beta = -0.06$). (For interpretation of the references to color in this figure legend, the reader is referred to the web version of this book.)

obtained by using an iterative algorithm. These parameters can be used as descriptors. There is also a slightly modified version of them:

- **Initial area under the curve until a specified time τ**

$$\mathrm{iAUC} = A \cdot \left(\frac{1 - e^{-\beta\tau}}{\beta} + \frac{e^{-(\alpha+\beta)\tau} - 1}{\alpha + \beta} \right)$$

- **Initial slope**

$$\mathrm{iSlope} = A\alpha$$

- **Time to peak enhancement**

$$T_{\mathrm{peak}} = \frac{1}{\alpha \log(1 + \frac{\alpha}{\beta})}$$

- **Curvature at the peak**

$$\kappa_{\mathrm{peak}} = -A\alpha\beta$$

Besides these features, the parameters A, α, and β can be also used as features.

13.4.2.3 Simultaneous Morphology and Dynamics Representations

The scaling index method [149] is a technique that is based on both morphology and kinetics. It represents the local structure around a given point. In the context of breast

MRI, such a point consists of the sagittal, coronal, and transverse position of a tumor voxel and its third time scan gray value, and the scaling index serves as an approximation of the dimension of local point distributions.

Mathematically, the scaling index represents the 2-D image as a set of points in a three-dimensional state space defined by the coordinates x, y, z and the gray value $f(x, y, z)$. For every point P_i with coordinates (x_i, y_i, z_i) the number of points in a sphere with radius r_1 and a sphere with radius r_2 is determined and the scaling index α_i is computed based on the following equation:

$$\alpha_i = \frac{(\log N(\mathbf{P_i}, r_2) - \log N(\mathbf{P_i}, r_1))}{(\log r_2 - \log r_1)} \tag{13.19}$$

where $N(\mathbf{P_i}, r)$ is the number of points located within an n-dimensional sphere of radius r centered at $\mathbf{P_i}$. As radii, the bounds of the tumor shape are chosen. Thus, the obtained scaling index is a measure for the local dimensionality of the tumor and thus quantifies its morphological and dynamical features. There is a correlation between the scaling index and the structural nature: $\alpha = 0$ for clumpy structures, $\alpha = 1$ for points embedded in straight lines, and $\alpha = 2$ for points in a flat distribution.

For each of the three time scans ($i \in \{1, 3, 5\}$), the standard deviation and entropy are determined and can be used as a feature to capture the heterogeneous behavior of the enhancement in a tumor.

13.5. AUTOMATED DETECTION OF SMALL LESIONS IN BREAST MRI BASED ON MORPHOLOGICAL, KINETIC, AND SPATIO-TEMPORAL FEATURES

The small lesion evaluation is based on a multistep system that includes a reduction of motion artifacts based on a novel nonrigid registration method, an extraction of morphologic features, dynamic enhancement patterns as well as mixed features for diagnostic feature selection and performance of lesion evaluation [348,349]. Figure 13.13 visualizes the proposed automated system for small lesion detection.

13.5.1 Patients and MR Imaging

A total of 40 patients, all female having an age range 42–73 years, with indeterminate small mammographic breast lesions were examined. All patients were consecutively selected after clinical examinations, mammography in standard projections (cranio-caudal and oblique medio-lateral projections) and ultrasound. Only lesions BIRADS 3 and 4 were selected where at least one of the following criteria was present: nonpalpable lesion, previous surgery with intense scarring, or location difficult for biopsy (close to chest wall). All patients had histopathologically confirmed diagnosis from needle aspiration/excision biopsy and surgical removal. Breast cancer was diagnosed in 17 out of the total 31 cases. The average size of both benign and malignant tumors was less than 1.1 cm.

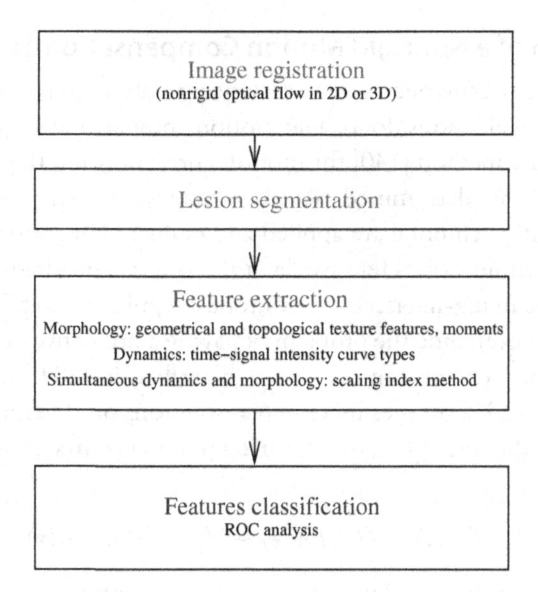

Figure 13.13 Diagram of a computer-assisted system for the evaluation of small contrast enhancing lesions.

MRI was performed with a 1.5 T system (Magnetom Vision, Siemens, Erlangen, Germany) with two different protocols equipped with a dedicated surface coil to enable simultaneous imaging of both breasts. The patients were placed in a prone position. First, transversal images were acquired with a STIR (short TI inversion recovery) sequence (TR = 5600 ms, TE = 60 ms, FA = 90°, IT = 150 ms, matrix size 256 × 256 pixels, slice thickness 4 mm). Then a dynamic T1-weighted gradient echo sequence (3-D fast low angle shot sequence) was performed (TR = 11 ms and TR = 9 ms, TE = 5 ms, FA = 25°) in transversal slice orientation with a matrix size of 256 × 256 pixels and an effective slice thickness of 4 mm or 2 mm.

The dynamic study consisted of six measurements with an interval of 83 s. The first frame was acquired before injection of paramagnetic contrast agent (gadopentetate dimeglumine, 0.1 mmol/kg body weight, Magnevist^{TM}, Schering, Berlin, Germany) immediately followed by the five other measurements. The initial localization of suspicious breast lesions was performed by computing difference images, i.e., subtracting the image data of the first from the fourth acquisition. As a preprocessing step to clustering, each raw gray level time-series $S(\tau), \tau \in \{1, \ldots, 6\}$ was transformed into a signal time-series of relative signal enhancement $x(\tau)$ for each voxel, the precontrast scan at $\tau = 1$ serving as reference, in other words $x(\tau) = \frac{x(\tau)-x(1)}{x(1)}$. Thus, it is ensured that the proposed method is less sensitive to changing between different MR scanners and/or protocols.

13.5.2 Evaluation of a Nonrigid Motion Compensation Technique

Motion correction algorithms become a necessary correction tool in order to improve the diagnostic value for small lesions (foci). The motion compensation algorithm is based on the Horn and Schunck method [140] for motion correction and the optimal parameters for lesion classification are determined. Several novel lesion descriptors such as morphologic, kinetic, and spatio-temporal are applied and evaluated in context with benign and malignant lesion discrimination. Here we favor the original quadratic formulation, since we explicitly need the filling-in effect of a nonrobust regularizer to fill in the information in masked regions. To overcome the problem of having a nonconvex energy in (13.5), the coarse-to-fine warping scheme can be used as described in [284], which linearizes the data term as in [140] and computes incremental solutions on different image scales. For this, we approximate the image f at the distorted point $(\mathbf{x}+\mathbf{u}(\mathbf{x}), t+k)$ by a first-order Taylor approximation:

$$f(\mathbf{x}+\mathbf{u}(\mathbf{x}), t+k) - f(\mathbf{x}, t) \approx f(\mathbf{x}, t+k) + \nabla f(\mathbf{x}, t+k)^{\top}\mathbf{u}(\mathbf{x}) - f(\mathbf{x}, t) \tag{13.20}$$

Alternatively, we can develop the Taylor series at time t, getting

$$f(\mathbf{x}+\mathbf{u}(\mathbf{x}), t+k) - f(\mathbf{x}, t) \approx f(\mathbf{x}, t) + \nabla f(\mathbf{x}, t)^{\top}\mathbf{u}(\mathbf{x}) + \frac{\partial}{\partial t} f(\mathbf{x}, t)k - f(\mathbf{x}, t) \tag{13.21}$$

Since the term $f(\mathbf{x}, t)$ cancels and the temporal derivative is again approximated by the difference of the two images at point $\mathbf{x}$, the only difference between eqs. (13.20) and (13.21) is the time at which the spatial derivative ∇f is being computed. In optical flow computation one usually uses the arithmetic mean $\frac{1}{2}(\nabla f(\mathbf{x}, t+k) + \nabla f(\mathbf{x}, t))$. For the purpose of registration, the scalar factor k can be neglected since it is arbitrary whether one computes a motion for a $k \neq 1$ and then scales the motion later on with k when registering the image, or simply sets k to 1. Incorporating this in the energy formulation and leaving out the indices for better readability, the linearized energy functional then reads as

$$E_{lin}(\mathbf{u}) = \int_{\Omega} m \left(\frac{\partial}{\partial t} f + \nabla f^{\top} \mathbf{u} \right)^2 + \alpha \left(|\nabla u_1|^2 + |\nabla u_2|^2 + |\nabla u_3|^2 \right) d\mathbf{x} \tag{13.22}$$

This functional is convex in $\mathbf{u}$, and a minimizer can be found by solving its Euler-Lagrange equations $\forall \mathbf{x} \in \Omega$:

$$0 = m \left(\frac{\partial f}{\partial x}\frac{\partial f}{\partial x} u_1 + \frac{\partial f}{\partial x}\frac{\partial f}{\partial y} u_2 + \frac{\partial f}{\partial x}\frac{\partial f}{\partial z} u_3 + \frac{\partial f}{\partial x}\frac{\partial f}{\partial t} \right) - \alpha \Delta u_1 \tag{13.23}$$

$$0 = m \left(\frac{\partial f}{\partial x}\frac{\partial f}{\partial y} u_1 + \frac{\partial f}{\partial y}\frac{\partial f}{\partial y} u_2 + \frac{\partial f}{\partial y}\frac{\partial f}{\partial z} u_3 + \frac{\partial f}{\partial y}\frac{\partial f}{\partial t} \right) - \alpha \Delta u_2 \tag{13.24}$$

$$0 = m \left(\frac{\partial f}{\partial x}\frac{\partial f}{\partial z} u_1 + \frac{\partial f}{\partial y}\frac{\partial f}{\partial z} u_2 + \frac{\partial f}{\partial z}\frac{\partial f}{\partial z} u_3 + \frac{\partial f}{\partial z}\frac{\partial f}{\partial t} \right) - \alpha \Delta u_3 \tag{13.25}$$

Since the linearization in eq. (13.20) or (13.21) is only valid for small motions in the subpixel range, a typical strategy to overcome the problem of large motions is to downsample the MR images to a coarse resolution, compute an approximate motion on the coarse resolution, interpolate this motion to the next finer resolution, register the second image with the approximate motion, compute the incremental motion from the first image to the registered second image, add the incremental motion to the approximate motion, and repeat this iteration up to the original resolution.

13.5.2.1 Motion Compensation Results

The applicability of the previously described motion compensation algorithm under different motion compensation parameters and features' sets is here evaluated. Table 13.1 describes the motion compensation parameters used in the subsequent evaluations.

The effect of the motion compensation based on motion compensation parameters such as the amount of presmoothing and the regularization parameter was analyzed based on different combinations of feature groups and ROC analysis.

As a classification method to evaluate the effect of motion compensation for small lesions the Fisher's linear discriminant analysis was chosen. Different features were chosen as descriptors for small lesions: contour, kinetic, and spatio-temporal features such as the scaling index method.

The contour features show for almost all motion compensation parameters high ROC values as shown in Table 13.2. Without motion compensation, the entropy as well as radius mean and maximum yield the best results. The radius minimum followed by the compactness show a significant improvement for motion compensation in 3-D directions.

The slope chosen as a kinetic feature is derived from the first-order approximation of relative signal intensity enhancement from the last three scans. Table 13.3 shows that both 2-D as well as 3-D motion compensation yield almost equally good results.

The scaling index is described in eq. (13.19) and the two radii r_1 and r_2 are chosen in the size of tumor structures, $r_1 = 3$ mm and $r_2 = 6$ mm. The maximum, mean, standard deviation, and entropy of the set of scaling indices, computed from tumor points as in

Table 13.1 Motion compensation parameters for two or three directions. σ represents the standard deviation for presmoothing and α the regularization parameter.

01	No motion compensation		
02	3 directions, $\sigma = 1$, $\alpha = 100$	**06**	2 directions, $\sigma = 1$, $\alpha = 100$
03	3 directions, $\sigma = 1$, $\alpha = 500$	**07**	2 directions, $\sigma = 1$, $\alpha = 500$
04	3 directions, $\sigma = 3$, $\alpha = 100$	**08**	2 directions, $\sigma = 3$, $\alpha = 100$
05	3 directions, $\sigma = 13$, $\alpha = 500$	**09**	2 directions, $\sigma = 13$, $\alpha = 500$

Table 13.2 Areas under the ROC curves for contour features using FLDA. The rows represent the motion compensation as given by Table 13.1. Numbers in boldface show the best results.

Feature Type	Area under the ROC curve (%)								
	01	02	03	04	05	06	07	08	09
Radius Min.	70.2	**85.1**	79.0	72.9	63.9	80.3	66.8	74.4	71.2
Radius Max.	**83.4**	76.3	81.7	84.0	84.2	80.7	80.0	79.6	83.4
Radius Mean	**83.2**	80.0	83.0	83.4	79.8	81.1	76.7	82.6	84.2
Radius St. Dev.	82.4	70.4	76.1	80.3	80.7	76.9	79.2	75.0	76.9
Radius Entropy	83.0	76.7	79.6	84.0	74.8	79.6	75.0	80.7	80.9
Azimuth Entropy	80.7	81.9	77.7	79.6	77.5	81.9	79.0	78.4	79.6
Compactness	69.5	**77.1**	73.7	75.6	67.9	68.9	68.7	71.4	65.8

Table 13.3 Areas under the ROC curves for dynamic features (slope) using FLDA. The rows represent the motion compensation as given by Table 13.1. Numbers in boldface show the best results.

Feature Type	Area under the ROC curve (%)								
	01	02	03	04	05	06	07	08	09
Slope	70.4	75.6	74.2	75.0	75.0	73.9	75.8	72.7	75.4

Table 13.4 Areas under the ROC curves for scaling index (SI) method using FLDA. The rows represent the motion compensation as given by Table 13.1. Numbers in boldface show the best results.

Feature Type	Area under the ROC curve (%)								
	01	02	03	04	05	06	07	08	09
SI Max.	74.6	66.6	64.3	69.3	62.8	**80.3**	69.1	60.9	62.8
SI Mean	**80.3**	79.8	79.6	80.0	78.6	80.5	77.1	79.6	77.5
SI St. Dev.	52.7	**81.5**	73.3	70.8	74.4	61.1	67.4	79.2	72.5
SI Entropy	70.8	71.4	75.0	72.7	71.6	68.9	65.1	75.0	**76.5**

(13.15) – (13.17), were used as features of the tumor. The minimum was neglected, since for almost every tumor it was 0, due to isolated points.

Table 13.4 shows the ROC values for the scaling index method for different motion compensation parameters shown in Table 13.1. The scaling index mean value yields the highest results without motion compensation and for both 2-D and 3-D motion compensation.

Table 13.5 Areas under the ROC curves for spatio-temporal features using FLDA. The rows represent the motion compensation as given by Table 13.1.

Feature	Area under the ROC curve (%)								
Type	01	02	03	04	05	06	07	08	09
RSIE St. Dev. (3)	56.7	52.5	55.0	52.3	48.1	52.5	55.5	58.4	55.3
RSIE St. Dev. (4)	64.9	65.8	68.5	62.8	63.7	66.6	66.8	69.5	64.5
RSIE St. Dev. (5)	64.7	70.6	70.0	64.9	68.3	69.7	65.8	66.8	69.5
RSIE Entropy (3)	**80.9**	84.2	**85.3**	83.0	79.4	84.5	81.3	81.5	83.4
RSIE Entropy (4)	77.9	87.4	81.9	80.7	76.7	83.8	78.8	77.3	78.4
RSIE Entropy (5)	**74.6**	79.8	81.7	81.7	73.3	**81.5**	76.3	76.9	73.5
Contour RSIE Mean (3)	54.4	51.7	52.7	52.3	54.0	55.7	52.7	52.7	55.5
Contour RSIE Mean (4)	63.7	56.9	62.8	59.2	61.1	59.5	59.5	59.0	59.9
Contour RSIE Mean (5)	62.2	64.9	60.3	68.5	68.7	62.6	63.9	62.6	66.6
Contour RSIE St. Dev. (3)	55.3	57.4	58.0	52.1	55.7	58.0	55.0	57.4	57.8
Contour RSIE St. Dev. (4)	58.4	59.9	57.1	56.7	61.3	60.7	58.6	56.9	62.4
Contour RSIE St. Dev. (5)	63.7	63.2	67.6	60.5	65.1	58.6	58.2	66.2	64.3
Contour RSIE Entropy (3)	77.5	81.3	77.1	74.2	76.9	77.1	74.8	75.4	74.4
Contour RSIE Entropy (4)	83.2	84.5	82.8	79.8	77.9	84.9	77.5	81.3	79.6
Contour RSIE Entropy (5)	80.5	81.3	78.2	80.7	72.9	79.6	78.2	77.5	78.4

The performance results for the spatio-temporal features related to both contour and tumor relative signal intensity enhancement are shown in Table 13.5 for the third, fourth, and fifth scans. For both the tumor and contour, the entropy showed the best results in the ROC analysis: the third scan for the tumor entropy and the fourth for the contour entropy for both uncompensated and compensated motion.

The optimal motion correction results were achieved for motion compensation in two directions for mostly small standard deviations of the Gaussian kernel and smoothing parameter. Consistent with the only study known for evaluating the effect of motion correction algorithms [336], the proposed motion compensation technique achieved good results for weak motion artifacts.

The performed ROC analysis shows that an integrated motion compensation step in a CAD system represents a valuable tool for supporting radiological diagnosis in dynamic breast MR imaging.

13.5.3 Evaluation of Morphological and Kinetic Features

A computer-aided diagnosis system based on feature extraction and classification is described in [309]. The features shown in Table 13.6 range from local and global morphological descriptors and the slope of RSIE as a kinetic descriptor and were tested for different classifiers shown in Table 13.7.

Table 13.6 Numerical code of features.

1.	Minkowski Functionals (MF)
1.1	Volume
1.2	Surface
1.3	Curvature
1.4	Euler Characteristic
2	Normed Minkowsi Functionals (independent of tumor size)
2.1	Surface
2.2	Curvature
2.3	Euler Characteristic
3	Krawtchouk Moments (Reduction to 28 dimensions)
3.1	Standard Krawtchouk Moments
3.2	Radial Krawtchouk Moments
4	Slope of RSIE

Table 13.7 Classifiers employed for lesion classification.

LDA	Linear Discriminant Analysis.
NLDA	Naive Bayes Linear Discriminant Analysis.
QDA	Quadratic Discriminant Analysis.
NQDA	Naive Bayes Quadratic Discriminant Analysis.
FLD	Fisher's Linear Discriminant Analysis.
PK	SVM Classification with a Polynomial Kernel.
RBF	SVM Classification with a Radial Basis Function.
PUK	SVM Classification with a Pearson VII Universal Function Kernel.
AUC	Area under the ROC curve.

Table 13.8 shows the results for different classifiers and single features when applied to tumor classification. Normed Minkowski functionals (independent of tumor size) such as the Euler characteristic achieve the highest AUC-value among all other features. Thus the descriptive power of this simple morphological parameter is almost independent of the classifier's type. The regular Krawtchouk moment scores higher than the radial invariant one and we assume that this occurs due to the re-discretization error and the usage of only one Krawtchouk polynomial. The area under the curve—representing the kinetic features—scored lower than some of the morphological features confirming that non-mass-enhancing lesions cannot be correctly captured by kinetics only.

Summarizing, morphological feature descriptors such as the Minkowski functional appear to be more adequate than kinetic descriptors for diagnostically challenging lesions.

Table 13.8 Area-under-the-curve (AUC) for the classifiers applied to the single features from Table 13.6.

	LDA	NLDA	QDA	NQDA	FLD	PK	RBF	PUK
1.1	0.74	0.74	0.72	0.72	0.79	0.78	0.78	0.77
1.2	0.75	0.75	0.72	0.72	0.78	0.75	0.75	0.72
1.3	0.75	0.75	0.76	0.76	0.75	0.79	0.79	0.78
1.4	0.78	0.78	0.80	0.80	0.82	0.80	0.80	0.80
2.1	0.70	0.70	0.70	0.70	0.75	0.72	0.72	0.69
2.2	0.78	0.78	0.81	0.81	0.82	0.82	0.82	0.79
2.3	0.85	0.85	0.87	0.87	0.84	0.85	0.85	0.84
3.1	0.82	0.78	0.78	0.75	0.66	0.82	0.54	0.90
3.2	0.64	0.64	0.58	0.57	0.58	0.66	0.54	0.61
4	0.76	0.76	0.75	0.75	0.73	0.75	0.77	0.76

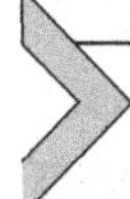

13.6. AUTOMATED ANALYSIS OF NON-MASS-ENHANCING LESIONS IN BREAST MRI BASED ON MORPHOLOGICAL, KINETIC, AND SPATIO-TEMPORAL FEATURES

The most comprehensive computer-aided diagnosis system for non-mass-enhancing lesions in the literature is presented [136] consisting of motion compensation, segmentation, feature extraction, and classification. These types 8 of lesions are diagnostically challenging since typical kinetic or morphologic descriptors are not so far known. The discriminative power of the novel joint spatio-temporal technique, the Zernike velocity moments, is evaluated versus single kinetic or shape descriptors for the diagnosis of these lesions in combination with or without motion compensation. The impact of nonrigid motion compensation on a correct diagnosis is additionally analyzed.

13.6.1 Patients and MR Imaging

The database for non-mass-enhancing lesions includes a total of 84 patients images, all female, with non-mass-enhancing tumors. All patients had histopathologically confirmed diagnosis from needle aspiration/excision biopsy and surgical removal. Histologic findings were malignant in 61 and benign in 23 lesions.

MRI was performed with a 1.5 T system (Magnetom Vision, Siemens, Erlangen, Germany) equipped with a dedicated surface coil to enable simultaneous imaging of both breasts for both types of lesions. The patients were placed in a prone position.

Transversal images were acquired with a STIR (short TI inversion recovery) sequence (TR = 5600 ms, TE = 60 ms, FA = 90°, IT = 150 ms, matrix size 228 × 182 pixels, slice thickness 3 mm). Then a dynamic T1-weighted gradient echo sequence (3-D fast low angle shot sequence) was performed (TR = 4.9 ms, TE = 1.83 ms, FA = 12°) in

Table 13.9 Classifiers employed for lesion classification.

SVM Kernel 1	SVM Classification with a Linear Kernel
SVM Kernel 2	SVM Classification with a Polynomial Kernel
SVM Kernel 3	SVM Classification with Radial Basis Kernel
SVM Kernel 4	SVM Classification with Sigmoidal Kernel

transversal slice orientation with a matrix size of 352 × 352 pixels and an effective slice thickness of 1 mm. The dynamic study consisted of five measurements with an interval of 1.4 min. The first frame was acquired before injection of paramagnetic contrast agent (gadopentetate dimeglumine, 0.1 mmol/kg body weight, Magnevist^{TM}, Schering, Berlin, Germany) immediately followed by the four other measurements.

13.6.2 Detection and Classification Results

Both the quantitative and qualitative effect of the previously introduced features for non-mass-enhancing lesions was analyzed in a computer-aided diagnosis system. The classifier is a SVM with different kernels as described in Table 13.9. The area under the ROC curve (AUC) served as a quantitative evaluation measure for the proposed CAD system.

Figure 13.14 gives an overview of the classification results for both motion-compensated data (left bars) as well as uncompensated data (right bars). The most important fact is that motion compensation improves in most cases the AUCs suggesting that motion artifacts play an important role in correct diagnosis of non-mass-enhancing lesions. The dynamical features, such as parameters extracted from the approximation of the RSIE curves, see Fig. 13.14(a), yield the best results while morphological features like the writhe number, see Fig. 13.14(e) are close to the dynamical features. The averaged Zernike descriptors provide a lower classification rate, however it is higher than slope of the mean values as shown in Fig. 13.14(d).

A recent study for mass-like-enhancing lesions [1] has shown that both morphological and kinetic features outperform spatio-temporal features. Their morphological features were the most discriminative suggesting that the lesion's morphology—in concordance with clinical practice—determines the further evaluation based on a needle biopsy. Kinetic features, on the other hand, seem to be more discriminative in case of non-mass-like enhancing lesions followed by the morphological ones.

Summarizing, it is shown that motion compensation proved in most cases to be diagnostically relevant. The best discriminative features are based on the dynamical properties of an approximation of the RSIE curve suggesting that kinetics such as washin and washout parameters plays a key role in correctly diagnosing non-mass-enhancing lesions and confirms existing studies. The next best feature set is represented by the writhe number computed for vertices along the surface of the tumor and describing the surface

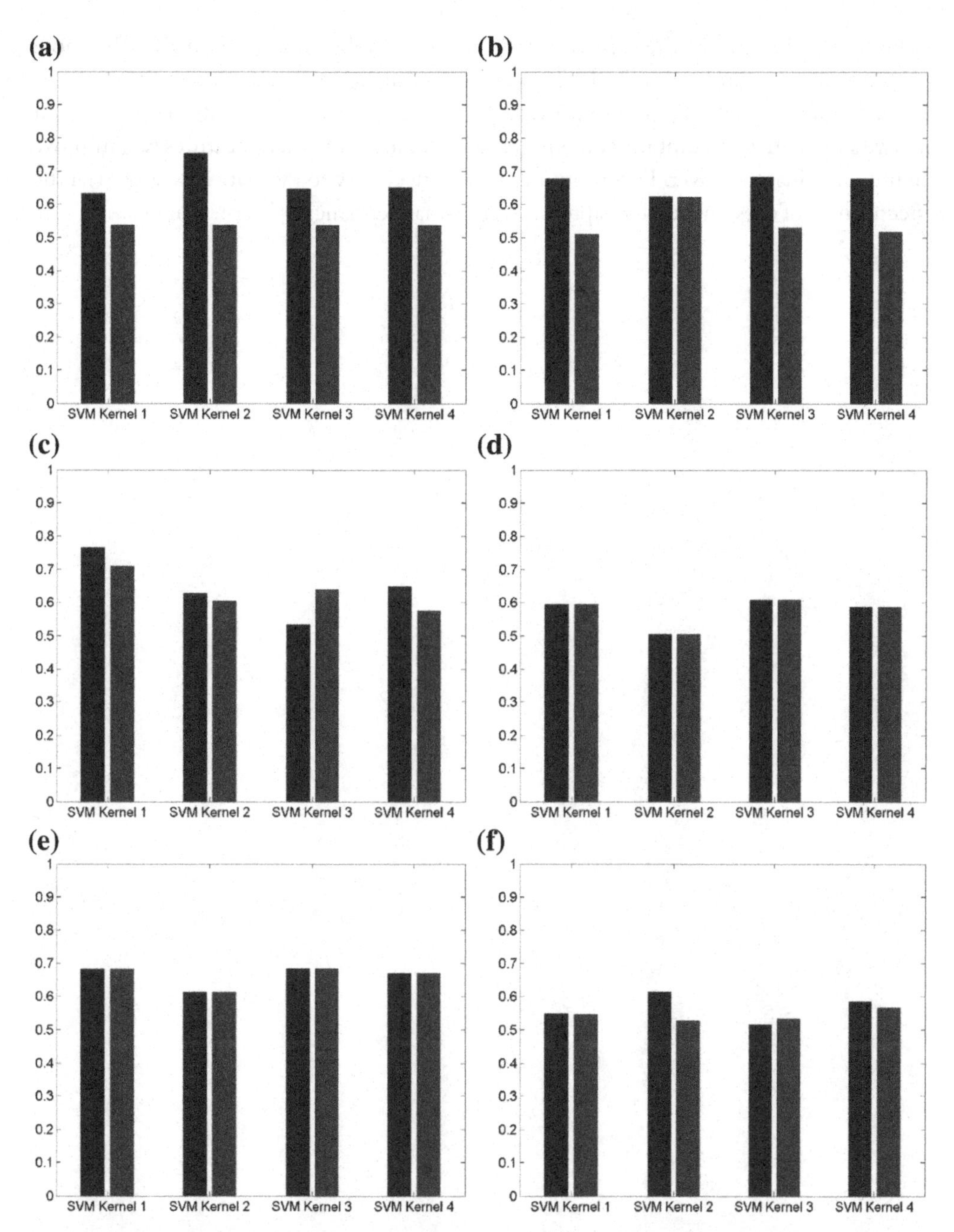

Figure 13.14 AUCs of SVM applied to kinetic and morphologic features, separately, using four different kernels and both motion compensated (blue bars) as well as original data (red bars) to compute the features (a) Regression w. exp. func. (b) Krawtchouk moments. (c) (a) Regression w. lin. func. (d) Morphology. (e) Writhe number. (f) Zernike vel. moments. (For interpretation of the references to color in this figure legend, the reader is referred to the web version of this book.)

asymmetries. Intuitively, the results for the writhe number have shown that the more twisted the surface of a lesion is, the greater the likelihood for malignancy is. The spatio-temporal moments, alone, seem to provide an independent characterization of the tumor and we suspect that in combination with morphological and kinetic features will improve their discriminative power. Future studies will be necessary to evaluate in a large trial the effectiveness of these novel descriptors for non-mass-enhancing lesion diagnosis.

Glossary

AFNM	Adaptive fuzzy *n*-means
AFNS	Adaptive fuzzy *n*-shells
AO	Alternating optimization
APEX	Adaptive principal components analysis using lateral inhibition
AR	Autoregressive
AUC	Area under the ROC curve
CAD	Computer-aided design
CNN	Convolution neural network
CNR	Contrast-to-noise ratio
CT	Computer tomography
CWT	Continuous wavelet transform
DCE	Dynamic contrast enhanced
DCT	Discrete cosine transform
DNA	Desoxyribonucleic acid
DST	Discrete sine transform
DWT	Discrete wavelet transform
fMRI	Functional magnetic resonance imaging
fn	False negative
fp	False positive
FALVQ	Fuzzy algorithm for learning vector quantization
FCA	Fuzzy clustering analysis
FIR	Finite impulse response
FNM	Fuzzy *n*-means
FT	Fourier transform
GA	Genetic algorithm
GAFNM	Generalized adaptive fuzzy *n*-means
GAFNS	Generalized adaptive fuzzy *n*-shells
GFNM	Generalized fuzzy *n*-means
GFNS	Generalized fuzzy *n*-shells
HPNN	Hierarchical pyramid neural network
ICA	Independent component analysis
Infomax	Information maximization
JPEG	Joint photographic experts group
KNN	K nearest neighbor
LC	Linear classifier
LBG	Linde-Buzo-Gray

LMS	Least mean squares
LPC	Linear predictive coding
LTM	Long-term memory
LVQ	Learning vector quantization
MEG	Magnetoenchephalographic
MIAS	Mammographic Image Analysis Society
MLP	Multilayer perceptron
MPEG	Motion Photographic Experts Group
MRI	Magnetic resonance imaging
NMSE	Normalized mean-squared error
NSNR	Normalized signal-to-noise ratio
PCA	Principal component analysis
PET	Positron emission tomography
PR-QMF	Perfect reconstruction quadrature mirror filter
PSNR	Peak signal-to-noise ratio
QC	Quadratic classifier
RCE	Restricted Coulomb energy
ROC	Receiver operating characteristics
ROI	Region of interest
SPECT	Single-photon-emission-computer-tomography
STM	Short-term memorytn True negative
tp	True-positive
TSI	Time signal intensity
US	Ultrasound
WT	Wavelet transform

References

[1] S. Agliozzo, M. De Luca, C. Bracco, A. Vignati, V. Giannini, L. Martincich, A. Bert, F. Sardanelli, D. Regge, Computer-aided diagnosis for contrast-enhanced breast MRI of mass-like lesions using a multiparametric model combining a selection of morphological, kinetic and spatio-temporal features, Medical Physics 39 (2012) 3102–3109.

[2] T.S. Ahearn, R.T. Staff, T.W. Redpath, S.I.K. Semple, The use of the Levenberg-Marquardt curve-fitting algorithm in pharmacokinetic modelling of DCE-MRI, Physics in Medicine and Biology 50 (9) (2005) N85–N92.

[3] A. Ahumada, Computational image quality metrics: a review, Society for Information Display Digest of Technical Papers, 1993, pp. 305–308.

[4] A.N. Akansu, R.A. Haddad, Multiresolution Signal Decomposition, Academic Press, 1992.

[5] M. Akay, Time Frequency and Wavelets in Biomedical Signal Processing, IEEE Press, 1998.

[6] A. Aldroubi, M. Unser, Wavelet in Medicine and Biology, CRC Press, 1996.

[7] D. Alkon, K. Blackwell, G. Barbor, A. Rigle, T. Vogel, Pattern-recognition by an artificial network derived from biological neuronal systems, Biological Cybernetics 62 (1990) 363–379.

[8] L.J.S. Allen, An Introduction to Stochastic Processes with Applications to Biology, second ed., CRC Press, Boca Raton, 2011.

[9] S. Amari, Competitive and cooperative aspects in dynamics of neural excitation and self-organization, Competition and Cooperation in Neural Networks 20 (1982) 1–28.

[10] S. Amari, Field theory of self-organizing neural nets, IEEE Transactions on Systems, Man and Cybernetics 13 (9 & 10) (1983) 741–748.

[11] S. Amari, A. Cichocki, H.H. Yang, A new learning algorithm for blind signal separation, in: NIPS96, vol. 8, 1996, pp. 757–763.

[12] S. Amartur, D. Piraino, Y. Takefuji, Optimization neural networks for the segmentation of magnetic resonance images, IEEE Transaction on Medical Imaging 11 (1992) 215–220.

[13] H. Andrews, Mathematical Techniques in Pattern Recognition, John Wiley Verlag, 1972.

[14] Y. Anzai, S. Minoshima, G. Wolf, R. Wahl, Head and neck cancer: detection of recurrence with 3D principal component analysis at dynamic FDG pet, Radiology 212 (1999) 285–290.

[15] H. Attias, Independent factor analysis, Neural Computation 11 (1999) 803–851.

[16] L. Axel, Cerebral blood flow determination by rapid-sequence computed tomography, Radiology 137 (1980) 679–686.

[17] W. Backfrieder, R. Baumgartner, M. Samal, E. Moser, H. Bergmann, Quantification of intensity variations in functional MR images using rotated principal components, Physics in Medicine and Biology 41 (1996) 1425–1438.

[18] P. Bandettini, E. Wong, R. Hinks, R. Tikofski, J. Hyde, Time course EPI of human brain function during task activation, Magnetic Resonance in Medicine 25 (1992) 390–397.

[19] B. Batchelor, Practical Approach to Pattern Classification, Plenum Press Verlag, 1974.

[20] R. Baumgartner, L. Ryder, W. Richter, R. Summers, M. Jarmasz, R. Somorjai, Comparison of two exploratory data analysis methods for fMRI: fuzzy clustering versus principal component analysis, Magnetic Resonance Imaging 18 (2000) 89–94.

[21] A. Beghdadi, A. Le Negrate, Contrast enhancement technique based on local detection of edges, Computer Vision, Graphics, and Image Processing 46 (1989) 162–174.

[22] S. Behrens, H. Laue, T. Boehler, B. Kuemmerlen, H. Hahn, H.O. Peitgen, Computer assistance for MR based diagnosis of breast cancer: present and future challenges, Computerized Medical Imaging and Graphics 31 (2007) 236–247.

Pattern Recognition and Signal Analysis in Medical Imaging
http://dx.doi.org/10.1016/B978-0-12-409545-8.00024-8

[23] A. Bell, T. Sejnowski, Fast blind separation based on information theory, in: International Symposium on Nonlinear Theory and Applications, 1995, pp. 43–47.

[24] A.J. Bell, T.J. Sejnowski, An information-maximization approach to blind separation and blind deconvolution, Neural Computation 7 (1995) 1129–1159.

[25] J.E. Besag, On the statistical analysis of dirty pictures, Journal of the Royal Statistical Society B 48 (3) (1986) 259–302.

[26] J.E. Besag, D.M. Higdon, Bayesian analysis of agricultural field experiments, Journal of the Royal Statistical Society: Series B 61 (4) (1999) 691–746.

[27] J.E. Besag, J. York, A. Mollie, Bayesian image restoration with two applications in spatial statistics (with discussion), Annals of the Institute of Statistical Mathematics 43 (1991) 1–59.

[28] B. Betts, A. Aiyer, J. Li, D. Ikeda, R. Birdwell, R.M. Gray, R.A. Olshen, Management and lesion detection effects of lossy image compression on digitized mammograms, in: Proceedings of the Fourth International Workshop on Digital Mammography, vol. 8, 1998.

[29] J. Bezdek, Pattern Recognition with Fuzzy Objective Function Algorithms, Plenum Press, 1981.

[30] J. Bezdek, L. Hall, L. Clarke, Review of MR image segmentation techniques using pattern recognition, Medical Physics 20 (1993) 1033–1048.

[31] C. Biller, Adaptive Bayesian regression splines in semiparametric generalized linear models, Journal of Computational and Graphical Statistics 9 (1) (2000) 122.

[32] G. Blomqvist, On the construction of functional maps in positron emission tomography, Journal of Cerebral Blood Flow and Metabolism 4 (4) (1984) 629–632.

[33] J. Boxerman, P. Bandettini, K. Kwong, J. Baker, The intravascular contribution to fMRI signal change: Monte Carlo modeling and diffusion-weighted studies in vivo, Magnetic Resonance in Medicine 34 (1995) 4–10.

[34] A. Brezger, L. Fahrmeir, A. Hennerfeind, Adaptive Gaussian Markov random fields with applications in human brain mapping, Journal of the Royal Statistical Society: Series C (Applied Statistics) 56 (3) (2007) 327–345.

[35] A. Brezger, T. Kneib, S. Lang, BayesX: analysing Bayesian structured additive regression models, Journal of Statistical Software 14 (11) (2005) 1–22.

[36] P. Brodatz, Textures: A Photographic Album for Artists and Designers, Dover Publications, 1966.

[37] R.L. Brooks, On coloring the points of a network, Proceedings of the Cambridge Philosophical Society 37 (1941) 194–197.

[38] S. Brooks, A. Gelman, G. Jones, X.L. Meng, Handbook of Markov Chain Monte Carlo, Chapman and Hall, Boca Raton, 2011.

[39] D. Broomhead, D. Lowe, Multivariable function interpolation and adaptive networks, Complex Systems 2 (1988) 321–355.

[40] T. Brox, A. Bruhn, N. Papenberg, J. Weickert, High accuracy optical flow estimation based on a theory for warping, Lecture Notes in Computer Science 3024 (2006) 26–36.

[41] D.L. Buckley, R.W. Kerslake, S.J. Blackband, A. Horsman, Quantitative analysis of multi-slice Gd-DTPA enhanced dynamic MR images using an automated simplex minimization procedure, Magnetic Resonance in Medicine 32 (5) (1994) 646–651.

[42] D.L. Buckley, Geoffrey J.M. Parker, Measuring contrast agent concentration in T1-weighted dynamic contrast-enhanced MRI, in: Alan Jackson, David Buckley, Geoffrey Parker (Eds.), Dynamic Contrast-Enhanced Magnetic Resonance Imaging in Oncology, Springer, 2005, pp. 69–79, (Chapter 5).

[43] H. Bunke, A. Sanfeliu, Syntactic and Structural Pattern Recognition: Theory and Applications, World Scientific, 1990.

[44] C.S. Burrus, R.A. Gopinath, H. Guo, Introduction to Wavelets and Wavelet Transform, Prentice Hall, 1998.

[45] P.J. Burt, E.H. Adelson, The Laplacian pyramid as a compact image code, IEEE Transaction on Communication 18 (1983) 532–540.

[46] R.B. Calinski, J. Harabasz, A dendrite method for cluster analysis, Psychometrika 3 (1974) 1–27.

[47] J. Canny, A computational approach to edge detection, IEEE Transaction on PAMI 8 (1986) 679–698.

[48] N. Canterakis, 3D zernike moments and zernike affine invariants for 3D image analysis and recognition, in: 11th Scandinavian Conference on Image Analysis, vol. 1, 1999, pp. 85–93.

[49] J.F. Cardoso, Blind signal seperation: statistical principles, Proceedings of IEEE 9 (1998) 2009–2025.

[50] J.F. Cardoso, Multidimensional independent component analysis, in: Proceedings of IEEE ICASSP, Seattle, vol. 4, 1998, pp. 1941–1944.

[51] R.E. Carson, S.C. Huang, M.V. Green, Weighted integration method for local cerebral blood flow measurements with positron emission tomography, Journal of Cerebral Blood Flow and Metabolism 6 (2) (1986) 245–258.

[52] K.R. Castleman, Digital Image Processing, Prentice Hall, 1996.

[53] M.D. Cerqueira, N.J. Weissman, V. Dilsizian, A.K. Jacobs, S. Kaul, Standardized myocardial segmentation and nomenclature for tomographic imaging of the heart: a statement for healthcare professionals from the cardiac imaging committee of the council on clinical cardiology of the American heart association, Circulation 105 (4) (2002) 539–542.

[54] T. Chan, L. Vese, Active contours without edges, IEEE Transactions on Image Processing 10 (2001) 266–277.

[55] W. Chang, H. Soliman, A. Sung, A vector quantization neural network to compress still monochromatic images, in: Proceedings of the IEEE Joint International Conference on Neural Networks in Orlando, 1994, pp. 4163–4168.

[56] C. Chatterjee, V. Roychowdhhurry, E. Chong, On relative convergence properties of principal component algorithms, IEEE Transactions on Neural Networks 9 (1998) 319–329.

[57] C. Chen, E. Tsao, W. Lin, Medical image segmentation by a constraint satisfaction neural network, IEEE Transaction on Nuclear Science 38 (1991) 678–686.

[58] W. Chen, M. Giger, G. Newstead, U. Bick, Automatic identification and classification of characteristic kinetic curves of breast lesions on DCE-MRI, Medical Physics 33 (2006) 2878–2887.

[59] Z. Cho, J. Jones, M. Singh, Foundations of Medical Imaging, John Wiley, 1993.

[60] N. Chomsky, Aspects of the Theory of Syntax, MIT Press, 1965.

[61] Leon O. Chua, Cellular neural networks: theory, IEEE Transactions on Circuit and Systems 35 (1988) 1257–1272.

[62] K. Chuang, M. Chiu, C. Lin, J. Chen, Model-free functional MRIanalysis using Kohonen clustering neural network and fuzzy c-means, IEEE Transaction on Medical Imaging 18 (1999) 1117–1128.

[63] E. Ciaccio, S. Dunn, M. Akay, Biosignal pattern recognition and interpretation systems: part I, IEEE Engineering in Medicine and Biology 13 (1993) 89–97.

[64] P. Clifford, Markov random fields in statistics, in: G.R. Grimmett, D.J.A. Welsh (Eds.), Disorder in Physical Systems: A Volume in Honour of John M. Hammersley, Oxford University Press, 1990, pp. 19–32.

[65] A.M. Cohen, S. Grossberg, Absolute stability of global pattern formation and parallel memory storage by competitive neural networks, IEEE Transactions on Systems, Man and Cybernetics SMC-13 (1983) 815–826.

[66] P. Comon, Independent component analysis, a new concept? Signal Processing 36 (1994) 287–314.

[67] P.C. Cosman, R.M. Gray, R.A. Olshen, Evaluating quality of compressed medical images: Snr, subjective rating, and diagnostic accuracy, Proceedings of IEEE 82 (1994) 919–932.

[68] P.C. Cosman, R.M. Gray, M. Vetterli, Vector quantization of image subbands: a survey, IEEE Transaction on Image Processing 5 (1996) 202–225.

[69] N. Cristiani, J. Shawe-Taylor, An Introduction to Support Vector Machines and Other Kernel-based Learning Methods, Cambridge Press, 2000.

[70] V.J. Cunningham, T. Jones, Spectral analysis of dynamic PET studies, Journal of Cerebral Blood Flow and Metabolism 13 (1993) 15–23.

[71] G. Hinton, D. Rumelhart, J. McClelland, A General Framework for Parallel Distributed Processing, Cambridge Press, 1986.

[72] R. Dave, Fuzzy shell clustering and appilcations to circle detection in digital images, International Journal of General Systems 16 (1990) 343–355.

[73] R. Dave, K. Bhaswan, Adaptive fuzzy c-shells clustering and detection of ellipses, IEEE Transactions on Neural Networks 3 (1992) 643–662.

[74] S. Davis, M. Fisher, S. Warach, Magnetic Resonance Imaging in Stroke, Cambridge University Press, Cambridge, 1993.
[75] F. De Pasquale, P. Barone, G. Sebastiani, J. Stander, Bayesian analysis of dynamic magnetic resonance breast images, Applied Statistics 53 (3) (2004) 475–493.
[76] C. Von der Malsburg, Self-organization of orientation sensitive cells in striata cortex, Kybernetik 14 (1973) 85–100.
[77] A.P. Dhawan, Y. Chitre, C. Kaiser, M. Moskowitz, Analysis of mammographic microcalcifications using gray-level image structure features, IEEE Transaction on Medical Imaging 15 (1996) 246–259.
[78] A.P. Dhawan, E. LeRoyer, Mammographic feature enhancement by computerized image processing, Computer Methods and Programs in Biomedicine (1988) 23–25.
[79] K.I. Diamantaras, S.Y. Kung, Principal Component Neural Networks, John Wiley, 1996.
[80] S. Dickson, Investigation of the use of neural networks for computerised medical image analysis (Ph.D. thesis), University of Bristol, 1998.
[81] S. Dodel, J. Herrmann, T. Geisel, Localization of brain activity-blind separation for fMRI data, Neurocomputing 33 (2000) 701–708.
[82] D.L. Donoho, De-noising by soft-thresholding, IEEE Transaction on Information Theory 41 (1995) 613–627.
[83] D.L. Donoho, I.M. Johnstone, De-noising by soft-thresholding, Biometrika 81 (1994) 425–455.
[84] R. Duda, P. Hart, Pattern Classification and Scene Analysis, John Wiley Verlag, 1973.
[85] D. Dumitrescu, B. Lazzerini, L. Jain, Fuzzy Sets and their Application to Clustering and Training, CRC Press, 2000.
[86] P.H.C. Eilers, B.D. Marx, Flexible smoothing with B-splines and penalties (with comments and rejoinder), Statistical Science 11 (2) (1996) 89–121.
[87] G. Ertas, O. Gulcur, O. Osman, O. Ucan, M. Tunaci, M. Dursun, Breast MR segmentation and lesion detection with cellular neural networks and 3D template matching, Computers in Biology and Medicine 38 (2008) 116–126.
[88] A.M. Eskiciouglu, P.S. Fisher, A survey of quality measures for gray scale image compression, in: Proceedings of Space and Earth Data Compression Workshop, 1993, pp. 49–61.
[89] L. Fahrmeir, T. Kneib, Bayesian Smoothing and Regression for Longitudinal, Spatial and Event History Data, Oxford University Press, Oxford, 2010.
[90] L. Fahrmeir, T. Kneib, S. Lang, Penalized structured additive regression for space-time data: a Bayesian perspective, Statistica Sinica 14 (2004) 715–745.
[91] J. Fan, Overcomplete wavelet representations with applications in image processing (Ph.D. thesis), University of Florida, 1997.
[92] M.A.T. Figueiredo, J.M.N. Leitao, Sequential and parallel image restoration: neural network implementations, IEEE Transactions on Neural Networks 3 (1994) 789–801.
[93] C.R. Fisel, J.L. Ackerman, R.B. Bruxton, MR contrast due to microscopically heterogeneous magnetic susceptibility: numerical simulations and applications to cerebral physiology, Magnetic Resonance in Medicine (1991) 336–347.
[94] H. Fisher, J. Hennig, Clustering of functional MR data, Proceedings of ISMRM Fourth Annual Meeting, vol. 96, 1996, pp. 1179–1183.
[95] D.B. Fogel, Evolutionary Computation, IEEE Press, 1995.
[96] J. Frahm, K. Merboldt, W. Hanicke, Functional MRI of human brain activation at high spatial resolution, Magnetic Resonance in Medicine 29 (1992) 139–144.
[97] K. Friston, A.P. Holmes, K.J. Worsley, J.P. Poline, C.D. Frith, R.S.J. Frackowiak, Statistical parametric maps in functional imaging: a general linear approach, Human Brain Mapping 2 (1995) 189–210.
[98] K.S. Fu, Syntactical Methods in Pattern Recognition, Academic Press, 1974.
[99] K.S. Fu, Syntactic Pattern Recognition with Applications, Prentice Hall, 1982.
[100] K. Fukunaga, Introduction to Statistical Pattern Recognition, Academic Press, 1990.
[101] F.B. Fuller, The writhing number of a space curve, Proceedings of the National Academy of Sciences of the United States of America 68 (1971) 815–819.
[102] D. Gabor, Theory of communication, Journal of the IEE 93 (1946) 429–457.

[103] D. Gamerman, H.F. Lopes, Markov Chain Monte Carlo: Stochastic Simulation for Bayesian Inference, Chapman and Hall, Boca Raton, 2006.

[104] C.R. Genovese, Nicole A. Lazar, T. Nichols, Thresholding of statistical maps in functional neuroimaging using the false discovery rate, NeuroImage 15 (4) (2002) 870–878.

[105] K. Gilhuijs, M. Giger, U. Bick, Computerized analysis of breast lesions in three dimensions using dynamic magnetic-resonance imaging, Medical Physics 25 (1998) 1647–1654.

[106] W.R. Gilks, S. Richardson, D.J. Spiegelhalter, Markov Chain Monte Carlo in Practice, Chapman and Hall, Boca Raton, 1996.

[107] M. Girolami, C. Fyfe, An extended exploratory pursuit network with linear and nonlinear anti-Hebbian lateral connections applied to the cocktail party problem, Neural Networks 10 (1997) 1607–1618.

[108] M. Girolami, C. Fyfe, Stochastic ICA contrast maximisation using Oja's nonlinear PCA algorithm, International Journal of Neural Systems 8 (1997) 661–679.

[109] T.A. Goldstein, M. Jerosch-Herold, B. Misselwitz, H. Zhang, R.J. Gropler, J. Zheng, Fast mapping of myocardial blood flow with MR first-pass perfusion imaging, Magnetic Resonance in Medicine 59 (6) (2008) 1394–1400.

[110] R.C. Gonzalez, R.E. Woods, Digital Image Processing, Prentice Hall, 2002.

[111] S.S. Gopal, B. Sahiner, H.P. Chan, N. Petrick, Neural network based segmentation using a priori image models, Proceedings of the IEEE Joint International Conference on Neural Networks in Houston, vol. 3, 1997, pp. 2456–2459.

[112] C. Gössl, D.P. Auer, L. Fahrmeir, Bayesian spatiotemporal inference in functional magnetic resonance imaging, Biometrics 57 (2) (2001) 554–562.

[113] C. Goutte, P. Toft, E. Rostrup, F. Nielsen, L.K. Hansen, On clustering fMRI series, NeuroImage 9 (1999) 298–310.

[114] H. Gries, Extracellular MRI contrast agents based on gadolinium, in: W. Krause (Ed.), Contrast Agents I. Magnetic Resonance Imaging, Springer, Heidelberg, 2002, pp. 1–24.

[115] G. Grimsby, R. Gray, A. Dueck, S. Carpenter, C. Stucky, H. Aspey, M. Giurescu, B. Pockaj, Is there concordance of invasive breast cancer pathologic tumor size with magnetic resonance imaging, American Journal of Surgery 198 (2009) 500–504.

[116] S. Grossberg, On learning and energy-entropy dependence in recurrent and nonrecurrent signed networks, Journal of Statistical Physics 1 (1969) 319–350.

[117] S. Grossberg, Adaptive pattern classification and universal recording, Biological Cybernetics 23 (1976) 121–134.

[118] S. Grossberg, Competition, decision and consensus, Journal of Mathematical Analysis and Applications 66 (1978) 470–493.

[119] H. Gudbjartsson, S. Patz, The Rician distribution of noisy MRI data, Magnetic Resonance in Medicine 34 (6) (2005) 910–914.

[120] L. Hadjiiski, B. Sahiner, H. Chan, Evaluating the effect of image preprocessing on an information-theoretic CAD system in mammography, Current Opinion in Obstetrics and Gynecology 18 (2006) 64–70.

[121] J.V. Hajnal, D.G. Hill, D.J. Hawkes (Eds.), Medical Image Registration, CRC Press, Boca Raton, 2001.

[122] L. Hall, A. Bensaid, L. Clarke, R. Velthuizen, M. Silbiger, J. Bezdek, A comparison of neural network and fuzzy clustering techniques in segmenting magnetic resonance images of the brain, IEEE Transactions on Neural Networks 5 (1992) 672–682.

[123] O. Haraldseth, R. Jones, T. Mullera, A. Fahlvik, A. Oksendal, Comparison of dysprosium BMA and superparamagnetic iron oxide particles as susceptibility contrast agents for perfusion imaging of regional cerebral ischemia in the rat, Journal of Magnetic Resonance in Imaging 6 (5) (1996) 714–717.

[124] R.M. Haralick, L. Shapiro, Computer and Robot Vision, Addison-Wesley, 1992.

[125] S. Haring, M. Viergever, N. Kok, Kohonen networks for multiscale image segmentation, Image and Vision Computing 12 (1994) 339–344.

[126] Eric J. Hartman, James D. Keeler, Jacek M. Kowalski, Layered neural networks with Gaussian hidden units as universal approximations, Neural Computation 2 (1990) 210–215.

[127] R. Hathaway, J. Bezdek, Optimization of clustering criteria by reformulation, IEEE Transactions on Fuzzy Systems 3 (1992) 241–245.

[128] C. Hayes, A.R. Padhani, M.O. Leach, Assessing changes in tumour vascular function using dynamic contrast-enhanced magnetic resonance imaging, NMR in Biomedicine 15 (2002) 154–163.

[129] S. Haykin, Neural Networks, Maxwell Macmillan Publishing Company, 1994.

[130] D.O. Hebb, The Organization of Behavior, John Wiley Verlag, 1949.

[131] S. Heim, Ludwig Fahrmeir, Paul H.C. Eilers, Brian D. Marx, 3D space-varying coefficient models with application to diffusion tensor imaging, Computational Statistics and Data Analysis 51 (12) (2007) 6212–6228.

[132] K.H. Herrmann, S. Wurdinger, D.R. Fischer, I. Krumbein, M. Schmitt, G. Hermosillo, K. Chaudhuri, A. Krishnan, M. Salganicoff, W.A. Kaiser, J.R. Reichenbach, Application and assessment of a robust elastic motion correction algorithm to dynamic MRI, European Radiology 17 (2007) 259–264.

[133] H. Herzog, Basic ideas and principles for quantifying regional blood flow with nuclear medical techniques, Nuklearmedizin 35 (5) (1996) 181–185.

[134] S. Heywang, A. Wolf, E. Pruss, MRI imaging of the breast: fast imaging sequences with and without GD-DTPA, Radiology 171 (1989) 95–103.

[135] A. Hill, A. Mehnert, S. Crozier, K. McMahon, Evaluating the accuracy and impact of registration in dynamic contrast-enhanced breast MRI, Concepts in Magnetic Resonance Part B 35B (2009) 106–120.

[136] S. Hoffmann, J.D. Shutler, M. Lobbes, B. Burgeth, A. Meyer-Baese, Automated analysis of diagnostically challenging lesions in breast MRI based on spatio-temporal moments and joint segmentation-motion compensation technique, EURASIP Journal on Advances in Signal Processing (2013), (page In print, 4).

[137] J.H. Holland, Adaptation in Natural and Artificial Systems, University of Michigan Press, 1975.

[138] J.J. Hopfield, Neural networks and physical systems with emergent collective computational abilities, Proceedings of the National Academy of Science 79 (1982) 2554–2558.

[139] J.J. Hopfield, D.W. Tank, Computing with neural circuits: a model, Science 233 (1986) 625–633.

[140] B. Horn, B. Schunck, Determining optical flow, Artificial Intelligence 17 (1981) 185–203.

[141] M. Horsfield, B. Morgan, Algorithms for calculation of kinetic parameters from T1-weighted dynamic contrast-enhanced magnetic resonance imaging, Journal of Magnetic Resonance Imaging 20 (4) (2004) 723–729.

[142] S.C. Huang, R.E. Carson, M.E. Phelps, Measurement of local blood flow and distribution volume with short-lived isotopes: a general input technique, Journal of Cerebral Blood Flow and Metabolism 2 (1) (1982) 99–108.

[143] P. Huber, Projection pursuit, Annals of Statistics 13 (1985) 435–475.

[144] A. Hyvarinen, Fast and robust fixed-point algorithms for independent component analysis, IEEE Transactions on Neural Networks 10 (1999) 626–634.

[145] A. Hyvarinen, P. Hoyer, Emergence of phase- and shift-invariant features by decomposition of natural images into independent feature subspaces, Neural Computation 12 (2000) 1705–1720.

[146] A. Hyvarinen, P. Hoyer, Topographic independent component analysis, Neural Computation 13 (2001) 1527–1558.

[147] A. Hyvarinen, E. Oja, Independent component analysis: algorithms and applications, Neural Networks 13 (2000) 411–430.

[148] H. Iida, P.M. Bloomfield, S. Miura, I. Kanno, M. Murakami, K. Uemura, M. Amano, K. Tanaka, Y. Hirose, S. Yamamoto, Effect of real-time weighted integration system for rapid calculation of functional images in clinical positron emission tomography, IEEE Transactions on Medical Imaging 14 (1) (1995) 116–121.

[149] F. Jamitzky, R. Stark, W. Bunk, S. Thalhammer, C. Räeth, T. Aschenbrenner, G. Morfill, W. Heckl, Scaling-index method as an image processing tool in scanning-probe microscopy, Ultramicroscopy 86 (2001) 241–246.

[150] S. Jansen, Ductal carcinoma in situ: detection, diagnosis, and characterization with magnetic resonance imaging, Seminars in Ultrasound, CT and MRI 32 (2011) 306–318.

[151] S.A. Jansen, X. Fan, G.S. Karczmar, H. Abe, R. Schmidt, M. Giger, G.M. Newstaed, DCEMRI of breast lesions: is kinetic analysis equally effective for both mass and nonmass-like enhancement? Medical Physics 35 (2008) 3102–3109.

[152] S.A. Jansen, A. Shimauchi, L. Zak, X. Fan, G.S. Karczmar, G.M. Newstaed, The diverse pathology and kinetics of mass, nonmass, and focus enhancement on MR imaging of the breast, Journal of Magnetic Resonance Imaging 33 (2011) 1382–1389.

[153] M. Jerosch-Herold, Perfusion reserve in asymptomatic individuals, International Journal of Cardiovascular Imaging 20 (6) (2004) 579–586.

[154] M. Jerosch-Herold, C. Swingen, R.T. Seethamraju, Myocardial blood flow quantification with MRI by model-independent deconvolution, Medical Physics 29 (5) (2002) 886.

[155] J.A. Johnson, T.A. Wilson, A model for capillary exchange, American Journal of Physiology 210 (6) (1966) 1299–1303.

[156] P.R. Johnston, R.M. Gulrajani, Selecting the corner in the L-curve approach to Tikhonov regularization, IEEE Transaction of Biomedical Engineering 47 (9) (2000) 1293–1296.

[157] M.I. Jordan, R.A. Jacobs, Hierarchical mixture of experts and the EM algorithm, Neural Computation 6 (1994) 181–214.

[158] C. Jutten, J. Herault, Blind separation of sources, Signal Processing 24 (1991) 1–10.

[159] A. Kalukin, M. Van Geet, R. Swennen, Principal component analysis of multienergy X-ray computed tomography of mineral samples, IEEE Transaction on Nuclear Science 47 (2000) 1729–1736.

[160] N. Kambhatla, S. Haykin, R. Dony, Image compression using KLT wavelets and an adaptive mixture of principal components model, Journal ofVLSI Signal Processing Systems 18 (3) (1998) 287–297.

[161] I. Kapouleas, Segmentation and feature extraction for magnetic resonance brain image analysis, in: ICPR90, vol. 1, 1990, pp. 583–590.

[162] N. Karayiannis, A methodology for constructing fuzzy algorithms for learning vector quantization, IEEE Transactions on Neural Networks 8 (1997) 505–518.

[163] N. Karayiannis, P. Pai, Fuzzy algorithms for learning vector quantization, IEEE Transactions on Neural Networks 7 (1996) 1196–1211.

[164] J.C. Kärcher, V.J. Schmid, Two tissue compartment model in DCE-MRI: a Bayesian approach, in: IEEE International Symposium on Biomedical Imaging. From Nano to Macro, IEEE, 2010, pp. 724–727.

[165] J. Karhunen, J. Jourtensalo, Representation and separation of signals using nonlinear PCA type learning, Neural Networks 7 (1994) 113–127.

[166] J. Karhunen, E. Oja, L. Wang, R. Vigario, J. Joutsensalo, A class of neural networks for independent component analysis, IEEE Transactions on Neural Networks 8 (1997) 486–504.

[167] N. Karssemeijer, Stochastic model for automated detection of calcification in digital mammogramms, Image and Vision Computing 10 (1992) 369–375.

[168] R.L. Kashyap, A. Khotanzad, A model based method for rotation invariant texture classification, IEEE Transaction on PAMI 8 (1986) 472–481.

[169] M. Kass, A. Witkin, D. Terzopoulos, Snakes: active contour model, International Journal on Computer Vision 1 (1988) 321–331.

[170] A.G. Katsaggelos, Digital Image Processing, Springer Verlag, 1991.

[171] W.P. Kegelmeyer, J.M. Pruneda, P.D. Bourland, Computer-aided mammographic screening for spiculated lesions, Radiology 191 (1994) 331–337.

[172] R.P. Kennan, J. Zhong, J.C. Gore, Intravascular susceptibility contrast mechanism in tissues, Magnetic Resonance in Medicine 31 (1) (1994) 9–21.

[173] D.N. Kennedy, P.A. Filipek, V.S. Caviness, Anatomic segmentation and volumetric calculations in nuclear magnetic resonance imaging, IEEE Transaction on Medical Imaging 1 (1989) 1–7.

[174] S.S. Kety, Blood-tissue exchange methods. Theory of blood-tissue exchange and its applications to measurement of blood flow, Methods in Medical Research 8 (1960) 223–227.

[175] H.K. Khalil, Nonlinear Systems, Prentice Hall, 1996.

[176] D. J. Kim, Y.W. Park, D. J. Park, A novel validity index for determination of the optimal number of clusters, IEICE Transactions on Information and Systems E84-D (2) (2001) 281–285.

[177] V. Kim, L. Yaroslavskii, Rank algorithms for picture processing, ComputingVision, Graphics, and Image Processing 35 (1986) 234–259.

[178] R. Kindermann, J.L. Snell, Markov random fields and their applications, American Mathematical Society, 1980.

[179] J. Kittler, Feature selection and extraction, Handbook of Pattern Recognition and Image Processing 1 (1986) 59–83.

[180] L. Knorr-Held, Bayesian modelling of inseparable space time variation in disease risk, Statistics in Medicine 19 (17–18) (2000) 2555–2567.

[181] R.A. Koeppe, J.E. Holden, W.R. Ip, Performance comparison of parameter estimation techniques for the quantitation of local cerebral blood flow by dynamic positron computed tomography, Journal of Cerebral Blood Flow and Metabolism 5 (2) (1985) 224–234.

[182] T.S. Koh, V. Zeman, J. Darko, T.Y. Lee, Michael Milosevic, M. Haider, P. Warde, I.W.T. Yeung, The inclusion of capillary distribution in the adiabatic tissue homogeneity model of blood flow, Physics in Medicine and Biology 46 (5) (2001) 1519–1538.

[183] T. Kohonen, Self-organized formation of topologically correct feature maps, Biological Cybernetics 43 (1982) 59–69.

[184] T. Kohonen, Self-Organization and Associative Memory, Springer Verlag, 1988.

[185] T. Kohonen, Emergence of invariant-feature detectors in the adaptive-subspace self-organizing map, Biological Cybernetics 75 (1996) 281–291.

[186] I. Koren, A. Laine, F. Taylor, Image fusion using steerable dyadic wavelet transform, in: ICIP95, vol. 3, 1995, pp. 1415–1418.

[187] I. Koren, A. Laine, F. Taylor, Enhancement via fusion of mammographic features, in: ICIP98, vol. 3, 1998, pp. 1415–1418.

[188] B. Kosko, Adaptive bidirectional associative memory, Applied Optics 26 (1987) 4947–4960.

[189] B. Kosko, Neural Networks and Fuzzy Systems, Prentice Hall, 1992.

[190] C. Kotropoulos, X. Magnisalis, I. Pitas, M.G. Strintzis, Nonlinear ultrasonic image processing based on signal-adaptive filters and self-organizing neural networks, IEEE Transaction on Image Processing 3 (1994) 65–77.

[191] C.K. Kuhl, P. Mielcareck, S. Klaschik, C. Leutner, E. Wardelmann, J. Gieseke, H. Schild, Dynamic breast MR imaging: are signal intensity time course data useful for differential diagnosis of enhancing lesions? Radiology 211 (1999) 101–110.

[192] M. Kupinski, M. Giger, Automated seeded lesion segmentation on digital mammograms, IEEE Transaction on Medical Imaging 17 (1998) 510–517.

[193] K. Kwong, Functional magnetic resonance imaging with echo planar imaging, Magnetic Resonance Quarterly 11 (1992) 1–20.

[194] K. Kwong, J. Belliveau, D. Chesler, Dynamic magnetic resonance imaging of human brain activity during primary sensor stimulation, Proceedings of the National Academy of Science 89 (1992) 5675–5679.

[195] A. Laine, I. Koren, J. Fan, F. Taylor, A steerable dyadic wavelet transform and interval wavelets for enhancement of digital mammography, in: SPIE Proceedings Series, vol. 2491, 1995, pp. 736–749.

[196] A.F. Laine, J. Fan, W. Yang, Wavelets for contrast enhancement of digital mammography, IEEE Engineering in Medicine and Biology 15 (1995) 536–550.

[197] A.F. Laine, A. Meyer-Base, W. Huda, J. Honeyman, B. Steinbach, Local Enhancement of Masses via Multiscale Analysis, Technical Report No. 4, University of Florida, 1996.

[198] A.F. Laine, S. Schuler, J. Fan, W. Huda, Mammographic feature enhancement by multiscale analysis, IEEE Transaction on Medical Imaging 13 (1994) 725–740.

[199] S. Lang, A. Brezger, Bayesian P-splines, Journal of Computational and Graphical Statistics 13 (2004) 183–212.

[200] N. Lange, Statistical approaches to human brain mapping by functional magnetic resonance imaging, Statistics in Medicine 15 (4) (1996) 389–428.

[201] O. Lange, A. Meyer-Baese, A. Wismueller, M. Hurdal, Analysis of dynamic cerebral contrast-enhanced perfusion MRI time-series based on unsupervised clustering methods, in: SPIE's 18th Annual International Symposium on Aerospace/Defense Sensing, Simulation and Controls, vol. 5818, 2005, pp. 26–37.

[202] H. Larsson, P. Tofts, Measurement of the blood-brain barrier permeability and leakage space using dynamic Gd-DTPA scanning – a comparison of methods, Magnetic Resonance in Medicine 24 (1) (1992) 174–176.
[203] B.W. Henrik, Larsson M. Stubgaard, J.L. Frederiksen, M. Jensen, O. Henriksen, O.B. Paulson, Quantitation of blood-brain barrier defect by magnetic resonance imaging and gadolinium-DTPA in patients with multiple sclerosis and brain tumors, Magnetic Resonance in Medicine 16 (1) (1990) 117–131.
[204] N.A. Lassen, W. Perl, Tracer Kinetic Methods in Medical Physiology, Raven Press, New York, 1979.
[205] A. Lauric, E. Miller, M. Baharoglu, A. Malek, 3D shape analysis of intracranial aneurysms using the writhe number as a discriminant for rupture, Annals of Biomedical Engineering 39 (2011) 1457–1469.
[206] A. Lauric, E. Miller, S. Frisken, A. Malek, Automated detection of intracranial aneurysms based on parent vessel 3D analysis, Medical Image Analysis 14 (2010) 149–159.
[207] K. Laws, Rapid texture identification, SPIE Proceedings Series, vol. 238, 1980, pp. 376–380.
[208] K.I. Laws, Textured image segmentation (Ph.D. thesis), University of Southern California, 1980.
[209] R.S. Ledley, L.B. Lusted, Reasoning foundations of medical diagnosis, MD Computation 8 (1991) 300–315.
[210] S. Lee, J. Kim, Z. Yang, Y. Jung, W. Moon, Multilevel analysis of spatiotemporal association features for differentiation of tumor enhancement pattern in breast DCE-MRI, Medical Physics 37 (2010) 3940–3956.
[211] Sukhan Lee, Rhee M. Kil, A gaussian potential function network with hierarchically self-organizing learning, Neural Networks 4 (1991) 207–224.
[212] T.W. Lee, M. Girolami, T.J. Sejnowski, Independent component analysis using an extended infomax algorithm for mixed sub-Gaussian and super-Gaussian sources, Neural Computation 11 (1999) 409–433.
[213] C. Leondes, Image Processing and Pattern Recognition, Academic Press, 1998.
[214] M.D. Levine, Vision in Man and Machine, McGraw-Hill, 1985.
[215] H. Li, Y. Wang, S. Lo, M. Freedman, Computerized radiographic mass detection—part1, IEEE Transaction on Medical Imaging 20 (2001) 289–301.
[216] J. Li, N. Chaddha, R.M. Gray, Multiresolution tree structured vector quantization, in: Asilomar Conference on Signals, Systems and Computer, vol. 11, 1996.
[217] S. Li, Markov Random Field Modeling in Image Analysis, Springer, London, 2009.
[218] Y. Linde, A. Buzo, R.M. Gray, An algorithm for vector quantizer design, IEEE Transactions on Communications 28 (1980) 84–95.
[219] R.P. Lipmann, An introduction to computing with neural networks, IEEE ASSP Magazine 4 (1987) 4–22.
[220] G. Liu, H. Rugo, G. Wilding, T. McShane, J. Evelhoch, C. Ng, E. Jackson, F. Kelcz, B. Yeh, F. Lee Jr, C. Charnsangavej, J. Park, E. Ashton, H. Steinfeldt, Y. Pithavala, S. Reich, R. Herbst, Dynamic contrast-enhanced magnetic resonance imaging as a pharmacodynamic measure of response after acute dosing of AG-013736, an oral angiogenesis inhibitor, in patients with advanced solid tumors: results from a phase I study, Journal of Clinical Oncology 24 (2005) 5464–5471.
[221] S.B. Lo, H.P. Chan, J. Lin, M.T. Freedman, S.K. Mun, Artificial convolution neural network for medical image pattern recognition, Neural Networks 8 (1995) 1201–1214.
[222] S.B. Lo, J. Lin, M.T. Freedman, S.K. Mun, Application of artificial neural networks to medical image pattern recognition: detection of clustered microcalcifications on mammograms and lung cancer on chest radiographs, Journal of VLSI Signal Processing 18 (1998) 263–274.
[223] C. Looney, Pattern Recognition Using Neural networks, Oxford University Press, 1997.
[224] J. Lu, D. Healy, J. Weaver, Contrast enhancement of medical images using multiscale edge representation, Optical Engineering 33 (1994) 2151–2161.
[225] E. Lucht, S. Delorme, G. Brix, Neural network-based segmentation of dynamic (MR) mammography images, Magnetic Resonance Imaging 20 (2002) 89–94.
[226] R. Lucht, S. Delorme, J. Heiss, M. Knopp, M.A. Weber, J. Griebel, G. Brix, Classification of signal-time curves obtained by dynamic-magnetic resonance mammography, Investigative Radiology 40 (2005) 442–447.

[227] D. Lunn, D. Spiegelhalter, A. Thomas, N. Best, The BUGS project: evolution, critique and future directions, Statistics in Medicine 28 (2009) 3049–3067.

[228] A. Macovski, Medical Imaging Systems, Prentice Hall, 1983.

[229] A. Mademlis, A. Axenopoulos, P. Daras, D. Tzovaras, M. Strintzis, 3D content-based search based on 3D Krawtchouk moments, in: Proceedings of the Third International Symposium on 3D Data Processing, Visualization and Transmission, vol. 1, 2006, pp. 743–749.

[230] J. Makhoul, Linear prediction: a tutorial review, Proceedings of the IEEE 63 (1975) 561–580.

[231] S. Mallat, Zero-crossings of a wavelet transform, IEEE Transaction on Information Theory 37 (1991) 1019–1033.

[232] S. Mallat, A Wavelet Tour of Signal Processing, Academic Press, 1997.

[233] S. Mallat, S. Zhong, Characterization of signals from multiscale edges, IEEE Transaction on PAMI 14 (1992) 710–732.

[234] J. Mao, A.K. Jain, Texture classification and segmentation using multiresolution simultaneous autoregressive models, Pattern Recognition 25 (1992) 173–188.

[235] J.A. Marshall, Adaptive perceptual pattern recognition by self-organizing neural networks, Neural Networks 8 (1995) 335–362.

[236] T. Martinetz, S. Berkovich, K. Schulten, Neural gas network for vector quantization and its application to time-series prediction, IEEE Transactions on Neural Networks 4 (1993) 558–569.

[237] B. Marx, P. Eilers, Direct generalized additive modeling with penalized likelihood, Computational Statistics and Data Analysis 28 (2) (1998) 193–209.

[238] K. Matsuoka, M. Kawamoto, A neural network that self-organizes to perform three operations related to principal component analysis, Neural Networks 7 (1994) 753–765.

[239] M. McKeown, T. Jung, S. Makeig, G. Brown, T. Jung, S. Kindermann, A. Bell, T. Sejnowski, Analysis of fMRI data by blind separation into independent spatial components, Human Brain Mapping 6 (1998) 160–188.

[240] M. McKeown, T. Jung, S. Makeig, G. Brown, T. Jung, S. Kindermann, A. Bell, T. Sejnowski, Spatially independent activity patterns in functional magnetic resonance imaging data during the stroop color-naming task, Proceedings of National Academic Science 95 (1998) 803–810.

[241] L. Arbash Meinel, T. Buelow, D. Huo, A. Shimauci, U. Kose, J. Buurman, G. Newstead, Robust segmentation of mass-lesions in contrast-enhanced dynamic breast MR images, Journal of Magnetic Resonance Imaging 32 (2010) 110–119.

[242] L. Arbash Meinel, A. Stolpen, K. Berbaum, L. Fajardo, J. Reinhardt, Breast MRI lesion classification: improved performance of human readers with a backpropagation network computer-aided diagnosis (CAD) system, Journal of Magnetic Resonance Imaging 25 (2007) 89–95.

[243] W. Meisel, Computer-Oriented Approaches to Pattern Recognition, Academic Press, 1972.

[244] N. Metropolis, A. Rosenbluth, M. Rosenbluth, A. Teller, E. Teller, Equation of state calculations by fast computing machines, Journal of Chemical Physics 21 (6) (1953) 1087.

[245] C.E. Metz, Roc methodology in radiologic imaging, Investigative Radiology 21 (1986) 720–733.

[246] A. Meyer-Baese, O. Lange, A. Wismueller, M. Hurdal, Analysis of dynamic susceptibility contrast MRI time-series based on unsupervised clustering methods, in: Press IEEE Transactions on Information Technology in Biomedicine, 2007, pp. 563–573.

[247] A. Meyer-Base, On the existence and stability of solutions in self-organizing cortical maps, IEICE Transactions on Fundamentals of Electronics, Communications and Computer Sciences E82-A (9) (1999) 1883–1887.

[248] A. Meyer-Bäse, Pattern Recognition for Medical Imaging, Elsevier Science/Academic Press, 2003.

[249] A. Meyer-Bäse, F. Ohl, H. Scheich, Singular perturbation analysis of competitive neural networks with different time-scales, Neural Computation (1996) 545–563.

[250] A. Meyer-Bäse, T. Schlossbauer, O. Lange, A. Wismüller, Small lesions evaluation based on unsupervised cluster analysis of signal-intensity time courses in dynamic breast MRI, International Journal of Biomedical Imaging (2010) ID 326924.

[251] Z. Michalewicz, Genetic Algorithms, Springer Verlag, 1995.

[252] K. Michielsen, H. de Raedt, Integral-Geometry Morphological Image Analysis, Elsevier, 2001.

[253] K. Michielsen, H. De Raedt, Morphological image analysis, Computer Physics Communications 132 (2000) 94–103.

[254] L. Miclet, Structural Methods in Pattern Recognition, Springer Verlag, 1986.

[255] J. Moody, C. Darken, Fast learning in networks of locally-tuned processing units, Neural Computation 1 (1989) 281–295.

[256] M.E. Moseley, Z. Vexler, H.S. Asgari, Comparison of Gd- and Dy-chelates for T2* contrast-enhanced imaging, Magnetic Resonance in Medicine (1991) 259–264.

[257] K. Murase, Efficient method for calculating kinetic parameters using T1-weighted dynamic contrast-enhanced magnetic resonance imaging, Magnetic Resonance in Medicine 51 (4) (2004) 858–862.

[258] B. Nabet, R. Pinter, Sensory Neural Networks, CRC Press, 1991.

[259] A.N. Netravali, B.G. Haskell, Digital Pictures: Representation and Compression, Plenum Press Verlag, 1988.

[260] D. Newell, K. Nie, J. Chen, C. Hsu, H. Yu, O. Nalcioglu, M. Su, Selection of diagnostic features on breast MRI to differentiate between malignant and benign lesions using computer-aided diagnostics: differences in lesions presenting as mass and non-mass-like enhancement, European Radiology 20 (2010) 771–781.

[261] S. Ngan, X. Hu, Analysis of fMRI imaging data using self-organizing mapping with spatial connectivity, Magnetic Resonance in Medicine 41 (1999) 939–946.

[262] N.J. Nilsson, Learning Machines: Foundations of Trainable Pattern-Classifying Systems, McGraw-Hill, 1965.

[263] R.M. Nishikawa, M.L. Giger, K. Doi, C.J. Vyborny, R.A. Schmidt, C.E. Metz, Y. Wu, F.-F. Yin, Y. Jiang, Z. Huo, P. Lu, W. Zhang, T. Ema, U. Bick, J. Papaioannou, R.H. Nagel, Computer-aided detection and diagnosis of masses and clustered microcalcifications from digital mammograms, SPIE Proceedings Series, vol. 1905, 1993, pp. 422–432.

[264] M. Novotni, R. Klein, Shape retrieval using 3D zernike descriptors, Computer Aided Design 36 (2004) 1047–1062.

[265] I.M. Obdeijn, C. Loo, A. Rijnsburger, M. Wasser, E. Bergers, T. Kok, J. Klijn, C. Boetes, Assessment of false-negative cases of breast MR imaging in women with a familial or genetic predisposition, Breast Cancer Research and Treatment 119 (2010) 399–407.

[266] S. Ogawa, T. Lee, B. Barrere, The sensitivity of magnetic resonance image signals of a rat brain to changes in the cerebral venous blood oxygenation activation, Magnetic Resonance in Medicine 29 (1993) 205–210.

[267] S. Ogawa, D. Tank, R. Menon, Intrinsic signal changes accompanying sensory stimulation: functional brain mapping with magnetic resonance imaging, Proceedings of the National Academy of Science 89 (1992) 5951–5955.

[268] M. Ogiela, R. Tadeusiewicz, Syntactic reasoning and pattern recognition for analysis of coronary artery images, Artificial Intelligence in Medicine 670 (2002) 1–15.

[269] M.R. Ogiela, R. Tadeusiewicz, Advances in syntactic imaging techniques for perception of medical images, Journal of Imaging Science 49 (2001) 113–120.

[270] M.R. Ogiela, R. Tadeusiewicz, New aspects of using the structured graph-grammar based technique for recognition of selected medical images, Journal of Digital Imaging 14 (2001) 231–232.

[271] E. Oja, A simplified neural model as a principal component analyzer, Journal of Mathematical Biology 15 (1982) 267–273.

[272] E. Oja, Neural networks, principal components, and subspaces, International Journal of Neural Systems 1 (1989) 61–68.

[273] E. Oja, Self-organizing maps and computer vision, Neural Networks for Perception 1 (1992) 368–385.

[274] E. Oja, H. Ogawa, J. Wangviwattana, Learning in nonlinear constrained Hebbian networks, Artificial Neural Networks 1 (1991) 385–390.

[275] Matthew R. Orton, David J. Collins, Simon Walker-Samuel, James A. D'Arcy, David J. Hawkes, David Atkinson, Martin O. Leach, Bayesian estimation of pharmacokinetic parameters for DCE-MRI with a robust treatment of enhancement onset time, Physics in Medicine and Biology 52 (9) (2007) 2393–2408.

[276] L. Østergaard, A.G. Sorensen, K.K. Kwong, R.M. Weisskopf, C. Gyldensted, B.R. Rosen, High resolution measurement of cerebral blood flow using intravascular tracer bolus passages. Part II: experimental comparison and preliminary results, Magnetic Resonance in Medicine 36 (1996) 726–736.

[277] T. Otto, A. Meyer-Bäese, M. Hurdal, D. Sumners, D. Auer, A. Wismüller, Model-free functional MRI analysis using transformation based methods, in: Proceeding of SPIE, vol. 5102, 2003, pp. 156–167.

[278] A. Padhani, J. Yarnold, J. Regan, J. Husband, Dynamic MRI of breast hardness following radiation treatment, Journal of Magnetic Resonance Imaging 17 (2003) 427–434.

[279] J.K. Paik, A.K. Katsagellos, Image restoration using a modified hopfield network, IEEE Transaction on Image Processing 1 (1992) 49–63.

[280] N. Pal, J. Bezdek, E. Tsao, Generalized clustering networks and Kohonen's self-organizing scheme, IEEE Transactions on Neural Networks 4 (1993) 549–557.

[281] S. Pal, A. Ghosh, M. Kundu, Soft Computing for Image Processing, Springer, 2000.

[282] S. Pal, S. Mitra, Neuro-Fuzzy Pattern Recognition, John Wiley, 1999.

[283] J. Panting, P. Gatehouse, G.-Z. Yang, F. Grothues, D. Firmin, P. Collins, D. Pennell, Abnormal subendocardial perfusion in cardiac syndrome X detected by cardiovascular magnetic resonance imaging, New England Journal of Medicine 346 (25) (2002) 1948–1953.

[284] N. Papenberg, A. Bruhn, T. Brox, S. Didas, J. Weickert, Highly accurate optic flow computation with theoretically justified warping, International Journal of Computer Vision 67 (2006) 141–158.

[285] A. Papoulis, Probability, Random Variables, and Stochastic Processes, McGraw-Hill, 1986.

[286] J. Park, I. Sandberg, Universal approximation using radial-basis-function networks, Neural Computation 3 (1991) 247–257.

[287] G. Parker, A. Padhani, T1-w DCE-MRI: T1-weighted dynamic contrast-enhanced MRI, in: P. Tofts (Ed.), Quantitative MRI of the Brain: Measuring Changes Caused by Disease, John Wiley and Sons, Chichester, 2004.

[288] F. Pedersen, M. Bergstrom, E. Bengtsson, B. Langstrom, Principal component analysis of dynamic positron emission tomography images, European Journal of Nuclear Medicine 21 (1994) 1285–1292.

[289] W.D. Penny, N.J. Trujillo-Barreto, K. Friston, Bayesian fMRI time series analysis with spatial priors, NeuroImage 24 (2) (2005) 350–362.

[290] H. Penzkofer, Entwicklung von Methoden zur magnetresonanztomographischen Bestimmung der myokardialen und zerebralen Perfusion (Ph.D. thesis), LMU Munich, 1998.

[291] S.M. Perlmutter, P.C. Cosman, R.M. Gray, R.A. Olshen, D. Ikeda, C.N. Adams, B.J. Betts, M. Williams, K.O. Perlmutter, J. Li, A. Aiyer, L. Fajardo, R. Birdwell, B.L. Daniel, Image quality in lossy compressed digital mammograms, Signal Processing 59 (2) (1997) 180–210.

[292] N. Petrick, H. Chan, B. Sahiner, M. Helvie, M. Goodsitt, D. Adler, Computer-aided breast mass detection: false positive reducing using breast tissue composition, Excerpta Medica 1119 (1996) 373–378.

[293] D.T. Pham, P. Garat, Blind separation of mixture of independent sources through a quasimaximum likelihood approach, IEEE Transactions on Signal Processing 45 (1997) 1712–1725.

[294] E. Pietka, A. Gertych, K. Witko, Informatics infrastructure of CAD system, Computerized Medical Imaging and Graphics 29 (2005) 157–169.

[295] John Platt, A resource-allocating network for function interpolation, Neural Computation 3 (1991) 213–225.

[296] T. Poggio, F. Girosi, Extensions of a theory of networks for approximations and learning: outliers and negative examples, Touretky's Connectionist Summer School 3 (1990) 750–756.

[297] Tomaso Poggio, Federico Girosi, Networks and the best approximation property, Biological Cybernetics 63 (1990) 169–176.

[298] Tomaso Poggio, Federico Girosi, Networks for approximation and learning, Proceedings of the IEEE 78 (1990) 1481–1497.

[299] Tomaso Poggio, Michael Jones, Federico Girosi, Regularization theory and neural networks architectures, Neural Computation 7 (1995) 219–269.

[300] R. Port, M. Knopp, U. Hoffmann, S. Milker-Zabel, G. Brix, Multicompartment analysis of gadolinium chelate kinetics: blood-tissue exchange in mammary tumors as monitored by dynamic MR imaging, Journal of Magnetic Resonance Imaging 10 (1999) 233–241.
[301] W.K. Pratt, Digital Image Processing, John Wiley, 1978.
[302] F.P. Preparata, M.I. Shamos, Computational Geometry: An Introduction, Springer Verlag, 1988.
[303] W.H. Press, S.A. Teukolsky, W.T. Vetterling, B.P. Flannery, Numerical Recipes in C, Cambridge University Press, Cambridge, 1992.
[304] P. Pudil, J. Novovicova, J. Kittler, Floating search methods in feature selection, Pattern Recognition Letters 15 (1994) 1119–1125.
[305] W. Qian, H. Li, M. Kallergi, D. Song, L. Clarke, Adaptive neural network for nuclear medicine image restoration, Journal of VLSI Signal Processing Systems 18 (3) (1998) 297–315.
[306] L.R. Rabiner, R. Schafer, Digital representation of speech signals, Proceedings of the IEEE 63 (1975) 662–677.
[307] W. Reith, S. Heiland, G. Erb, T. Brenner, M. Forsting, K. Sartor, Dynamic contrast-enhanced T2*-weighted MRI in patients with cerebrovascular disease, Neuroradiology (1997) 250–257.
[308] K.A. Rempp, G. Brix, F. Wenz, C.R. Becker, F. Gückel, W.J. Lorenz, Quantification of regional cerebral blood flow and volume with dynamic susceptibility contrast-enhanced MR imaging, Radiology 193 (1994) 637–641.
[309] F. Retter, C. Plant, B. Burgeth, G. Botilla, T. Schlossbauer, A. Meyer-Baese, Computer-aided diagnosis for diagnostically challenging breast lesions in DCE-MRI based on image registration and integration of morphologic and dynamic characteristics, EURASIP Journal on Advances in Signal Processing (2013), page In print.
[310] G.X. Ritter, J.N. Wilson, Handbook of Computer Vision Algorithms in Image Algebra, CRC Press, 1996.
[311] G.P. Robinson, A.C.F. Colchester, L.D. Griffin, Model-based recognition of anatomical objects from medical images, Image and Vision Computing 12 (1994) 499–507.
[312] B.R. Rosen, J.W. Belliveau, J.M. Vevea, T.J. Brady, Perfusion imaging with NMR contrast agents, Magnetic Resonance in Medicine 14 (1990) 249–265.
[313] E. Rosen, S. Smith-Foley, W. DeMartini, P. Eby, S. Peacock, C. Lehman, Bi-rads MRI enhancement characteristics of ductal carcinoma in situ, Breast Journal 13 (2007) 545–550.
[314] F. Rosenblatt, The perceptron: a probabilistic model for information storage and organization in the brain, Psychological Review 65 (1958) 386–408.
[315] Z. Roth, Y. Baram, Multidimensional density shaping by sigmoids, IEEE Transactions on Neural Networks 7 (1996) 1291–1298.
[316] H. Rue, L. Held, Gaussian Markov Random Fields: Theory and Applications (Monographs on Statistics and Applied Probability), Chapman and Hall, 2005.
[317] H. Rue, S. Martino, N. Chopin, Approximate Bayesian inference for latent Gaussian models by using integrated nested Laplace approximations, Journal of the Royal Statistical Society. Series B (Statistical Methodology) 71 (2) (2009) 319–392.
[318] S.J. Russell, P. Norvig, Artificial Intelligence: A Modern Approach, Prentice Hall, 2010.
[319] Ali Saberi, Hassan Khalil, Quadratic-type Lyapunov functions for singularly perturbed systems, IEEE Transactions on Automatic Control (1984) 542–550.
[320] E. Säckinger, B.E. Boser, J. Bromley, Y. LeCun, L.D. Jackel, Application of an ANNA neural network chip to high-speed character recognition, IEEE Transactions on Neural Networks 3 (1992) 498–505.
[321] A. Saha, D.S. Christian, D.S. Tang, C.L. Wu, Oriented non-radial basis functions for image coding and analysis, Touretky's Connectionist Summer School 2 (1991) 728–734.
[322] B. Sahiner, H.P. Chan, N. Petrick, D. Wei, M.A. Helvie, D. Adler, M.M. Goodsitt, Classification of mass and normal breast tissue: a convolution neural network classifier with spatial domain and texture images, IEEE Transaction on Medical Imaging 15 (1996) 598–610.
[323] A. Said, W.A. Pearlman, A new fast and efficient image codec based on set partitioning in hierarchical trees, IEEE Transactions on Circuits and Systems for Video Technology 6 (1996) 243–250.
[324] S. Saito, K. Nakato, Fundamentals of Speech Signal Processing, Academic Press, 1985.

[325] P. Sajda, C.D. Spence, J.C. Pearson, R.M. Nishikawa, Exploiting context in mammograms: a hierarchical neural network for detecting microcalcifications, SPIE Proceedings Series, vol. 2710, 1996, pp. 733–742.

[326] N. Sakamoto, M. Tozaki, K. Higa, Y. Tsunoda, T. Ogawa, S. Abe, S. Ozaki, M. Sakamoto, T. Tsuruhara, N. Kawano, T. Suzuki, N. Yamashiro, E. Fukuma, Categorization of non-mass-like breast lesions detected by MRI, Breast Cancer 15 (2008) 241–246.

[327] T. Sanger, Optimal unsupervised learning in a single-layer linear feedforward neural network, Neural Networks 12 (1989) 459–473.

[328] G. Scarth, M. McIntrye, B. Wowk, R. Samorjai, Detection novelty in functional imaging using fuzzy clustering, in: Proceedings SMR Third Annual Meeting, vol. 95, 1995, pp. 238–242.

[329] R. Schalkoff, Pattern Recognition, John Wiley, 1992.

[330] V.J. Schmid, P.D. Gatehouse, G.-Z. Yang, Attenuation resilient AIF estimation based on hierarchical Bayesian modelling for first pass myocardial perfusion MRI, in: Nicholas Ayache, Sebastien Ourselin, Anthony Maeder (Eds.), Medical Image Computing and Computer-Assisted Intervention, Springer, Berlin, 2007, pp. 393–400.

[331] V.J. Schmid, G.-Z. Yang, Spatio-temporal modelling of first-pass perfusion cardiovascular MRI, in: World Congress on Medical Physics and Biomedical Engineering, September 7–12, 2009, Springer, Munich, Germany, 2008, pp. 45–48.

[332] Volker J. Schmid, Voxel-based adaptive spatio-temporal modelling of perfusion cardiovascular MRI, IEEE Transactions on Medical Imaging 30 (7) (2011) 1305–1313.

[333] Volker J. Schmid, Brandon Whitcher, Anwar R. Padhani, N. Jane Taylor, Guang-Zhong Yang, Bayesian methods for pharmacokinetic models in dynamic contrast-enhanced magnetic resonance imaging, IEEE Transactions on Medical Imaging 25 (12) (2006) 1627–1636.

[334] M.D. Schnall, S. Rosten, S. Englander, S. Orel, L. Nunes, A combined architectural and kinetic interpretation model for breast MR images, Academic Radiology 8 (2001) 591–597.

[335] B. Scholkopf, Support Vector Learning, R. Oldenbourg Verlag, 1997.

[336] U. Schwarz-Boeger, M. Mueller, G. Schimpfle, N. Harbeck, G. Zahlmann, M. Schmidt, C. Geppert, S. Heywang-Koebrunner, Moco – comparison of two different algorithms for motion correction in breast MRI, Onkologie 31 (2008) 141–158.

[337] J. Shapiro, Embedded image coding using zerotrees of wavelet coefficients, IEEE Transactions on Signal Processing 41 (1993) 3445–3462.

[338] L.. Shapiro, G. Stockman, Computer Vision, Prentice-Hall, 2001.

[339] J. Shi, B. Sahiner, H. Chan, C. Paramagul, L. Hadjinski, M. Helvie, T. Chenevert, Treatment response assessment of breast masses on dynamic contrast-enhanced magnetic resonance scans using fuzzy c-means clustering and level set segmentation, Medical Physics 36 (2009) 5052–5063.

[340] J.D. Shutler, M.S. Nixon, Zernike velocity moments for sequence-based description of moving features, Image and Vision Computing 24 (2006) 343–356.

[341] W. Siedlecki, J. Sklansky, A note on genetic algorithms for large-scale feature selection, Pattern Recognition Letters 10 (1989) 335–347.

[342] J.C. Sommer, Regularized estimation and model selection in compartment models, Dr. Hut, Munich, 2013.

[343] J. Sommer, J. Gertheiss, V.J. Schmid, Spatially regularized estimation for the analysis of DCE-MRI data, Statistics in Medicine (2014), http://dx.doi.org/10.1002/sim.5997, in press.

[344] J. Sommer, V.J. Schmid, Spatial two-tissue compartment model for dynamic contrast-enhanced magnetic resonance imaging, Journal of the Royal Statistical Society, Series C - Applied Statistics (2014). In print.

[345] S.P. Sourbron, D.L. Buckley, Tracer kinetic modelling in MRI: estimating perfusion and capillary permeability, Physics in Medicine and Biology 57 (2) (2012) R1–R33.

[346] K.S. St. Lawrence, T.Y. Lee, An adiabatic approximation to the tissue homogeneity model for water exchange in the brain: I. Theoretical derivation, Journal of Cerebral Blood Flow and Metabolism 18 (12) (1998) 1365–1377.

[347] S.D. Stearns, D.R. Hush, Digital Signal Analysis, Prentice-Hall, 1990.

[348] F. Steinbruecker, A. Meyer-Baese, C. Plant, T. Schlossbauer, U. Meyer-Baese, Selection of spatiotemporal features in breast MRI to differentiate between malignant and benign small lesions using computer-aided diagnosis, Advances in Artificial Neural Systems (2012) 919281.

[349] F. Steinbruecker, A. Meyer-Baese, T. Schlossbauer, D. Cremers, Evaluation of a nonrigid motion compensation technique based on spatiotemporal features for small lesion detection in breast MRI, Advances in Artificial Neural Systems (2012) 808602.

[350] G.N. Stewart, Researches on the circulation time in organs and on the influences which affect it, Journal of Physiology 15 (1-2) (1994) 1–89.

[351] C. Stippich, Introduction to presurgical functional MRI, in: Christoph Stippich (Ed.), Clinical Functional MRI, Springer, Berlin Heidelberg, 2007, pp. 1–7 (Chapter 1).

[352] M. Stone, Cross-validatory choice and assessment of statistical predictions, Journal of the Royal Statistical Society (1974) 111–133.

[353] M. Stoutjesdijk, J. Fuetterer, C. Boetes, L. van Dienand, G. Jaeger, J. Barentsz, Variability in the description of morphologic and contrast enhancement characteristics of breast lesions on magnetic resonance imaging, Investigative Radiology 40 (2005) 355–362.

[354] M. Stoutjesdijk, J. Veltman, H. Huisman, N. Karssemeijer, J. Barentsz, J. Blickman, C. Boetes, Automated analysis of contrast enhancement in breast MRI lesions using mean shift clustering for roi selection, Journal Magnetic Resonance Imaging 26 (2007) 606–614.

[355] M. Strintzis, Optimal pyramidal and subband decomposition for hierarchical coding of noisy and quantized images, IEEE Transaction on Image Processing 7 (1998) 155–166.

[356] A. Sudjianto, M. Hassoun, G. Wasserman, Extension of principal component analysis for nonlinear feature extraction, SPIE Proceedings Series, vol. 3210, 1996, pp. 1433–1438.

[357] Y. Sun, J.G. Li, S.Y. Wu, Improvement on performance of modified hopfield neural network for image restoration, IEEE Transaction on Image Processing 4 (1995) 688–692.

[358] J. Sychra, P. Bandettini, N. Bhattacharya, Q. Lin, Synthetic images by subspace transforms I. Principal components images and related filters, Medical Physics 21 (1994) 193–201.

[359] B. Szabo, P. Aspelin, M. Wiberg, B. Bone, Dynamic MR imaging of the breast – analysis of kinetic and morphologic diagnostic criteria, Acta Radiologica 44 (2003) 379–386.

[360] B. Szabo, M. Wilberg, B. Bone, P. Aspelin, Application of artificial neural networks to the analysis of dynamic MR imaging features to the breast, European Radiology 14 (2004) 1217–1225.

[361] L. Tabar, P.B. Dean, Teaching Atlas of Mammography, Thieme Inc., 1985.

[362] M.R. Teague, Image analysis via the general theory of moments, Journal of the Optical Society of America 70 (1979) 920–930.

[363] R Core Team, R: A language and environment for statistical computing, R Foundation for Statistical Computing, Vienna, Austria, 2013, ISBN 3-900051-07-0 http://www.R-project.org/.

[364] S. Theodoridis, K. Koutroumbas, Pattern Recognition, Academic Press, 1998.

[365] H.K. Thompson, C.F. Starmer, R.E. Whalen, D. McIntosh, Indicator transit time considered as a gamma variate, Circulation Research (1964) 502–515.

[366] A.N. Tikhonov, V.Y. Arsenin, Solutions of Ill-Posed Problems, W. H. Winston, 1977.

[367] A. Toet, Multiscale contrast enhancement with application to image fusion, Optical Engineering 31 (1992) 1026–1030.

[368] P. Tofts, G. Brix, D. Buckley, J. Evelhoch, E. Henderson, M. Knopp, H. Larsson, T. Lee, N.A. Mayr, G. Parker, E. Port, J. Taylor, R. Weiskoff, Estimating kinetic parameters from dynamic contrast-enhanced T1-weighted MRI of a diffusable tracer: standardized quantities and symbols, Journal of Magnetic Resonance Imaging 10 (1999) 223–232.

[369] P. Tofts, A. Kermode, Measurement of the blood-brain barrier permeability and leakage space using dynamic MR imaging – 1. Fundamental concepts, Magnetic Resonance in Medicine 17 (1991) 357–367.

[370] G. Tourassi, B. Harrawood, S. Singh, J. Lo, Information-theoretic cad system in mammography: entropy-based indexing for computational efficiency and robust performance, Medical Physics 34 (2007) 3193–3204.

[371] G. Tourassi, R. Ike, S. Singh, B. Harrawood, Evaluating the effect of image preprocessing on an information-theoretic cad system in mammography, Academic Radiology 15 (2008) 626–634.

[372] G. Tourassi, R. Vargas-Voracek, D. Catarious, Computer-assisted detection of mammographic masses: a template matching scheme based on mutual information, Medical Physics 30 (2003) 2123–2130.

[373] P. Trahanias, E. Skordalakis, Syntactic pattern recognition of the ECG, IEEE Transactions on Pattern Analysis and Machine Intelligence 12 (1990) 648–657.

[374] O. Tsujii, A. Hasegawa, C. Wu, B. Lo, M. Freedman, S. Mun, Classification of microcalcifications in digital mammograms for the diagnosis of breast cancer, SPIE Proceedings Series, vol. 2710, 1996, pp. 794–804.

[375] D. Tzovaras, M. Strintzis, Use of nonlinear principal component analysis and vector quantization for image coding, IEEE Transaction on Image Processing 7 (1998) 1218–1223.

[376] T. Vag, P. Baltzer, M. Dietzel, R. Zoubi, M. Gajda, O. Camara, W. Kaiser, Kinetic analysis of lesions without mass effect on breast MRI using manual and computer-assisted methods of dynamic MR imaging features to the breast, European Radiology 21 (2011) 893–898.

[377] P.P. Vaidyanathan, Multirate Systems and Filterbanks, Prentice Hall, 1993.

[378] A.P. Schouten van der Velden, C. Boetes, P. Bult, T. Wobbes, Variability in the description of morphologic and contrast enhancement characteristics of breast lesions on magnetic resonance imaging, American Journal of Surgery 192 (2006) 172–178.

[379] V.N. Vapnik, The Nature of Statistical Learning Theory, Springer, 2000.

[380] R. Vigario, V. Jousmaki, M. Hamalainen, R. Hari, E. Oja, Independent component analysis for identification of artifacts in magnetoencephalographic recordings, in: NIPS98, vol. 10, 1998, pp. 229–235.

[381] A. Villringer, B.R. Rosen, J.W. Belliveauand, J.L. Ackerman, R.B. Lauffer, R.B. Buxton, Y-S. Chao, V.J. Wedeen, T.J. Brady, Dynamic imaging of lanthanide chelates in normal brain: changes in signal intensity due to susceptibility effects, Magnetic Resonance in Medicine (1988) 164–174.

[382] D. Walnut, An Introduction to Wavelet Analysis, Birkhäuser, 2001.

[383] D.C. Wang, A.H. Vagnucci, C.C. Li, Digital image enhancement: a survey, IEEE Transaction on Information Theory 24 (1983) 363–381.

[384] S. Webb, The Physics of Medical Imaging, Adam Hilger, 1990.

[385] R.M. Weisskoff, D. Chesler, J.L. Boxerman, B.R. Rosen, Pitfalls in MR measurement of tissue blood flow with intravascular tracers: which mean transit time? Magnetic Resonance in Medicine 29 (1993) 553–559.

[386] B. Whitcher, V.J. Schmid, Quantitative analysis of dynamic contrast-enhanced and diffusion-weighted magnetic resonance imaging for oncology in R, Journal of Statistical Software 44 (5) (2011) 1–29.

[387] D. Whitley, Genetic algorithm tutorial, Statistics and Computing 4 (1994) 65–85.

[388] N. Wilke, C. Simm, J. Zhang, J. Ellermann, X. Ya, H. Merkle, G. Path, H. Lüdemann, R. Bache, K. Ugurbil, Contrast-enhanced first-pass myocardial perfusion imaging: correlation between myocardial bloodflow in dogs at rest and during hyperemia, Magnetic Resonance in Medicine (4) (1993) 485–497.

[389] Q. Williams, J. Noble, A. Ehlgen, H. Becher, Tissue perfusion diagnostic classification using a spatio-temporal analysis of contrast ultrasound image sequences – information processing in medical imaging – lecture notes in computer science, in: G. Christensen, M. Sonka (Eds.), Information Processing in Medical Imaging, Springer, Berlin, 2005, pp. 222–233.

[390] D.J. Willshaw, C. von der Malsburg, How patterned neural connections can be set up by self-organization, Proceedings of the Royal Society London B 194 (1976) 431–445.

[391] A. Wismüller, A. Meyer-Bäse, O. Lange, M. Reiser, G. Leinsinger, Cluster analysis of dynamic cerebral contrast-enhanced perfusion MRI time-series, IEEE Transactions on Medical Imaging 25 (1) (2006) 62–73.

[392] J.W. Woods, Subband Image Coding, Kluwer Academic Publishers, 1991.

[393] K.S. Woods, Automated image analysis techniques for digital mammography (Ph.D. thesis), University of South Florida, 1994.

[394] K.S. Woods, J.L. Solka, C.E. Priebe, C.C. Doss, K.W. Boyer, L.P. Clarke, Comparative evaluation of pattern recognition techniques for detection of microcalcifications, SPIE Proceedings Series, vol. 1905, 1993, pp. 841–852.

[395] R. Woods, S. Cherry, J. Mazziotta, Rapid automated algorithm for aligning and reslicing PET images, Journal of Computer Assisted Tomography 16 (1992) 620–633.

[396] M. Woolrich, M. Jenkinson, J.M. Brady, Stephen M. Smith, Fully Bayesian spatio-temporal modeling of FMRI data, IEEE Transactions on Medical Imaging 23 (2) (2004) 213–231.

[397] Y. Wu, M. Giger, K. Doi, R.M. Nishikawa, Computerized detection of clustered microcalcifications in digital mammograms: applications of artificial neural networks, Medical Physics 19 (1992) 555–560.

[398] H. Yabuuchi, Y. Matsuo, T. Kamitani, T. Setoguchi, T. Okafuji, H. Soeda, S. Sakai, M. Hatekenata, M. Kubo, E. Tokunaga, H. Yamamoto, H. Honda, Non-mass-like enhancement on contrast-enhanced breast MRI imaging: lesion characterization using combination of dynamic contrast-enhanced and diffusion-weighted MR images, European Journal of Radiology 75 (2010) 126–132.

[399] W. Yan, U. Helmke, J. Moore, Global analysis of Oja's flow for neural networks, IEEE Transactions on Neural Networks 5 (1994) 674–883.

[400] P. Yap, R. Paramesran, S. Ong, Image analysis by Krawtchouk moments, IEEE Transactions on Image Processing 12 (2003) 1367–1377.

[401] T.Y. Young, K.S. Fu, Handbook of Pattern Recognition and Image Processing, Academic Press, 1986.

[402] E. Yousef, R. Duchesneau, R. Alfidi, Magnetic resonance imaging of the breast, Radiology 150 (1984) 761–766.

[403] S. Yu, L. Guan, A CAD system for the automatic detection of clustered microcalcifications in digitized mammogram films, IEEE Transaction on Medical Imaging 19 (2000) 115–126.

[404] L. Zadeh, Fuzzy sets, Information and Control 8 (1965) 338–353.

[405] W. Zhang, M. Giger, K. Doi, Y. Wu, R. Nishikawa, R. Schmidt, Computerized detection of masses in digital mammograms using a shift-invariant artificial neural network, Medical Physics 21 (1994) 517–524.

[406] W. Zhang, Y. Wu, M. Giger, K. Doi, R.M. Nishikawa, Computerized detection of clustered microcalcifications in digital mammograms using a shift-invariant neural network, Medical Physics 21 (1994) 517–524.

[407] Y. Zheng, S. Englander, S. Baloch, E. Zacharaki, Y. Fan, M. Schnall, D. Shen, Step: spatiotemporal enhancement pattern for MR-based breast tumor diagnosis, Medical Physics 36 (2009) 3192–3204.

[408] Y. Zhu, H. Yan, Computerized tumor boundary detection using a hopfield neural network, in: Proceedings of the IEEE Joint International Conference on Neural Networks in Houston, vol. 3, 1997, pp. 2467–2472.

[409] K.L. Zierler, Theoretical basis of indicator-dilution methods for measuring flow and volume, Circulation Research 10 (3) (1965) 393–407.

[410] X. Zong, A. Meyer-Bäse, A. Laine, Multiscale segmentation through a radial basis neural network, in: IEEE International Conference on Image Processing, 1997, pp. 400–403.

INDEX

N

O

P

Q

R

S

T

U

V

W

X

Printed in the United States
By Bookmasters